Student's Solutions Manu

D0508246

to accompany

Beginning and Intermediate Algebra

Second Edition

Sherri Messersmith

College of DuPage

McGraw-Hill Higher Education

Boston Burr Ridge, IL Dubuque, IA New York San Francisco St. Louis
Bangkok Bogotá Caracas Kuala Lumpur Lisbon London Madrid Mexico City
Milan Montreal New Delhi Santiago Seoul Singapore Sydney Taipei Toronto

The McGraw-Hill Companies

McGraw-Hill Higher Education

Student's Solutions Manual to accompany
BEGINNING AND INTERMEDIATE ALGEBRA, SECOND EDITION
SHERRI MESSERSMITH

Published by McGraw-Hill Higher Education, an imprint of The McGraw-Hill Companies, Inc., 1221 Avenue of the Americas, New York, NY 10020.

This book is printed on acid-free paper.

2 3 4 5 6 7 8 9 0 CCW/CCW 0 9

ISBN: 978-0-07-322758-0
MHID: 0-07-322758-7

www.mhhe.com

Table of Contents

Chapter 1: The Real Number System and Geometry

Section 1.1: Exercises

1) a) $\frac{2}{5}$

b) $\frac{4}{6}=\frac{2}{3}$

c) $\frac{4}{4}=1$

3) $\frac{1}{2}$ 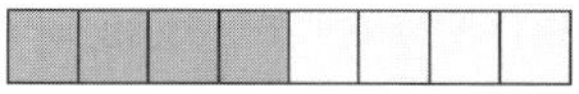

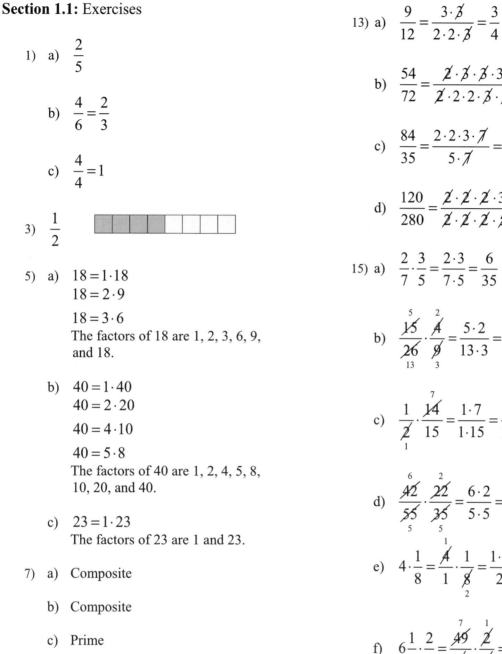

5) a) $18=1\cdot18$
$18=2\cdot9$
$18=3\cdot6$
The factors of 18 are 1, 2, 3, 6, 9, and 18.

b) $40=1\cdot40$
$40=2\cdot20$
$40=4\cdot10$
$40=5\cdot8$
The factors of 40 are 1, 2, 4, 5, 8, 10, 20, and 40.

c) $23=1\cdot23$
The factors of 23 are 1 and 23.

7) a) Composite

b) Composite

c) Prime

9) Composite. It is divisible by 2 and has other factors as well.

11) a) $18=2\cdot3\cdot3$

b) $54=2\cdot3\cdot3\cdot3$

c) $42=2\cdot3\cdot7$

d) $150=2\cdot3\cdot5\cdot5$

13) a) $\frac{9}{12}=\frac{3\cdot\cancel{3}}{2\cdot2\cdot\cancel{3}}=\frac{3}{4}$

b) $\frac{54}{72}=\frac{\cancel{2}\cdot\cancel{3}\cdot\cancel{3}\cdot3}{\cancel{2}\cdot2\cdot2\cdot\cancel{3}\cdot\cancel{3}}=\frac{3}{4}$

c) $\frac{84}{35}=\frac{2\cdot2\cdot3\cdot\cancel{7}}{5\cdot\cancel{7}}=\frac{12}{5}$

d) $\frac{120}{280}=\frac{\cancel{2}\cdot\cancel{2}\cdot\cancel{2}\cdot3\cdot\cancel{5}}{\cancel{2}\cdot\cancel{2}\cdot\cancel{2}\cdot\cancel{5}\cdot7}=\frac{3}{7}$

15) a) $\frac{2}{7}\cdot\frac{3}{5}=\frac{2\cdot3}{7\cdot5}=\frac{6}{35}$

b) $\frac{\overset{5}{\cancel{15}}}{\underset{13}{\cancel{26}}}\cdot\frac{\overset{2}{\cancel{4}}}{\underset{3}{\cancel{9}}}=\frac{5\cdot2}{13\cdot3}=\frac{10}{39}$

c) $\frac{1}{\underset{1}{\cancel{2}}}\cdot\frac{\overset{7}{\cancel{14}}}{15}=\frac{1\cdot7}{1\cdot15}=\frac{7}{15}$

d) $\frac{\overset{6}{\cancel{42}}}{\underset{5}{\cancel{55}}}\cdot\frac{\overset{2}{\cancel{22}}}{\underset{5}{\cancel{35}}}=\frac{6\cdot2}{5\cdot5}=\frac{12}{25}$

e) $4\cdot\frac{1}{8}=\frac{\overset{1}{\cancel{4}}}{1}\cdot\frac{1}{\underset{2}{\cancel{8}}}=\frac{1\cdot1}{2}=\frac{1}{2}$

f) $6\frac{1}{8}\cdot\frac{2}{7}=\frac{\overset{7}{\cancel{49}}}{\underset{4}{\cancel{8}}}\cdot\frac{\overset{1}{\cancel{2}}}{\underset{1}{\cancel{7}}}=\frac{7\cdot1}{4\cdot1}=\frac{7}{4}$ or $1\frac{3}{4}$

17) She multiplied the whole numbers and multiplied the fractions. She should have converted the mixed numbers to improper fractions before multiplying.
$5\frac{1}{2}\cdot2\frac{1}{3}=\frac{11}{2}\cdot\frac{7}{3}=\frac{11\cdot7}{2\cdot3}=\frac{77}{6}$ or $12\frac{5}{6}$

19) a) $\dfrac{1}{42}\div\dfrac{2}{7}=\dfrac{1}{\cancel{42}_6}\cdot\dfrac{\cancel{7}^1}{2}=\dfrac{1\cdot1}{6\cdot2}=\dfrac{1}{12}$

b) $\dfrac{3}{11}\div\dfrac{4}{5}=\dfrac{3}{11}\cdot\dfrac{5}{4}=\dfrac{3\cdot5}{11\cdot4}=\dfrac{15}{44}$

c) $\dfrac{18}{35}\div\dfrac{9}{10}=\dfrac{\cancel{18}^2}{\cancel{35}_7}\cdot\dfrac{\cancel{10}^2}{\cancel{9}_1}=\dfrac{2\cdot2}{7\cdot1}=\dfrac{4}{7}$

d) $\dfrac{14}{15}\div\dfrac{2}{15}=\dfrac{\cancel{14}^7}{\cancel{15}_1}\cdot\dfrac{\cancel{15}^1}{\cancel{2}_1}=\dfrac{7\cdot1}{1\cdot1}=7$

e) $6\dfrac{2}{5}\div1\dfrac{13}{15}=\dfrac{32}{5}\div\dfrac{28}{15}$

$=\dfrac{\cancel{32}^8}{\cancel{5}_1}\cdot\dfrac{\cancel{15}^3}{\cancel{28}_7}$

$=\dfrac{8\cdot3}{1\cdot7}$

$=\dfrac{24}{7}$ or $3\dfrac{3}{7}$

f) $\dfrac{4}{7}\div8=\dfrac{\cancel{4}^1}{7}\cdot\dfrac{1}{\cancel{8}_2}=\dfrac{1\cdot1}{7\cdot2}=\dfrac{1}{14}$

21) $10=2\cdot5,\ 15=3\cdot5$

LCM of 10 and 15 $=2\cdot3\cdot5=30$

23) a) $10=2\cdot5,\ 30=2\cdot3\cdot5$

LCD of $\dfrac{9}{10}$ and $\dfrac{11}{30}=2\cdot3\cdot5$

$=30$

b) $8=2\cdot2\cdot2,\ 12=2\cdot2\cdot3$

LCD of $\dfrac{7}{8}$ and $\dfrac{5}{12}=2\cdot2\cdot2\cdot3$

$=24$

c) $9=3\cdot3,\ 6=2\cdot3,\ 4=2\cdot2$

LCD of $\dfrac{4}{9},\dfrac{1}{6},$ and $\dfrac{3}{4}=2\cdot2\cdot3\cdot3$

$=36$

25) a) $\dfrac{6}{11}+\dfrac{2}{11}=\dfrac{6+2}{11}=\dfrac{8}{11}$

b) $\dfrac{19}{20}-\dfrac{7}{20}=\dfrac{19-7}{20}=\dfrac{12}{20}=\dfrac{3}{5}$

c) $\dfrac{4}{25}+\dfrac{2}{25}+\dfrac{9}{25}=\dfrac{4+2+9}{25}$

$=\dfrac{15}{25}$

$=\dfrac{3}{5}$

d) $\dfrac{2}{9}+\dfrac{1}{6}=\dfrac{4}{18}+\dfrac{3}{18}=\dfrac{4+3}{18}=\dfrac{7}{18}$

e) $\dfrac{3}{5}+\dfrac{11}{30}=\dfrac{18}{30}+\dfrac{11}{30}=\dfrac{18+11}{30}=\dfrac{29}{30}$

f) $\dfrac{13}{18}-\dfrac{2}{3}=\dfrac{13}{18}-\dfrac{12}{18}=\dfrac{13-12}{18}=\dfrac{1}{18}$

g) $\dfrac{4}{7}+\dfrac{5}{9}=\dfrac{36}{63}+\dfrac{35}{63}=\dfrac{36+35}{63}=\dfrac{71}{63}$

h) $\dfrac{5}{6}-\dfrac{1}{4}=\dfrac{10}{12}-\dfrac{3}{12}=\dfrac{10-3}{12}=\dfrac{7}{12}$

i) $\frac{3}{10}+\frac{7}{20}+\frac{3}{4}=\frac{6}{20}+\frac{7}{20}+\frac{15}{20}$
$=\frac{6+7+15}{20}$
$=\frac{28}{20}$
$=\frac{7}{5}$

j) $\frac{1}{6}+\frac{2}{9}+\frac{10}{27}=\frac{9}{54}+\frac{12}{54}+\frac{20}{54}$
$=\frac{9+12+20}{54}$
$=\frac{41}{54}$

27) a) $8\frac{5}{11}+6\frac{2}{11}=14\frac{5+2}{11}=14\frac{7}{11}$

b) $2\frac{1}{10}+9\frac{3}{10}=11\frac{1+3}{10}$
$=11\frac{4}{10}=11\frac{2}{5}$

c) $7\frac{11}{12}-1\frac{5}{12}=6\frac{11-5}{12}=6\frac{6}{12}=6\frac{1}{2}$

d) $3\frac{1}{5}+2\frac{1}{4}=3\frac{4}{20}+2\frac{5}{20}$
$=5\frac{4+5}{20}$
$=5\frac{9}{20}$

e) $5\frac{2}{3}-4\frac{4}{15}=5\frac{10}{15}-4\frac{4}{15}$
$=1\frac{10-4}{15}$
$=1\frac{6}{15}$
$=1\frac{2}{5}$

f) $9\frac{5}{8}-5\frac{3}{10}=9\frac{25}{40}-5\frac{12}{40}$
$=4\frac{25-12}{40}$
$=4\frac{13}{40}$

g) $4\frac{3}{7}+6\frac{3}{4}=4\frac{12}{28}+6\frac{21}{28}$
$=10\frac{12+21}{28}$
$=10\frac{33}{28}$
$=11\frac{5}{28}$

h) $7\frac{13}{20}+\frac{4}{5}=7\frac{13}{20}+\frac{16}{20}$
$=7\frac{13+16}{20}$
$=7\frac{29}{20}$
$=8\frac{9}{20}$

29) $7 \div 1\frac{2}{3} = 7 \div \frac{5}{3} = \frac{7}{1} \cdot \frac{3}{5} = \frac{21}{5} = 4\frac{1}{5}$

Alex can make 4 whole bears.
Determine amount of fabric used to make 4 bears: $4 \cdot 1\frac{2}{3} = 4 \cdot \frac{5}{3} = \frac{20}{3}$ yds.

Amount of fabric remaining

$= 7 \text{ yds} - \frac{20}{3} \text{ yds}$

$= \frac{21}{3} \text{ yds} - \frac{20}{3} \text{ yds}$

$= \frac{1}{3}$ yd left over

31) $99 \cdot \frac{4}{11} = \frac{\overset{9}{\cancel{99}}}{1} \cdot \frac{4}{\underset{1}{\cancel{11}}} = 36$ hits

33) Add the width of the frame twice to each dimension since it is being added to both sides of the picture.

$18\frac{3}{8} + 2\frac{1}{8}\text{in} + 2\frac{1}{8} = 22\frac{3+1+1}{8}$

$= 22\frac{5}{8}$

$12\frac{1}{4} + 2\frac{1}{8} + 2\frac{1}{8} = 2\frac{2}{8} + 2\frac{1}{8} + 2\frac{1}{8}$

$= 16\frac{2+1+1}{8}$

$= 16\frac{4}{8}$

$= 16\frac{1}{2}$

$16\frac{1}{2}$ in by $22\frac{5}{8}$ in

35) $\frac{2}{3} + 1\frac{1}{4} + 1\frac{1}{2} = \frac{8}{12} + 1\frac{3}{12} + 1\frac{6}{12}$

$= 2\frac{8+3+6}{12}$

$= 2\frac{17}{12}$

$= 3\frac{5}{12}$ cups

37) $16\frac{3}{4} - 11\frac{3}{5} = 16\frac{15}{20} - 11\frac{12}{20}$

$= 5\frac{15-12}{20}$

$= 5\frac{3}{20}$ gallons

39) $42 \cdot \frac{5}{6} = \frac{\overset{7}{\cancel{42}}}{1} \cdot \frac{5}{\underset{1}{\cancel{11}}} = 35$ problems

41) Add the total length welded so far:

$14\frac{1}{6} + 10\frac{3}{4} = 14\frac{2}{12} + 10\frac{9}{12}$

$= 24\frac{2+9}{12} = 24\frac{11}{12}$

Subtract the total from the desired length:

$32\frac{7}{8} - 24\frac{11}{12} = 32\frac{21}{24} - 24\frac{22}{24}$

$= 31\frac{45}{24} - 24\frac{22}{24}$

$= 7\frac{45-22}{24} = 7\frac{23}{24}$ inches.

43) $400 \cdot \frac{3}{5} = \frac{\overset{8}{\cancel{400}}}{1} \cdot \frac{3}{\underset{1}{\cancel{5}}} = 240$ students

Section 1.2: Exercises

1) a) base: 6; exponent: 4

b) base: 2; exponent: 3

c) base: $\frac{9}{8}$; exponent: 5

3) a) $9 \cdot 9 \cdot 9 \cdot 9 = 9^4$

b) $2 \cdot 2 \cdot 2 \cdot 2 \cdot 2 \cdot 2 \cdot 2 \cdot 2 = 2^8$

c) $\frac{1}{4} \cdot \frac{1}{4} \cdot \frac{1}{4} = \left(\frac{1}{4}\right)^3$

5) a) $8^2 = 8 \cdot 8 = 64$

b) $(11)^2 = 11 \cdot 11 = 121$

c) $2^4 = 2 \cdot 2 \cdot 2 \cdot 2 = 16$

d) $5^3 = 5 \cdot 5 \cdot 5 = 125$

e) $3^4 = 3 \cdot 3 \cdot 3 \cdot 3 = 81$

f) $(12)^2 = 12 \cdot 12 = 144$

g) $1^2 = 1 \cdot 1 = 1$

h) $\left(\frac{3}{10}\right)^2 = \frac{3}{10} \cdot \frac{3}{10} = \frac{9}{100}$

i) $\left(\frac{1}{2}\right)^6 = \frac{1}{2} \cdot \frac{1}{2} \cdot \frac{1}{2} \cdot \frac{1}{2} \cdot \frac{1}{2} \cdot \frac{1}{2} = \frac{1}{64}$

j) $(0.3)^2 = (0.3) \cdot (0.3) = 0.09$

7) $(0.5)^2 = 0.5 \cdot 0.5 = 0.25$ or

$(0.5)^2 = \left(\frac{1}{2}\right)^2 = \frac{1}{4}$

9) Answers may vary.

11) $17 - 2 + 4 = 15 + 4 = 19$

13) $50 \div 10 + 15 = 5 + 15 = 20$

15) $$\begin{aligned} 20 - 3 \cdot 2 + 9 &= 20 - 6 + 9 \\ &= 14 + 9 \\ &= 23 \end{aligned}$$

17) $8 + 12 \cdot \frac{3}{4} = 8 + 9 = 17$

19) $$\begin{aligned} \frac{3}{4} \cdot \frac{1}{6} + \frac{1}{2} \cdot \frac{1}{3} &= \frac{3}{24} + \frac{1}{6} \\ &= \frac{3}{24} + \frac{4}{24} \\ &= \frac{7}{24} \end{aligned}$$

21) $$\begin{aligned} 2 \cdot \frac{3}{4} - \left(\frac{2}{3}\right)^2 &= 2 \cdot \frac{3}{4} - \frac{4}{9} \\ &= \frac{3}{2} - \frac{4}{9} \\ &= \frac{27}{18} - \frac{8}{18} \\ &= \frac{19}{18} \end{aligned}$$

23) $$\begin{aligned} 25 - 11 \cdot 2 + 1 &= 25 - 22 + 1 \\ &= 3 + 1 \\ &= 4 \end{aligned}$$

25) $$\begin{aligned} 15 - 3(6-4)^2 &= 15 - 3(2)^2 \\ &= 15 - 3(4) \\ &= 15 - 12 \\ &= 3 \end{aligned}$$

27) $60 \div 15 + 5 \cdot 3 = 4 + 15 = 19$

29) $6[21\div(3+4)]-9$
$=6[21\div(7)]-9$
$=6[3]-9$
$=18-9=9$

31) $4+3\left[(1+3)^3\div(10-2)\right]$
$=4+3\left[(4)^3\div(8)\right]$
$=4+3[64\div(8)]$
$=4+3[8]$
$=4+24=28$

33) $\dfrac{12(5+1)}{2\cdot5-1}=\dfrac{12(6)}{10-1}$
$=\dfrac{72}{9}$
$=8$

35) $\dfrac{4(7-2)^2}{(12)^2-8\cdot3}=\dfrac{4(5)^2}{144-8\cdot3}$
$=\dfrac{4(25)}{144-24}$
$=\dfrac{100}{120}$
$=\dfrac{5}{6}$

Section 1.3: Exercises

1) Right

3) Obtuse

5) Supplementary; complementary

7) $90^\circ-24^\circ=66^\circ$

9) $90^\circ-45^\circ=45^\circ$

11) $180^\circ-100^\circ=80^\circ$

13) $180^\circ-71^\circ=109^\circ$

15)

Angle Measure	Reason
$m\angle A=180^\circ-29^\circ=151^\circ$	supplementary
$m\angle B=29^\circ$	vertical angles
$m\angle C=m\angle A=151^\circ$	vertical angles

17) 180

19) the sum of the two known angles
$90^\circ+41^\circ=131^\circ$
the measure of the unknown angle
$180^\circ-131^\circ=49^\circ$
The triangle is right since it contains one right angle.

21) the sum of the two known angles
$70^\circ+33^\circ=103^\circ$
the measure of the unknown angle
$180^\circ-103^\circ=77^\circ$
The triangle is acute since all three angles are acute.

23) Isosceles

25) Equilateral

27) True

29) $A=s^2=(9\text{ ft})^2=81\text{ ft}^2$

$P=4s=4(9\text{ ft})=36\text{ ft}$

31) $A=\dfrac{1}{2}bh$
$=\dfrac{1}{2}(6\text{ mm})(3\text{ mm})$
$=9\text{ mm}^2$

$$\begin{aligned} P &= a+b+c \\ &= 3.6 \text{ mm} + 6 \text{ mm} + 5 \text{ mm} \\ &= 14.6 \text{ mm} \end{aligned}$$

33) $$\begin{aligned} A &= lw \\ &= \left(3\frac{1}{2} \text{ mi}\right)\left(1\frac{7}{8} \text{ mi}\right) \\ &= \left(\frac{7}{2} \text{ mi}\right)\left(\frac{15}{8} \text{ mi}\right) \\ &= 6\frac{9}{16} \text{ mi}^2 \end{aligned}$$

$$\begin{aligned} P &= 2l+2w \\ &= 2\left(3\frac{1}{2} \text{ mi}\right)+2\left(1\frac{7}{8} \text{ mi}\right) \\ &= 7 \text{ mi} + \frac{15}{4} \text{ mi} \\ &= 10\frac{3}{4} \text{ miles} \end{aligned}$$

35) $$\begin{aligned} A &= \frac{1}{2}bh \\ &= \frac{1}{2}(12 \text{ cm})(5 \text{ cm}) \\ &= 30 \text{ cm}^2 \end{aligned}$$

$$\begin{aligned} P &= a+b+c \\ &= 5 \text{ cm} + 13 \text{ cm} + 12 \text{ cm} \\ &= 30 \text{ cm} \end{aligned}$$

37) a) $$\begin{aligned} A &= \pi r^2 \\ &= \pi(5 \text{ in})^2 \\ &= 25\pi \text{ in}^2 \\ &\approx 78.5 \text{ in}^2 \end{aligned}$$

b) $$\begin{aligned} C &= 2\pi r \\ &= 2\pi(5 \text{ in}) \\ &= 10\pi \text{ in} \\ &\approx 31.4 \text{ in} \end{aligned}$$

39) a) $$\begin{aligned} A &= \pi r^2 \\ &= \pi(10 \text{ cm})^2 \\ &= 100\pi \text{ cm}^2 \\ &\approx 314 \text{ cm}^2 \end{aligned}$$

b) $$\begin{aligned} C &= 2\pi r \\ &= 2\pi(10 \text{ cm}) \\ &= 20\pi \text{ cm} \\ &\approx 62.8 \text{ cm} \end{aligned}$$

41) $$\begin{aligned} A &= \pi r^2 \\ &= \pi(7.2 \text{ ft})^2 \\ &= 51.84\pi \text{ ft}^2 \end{aligned}$$

$$\begin{aligned} C &= 2\pi r \\ &= 2\pi(7.2 \text{ ft}) \\ &= 14.4\pi \text{ ft} \end{aligned}$$

43) $$\begin{aligned} A &= \pi r^2 \\ &= \pi(8 \text{ yd})^2 \\ &= 64\pi \text{ yd}^2 \end{aligned}$$

$$\begin{aligned} C &= 2\pi r \\ &= 2\pi(8 \text{ yd}) \\ &= 16\pi \text{ yd} \end{aligned}$$

45) a) $A = (4\text{ ft})(7\text{ ft}) + (5\text{ ft})(6\text{ ft})$

$= 28\text{ ft}^2 + 30\text{ ft}^2$

$= 58\text{ ft}^2$

$P = 4\text{ ft} + 7\text{ ft} + 2\text{ ft}$

$+5\text{ ft} + 6\text{ ft} + 12\text{ ft}$

$= 36\text{ ft}$

b) $A = (8)(10) + (3)(14)$

$= 80 + 42$

$= 112\text{ cm}^2$

$P = (10 + 14 + 3 + 4 + 7 + 10)\text{ cm}$

$= 48\text{ cm}$

c) $A = (4.4)(17.3) + (8.6)(6.1)$

$+ (6.2)(4.1)$

$= 76.12 + 52.46 + 25.42$

$= 154\text{ in}^2$

$P = 17.3\text{ in} + 17.3\text{ in} + 4.4\text{ in}$

$+ 4.4\text{ in} + 8.6\text{ in} + 8.6\text{ in}$

$= 60.6\text{ in}$

47) $A = (15\text{ in})(20\text{ in}) - (13\text{ in})(18\text{ in})$

$= 300\text{ in}^2 - 234\text{ in}^2$

$= 66\text{ in}^2$

49) $A = (8.5\text{ m})^2 - 3.14(3\text{ m})^2$

$= 72.25\text{ m}^2 - 28.26\text{ m}^2$

$= 43.99\text{ m}^2$

51) $V = lwh = (7\text{ m})(5\text{ m})(2\text{ m}) = 70\text{ m}^3$

53) $V = \frac{4}{3}\pi r^3$

$= \frac{4}{3}\pi(3\text{ in})^3$

$= \frac{4}{3}\pi(27\text{ in}^3)$

$= 36\pi\text{ in}^3$

55) $V = \frac{4}{3}\pi r^3$

$= \frac{4}{3}\pi(4\text{ ft})^3$

$= \frac{4}{3}\pi(64\text{ ft}^3)$

$= \frac{256}{3}\pi\text{ ft}^3$

57) $V = \pi r^2 h$

$= \pi(3\text{ cm})^2(5.8\text{ cm})$

$= 52.2\pi\text{ cm}^3$

59) a) $A = (18\text{ ft})(15\text{ ft}) = 270\text{ ft}^2$

b) $270\ \cancel{\text{ft}^2} \cdot \left(\frac{\$2.30}{\cancel{\text{ft}^2}}\right) = \$621$

No, it would cost $621 to use this carpet.

61) $V = \pi(3\text{ cm})^2(15\text{ cm})$

$\approx (3.14)(9\text{ cm}^2)(15\text{ cm})$

$\approx 423.9\text{ cm}^3$

63) a) $C = 2\pi(7\text{ in}) = 14(3.14)\text{ in}$

$\approx 43.96\text{ in}$

b) $A = \pi(7\text{ in})^2 \approx 3.14(49\text{ in}^2)$

$\approx 153.86\text{ in}^2$

65) $V=(60\text{ ft})(19\text{ ft})(50\text{ ft})=57000\text{ ft}^3$

Multiply to convert to gallons.

$$57000\text{ ft}^3\cdot\left(\frac{7.48\text{ gallons}}{1\text{ ft}^3}\right)$$
$$=426,360\text{ gallons}$$

67) $A=(15\text{ ft})(50\text{ ft})+(8\text{ ft})^2$
$$=750\text{ ft}^2+64\text{ ft}^2$$
$$=814\text{ ft}^2$$

Multiply by the rate to determine if Micah can afford the office.

$$814\text{ ft}^2\cdot\left(\frac{\$1.50}{\text{ft}^2}\right)=\$1221$$

Yes, Micah can afford the office.

69) $C=2\pi(4.7\text{ in})$
$$\approx 9.4\text{ in}(3.14)$$
$$\approx 29.516\text{ in}$$
$$\approx 29.5\text{ in}$$

71) $P=2(9\text{ ft})+2(5\text{ ft})$
$$=18\text{ ft}+10\text{ ft}=28\text{ ft}$$

$$28\text{ ft}\cdot\left(\frac{\$160}{\text{ft}}\right)=\$44.80$$

73) $V=\frac{1}{3}\pi(2\text{ ft})^2(6\text{ ft})$
$$=\frac{1}{3}\pi(4\text{ ft}^2)(6\text{ ft})$$
$$\approx(3.14)(4\text{ ft}^2)(2\text{ ft})$$
$$\approx 25.12\text{ ft}^3$$

Section 1.4: Exercises

1) a) $6,0$

b) $-14,6,0$

c) $\sqrt{19}$

d) 6

e) $-14,6,\frac{2}{5},\sqrt{19},0,3.\overline{28},-1\frac{3}{7},0.95$

f) $-14,6,\frac{2}{5},\sqrt{19},0,3.\overline{28},-1\frac{3}{7},0.95$

3) True

5) False

7) True

9)

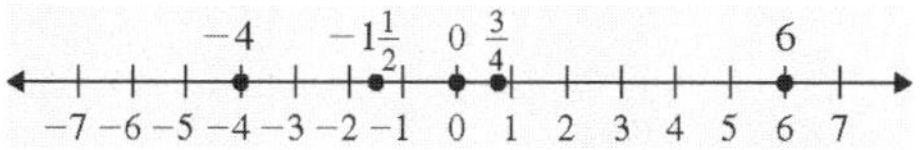

11)

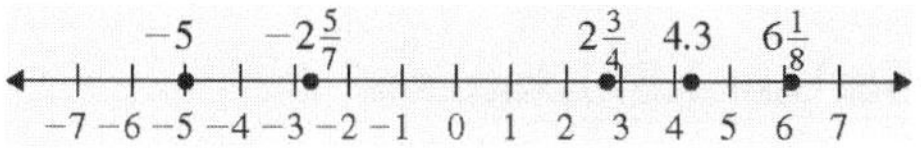

13)

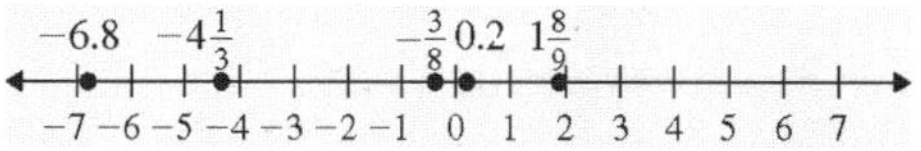

15) $|-13|=13$

17) $\left|\frac{3}{2}\right|=\frac{3}{2}$

19) $-|10|=-10$

21) $-|-19|=-19$

23) -11

25) 7

27) 4.2

29) $-10,\ -2,\ 0,\ \frac{9}{10},\ 3.8,\ 7$

31) $-9, -4\frac{1}{2}, -0.3, \frac{1}{4}, \frac{5}{8}, 1$

33) True

35) True

37) False

39) False

41) True

43) -27

45) 1.1%

47) $-4{,}371$

49) $3{,}000$

Section 1.5: Exercises

1) Answers may vary.

3) Answers may vary.

5) $6-11=-5$

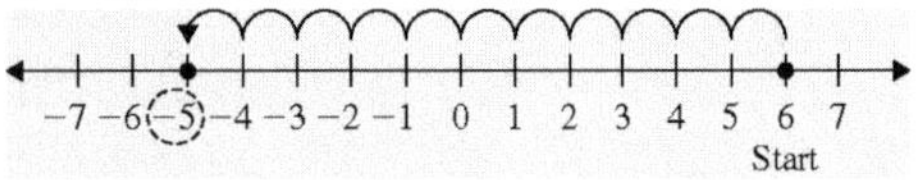

7) $-2+(-7)=-9$

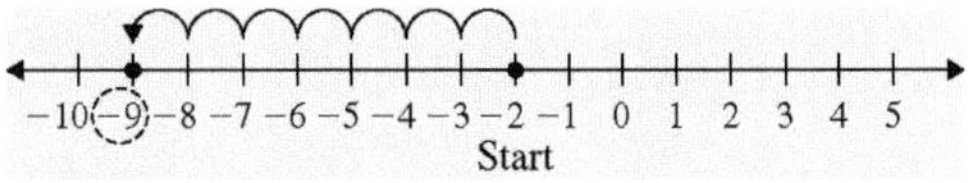

9) $9+(-13)=-4$

11) $-2-12=-14$

13) $-25+38=13$

15) $-1-(-19)=-1+19=18$

17) $-794-657=-794+(-657)=-1451$

19) $-\frac{3}{10}+\frac{4}{5}=-\frac{3}{10}+\frac{8}{10}=\frac{5}{10}=\frac{1}{2}$

21) $-\frac{5}{8}-\frac{2}{3}=-\frac{15}{24}+\left(-\frac{16}{24}\right)=-\frac{31}{24}$

23) $-\frac{11}{12}-\left(-\frac{5}{9}\right)=-\frac{33}{36}+\frac{20}{36}=-\frac{13}{36}$

25) $7.3-11.2=7.3+(-11.2)=-3.9$

27) $-5.09-(-12.4)=-5.09+12.4=7.31$

29) $9-(5-11)=9-(-6)=9+6=15$

31) $-1+(-6-4)=-1+(-10)=-11$

33) $(-3-1)-(-8+6)=(-4)-(-2)$
$=-4+2$
$=-2$

35) $-16+4+3-10=-12+3-10$
$=-9-10$
$=-19$

37) $5-(-30)-14+2=5+30-14+2$
$=35-14+2$
$=21+2$
$=23$

39) $\frac{4}{9}-\left(\frac{2}{3}+\frac{5}{6}\right)=\frac{8}{18}-\left(\frac{12}{18}+\frac{15}{18}\right)$
$=\frac{8}{18}-\left(\frac{27}{18}\right)$
$=-\frac{19}{18}$

41) $\left(\frac{1}{8}-\frac{1}{2}\right)+\left(\frac{3}{4}-\frac{1}{6}\right)$
$=\left(\frac{3}{24}-\frac{12}{24}\right)+\left(\frac{18}{24}-\frac{4}{24}\right)$
$=-\frac{9}{24}+\frac{14}{24}$
$=\frac{5}{24}$

43) $(2.7+3.8)-(1.4-6.9)=6.5-(-5.5)$
$=6.5+5.5$
$=12$

45) $|7-11|+|6+(-13)|=|-4|+|-7|$
$=4+7$
$=11$

47) $-|-2-(-3)|-2|-5+8|$
$=-|-2+3|-2|3|$
$=-|1|-2(3)$
$=-1-6$
$=-7$

49) True

51) False

53) True

55) $29{,}028-(-36{,}201)$
$=29{,}028+36{,}201$
$=65{,}229$
There is a 65,229 ft difference between Mt. Everest and the Mariana Trench.

57) $51{,}700-51{,}081=619$
The median income for a male with a bachelor's degree increased by $619 from 2004 to 2005.

59) $-79.8+213.8=134$
The highest temperature on record in the U.S. is 134° F.

61) $7+4+1+6-10=11+1+6-10$
$=12+6-10$
$=18-10$
$=8$
The Patriots' net yardage on this offensive drive was 8 yards.

63) a) $19.5-17.6=1.9$; 1900

b) $18.3-19.5=-1.2$; -1200

c) $19.8-18.3=1.5$; 1500

d) $17.7-19.8=-2.1$; -2100

65) $5+7$; 12

67) $10-16$; -6

69) $9-(-8)$; 17

71) $-21+13$; -8

73) $-20+30$; 10

75) $23-19$; 4

77) $(-5+11)-18$; $6-18$; -12

Section 1.6: Exercises

1) Positive

3) Negative

5) $-5\cdot 9=-45$

7) $-14\cdot(-3)=42$

9) $-2\cdot5\cdot(-3)=-10\cdot(-3)=30$

11) $\frac{7}{9}\cdot\left(-\frac{6}{5}\right)=\frac{7}{\cancel{9}_3}\cdot\left(-\frac{\cancel{6}^2}{5}\right)=-\frac{14}{15}$

13) $(-0.25)(1.2)=-0.3$

15) $8\cdot(-2)\cdot(-4)\cdot(-1)=-16\cdot4=-64$

17) $(-8)\cdot(-9)\cdot0\cdot\left(-\frac{1}{4}\right)\cdot(-2)=0$

19) Negative

21) Positive

23) $(-6)^2=36$

25) $-5^3=-125$

27) $(-3)^2=9$

29) $-7^2=-49$

31) $-2^5=-32$

33) $-42\div(-6)=7$

35) $\frac{56}{-7}=-8$

37) $\frac{-3.6}{0.9}=-4$

39) $-\frac{12}{13}\div\left(-\frac{6}{5}\right)=-\frac{\cancel{12}^2}{13}\cdot\left(-\frac{5}{\cancel{6}}\right)=\frac{10}{13}$

41) $\frac{0}{-4}=0$

43) $\frac{360}{-280}=-\frac{360\div40}{280\div40}=-\frac{9}{7}=-1\frac{2}{7}$

45) $7+8(-5)=7-40=-33$

47) $(9-14)^2-(-3)(6)=(-5)^2-(-3)(6)$
$=25-(-3)(6)$
$=25+18$
$=43$

49) $10-2(1-4)^3\div9=10-2(-3)^3\div9$
$=10-2(-27)\div9$
$=10+54\div9$
$=10+6=16$

51) $\left(-\frac{3}{4}\right)(8)-2[7-(-3)(-6)]$
$=-6-2[7-18]$
$=-6-2[-11]$
$=-6+22=16$

53) $\frac{-46-3(-12)}{(-5)(-2)(-4)}=\frac{-46+36}{-40}$
$=\frac{-10}{-40}=\frac{1}{4}$

55) $-12\cdot6;\ -72$

57) $9+(-7)(-5);\ 9+35;\ 44$

59) $\frac{63}{-9}+7;\ -7+7;\ 0$

61) $(-4)(-8)-19;\ 32-19;\ 13$

63) $\frac{-100}{4}-(-7+2);\ -25-(-7+2);$
$-25-(-5);\ -25+5;\ -20$

65) $2[18+(-31)]$; $2[-13]$; -26

67) $\frac{2}{3}(-27)$; $\frac{2}{\cancel{3}}\left(-\overset{9}{\cancel{27}}\right)$; -18

69) $12(-5)+\frac{1}{2}(36)$; $-60+18$; -42

Section 1.7: Exercises

1) The constant is 4.

Term	Coeff.
$7p^2$	7
$-6p$	-6
4	4

3) The constant is 11.

Term	Coeff.
x^2y^2	1
$2xy$	2
$-y$	-1
11	11

5) The constant is -1.

Term	Coeff.
$-2g^5$	-2
$\frac{g^4}{5}$	$\frac{1}{5}$
$3.8g^2$	3.8
g	1
-1	-1

7) a) $4c+3$ when $c=2$

$$=4(2)+3$$
$$=8+3$$
$$=11$$

b) $4c+3$ when $c=-5$

$$=4(-5)+3$$
$$=-20+3$$
$$=-17$$

9) a) $2j^2+3j-7$ when $j=4$

$$=2(4)^2+3(4)-7$$
$$=2(16)+3(4)-7$$
$$=32+12-7$$
$$=37$$

b) $2j^2+3j-7$ when $j=-5$

$$=2(-5)^2+3(-5)-7$$
$$=2(25)+3(-5)-7$$
$$=50-15-7$$
$$=28$$

11) $8x+y=8(-2)+7=-16+7=-9$

13) $x^2+xy+10=(-2)^2+(-2)(7)+10$
$$=4+(-14)+10=0$$

15) $z^3-x^3=(-3)^3-(-2)^3$
$$=-27-(-8)$$
$$=-27+8=-19$$

17) $\frac{2x}{y+z}=\frac{2(-2)}{(7)+(-3)}=\frac{-4}{4}=-1$

19) $\frac{x^2-y^2}{2z^2+y}=\frac{(-2)^2-(7)^2}{2(-3)^2+(7)}$
$$=\frac{4-49}{2(9)+(7)}$$
$$=\frac{-45}{18+(7)}$$
$$=\frac{-45}{25}=-\frac{9}{5}$$

21) 0

23) $\frac{1}{6}$

25) Associative

27) Commutative

29) Associative

31) Distributive

33) Identity

35) Distributive

37) $7(u+v)=7u+7v$

39) $k+4=4+k$

41) $m+0=m$

43) $(-5+3)+6=-5+(3+6)$

45) No. Subtraction is not commutative.

47) $5(4+3)=5\cdot4+5\cdot3=20+15=35$

49) $-2(5+7)=(-2)5+(-2)7$
$=-10+(-14)$
$=-24$

51) $4(11-3)=4\cdot11+4\cdot(-3)$
$=44+(-12)$
$=32$

53) $-(9-5)=-9+5=-4$

55) $(8-2)4=8\cdot4+(-2)\cdot4$
$=32+(-8)$
$=24$

57) $2(-6+5+3)=2\cdot(-6)+2\cdot5+2\cdot3$
$=-12+10+6$
$=4$

59) $9(g+6)=9g+9\cdot6=9g+54$

61) $4(t-5)=4t+4\cdot(-5)=4t-20$

63) $-5(z+3)=-5z+(-5)\cdot3=-5z-15$

65) $-8(u-4)=-8u+(-8)\cdot(-4)$
$=-8u+32$

67) $-(v-6)=-v+6$

69) $10(m+5n-3)$
$=10m+10\cdot5n+10\cdot(-3)$
$=10m+50n-30$

71) $-(-8c+9d-14)=8c-9d+14$

Chapter 1 Review

1) $16=1\cdot16$
$16=2\cdot8$
$16=4\cdot4$
The factors of 16 are 1, 2, 4, 8, 16.

3) $28=2\cdot2\cdot7$

5) $\frac{12}{30}=\frac{12\div6}{30\div6}=\frac{2}{5}$

7) $\frac{4}{11}\cdot\frac{3}{5}=\frac{12}{55}$

9) $\frac{5}{8}\div\frac{3}{10}=\frac{5}{8}\cdot\frac{10}{3}=\frac{5}{\cancel{8}_4}\cdot\frac{\cancel{10}^5}{3}=\frac{25}{12}=2\frac{1}{12}$

11) $4\frac{2}{3}\cdot 1\frac{1}{8}=\frac{14}{3}\cdot\frac{9}{8}=\frac{\overset{7}{\cancel{14}}}{\cancel{3}}\cdot\frac{\overset{3}{\cancel{9}}}{\underset{4}{\cancel{8}}}=\frac{21}{4}=5\frac{1}{4}$

13) $\frac{2}{9}+\frac{4}{9}=\frac{6}{9}=\frac{2}{3}$

15) $\frac{9}{40}+\frac{7}{16}=\frac{18}{80}+\frac{35}{80}=\frac{53}{80}$

17) $\frac{21}{25}-\frac{11}{25}=\frac{10}{25}=\frac{2}{5}$

19) $3\frac{2}{9}+5\frac{3}{8}=3\frac{16}{72}+5\frac{27}{72}$
$=8\frac{16+27}{72}=8\frac{43}{72}$

21) $2\cdot 1\frac{7}{8}=\cancel{2}\cdot\frac{15}{\underset{4}{\cancel{8}}}=\frac{15}{4}=3\frac{3}{4}$ yd

23) $2^6=2\cdot 2\cdot 2\cdot 2\cdot 2\cdot 2=64$

25) $(0.6)^2=(0.6)\cdot(0.6)=0.36$

27) $8\cdot 3+20\div 4=24+5=29$

29) $90^\circ-51^\circ=39^\circ$

31) the sum of the two known angles
$78^\circ+46^\circ=124^\circ$
the measure of the unknown angle
$180^\circ-124^\circ=56^\circ$
The triangle is acute since all three angles are acute.

33) $A=\frac{1}{2}bh=\frac{1}{2}(18\text{ in})(5\text{ in})=45\text{ in}^2$

$P=18\text{ in}+13\text{ in}+7.8\text{ in}=38.8\text{ in}$

35) $A=(23\text{ m})(13\text{ m})+(11\text{ m})(7\text{ m})$
$=299\text{ m}^2+77\text{ m}^2$
$=376\text{ m}^2$

$P=2(23\text{ m})+2(20\text{ m})$
$=46\text{ m}+40\text{ m}$
$=86\text{ m}$

37) a) $A=\pi r^2$
$=\pi(2.5\text{ m})^2$
$=6.25\pi\text{ m}^2$
$\approx 19.625\text{ m}^2$

b) $C=2\pi r$
$=2\pi(2.5\text{ m})$
$=5\pi\text{ m}$
$\approx 15.7\text{ cm}$

39) $V=lwh$
$=(13\text{ in})(9\text{ in})(8\text{ in})$
$=936\text{ in}^3$

41) $V=\frac{1}{3}\pi r^2h$
$=\frac{1}{3}\pi(5\text{ ft})^2(16\text{ ft})$
$=\frac{1}{3}\pi(25\text{ ft}^2)(16\text{ ft})$
$=\frac{1}{3}\pi(400\text{ ft}^3)=\frac{400}{3}\pi\text{ ft}^3$

43) $V=\pi r^2h$
$=\pi(1.5\text{ in})^2(1\text{ in})$
$=\pi(2.25\text{ in}^2)(1\text{ in})$
$=2.25\pi\text{ in}^3$

45) 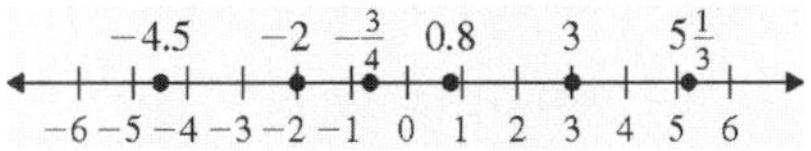

47) $-18+4=-14$

49) $$-\frac{5}{8}+\left(-\frac{2}{3}\right)=-\frac{15}{24}+\left(-\frac{16}{24}\right)$$
$$=-\frac{31}{24}$$
$$=-1\frac{7}{24}$$

51) $-66^{\circ}+49^{\circ}=-17^{\circ}$ F

53) $\left(-\frac{2}{3}\right)(15)=\left(-\frac{2}{\cancel{3}}\right)\left(\overset{5}{\cancel{15}}\right)=-10$

55) $(-3)(-5)(-2)=-30$

57) $-54\div 6=-9$

59) $\frac{38}{-44}=-\frac{19}{22}$

61) $-\frac{8}{9}\div(-4)=-\frac{\overset{2}{\cancel{8}}}{9}\cdot\left(-\frac{1}{\cancel{4}}\right)=\frac{2}{9}$

63) $(-5)^2=25$

65) $(-1)^9=-1$

67) $64\div(-8)+6=-8+6=-2$

69) $$-11-3\cdot 9+(-2)^1=-11-3\cdot 9+(-2)$$
$$=-11-27+(-2)$$
$$=-40$$

71) $\frac{-120}{-3}$; 40

73) $(-4)\cdot 7-15;\ -28-15;\ -43$

75) The constant is 11.

Term	Coeff.
c^4	1
$12c^3$	12
$-c^2$	−1
$-3.8c$	−3.8
11	11

77) $\frac{t-6s}{s^2-t^2}$ when $s=-4$ and $t=5$
$$=\frac{(5)-6(-4)}{(-4)^2-(5)^2}$$
$$=\frac{5+24}{16-25}$$
$$=\frac{29}{-9}$$
$$=-\frac{29}{9}=-3\frac{2}{9}$$

79) Associative

81) Commutative

83) $3(10-6)=3\cdot 10-3\cdot 6=30-18=12$

85) $-(12+5)=-12-5=-17$

Chapter 1 Test

1) $210=2\cdot 3\cdot 5\cdot 7$

3) $\frac{9}{14}\cdot\frac{7}{24}=\frac{\overset{3}{\cancel{9}}}{\underset{2}{\cancel{14}}}\cdot\frac{\cancel{7}}{\underset{8}{\cancel{24}}}=\frac{3}{16}$

5) $5\frac{1}{4}-2\frac{1}{6}=5\frac{3}{12}-2\frac{2}{12}=3\frac{3-2}{12}=3\frac{1}{12}$

7) $\frac{4}{7}-\frac{5}{6}=\frac{24}{42}-\frac{35}{42}=-\frac{11}{42}$

9) $25+15\div15=25+3=28$

11) $-8\cdot(-6)=48$

13) $30-5\left[-10+(2-6)^2\right]$

$=30-5\left[-10+(-4)^2\right]$

$=30-5[-10+16]$

$=30-5[6]$

$=30-30$

$=0$

15) $14,505\text{ ft}-(-282\text{ ft})=14,505\text{ ft}+282\text{ ft}=14,787\text{ ft}$

17) $-3^4=-81$

19) $|2-12|-4|6-1|=|-10|-4|5|$

$=10-20=-10$

21) the sum of the two known angles

$112^\circ+42^\circ=154^\circ$

the measure of the unknown angle

$180^\circ-154^\circ=26^\circ$

The triangle is obtuse since it contains one obtuse angle.

23) a) $A=lw=(10\text{ in})(4\text{ in})=40\text{ in}^2$

$P=2l+2w$

$=2(10\text{ in})+2(4\text{ in})$

$=20\text{ in}+8\text{ in}=28\text{ in}$

b) $A=b\cdot h+\frac{1}{2}\cdot b\cdot h$

$=(12\text{ yd})(6\text{ yd})+\frac{1}{2}(3\text{ yd})(4\text{ yd})$

$=72\text{ yd}^2+6\text{ yd}^2=78\text{ yd}^2$

$P=(9+12+6+8+5)\text{ yd}$

$=40\text{ yd}$

25) a) $C=2\pi r=2\pi(1.5\text{ in})=3\pi\text{ in}$

b) $V=\frac{4}{3}\pi r^3=\frac{4}{3}\pi(1.5\text{ in})^3$

$=\frac{4}{3}\pi(3.375\text{ in}^3)=4.5\pi\text{ in}^3$

27)

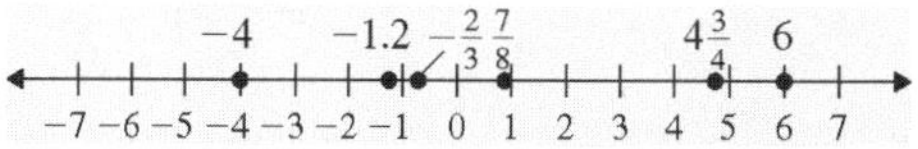

29) The constant is -14.

Term	Coeff.
$5a^2b^2$	5
$2a^2b$	2
$\frac{8}{9}ab$	$\frac{8}{9}$
a	1
$-\frac{b}{7}$	$-\frac{1}{7}$
-14	-14

31) a) Distributive

b) Inverse

c) Commutative

Section 2.1: Exercises

1) No; the exponents are different.

3) Yes; both are x^4y^3- terms.

5) $10p+9+14p-2=10p+14p+9-2$
$=24p+7$

7) $-18y^2-2y^2+19+y^2-2+13$
$=-18y^2-2y^2+y^2+19-2+13$
$=-19y^2+30$

9) $\frac{4}{9}+3r-\frac{2}{3}+\frac{1}{5}r=3r+\frac{1}{5}r+\frac{4}{9}-\frac{2}{3}$
$=\frac{15}{5}r+\frac{1}{5}r+\frac{4}{9}-\frac{6}{9}$
$=\frac{16}{5}r-\frac{2}{9}$

11) $2(3w+5)+w=6w+10+w$
$=6w+w+10$
$=7w+10$

13) $9-4(3-x)-4x+3$
$=9-12+4x-4x+3$
$=4x-4x+9-12+3$
$=0x+0=0$

15) $3g-(8g+3)+5=3g-8g-3+5$
$=-5g+2$

17) $-5(t-2)-(10-2t)$
$=-5t+10-10+2t$
$=-5t+2t+10-10$
$=-3t$

19) $3[2(5x+7)-11]+4(7-x)$
$=3[10x+14-11]+28-4x$
$=3[10x+3]+28-4x$
$=30x+9+28-4x$
$=30x-4x+9+28$
$=26x+37$

21) $\frac{4}{5}(2z+10)-\frac{1}{2}(z+3)$
$=\frac{8}{5}z+8-\frac{1}{2}z-\frac{3}{2}$
$=\frac{8}{5}z-\frac{1}{2}z+8-\frac{3}{2}$
$=\frac{16}{10}z-\frac{5}{10}z+\frac{16}{2}-\frac{3}{2}$
$=\frac{11}{10}z+\frac{13}{2}$

23) $1+\frac{3}{4}(10t-3)+\frac{5}{8}\left(t+\frac{1}{10}\right)$
$=1+\frac{15}{2}t-\frac{9}{4}+\frac{5}{8}t+\frac{1}{16}$
$=\frac{15}{2}t+\frac{5}{8}t+1-\frac{9}{4}+\frac{1}{16}$
$=\frac{60}{8}t+\frac{5}{8}t+\frac{16}{16}-\frac{36}{16}+\frac{1}{16}$
$=\frac{65}{8}t-\frac{19}{16}$

25) $2.5(x-4)-1.2(3x+8)$
$=2.5x-10-3.6x-9.6$
$=2.5x-3.6x-10-9.6$
$=-1.1x-19.6$

27) $18+x$

29) $x-6$

31) $x-3$

33) $12+2x$

35) $(3+2x)-7=2x+3-7=2x-4$

37) $(x+15)-5=x+15-5=x+10$

Section 2.2a: Exercises

1) $9\cdot 9\cdot 9\cdot 9=9^4$

3) $\left(\frac{1}{7}\right)\left(\frac{1}{7}\right)\left(\frac{1}{7}\right)\left(\frac{1}{7}\right)\left(\frac{1}{7}\right)=\left(\frac{1}{7}\right)^5$

5) $(-5)(-5)(-5)(-5)(-5)(-5)(-5)$
$=(-5)^7$

7) $(-3y)(-3y)(-3y)(-3y)$
$\cdot(-3y)(-3y)(-3y)(-3y)$
$=(-3y)^8$

9) base: 6; exponent: 8

11) base: 0.05; exponent: 7

13) base: -8; exponent: 5

15) base: $9x$; exponent: 8

17) base: $-11a$; exponent: 2

19) base: p; exponent: 4

21) base: y; exponent: 2

23) $(3+4)^2=49$, $3^2+4^2=25$. They are not equivalent because when evaluating $(3+4)^2$, first add $3+4$ to get 7, then square the 7.

25) Answers may vary.

27) No. $3t^4=3\cdot t^4$,
$(3t)^4=3^4\cdot t^4=81t^4$

29) $2^5=32$

31) $(11)^2=121$

33) $(-2)^4=16$

35) $-3^4=-81$

37) $-2^3=-8$

39) $\left(\frac{1}{5}\right)^3=\frac{1}{125}$

41) $2^2\cdot 2^4=2^{2+4}=2^6=64$

43) $3^2\cdot 3^2=3^{2+2}=3^4=81$

45) $5^2\cdot 2^3=25\cdot 8=200$

47) $\left(\frac{1}{2}\right)^4\cdot\left(\frac{1}{2}\right)=\left(\frac{1}{2}\right)^{4+1}=\left(\frac{1}{2}\right)^5=\frac{1}{32}$

49) $8^3\cdot 8^9=8^{3+9}=8^{12}$

51) $7^5\cdot 7\cdot 7^4=7^{5+1+4}=7^{10}$

53) $(-4)^2\cdot(-4)^3\cdot(-4)^2=(-4)^{2+3+2}$
$=(-4)^7$

55) $a^2\cdot a^3=a^{2+3}=a^5$

57) $k\cdot k^2\cdot k^3=k^{1+2+3}=k^6$

59) $8y^3\cdot y^2=8\left(y^3\cdot y^2\right)=8y^5$

61) $(9m^4)(6m^{11}) = (9\cdot 6)(m^4 \cdot m^{11})$
$= 54m^{15}$

63) $(-6r)(7r^4) = (-6\cdot 7)(r\cdot r^4) = -42r^5$

65) $(-7t^6)(t^3)(-4t^7)$
$= [-7\cdot 1\cdot(-4)](t^6\cdot t^3\cdot t^7)$
$= 28t^{16}$

67) $\left(\frac{5}{3}x^2\right)(12x)(-2x^3)$
$= \left[\frac{5}{3}\cdot 12\cdot(-2)\right](x^2\cdot x\cdot x^3)$
$= -40x^6$

69) $\left(\frac{8}{21}b\right)(-6b^8)\left(-\frac{7}{2}b^6\right)$
$= \left[\frac{8}{21}\cdot(-6)\cdot\left(-\frac{7}{2}\right)\right](b\cdot b^8\cdot b^6)$
$= 8b^{15}$

71) $(x^4)^3 = x^{4\cdot 3} = x^{12}$

73) $(t^6)^7 = t^{6\cdot 7} = t^{42}$

75) $(2^3)^2 = 2^{3\cdot 2} = 2^6 = 64$

77) $(-5^3)^2 = (-5)^{3\cdot 2} = (-5)^6$

79) $\left(\frac{1}{2}\right)^5 = \frac{1}{2^5} = \frac{1}{32}$

81) $\left(\frac{4}{y}\right)^3 = \frac{4^3}{y^3} = \frac{64}{y^3}$

83) $\left(\frac{d}{c}\right)^8 = \frac{d^8}{c^8}$

85) $(5z)^3 = 5^3z^3 = 125z^3$

87) $(-3p)^4 = (-3)^4\cdot p^4 = 81p^4$

89) $(-4ab)^3 = (-4)^3\cdot a^3\cdot b^3 = -64a^3b^3$

91) $6(xy)^3 = 6\cdot x^3\cdot y^3 = 6x^3y^3$

93) $-9(tu)^4 = -9\cdot t^4\cdot u^4 = -9t^4u^4$

95) a) $A = lw$
$= (3w)(w)$
$= 3(w\cdot w)$
$= 3w^2$ sq. units

$P = 2l + 2w$
$= 2(3w) + 2(w)$
$= 6w + 2w$
$= 8w$ units

b) $A = lw$
$= (5k^3)(k^2)$
$= 5(k^3\cdot k^2)$
$= 5k^5$ sq. units

$P = 2l + 2w$
$= 2(5k^3) + 2(k^2)$
$= 10k^3 + 2k^2$ units

97) $A=\frac{1}{2}bh$

$=\frac{1}{2}(x)\left(\frac{3}{4}x\right)$

$=\left(\frac{1}{2}\cdot\frac{3}{4}\right)(x\cdot x)$

$=\frac{3}{8}x^2$ sq. units

Section 2.2b: Exercises

1) operations

3) $\left(k^9\right)^2\left(k^3\right)^2=\left(k^{18}\right)\left(k^6\right)=k^{24}$

5) $\left(5z^4\right)^2\left(2z^6\right)^3=\left(5^2\right)\left(z^4\right)^2\left(2^3\right)\left(z^6\right)^3$

$=\left(25z^8\right)\left(8z^{18}\right)$

$=200z^{26}$

7) $9pq\left(-p^{10}q^3\right)^5$

$=9pq\cdot(-1)^5\left(p^{10}\right)^5\left(q^3\right)^5$

$=9pq\cdot\left(-1p^{50}q^{15}\right)$

$=-9p^{51}q^{16}$

9) $(5+3)^2=8^2=64$

11) $\left(-4t^6u^2\right)^3\left(u^4\right)^5$

$=(-4)^3\left(t^6\right)^3\left(u^2\right)^3\cdot u^{20}$

$=\left(-64t^{18}u^6\right)\cdot u^{20}$

$=-64t^{18}u^{26}$

13) $8\left(6k^7l^2\right)^2=8\left(6^2\right)\left(k^7\right)^2\left(l^2\right)^2$

$=8(36)k^{14}l^4$

$=288k^{14}l^4$

15) $\left(\frac{3}{g^5}\right)^3\left(\frac{1}{6}\right)^2=\frac{3^3}{\left(g^5\right)^3}\cdot\frac{1}{6^2}=\frac{27}{g^{15}}\cdot\frac{1}{36}$

$=\frac{27}{36g^{15}}=\frac{\overset{3}{\cancel{27}}}{\underset{4}{\cancel{36}}\,g^{15}}$

$=\frac{3}{4g^{15}}$

17) $\left(\frac{7}{8}n^2\right)^2\left(-4n^9\right)^2$

$=\left(\frac{7^2}{8^2}\right)\left(n^2\right)^2(-4)^2\left(n^9\right)^2$

$=\frac{49}{64}\left(n^4\right)(16)\left(n^{18}\right)$

$=\frac{49}{\underset{4}{\cancel{64}}}\left(n^4\right)\left(\cancel{16}\right)\left(n^{18}\right)$

$=\frac{49}{4}n^{22}$

19) $h^4\left(10h^3\right)^2\left(-3h^9\right)^2$

$=h^4\left(10^2\right)\left(h^3\right)^2(-3)^2\left(h^9\right)^2$

$=h^4(100)\left(h^6\right)(9)\left(h^{18}\right)$

$=900h^{28}$

21) $3w^{11}\left(7w^2\right)^2\left(-w^6\right)^5$

$=3w^{11}\left(7^2\right)\left(w^2\right)^2(-1)^5\left(w^6\right)^5$

$=3w^{11}(49)\left(w^4\right)(-1)\left(w^{30}\right)$

$=-147w^{45}$

23) $\dfrac{(12x^3)^2}{(10y^5)^2}=\dfrac{(12^2)(x^3)^2}{(10^2)(y^5)^2}$
$=\dfrac{144x^6}{100y^{10}}$
$=\dfrac{36x^6}{25y^{10}}$

25) $\dfrac{(4d^9)^2}{(-2c^5)^6}=\dfrac{(4^2)(d^9)^2}{(-2)^6(c^5)^6}$
$=\dfrac{16d^{18}}{64c^{30}}$
$=\dfrac{d^{18}}{4c^{30}}$

27) $\dfrac{6(a^8b^3)^5}{(2c)^3}=\dfrac{6(a^8)^5(b^3)^5}{(2^3)(c^3)}$
$=\dfrac{6a^{40}b^{15}}{8c^3}$
$=\dfrac{3a^{40}b^{15}}{4c^3}$

29) $\dfrac{r^4(r^5)^7}{2t(11t^2)^2}=\dfrac{r^4(r^{35})}{2t(11^2)(t^2)^2}$
$=\dfrac{r^{39}}{2t(121t^4)}$
$=\dfrac{r^{39}}{242t^5}$

31) $\left(\dfrac{4}{9}x^3y\right)^2\left(\dfrac{3}{2}x^6y^4\right)^3$
$=\left(\dfrac{4^2}{9^2}\right)(x^3)^2(y^2)\left(\dfrac{3^3}{2^3}\right)(x^6)^3(y^4)^3$
$=\left(\dfrac{16}{81}x^6y^2\right)\left(\dfrac{27}{8}x^{18}y^{12}\right)$
$=\left(\dfrac{\cancel{16}^{\,2}}{\cancel{81}_{\,3}}x^6y^2\right)\left(\dfrac{\cancel{27}}{\cancel{8}}x^{18}y^{12}\right)$
$=\dfrac{2}{3}x^{24}y^{14}$

33) $\left(-\dfrac{2}{5}c^9d^2\right)^3\left(\dfrac{5}{4}cd^6\right)^2$
$=(-1)^3\left(\dfrac{2^3}{5^3}\right)(c^9)^3(d^2)^3\left(\dfrac{5^2}{4^2}\right)(c^2)(d^6)^2$
$=\left(-\dfrac{8}{125}c^{27}d^6\right)\left(\dfrac{25}{16}c^2d^{12}\right)$
$=\left(-\dfrac{\cancel{8}}{\cancel{125}_{\,5}}c^{27}d^6\right)\left(\dfrac{\cancel{25}}{\cancel{16}_{\,2}}c^2d^{12}\right)$
$=-\dfrac{1}{10}c^{29}d^{18}$

35) $\left(\dfrac{5x^5y^2}{z^4}\right)^3=\dfrac{(5^3)(x^5)^3(y^2)^3}{(z^4)^3}$
$=\dfrac{125x^{15}y^6}{z^{12}}$

37) $\left(-\dfrac{3t^4u^9}{2v^7}\right)^4=(-1)^4\cdot\dfrac{(3^4)(t^4)^4(u^9)^4}{(2^4)(v^7)^4}$
$=\dfrac{81t^{16}u^{36}}{16v^{28}}$

39) $\left(\frac{12w^5}{4x^3y^6}\right)^2 = \frac{(12^2)(w^5)^2}{(4^2)(x^3)^2(y^6)^2}$

$= \frac{144w^{10}}{16x^6y^{12}}$

$= \frac{9w^{10}}{x^6y^{12}}$

41) a) $P = 4(5l^2 \text{ units}) = 20l^2 \text{ units}$

b) $A = (5l^2 \text{ units})^2$

$= (5^2)(l^2)^2 \text{ sq. units}$

$= 25l^4 \text{ sq. units}$

43) a) $A = (x \text{ units})\left(\frac{3}{8}x \text{ units}\right)$

$= \frac{3}{8}x^2 \text{ sq. units}$

b) $P = 2(x \text{ units}) + 2\left(\frac{3}{8}x \text{ units}\right)$

$= 2x \text{ units} + \frac{3}{4}x \text{ units}$

$= \frac{8}{4}x \text{ units} + \frac{3}{4}x \text{ units}$

$= \frac{11}{4}x \text{ units}$

Section 2.3a: Exercises

1) False

3) True

5) $6^0 = 1$

7) $-4^0 = -1 \cdot 4^0 = -1 \cdot 1 = -1$

9) $0^4 = 0$

11) $(5)^0 + (-5)^0 = 1 + 1 = 2$

13) $6^{-2} = \left(\frac{1}{6}\right)^2 = \frac{1^2}{6^2} = \frac{1}{36}$

15) $2^{-4} = \left(\frac{1}{2}\right)^4 = \frac{1^4}{2^4} = \frac{1}{16}$

17) $5^{-3} = \left(\frac{1}{5}\right)^3 = \frac{1^3}{5^3} = \frac{1}{125}$

19) $\left(\frac{1}{8}\right)^{-2} = 8^2 = 64$

21) $\left(\frac{1}{2}\right)^{-5} = 2^5 = 32$

23) $\left(\frac{4}{3}\right)^{-3} = \left(\frac{3}{4}\right)^3 = \frac{3^3}{4^3} = \frac{27}{64}$

25) $\left(\frac{9}{7}\right)^{-2} = \left(\frac{7}{9}\right)^2 = \frac{7^2}{9^2} = \frac{49}{81}$

27) $\left(-\frac{1}{4}\right)^{-3} = \left(\frac{4}{-1}\right)^3 = \frac{4^3}{(-1)^3} = \frac{64}{-1} = -64$

29) $\left(-\frac{3}{8}\right)^{-2} = \left(\frac{8}{-3}\right)^2 = \frac{8^2}{(-3)^2} = \frac{64}{9}$

31) $-2^{-6} = -\left(\frac{1}{2}\right)^6 = -\frac{1^6}{2^6} = -\frac{1}{64}$

33) $-1^{-5} = -(1)^5 = -1$

35) $2^{-3}-4^{-2}=\left(\frac{1}{2}\right)^3-\left(\frac{1}{4}\right)^2$

$=\frac{1^3}{2^3}-\frac{1^2}{4^2}=\frac{1}{8}-\frac{1}{16}$

$=\frac{2}{16}-\frac{1}{16}=\frac{1}{16}$

37) $2^{-2}+3^{-2}=\left(\frac{1}{2}\right)^2+\left(\frac{1}{3}\right)^2$

$=\frac{1^2}{2^2}+\frac{1^2}{3^2}=\frac{1}{4}+\frac{1}{9}$

$=\frac{9}{36}+\frac{4}{36}=\frac{13}{36}$

39) $-9^{-2}+3^{-3}+(-7)^0$

$=-\left(\frac{1}{9}\right)^2+\left(\frac{1}{3}\right)^3+1$

$=-\frac{1^2}{9^2}+\frac{1^3}{3^3}+1=-\frac{1}{81}+\frac{1}{27}+1$

$=-\frac{1}{81}+\frac{3}{81}+\frac{81}{81}=\frac{83}{81}$

Section 2.3b: Exercises

1) a) w

b) n

c) $2p$

d) c

3) $r^0=1$

5) $-2k^0=-2\cdot 1=-2$

7) $x^0+(2x)^0=1+1=2$

9) $y^{-4}=\left(\frac{1}{y}\right)^4=\frac{1^4}{y^4}=\frac{1}{y^4}$

11) $p^{-1}=\left(\frac{1}{p}\right)^1=\frac{1}{p}$

13) $\left(\frac{a^{-10}}{b^{-3}}\right)=\frac{b^3}{a^{10}}$

15) $\frac{y^{-8}}{x^{-5}}=\frac{x^5}{y^8}$

17) $\frac{x^4}{10y^{-5}}=\frac{x^4y^5}{10}$

19) $9a^4b^{-3}=\frac{9a^4}{b^3}$

21) $\frac{1}{c^{-5}d^{-8}}=c^5d^8$

23) $\frac{8a^6b^{-1}}{5c^{-10}d}=\frac{8a^6c^{10}}{5bd}$

25) $\frac{2z^4}{x^{-7}y^{-6}}=2x^7y^6z^4$

27) $\left(\frac{a}{6}\right)^{-2}=\left(\frac{6}{a}\right)^2=\frac{6^2}{a^2}=\frac{36}{a^2}$

29) $\left(\frac{2n}{q}\right)^{-5}=\left(\frac{q}{2n}\right)^5=\frac{q^5}{2^5n^5}=\frac{q^5}{32n^5}$

31) $\left(\frac{12b}{cd}\right)^{-2}=\left(\frac{cd}{12b}\right)^2=\frac{c^2d^2}{12^2b^2}=\frac{c^2d^2}{144b^2}$

33) $-9k^{-2}=-9\cdot\frac{1}{k^2}=-\frac{9}{k^2}$

35) $3t^{-3}=3\cdot\frac{1}{t^3}=\frac{3}{t^3}$

37) $-m^{-9} = -1 \cdot \frac{1}{m^9} = -\frac{1}{m^9}$

39) $\left(\frac{1}{z}\right)^{-10} = z^{10}$

41) $\left(\frac{1}{j}\right)^{-1} = j$

43) $5\left(\frac{1}{n}\right)^{-2} = 5n^2$

45) $c\left(\frac{1}{d}\right)^{-3} = cd^3$

Section 2.4: Exercises

1) $\frac{d^{12}}{d^5} = d^{12-5} = d^7$

3) $\frac{m^7}{m^5} = m^{7-5} = m^2$

5) $\frac{9t^{11}}{t^6} = 9t^{11-6} = 9t^5$

7) $\frac{6^{15}}{6^{13}} = 6^{15-13} = 6^2 = 36$

9) $\frac{3^{11}}{3^7} = 3^{11-7} = 3^4 = 81$

11) $\frac{2^3}{2^7} = 2^{3-7} = 2^{-4} = \left(\frac{1}{2}\right)^4 = \frac{1}{16}$

13) $\frac{5^6}{5^9} = 5^{6-9} = 5^{-3} = \left(\frac{1}{5}\right)^3 = \frac{1}{125}$

15) $\frac{10k^4}{k} = 10k^{4-1} = 10k^3$

17) $\frac{20c^{11}}{30c^6} = \frac{\overset{2}{\cancel{20}}\, c^{11}}{\underset{3}{\cancel{30}}\, c^6} = \frac{2}{3}c^{11-6} = \frac{2}{3}c^5$

19) $\frac{z^2}{z^8} = z^{2-8} = z^{-6} = \frac{1}{z^6}$

21) $\frac{x^{-3}}{x^6} = x^{-3-6} = x^{-9} = \frac{1}{x^9}$

23) $\frac{r^{-5}}{r^{-3}} = r^{-5-(-3)} = r^{-5+3} = r^{-2} = \frac{1}{r^2}$

25) $\frac{a^{-1}}{a^9} = a^{-1-9} = a^{-10} = \frac{1}{a^{10}}$

27) $\frac{t^4}{t} = t^{4-1} = t^3$

29) $\frac{15w^2}{w^{10}} = 15w^{2-10} = 15w^{-8} = \frac{15}{w^8}$

31) $\frac{-6k}{k^4} = -6k^{1-4} = -6k^{-3} = -\frac{6}{k^3}$

33) $\frac{a^4b^9}{ab^2} = a^{4-1}b^{9-2} = a^3b^7$

35) $\frac{5m^{-1}n^{-6}}{15m^{-5}n^2} = \frac{\cancel{5}m^{-1}n^{-6}}{\underset{3}{\cancel{15}}\, m^{-5}n^2}$

$= \frac{1}{3}m^{-1-(-5)}n^{-6-2}$

$= \frac{1}{3}m^{-1+5}n^{-6-2}$

$= \frac{1}{3}m^4n^{-8} = \frac{m^4}{3n^8}$

37) $\dfrac{200x^8y^3}{20x^{10}y^{11}} = \dfrac{\overset{10}{\cancel{200}}\, x^8y^3}{\cancel{20}x^{10}y^{11}}$
$= 10x^{8-10}y^{3-11}$
$= 10x^{-2}y^{-8} = \dfrac{10}{x^2y^8}$

39) $\dfrac{6v^{-1}w}{54v^2w^{-5}} = \dfrac{\cancel{6}v^{-1}w}{\underset{9}{\cancel{54}}\, v^2w^{-5}}$
$= \dfrac{1}{9}v^{-1-2}w^{1-(-5)}$
$= \dfrac{1}{9}v^{-3}w^{1+5}$
$= \dfrac{1}{9}v^{-3}w^6 = \dfrac{w^6}{9v^3}$

41) $\dfrac{3c^5d^{-2}}{8cd^{-3}} = \dfrac{3}{8}c^{5-1}d^{-2-(-3)}$
$= \dfrac{3}{8}c^4d^{-2+3} = \dfrac{3}{8}c^4d$

43) $\dfrac{(x+y)^6}{(x+y)^2} = (x+y)^{6-2} = (x+y)^4$

45) $\dfrac{(c+d)^{-5}}{(c+d)^{-11}} = (c+d)^{-5-(-11)}$
$= (c+d)^{-5+11} = (c+d)^6$

Mid-Chapter Summary

1) $\left(\dfrac{2}{3}\right)^4 = \dfrac{2^4}{3^4} = \dfrac{16}{81}$

3) $\dfrac{6^9}{6^5 \cdot 6^4} = \dfrac{6^9}{6^{5+4}} = \dfrac{6^9}{6^9} = 6^{9-9} = 6^0 = 1$

5) $\left(\dfrac{10}{3}\right)^{-2} = \left(\dfrac{3}{10}\right)^2 = \dfrac{3^2}{10^2} = \dfrac{9}{100}$

7) $(9-4)^2 = 5^2 = 25$

9) $9^{-2} = \dfrac{1}{9^2} = \dfrac{1}{81}$

11) $\dfrac{3^7}{3^{11}} = 3^{7-11} = 3^{-4} = \dfrac{1}{3^4} = \dfrac{1}{81}$

13) $\left(-\dfrac{5}{3}\right)^7 \cdot \left(-\dfrac{5}{3}\right)^{-4} = \left(-\dfrac{5}{3}\right)^{7-4}$
$= \left(-\dfrac{5}{3}\right)^3 = -\dfrac{125}{27}$

15) $4^{-2} - 12^{-1} = \left(\dfrac{1}{4}\right)^2 - \dfrac{1}{12}$
$= \dfrac{1}{16} - \dfrac{1}{12}$
$= \dfrac{3}{48} - \dfrac{4}{48} = -\dfrac{1}{48}$

17) $-10\left(-3g^4\right)^3 = -10 \cdot (-27)g^{12}$
$= 270g^{12}$

19) $\dfrac{23t}{t^{11}} = 23t^{1-11} = 23t^{-10} = \dfrac{23}{t^{10}}$

21) $\left(\dfrac{2xy^4}{3x^{-9}y^{-2}}\right)^3 = \left(\dfrac{2}{3}x^{1-(-9)}y^{4-(-2)}\right)^3$
$= \left(\dfrac{2}{3}x^{1+9}y^{4+2}\right)^3$
$= \left(\dfrac{2}{3}x^{10}y^6\right)^3 = \dfrac{8}{27}x^{30}y^{18}$

23) $\left(\dfrac{7r^3}{s^8}\right)^{-2} = \left(\dfrac{s^8}{7r^3}\right)^2 = \dfrac{s^{16}}{49r^6}$

25) $\left(-k^4\right)^3 = -k^{12}$

27) $\left(-2m^5n^2\right)^5 = -32m^5n^{10}$

29) $\left(-\dfrac{9}{4}z^5\right)\left(\dfrac{2}{3}z^{-1}\right) = \left(-\dfrac{\overset{3}{\cancel{9}}}{\underset{2}{\cancel{4}}}z^5\right)\left(\dfrac{\cancel{2}}{\cancel{3}}z^{-1}\right)$

$= -\dfrac{3}{2}z^{5-1} = -\dfrac{3}{2}z^4$

31) $\left(\dfrac{a^5}{b^4}\right)^{-6} = \left(\dfrac{b^4}{a^5}\right)^6 = \dfrac{b^{24}}{a^{30}}$

33) $\left(-ab^3c^5\right)^2\left(\dfrac{a^4}{bc}\right)^3 = \left(a^2b^6c^{10}\right)\left(\dfrac{a^{12}}{b^3c^3}\right)$

$= a^{2+12}b^{6-3}c^{10-3}$

$= a^{14}b^3c^7$

35) $\left(\dfrac{48u^{-7}v^2}{36u^3v^{-5}}\right)^{-3} = \left(\dfrac{36u^3v^{-5}}{48u^{-7}v^2}\right)^3$

$= \left(\dfrac{\overset{3}{\cancel{36}}\,u^3v^{-5}}{\underset{4}{\cancel{48}}\,u^{-7}v^2}\right)^3$

$= \left(\dfrac{3}{4}u^{3-(-7)}v^{-5-2}\right)^3$

$= \left(\dfrac{3}{4}u^{3+7}v^{-7}\right)^3$

$= \left(\dfrac{3}{4}u^{10}v^{-7}\right)^3$

$= \dfrac{27}{64}u^{30}v^{-21} = \dfrac{27u^{30}}{64v^{21}}$

37) $\left(\dfrac{-3t^4u}{t^2u^{-4}}\right)^4 = \left(-3t^{4-2}u^{1-(-4)}\right)^4$

$= \left(-3t^2u^{1+4}\right)^4$

$= \left(-3t^2u^5\right)^4 = 81t^8u^{20}$

39) $\left(h^{-3}\right)^7 = h^{-21} = \dfrac{1}{h^{21}}$

41) $\left(\dfrac{h}{2}\right)^5 = \dfrac{h^5}{32}$

43) $-7c^8\left(-2c^2\right)^3 = -7c^8\cdot(-8)c^6 = 56c^{14}$

45) $\left(12a^7\right)^{-1}(6a)^2 = \dfrac{(6a)^2}{\left(12a^7\right)} = \dfrac{36a^2}{12a^7}$

$= \dfrac{\overset{3}{\cancel{36}}\,a^2}{\cancel{12}a^7} = 3a^{2-7}$

$= 3a^{-5} = \dfrac{3}{a^5}$

47) $\left(\dfrac{9}{20}d^5\right)\left(2d^{-3}\right)\left(\dfrac{4}{33}d^9\right)$

$= \left(\dfrac{\overset{3}{\cancel{9}}}{\underset{5}{\cancel{20}}}d^5\right)\left(2d^{-3}\right)\left(\dfrac{\cancel{4}}{\underset{11}{\cancel{33}}}d^9\right)$

$= \dfrac{6}{55}d^{5-3+9}$

$= \dfrac{6}{55}d^{11}$

49) $\left(\dfrac{56m^4n^8}{21m^4n^5}\right)^{-2} = \left(\dfrac{\overset{3}{\cancel{21}}\,m^4n^5}{\underset{8}{\cancel{56}}\,m^4n^8}\right)^2$

$= \left(\dfrac{3}{8}m^{4-4}n^{5-8}\right)^2$

$= \left(\dfrac{3}{8}m^0n^{-3}\right)^2$

$= \left(\dfrac{3}{8n^3}\right)^2 = \dfrac{9}{64n^6}$

51) $\left(p^{2n}\right)^5 = p^{10n}$

53) $y^{m} \cdot y^{10m} = y^{11m}$

55) $x^{5a} \cdot x^{-8a} = x^{-3a} = \dfrac{1}{x^{3a}}$

57) $\dfrac{21c^{2x}}{35c^{8x}} = \dfrac{\overset{3}{\cancel{21}}\, c^{2x}}{\underset{5}{\cancel{35}}\, c^{8x}} = \dfrac{3}{5}c^{2x-8x} = \dfrac{3}{5}c^{-6x} = \dfrac{3}{5c^{6x}}$

Section 2.5: Exercises

1) Yes

3) No

5) No

7) Yes

9) Answers may vary.

11) Answers may vary.

13) -6.8×10^{-5} : $-00006.8 = -0.000068$

15) -5.26×10^{4} : $-5.2600 = -52,600$

17) 8×10^{-6} : $000008. = 0.000008$

19) 6.021967×10^{5} : 6.021967
$= 602,196.7$

21) 3×10^{6} : $3.000000 = 3,000,000$

23) -7.44×10^{-4} : $-0007.44 = -0.000744$

25) $2110.5 = 2110.5 = 2.1105\times10^{3}$

27) $0.000096 = 0.000096 = 9.6\times10^{-5}$

29) $-7,000,000 = -7,000,000. = -7\times10^{6}$

31) $3400 = 3400. = 3.4\times10^{3}$

33) $0.0008 = 0.0008 = 8\times10^{-4}$

35) $-0.076 = -0.076 = -7.6\times10^{-2}$

37) $6000 = 6000. = 6\times10^{3}$

39) $380,8000,000$ kg $= 3.808\times10^{8}$ kg

41) 0.00000001 cm $= 1\times10^{-8}$ cm

43) $\dfrac{6\times10^{9}}{2\times10^{5}} = \dfrac{6}{2}\times\dfrac{10^{9}}{10^{5}} = 3\times10^{4} = 30,000$

45) $(2.3\times10^{3})(3\times10^{2})$
$= (2.3\times3)(10^{3}\times10^{2})$
$= 6.9\times10^{5}$
$= 690,000$

47) $\dfrac{8.4\times10^{12}}{-7\times10^{9}} = -\dfrac{8.4}{7}\times\dfrac{10^{12}}{10^{9}}$
$= -1.2\times10^{3} = -1200$

49) $(-1.5\times10^{-8})(4\times10^{6})$
$= (-1.5\times4)(10^{-8}\times10^{6})$
$= -6.0\times10^{-2}$
$= -0.06$

51) $\dfrac{-3\times10^{5}}{6\times10^{8}} = -\dfrac{3}{6}\times\dfrac{10^{5}}{10^{8}}$
$= -0.5\times10^{-3} = -0.0005$

53) $(9.75\times10^{4}) + (6.25\times10^{4})$
$= (9.75+6.25)10^{4}$
$= 16\times10^{4}$
$= 160,000$

55) $\left(3.19\times10^{-5}\right)+\left(9.2\times10^{-5}\right)$
$=\left(3.19+9.2\right)10^{-5}$
$=12.39\times10^{-5}$
$=0.0001239$

57) $365\left(1.44\times10^{7}\right)$
$=\left(365\cdot1.44\right)10^{7}$
$=525.6\times10^{7}$
$=5{,}256{,}000{,}000$ particles

59) $\dfrac{2.21\times10^{10}\text{ lb}}{1{,}300{,}000\text{ cow}}=\dfrac{2.21\times10^{10}\text{ lb}}{1.3\times10^{6}\text{ cow}}$
$=\dfrac{2.21}{1.3}\times\dfrac{10^{10}\text{ lb}}{10^{6}\text{ cow}}$
$=1.7\times10^{4}\dfrac{\text{lb}}{\text{cow}}$
$=17{,}000$ lb/cow

61) First determine the area of the photo.
$A=4\text{ in}\cdot6\text{ in}=24\text{ in}^2$
Then multiply by the rate.
$24\text{ in}^2\cdot\dfrac{1.1\times10^{6}\text{ droplets}}{\text{in}^2}$
$=2.4\times10\ \cancel{\text{in}^2}\cdot\dfrac{1.1\times10^{6}\text{ droplets}}{\cancel{\text{in}^2}}$
$=\left(2.4\times1.1\right)\left(10\times10^{6}\right)$ droplets
$=2.64\times10^{7}$ droplets
$=26{,}400{,}000$ droplets

Chapter 2 Review

1) $15y^2+8y-4+2y^2-11y+1$
$=15y^2+2y^2+8y-11y-4+1$
$=17y^2-3y-3$

3) $\dfrac{3}{2}(5n-4)+\dfrac{1}{4}(n+6)$
$=\dfrac{15}{2}n-6+\dfrac{1}{4}n+\dfrac{3}{2}$
$=\dfrac{15}{2}n+\dfrac{1}{4}n-6+\dfrac{3}{2}$
$=\dfrac{30}{4}n+\dfrac{1}{4}n-\dfrac{12}{2}+\dfrac{3}{2}$
$=\dfrac{31}{4}n-\dfrac{9}{2}$

5) $x-10$

7) $(x+8)-3=x+5$

9) a) $8\cdot8\cdot8\cdot8\cdot8\cdot8=8^6$

b) $(-7)(-7)(-7)(-7)=(-7)^4$

11) a) $2^3\cdot2^2=2^{3+2}=2^5=32$

b) $\left(\dfrac{1}{3}\right)^2\cdot\left(\dfrac{1}{3}\right)=\left(\dfrac{1}{3}\right)^{2+1}=\left(\dfrac{1}{3}\right)^3=\dfrac{1}{27}$

c) $\left(7^3\right)^4=7^{3\cdot4}=7^{12}$

d) $\left(k^5\right)^6=k^{5\cdot6}=k^{30}$

13) a) $(5y)^3=5^3y^3=125y^3$

b) $\left(-7m^4\right)\left(2m^{12}\right)=-14m^{4+12}$
$=-14m^{16}$

c) $\left(\dfrac{a}{b}\right)^6=\dfrac{a^6}{b^6}$

d) $6(xy)^2=6x^2y^2$

e) $\left(\frac{10}{9}c^4\right)(2c)\left(\frac{15}{4}c^3\right)$

$=\left(\frac{\overset{5}{\cancel{10}}}{\underset{3}{\cancel{9}}}c^4\right)(\cancel{2}c)\left(\frac{\overset{5}{\cancel{15}}}{\cancel{4}}c^3\right)$

$=\frac{25}{3}c^{4+1+3}$

$=\frac{25}{3}c^8$

15) a) $(z^5)^2(z^3)^4=z^{10}z^{12}=z^{22}$

b) $-2(3c^5d^8)^2=-2(9c^{10}d^{16})$

$=-18c^{10}d^{16}$

c) $(9-4)^3=5^3=125$

d) $\frac{(10t^3)^2}{(2u^7)^3}=\frac{100t^6}{8u^{21}}=\frac{25t^6}{2u^{21}}$

17) a) $8^0=1$

b) $-3^0=-1$

c) $9^{-1}=\frac{1}{9}$

d) $3^{-2}-2^{-2}=\left(\frac{1}{3}\right)^2-\left(\frac{1}{2}\right)^2$

$=\frac{1}{9}-\frac{1}{4}$

$=\frac{4}{36}-\frac{9}{36}=-\frac{5}{36}$

e) $\left(\frac{4}{5}\right)^{-3}=\left(\frac{5}{4}\right)^3=\frac{125}{64}$

19) a) $v^{-9}=\frac{1}{v^9}$

b) $\left(\frac{9}{c}\right)^{-2}=\left(\frac{c}{9}\right)^2=\frac{c^2}{81}$

c) $\left(\frac{1}{y}\right)^{-8}=y^8$

d) $-7k^{-9}=-\frac{7}{k^9}$

e) $\frac{19z^{-4}}{a^{-1}}=\frac{19a}{z^4}$

f) $20m^{-6}n^5=\frac{20n^5}{m^6}$

g) $\left(\frac{2j}{k}\right)^{-5}=\left(\frac{k}{2j}\right)^5=\frac{k^5}{32j^5}$

21) a) $\frac{3^8}{3^6}=3^{8-6}=3^2=9$

b) $\frac{r^{11}}{r^3}=r^{11-3}=r^8$

c) $\frac{48t^{-2}}{32t^3}=\frac{3}{2}t^{-2-3}=\frac{3}{2}t^{-5}=\frac{3}{2t^5}$

d) $\frac{21xy^2}{35x^{-6}y^3}=\frac{3}{5}x^{1-(-6)}y^{2-3}$

$=\frac{3}{5}x^7y^{-1}=\frac{3x^7}{5y}$

23) a) $(-3s^4t^5)^4=81s^{16}t^{20}$

b) $\frac{(2a^6)^5}{(4a^7)^2}=\frac{32a^{30}}{16a^{14}}=2a^{30-14}=2a^{16}$

c) $\left(\frac{z^4}{y^3}\right)^{-6} = \left(\frac{y^3}{z^4}\right)^6 = \frac{y^{18}}{z^{24}}$

d) $\left(-x^3y\right)^5\left(6x^{-2}y^3\right)^2$
$= \left(-x^{15}y^5\right)\left(36x^{-4}y^6\right)$
$= -36x^{11}y^{11}$

e) $\left(\frac{cd^{-4}}{c^8d^{-9}}\right)^5 = \left(c^{1-8}d^{-4-(-9)}\right)^5$
$= \left(c^{-7}d^5\right)^5$
$= c^{-35}d^{25}$
$= \frac{d^{25}}{c^{35}}$

f) $\left(\frac{14m^5n^5}{7m^4n}\right)^3 = \left(2m^{5-4}n^{5-1}\right)^3$
$= \left(2mn^4\right)^3$
$= 8m^3n^{12}$

g) $\left(\frac{3k^{-1}t}{5k^{-7}t^4}\right)^{-3} = \left(\frac{3}{5}k^{-1-(-7)}t^{1-4}\right)^{-3}$
$= \left(\frac{3}{5}k^6t^{-3}\right)^{-3}$
$= \left(\frac{3k^6}{5t^3}\right)^{-3}$
$= \left(\frac{5t^3}{3k^6}\right)^3 = \frac{125t^9}{27k^{18}}$

h) $\left(\frac{40}{21}x^{10}\right)\left(3x^{-12}\right)\left(\frac{49}{20}x^2\right)$
$= \left(\frac{\overset{2}{\cancel{40}}}{\cancel{21}}x^{10}\right)\left(\cancel{3}x^{-12}\right)\left(\frac{\overset{7}{\cancel{49}}}{\cancel{20}}x^2\right)$
$= 14x^{10-12+2} = 14x^0 = 14$

25) a) $y^{3k} \cdot y^{7k} = y^{3k+7k} = y^{10k}$

b) $\left(x^{5p}\right)^2 = x^{2 \cdot 5p} = x^{10p}$

c) $\frac{z^{12c}}{z^{5c}} = z^{12c-5c} = z^{7c}$

d) $\frac{t^{6d}}{t^{11d}} = t^{6d-11d} = t^{-5d} = \frac{1}{t^{5d}}$

27) $-4.185 \times 10^2 = -418.5$

29) $6.7 \times 10^{-4} = 0.00067$

31) $2 \times 10^4 = 20,000$

33) $0.0000575 = 5.75 \times 10^{-5}$

35) $32,000,000 = 3.2 \times 10^7$

37) $178,000 = 1.78 \times 10^5$

39) $0.0009315 = 9.315 \times 10^{-4}$

41) $\frac{8 \times 10^6}{2 \times 10^{13}} = \frac{8}{2} \times \frac{10^6}{10^{13}}$
$= 4 \times 10^{-7}$
$= 0.0000004$

43) $\left(9 \times 10^{-8}\right)\left(4 \times 10^7\right)$
$= (9 \times 4)\left(10^{-8} \times 10^7\right)$
$= 36 \times 10^{-1}$
$= 3.6$

45) $\frac{-3 \times 10^{10}}{-4 \times 10^6} = \frac{3}{4} \times \frac{10^{10}}{10^6}$
$= .75 \times 10^4$
$= 7500$

47) $\dfrac{2.4\times10^5}{8}=\dfrac{2.4}{8}\times10^5$
$=0.3\times10^5=30{,}000$ quills

49) $\dfrac{2.99\times10^{-23}\text{ g}}{\cancel{\text{molecule}}}\cdot100{,}000{,}000\ \cancel{\text{molecule}}$

$=2.99\times10^{-23}\cdot1\times10^8$ g
$=2.99\left(10^{-23}\times10^8\right)$ g
$=2.99\times10^{-15}$ g
$=0.00000000000000299$ g

Chapter 2 Test

1) a) $\left(-8k^2+3k-5\right)+\left(2k^2+k-9\right)$
$=-8k^2+2k^2+3k+k-5-9$
$=-6k^2+4k-14$

b) $\frac{4}{3}(6c-5)-\frac{1}{2}(4c+3)$
$=8c-\frac{20}{3}-2c-\frac{3}{2}$
$=8c-2c-\frac{20}{3}-\frac{3}{2}$
$=8c-2c-\frac{40}{6}-\frac{9}{6}=6c-\frac{49}{6}$

3) $3^4=81$

5) $2^{-5}=\left(\frac{1}{2}\right)^5=\frac{1}{32}$

7) $\left(-\frac{3}{4}\right)^3=-\frac{27}{64}$

9) $\left(5n^6\right)^3=125n^{18}$

11) $\dfrac{m^{10}}{m^4}=m^{10-4}=m^6$

13) $\left(\dfrac{-12t^{-6}u^8}{4t^5u^{-1}}\right)^{-3}=\left(-3t^{-6-5}u^{8-(-1)}\right)^{-3}$
$=\left(-3t^{-11}u^9\right)^{-3}$
$=\left(\dfrac{-3u^9}{t^{11}}\right)^{-3}$
$=\left(-\dfrac{t^{11}}{3u^9}\right)^3=-\dfrac{t^{33}}{27u^{27}}$

15) $t^{10k}\cdot t^{3k}=t^{10k+3k}=t^{13k}$

17) $0.000165=1.65\times10^{-4}$

19) $2.18\times10^7=21{,}800{,}000$

Cumulative Review: Chapters 1–2

1) $\dfrac{90}{150}=\dfrac{90\div30}{150\div30}=\dfrac{3}{5}$

3) $\dfrac{4}{15}\div\dfrac{20}{21}=\dfrac{4}{15}\cdot\dfrac{21}{20}=\dfrac{\cancel{4}}{\cancel{15}_5}\cdot\dfrac{\cancel{21}^7}{\cancel{20}_5}=\dfrac{7}{25}$

5) $-26+5-7=-21-7=-28$

7) $(5+1)^2-2[17+5(10-14)]$
$=(6)^2-2[17+5(-4)]$
$=36-2[17+(-20)]$
$=36-2[-3]$
$=36+6$
$=42$

9) $2p^2 - 11q$ when $p = 3$ and $q = -4$.

$$2(3)^2 - 11(-4) = 2(9) - 11(-4)$$
$$= 18 + 44$$
$$= 62$$

11) $5(t^2 + 7t - 3) - 2(4t^2 - t + 5)$

$$= 5t^2 + 35t - 15 - 8t^2 + 2t - 10$$
$$= 5t^2 - 8t^2 + 35t + 2t - 15 - 10$$
$$= -3t^2 + 37t - 25$$

13) $(4z^3)(-7z^5) = -28z^8$

15) $(-2a^{-6}b)^5 = -32a^{-30}b^5 = -\dfrac{32b^5}{a^{30}}$

17) $(6.2\times10^5)(9.4\times10^{-2})$

$$= (6.2\times9.4)(10^5\times10^{-2})$$
$$= 58.28\times10^3$$
$$= 58{,}280$$

Chapter 3: Linear Equations and Inequalities

Section 3.1 Exercises

1) Equation

3) Expression

5) No, it is an expression.

7) b, d

9) No.

$$a-4=-9$$
$$5-4=-9$$
$$1\neq-9$$

11) Yes.

$$-8p=12$$
$$-8\left(-\frac{3}{2}\right)=12$$
$$\overset{4}{\cancel{-8}}\left(\frac{3}{\cancel{-2}}\right)=12$$
$$12=12$$

13) Yes.

$$10-2(3y-1)+y=8$$
$$10-2(3(4)-1)+4=8$$
$$10-2(12-1)+4=8$$
$$10-2(11)+4=8$$
$$10-22+4=8$$
$$8=8$$

15)

$$r-6=11$$
$$r-6+6=11+6$$
$$r=17$$

The solution set is $\{17\}$.

17)

$$b+10=4$$
$$b+10-10=4-10$$
$$b=-6$$

The solution set is $\{-6\}$.

19)

$$-16=k-12$$
$$-16+12=k-12+12$$
$$-4=k$$

The solution set is $\{-4\}$.

21)

$$a+\frac{5}{8}=\frac{1}{2}$$
$$a+\frac{5}{8}-\frac{5}{8}=\frac{1}{2}-\frac{5}{8}$$
$$a=\frac{4}{8}-\frac{5}{8}$$
$$a=-\frac{1}{8}$$

The solution set is $\left\{-\frac{1}{8}\right\}$.

23) $3y=30$

$$\frac{\cancel{3}y}{\cancel{3}}=\frac{30}{3}$$
$$y=10$$

The solution set is $\{10\}$.

25) $-5z=35$

$$\frac{\cancel{-5}z}{\cancel{-5}}=\frac{35}{-5}$$
$$z=-7$$

The solution set is $\{-7\}$.

27) $-56=-7v$

$$\frac{-56}{-7}=\frac{\cancel{-7}v}{\cancel{-7}}$$
$$8=v$$

The solution set is $\{8\}$.

29) $\frac{a}{4}=12$

$4\cdot\frac{a}{4}=4\cdot12$

$1a=48$

$a=48$

The solution set is $\{48\}$.

31) $-6=\frac{k}{8}$

$-6\cdot8=\frac{k}{8}\cdot8$

$-48=1k$

$-48=k$

The solution set is $\{-48\}$.

33) $\frac{2}{3}g=-10$

$\frac{3}{2}\cdot\frac{2}{3}g=\frac{3}{2}\cdot-10$

$1g=\frac{3}{\cancel{2}}\cdot-\overset{5}{\cancel{10}}$

$g=-15$

The solution set is $\{-15\}$.

35) $-\frac{5}{3}d=-30$

$-\frac{3}{5}\cdot\left(-\frac{5}{3}\right)d=-\frac{3}{5}\cdot(-30)$

$1d=\frac{3}{\cancel{-5}}\cdot\left(\overset{6}{\cancel{-30}}\right)$

$d=18$

The solution set is $\{18\}$.

37) $\frac{11}{15}=\frac{1}{3}y$

$3\cdot\frac{11}{15}=3\cdot\frac{1}{3}y$

$\cancel{3}\cdot\frac{11}{\underset{5}{\cancel{15}}}=1y$

$\frac{11}{5}=y$

The solution set is $\left\{\frac{11}{5}\right\}$.

39) $0.5q=6$

$\frac{0.5q}{0.5}=\frac{6}{0.5}$

$q=12$

The solution set is $\{12\}$.

41) $-w=-7$

$\frac{-w}{-1}=\frac{-7}{-1}$

$w=7$

The solution set is $\{7\}$.

43) $-12d=0$

$\frac{-12d}{-12}=\frac{0}{-12}$

$d=0$

The solution set is $\{0\}$.

45) $3x-7=17$

$3x-7+7=17+7$

$3x=24$

$\frac{\cancel{3}x}{\cancel{3}}=\frac{24}{3}$

$x=8$

The solution set is $\{8\}$.

47) $7c+4=18$

$$7c+4-4=18-4$$
$$7c=14$$
$$\frac{7c}{7}=\frac{14}{7}$$
$$c=2$$

The solution set is $\{2\}$.

49) $8d-15=-15$

$$8d-15+15=-15+15$$
$$8d=0$$
$$\frac{8d}{8}=\frac{0}{8}$$
$$d=0$$

The solution set is $\{0\}$.

51) $-11=5t-9$

$$-11+9=5t-9+9$$
$$-2=5t$$
$$\frac{-2}{5}=\frac{5t}{5}$$
$$-\frac{2}{5}=t$$

The solution set is $\left\{-\frac{2}{5}\right\}$.

53) $10=3-7y$

$$10-3=3-3-7y$$
$$7=-7y$$
$$\frac{7}{-7}=\frac{-7y}{-7}$$
$$-1=y$$

The solution set is $\{-1\}$.

55) $\frac{4}{9}w-11=1$

$$\frac{4}{9}w-11+11=1+11$$
$$\frac{4}{9}w=12$$
$$\frac{9}{4}\cdot\frac{4}{9}w=\frac{9}{4}\cdot 12$$
$$1w=\frac{9}{4}\cdot \overset{3}{12}$$
$$w=27$$

The solution set is $\{27\}$.

57) $\frac{10}{7}m+3=1$

$$\frac{10}{7}m+3-3=1-3$$
$$\frac{10}{7}m=-2$$
$$\frac{7}{10}\cdot\frac{10}{7}m=\frac{7}{10}\cdot -2$$
$$1m=\frac{7}{\underset{5}{10}}\cdot -2$$
$$m=-\frac{7}{5}$$

The solution set is $\left\{-\frac{7}{5}\right\}$.

59) $-\frac{1}{6}z+\frac{1}{2}=\frac{3}{4}$

$-\frac{1}{6}z+\frac{1}{2}-\frac{1}{2}=\frac{3}{4}-\frac{1}{2}$

$-\frac{1}{6}z=\frac{3}{4}-\frac{2}{4}$

$-\frac{1}{6}z=\frac{1}{4}$

$-6\cdot\left(-\frac{1}{6}z\right)=-6\cdot\frac{1}{4}$

$1z=-\cancel{6}^{3}\cdot\frac{1}{\cancel{4}_{2}}$

$z=-\frac{3}{2}$

The solution set is $\left\{-\frac{3}{2}\right\}$.

61) $5-0.4p=2.6$

$5-5-0.4p=2.6-5$

$-0.4p=-2.4$

$\frac{\cancel{-0.4}p}{\cancel{-0.4}}=\frac{-2.4}{-0.4}$

$1p=6$

$p=6$

The solution set is $\{6\}$.

63) $10v+9-2v+16=1$

$10v-2v+9+16=1$

$8v+25=1$

$8v+25-25=1-25$

$8v=-24$

$\frac{\cancel{8}v}{\cancel{8}}=\frac{-24}{8}$

$v=-3$

The solution set is $\{-3\}$.

65) $5-3m+9m+10-7m=-4$

$-3m+9m-7m+5+10=-4$

$-m+15=-4$

$-m+15-15=-4-15$

$-m=-19$

$\frac{\cancel{-}m}{\cancel{-1}}=\frac{-19}{-1}$

$m=19$

The solution set is $\{19\}$.

67) $5=-12p+7+4p-12$

$5=-12p+4p+7-12$

$5=-8p-5$

$5+5=-8p-5+5$

$10=-8p$

$\frac{10}{-8}=\frac{\cancel{-8}p}{\cancel{-8}}$

$-\frac{5}{4}=p$

The solution set is $\left\{-\frac{5}{4}\right\}$.

69) $2(5x+3)-3x+4=-11$

$10x+6-3x+4=-11$

$10x-3x+6+4=-11$

$7x+10=-11$

$7x+10-10=-11-10$

$7x=-21$

$\frac{\cancel{7}x}{\cancel{7}}=\frac{-21}{7}$

$x=-3$

The solution set is $\{-3\}$.

71) $$\begin{aligned} -12 &= 7(2a-3)-(8a-9) \\ -12 &= 14a-21-8a+9 \\ -12 &= 14a-8a-21+9 \\ -12 &= 6a-12 \\ -12+12 &= 6a-12+12 \\ 0 &= 6a \\ \frac{0}{6} &= \frac{\cancel{6}a}{\cancel{6}} \\ 0 &= a \end{aligned}$$
The solution set is $\{0\}$.

Section 3.2 Exercises

1) $$\begin{aligned} \frac{9}{2}c-7 &= 29 \\ \frac{9}{2}c-7+7 &= 29+7 \\ \frac{9}{2}c &= 36 \\ \frac{2}{9}\cdot\frac{9}{2}c &= \frac{2}{\cancel{9}}\cdot\overset{4}{\cancel{36}} \\ c &= 8 \end{aligned}$$
The solution set is $\{8\}$.

3) $$\begin{aligned} 2y+7 &= 5y-2 \\ 2y-2y+7 &= 5y-2y-2 \\ 7 &= 3y-2 \\ 7+2 &= 3y-2+2 \\ 9 &= 3y \\ \frac{9}{3} &= \frac{3y}{3} \\ 3 &= y \end{aligned}$$
The solution set is $\{3\}$.

5) $$\begin{aligned} 6-7p &= 2p+33 \\ 6-7p+7p &= 2p+7p+33 \\ 6 &= 9p+33 \\ 6-33 &= 9p+33-33 \\ -27 &= 9p \\ \frac{-27}{9} &= \frac{9p}{9} \\ -3 &= p \end{aligned}$$
The solution set is $\{-3\}$.

7) $$\begin{aligned} -8x+6-2x+11 &= 3+3x-7x \\ -10x+17 &= 3-4x \\ -10x+10x+17 &= 3-4x+10x \\ 17 &= 3+6x \\ 17-3 &= 3-3+6x \\ 14 &= 6x \\ \frac{14}{6} &= \frac{6x}{6} \\ \frac{7}{3} &= x \end{aligned}$$
The solution set is $\left\{\frac{7}{3}\right\}$.

9) $$\begin{aligned} 4(2t+5)-7 &= 5(t+5) \\ 8t+20-7 &= 5t+25 \\ 8t+13 &= 5t+25 \\ 8t-5t+13 &= 5t-5t+25 \\ 3t+13 &= 25 \\ 3t+13-13 &= 25-13 \\ 3t &= 12 \\ \frac{3t}{3} &= \frac{12}{3} \\ t &= 4 \end{aligned}$$
The solution set is $\{4\}$.

11) $-9r+4r-11+2=3r+7-8r+9$

$$-5r-9=-5r+16$$
$$-5r+5r-9=-5r+5r+16$$
$$-9\neq 16$$

The solution set is $\varnothing$.

13) $j-15j+8=-3(4j-3)-2j-1$

$$j-15j+8=-12j+9-2j-1$$
$$-14j+8=-14j+8$$
$$-14j+14j+8=-14j+14j+8$$
$$8=8$$

The solution set is $\{\text{all real numbers}\}$.

15) $8(3t+4)=10t-3+7(2t+5)$

$$24t+32=10t-3+14t+35$$
$$24t+32=24t+32$$
$$24t-24t+32=24t-24t+32$$
$$32=32$$

The solution set is $\{\text{all real numbers}\}$.

17) $8-7(2-3w)-9w=4(5w-1)-3w-2$

$$8-14+21w-9w=20w-4-3w-2$$
$$-6+12w=17w-6$$
$$-6+12w-12w=17w-12w-6$$
$$-6=5w-6$$
$$-6+6=5w-6+6$$
$$0=5w$$
$$\frac{0}{5}=\frac{5w}{5}$$
$$0=w$$

The solution set is $\{0\}$.

19) $7y+2(1-4y)=8y-5(y+4)$

$$7y+2-8y=8y-5y-20$$
$$-y+2=3y-20$$
$$-y+y+2=3y+y-20$$
$$2=4y-20$$
$$2+20=4y-20+20$$
$$22=4y$$
$$\frac{22}{4}=\frac{4y}{4}$$
$$\frac{11}{2}=y$$

The solution set is $\left\{\frac{11}{2}\right\}$.

21) $\frac{1}{6}x+\frac{5}{4}=\frac{1}{2}x-\frac{5}{12}$

$$12\left(\frac{1}{6}x+\frac{5}{4}\right)=12\left(\frac{1}{2}x-\frac{5}{12}\right)$$
$$2x+15=6x-5$$
$$2x-2x+15=6x-2x-5$$
$$15=4x-5$$
$$15+5=4x-5+5$$
$$20=4x$$
$$\frac{20}{4}=\frac{4x}{4}$$
$$5=x$$

The solution set is $\{5\}$.

23) $\frac{2}{3}d-1=\frac{1}{5}d+\frac{2}{5}$

$$15\left(\frac{2}{3}d-1\right)=15\left(\frac{1}{5}d+\frac{2}{5}\right)$$
$$10d-15=3d+6$$
$$10d-3d-15=3d-3d+6$$
$$7d-15=6$$
$$7d-15+15=6+15$$
$$7d=21$$
$$\frac{7d}{7}=\frac{21}{7}$$
$$d=3$$

The solution set is $\{3\}$.

25) $\frac{m}{3}+\frac{1}{2}=\frac{2m}{3}+3$

$$6\left(\frac{m}{3}+\frac{1}{2}\right)=6\left(\frac{2m}{3}+3\right)$$
$$2m+3=4m+18$$
$$2m-2m+3=4m-2m+18$$
$$3=2m+18$$
$$3-18=2m+18-18$$
$$-15=2m$$
$$\frac{-15}{2}=\frac{2m}{2}$$
$$-\frac{15}{2}=m$$

The solution set is $\left\{-\frac{15}{2}\right\}$.

27) $\frac{1}{3}+\frac{1}{9}(k+5)-\frac{k}{4}=2$

$$36\left(\frac{1}{3}+\frac{1}{9}(k+5)-\frac{k}{4}\right)=36(2)$$
$$12+4(k+5)-9k=72$$
$$12+4k+20-9k=72$$
$$32-5k=72$$
$$32-32-5k=72-32$$
$$-5k=40$$
$$\frac{-5k}{-5}=\frac{40}{-5}$$
$$k=-8$$

The solution set is $\{-8\}$.

29) $0.05(t+8)-0.01t=0.6$

$$100(0.05(t+8)-0.01t)=100(0.6)$$
$$5(t+8)-1t=60$$
$$5t+40-t=60$$
$$4t+40-40=60-40$$
$$4t=20$$
$$\frac{4t}{4}=\frac{20}{4}$$
$$t=5$$

The solution set is $\{5\}$.

31)
$$0.1x+0.15(8-x)=0.125(8)$$
$$1000(0.1x+0.15(8-x))=1000(0.125(8))$$
$$100x+150(8-x)=125(8)$$
$$100x+1200-150x=1000$$
$$-50x+1200=1000$$
$$-50x+1200-1200=1000-1200$$
$$\frac{-50x}{-50}=\frac{-200}{-50}$$
$$x=4$$
The solution set is $\{4\}$.

33)
$$0.04s+0.03(s+200)=27$$
$$100(0.04s+0.03(s+200))=100(27)$$
$$4s+3(s+200)=2700$$
$$4s+3s+600=2700$$
$$7s+600=2700$$
$$7s+600-600=2700-600$$
$$7s=2100$$
$$\frac{7s}{7}=\frac{2100}{7}$$
$$s=300$$
The solution set is $\{300\}$.

35) Let x = a number
$$4+x=15$$
$$4-4+x=15-4$$
$$x=11$$
The number is 11.

37) Let x = a number
$$x-7=22$$
$$x-7+7=22+7$$
$$x=29$$
The number is 29.

39) Let x = a number

$$2x = -16$$
$$\frac{2x}{2} = \frac{-16}{2}$$
$$x = -8$$

The number is -8.

41) Let x = a number

$$\frac{2}{3}x = 10$$
$$\frac{3}{2} \cdot \frac{2}{3}x = \frac{3}{\cancel{2}} \cdot \cancel{10}^{5}$$
$$x = 15$$

The number is 15.

43) Let x = a number

$$7 + 2x = 35$$
$$7 - 7 + 2x = 35 - 7$$
$$2x = 28$$
$$\frac{2x}{2} = \frac{28}{2}$$
$$x = 14$$

The number is 14.

45) Let x = a number

$$3x - 8 = 40$$
$$3x - 8 + 8 = 40 + 8$$
$$3x = 48$$
$$\frac{3x}{3} = \frac{48}{3}$$
$$x = 16$$

The number is 16.

47) Let x = a number

$$\frac{1}{2}x + 10 = 3$$
$$\frac{1}{2}x + 10 - 10 = 3 - 10$$
$$\frac{1}{2}x = -7$$
$$2 \cdot \frac{1}{2}x = 2 \cdot (-7)$$
$$x = -14$$

The number is -14.

49) Let x = a number

$$2(x + 5) = 16$$
$$2x + 10 = 16$$
$$2x + 10 - 10 = 16 - 10$$
$$2x = 6$$
$$\frac{2x}{2} = \frac{6}{2}$$
$$x = 3$$

The number is 3.

51) Let x = a number

$$3x = 15 + \frac{1}{2}x$$
$$2(3x) = 2\left(15 + \frac{1}{2}x\right)$$
$$6x = 30 + x$$
$$6x - x = 30 + x - x$$
$$5x = 30$$
$$\frac{5x}{5} = \frac{30}{5}$$
$$x = 6$$

The number is 6.

53) Let $x =$ a number

$$x-6=5+2x$$
$$x-x-6=5+2x-x$$
$$-6=5+x$$
$$-6-5=5-5+x$$
$$-11=x$$

The number is -11.

Section 3.3: Exercises

1) $c+5$

3) $p-31$

5) $3w$

7) $14-x$

9) Let $x =$ the amount of sugar in Gatorade. Then the amount of sugar in Pepsi $= x+6.5$.

$$\begin{pmatrix}\text{amount in}\\ \text{Gatorade}\end{pmatrix}+\begin{pmatrix}\text{amount in}\\ \text{Pepsi}\end{pmatrix}=13.1$$
$$x \quad + \quad x+6.5 \quad =13.1$$
$$2x=6.6$$
$$x=3.3$$

$$\text{Pepsi}=x+6.5$$
$$=3.3+6.5=9.8$$

There are 3.3 teaspoons in Gatorade and 9.8 teaspoons in Pepsi.

11) $x =$ the number of medals Greece won. Then the amount of medals Thailand won $= \frac{1}{2}x$.

$$\begin{pmatrix}\text{medals won}\\ \text{by Greece}\end{pmatrix}+\begin{pmatrix}\text{medals won}\\ \text{by Thailand}\end{pmatrix}=24$$
$$x \quad + \quad \frac{1}{2}x \quad =24$$
$$\frac{3}{2}x=24$$
$$\frac{2}{3}\cdot\frac{3}{2}x=\frac{2}{3}\cdot 24$$
$$x=16$$

$$\text{Thailand}=\frac{1}{2}x$$
$$=\frac{1}{2}(16)=8$$

Greece won 16 medals, and Thailand won 8 medals at the 2004 olympics.

13) Let $x =$ length of the Ohio River.
Then the length of the Columbia River $= x - 70$.

$$\begin{pmatrix}\text{length of}\\ \text{Ohio}\\ \text{River}\end{pmatrix} + \begin{pmatrix}\text{Length of}\\ \text{Columbia}\\ \text{River}\end{pmatrix} = 2550$$

$$\begin{aligned} x \quad + \quad x - 70 &= 2550 \\ 2x &= 2620 \\ x &= 1310 \end{aligned}$$

$$\begin{aligned}\text{Columbia River} &= x - 70 \\ &= 1310 - 70 = 1240\end{aligned}$$

The length of the Ohio River is 1310 mi.
and the length of the Columbia River is 1240 mi.

15) Let $x =$ the length of the shorter piece. Then the length of the longer piece $= x + 14$.

$$(\text{shorter piece}) + (\text{longer piece}) = 36$$

$$\begin{aligned} x \quad + \quad x + 14 &= 36 \\ 2x &= 22 \\ x &= 11 \end{aligned}$$

$$\begin{aligned}\text{longer piece} &= x + 14 \\ &= 11 + 14 = 25\end{aligned}$$

The pipe will be cut into an 11 in. piece and a 25 in. piece.

17) Let $x =$ the length of the short jump rope. Then the length of the long jump rope $= 2x$.

$$\begin{pmatrix}\text{short jump}\\ \text{rope}\end{pmatrix} + \begin{pmatrix}\text{long jump}\\ \text{rope}\end{pmatrix} = 18$$

$$\begin{aligned} x \quad + \quad 2x &= 18 \\ 3x &= 18 \\ x &= 6 \end{aligned}$$

$$\begin{aligned}\text{long jump rope} &= 2x \\ &= 2(6) = 12\end{aligned}$$

The long jump rope will be 12 ft. and the short jump rope will be 6 ft.

19) $x = \text{first integer}$

$x+1 = \text{second integer}$

$x+2 = \text{third integer}$

$$\begin{pmatrix}\text{first}\\\text{integer}\end{pmatrix}+\begin{pmatrix}\text{second}\\\text{integer}\end{pmatrix}+\begin{pmatrix}\text{third}\\\text{integer}\end{pmatrix}=195$$

$$\begin{aligned} x \quad + \quad x+1 \quad + \quad x+2 &= 195\\ 3x+3 &= 195\\ 3x &= 192\\ x &= 64 \end{aligned}$$

$x = \text{first integer} = 64$

$x+1 = \text{second integer} = 65$

$x+2 = \text{third integer} = 66$

21) $x = \text{first even integer}$

$x+2 = \text{second even integer}$

$$2\begin{pmatrix}\text{first even}\\\text{integer}\end{pmatrix}=10+\begin{pmatrix}\text{second even}\\\text{integer}\end{pmatrix}$$

$$\begin{aligned} 2x &= 10+ \quad x+2\\ 2x &= 12+x\\ x &= 12 \end{aligned}$$

$x = \text{first even integer} = 12$

$x+2 = \text{second even integer} = 14$

23) $x = \text{first odd integer}$

$x+2 = \text{second odd integer}$

$x+4 = \text{third odd integer}$

$$\begin{pmatrix}\text{first odd}\\\text{integer}\end{pmatrix}+\begin{pmatrix}\text{second odd}\\\text{integer}\end{pmatrix}+\begin{pmatrix}\text{third odd}\\\text{integer}\end{pmatrix}=5+4\begin{pmatrix}\text{third}\\\text{odd integer}\end{pmatrix}$$

$$\begin{aligned} x \quad + \quad x+2 \quad + \quad x+4 &= 5+4(x+4)\\ 3x+6 &= 5+4x+16\\ 3x+6 &= 21+4x\\ 6 &= 21+x\\ -15 &= x \end{aligned}$$

$x = \text{first odd integer} = -15$

$x+2 = \text{second odd integer} = -13$

$x+4 = \text{third odd integer} = -11$

25) $x = \text{first page number}$
$x+1 = \text{second page number}$
$$\begin{pmatrix}\text{first page}\\ \text{number}\end{pmatrix}+\begin{pmatrix}\text{second page}\\ \text{number}\end{pmatrix}=345$$
$$x \quad + \quad x+1 \quad = 345$$
$$2x = 344$$
$$x = 172$$
$x = \text{first page number} = 172$
$x+1 = \text{second page number} = 173$

27) $\text{Hourly Cost} = \begin{pmatrix}\text{Price per}\\ \text{hour}\end{pmatrix}\begin{pmatrix}\text{Number}\\ \text{of hours}\end{pmatrix}$
$$= (\$6)(4)$$
$$= \$24$$
$\text{Total Cost} = \text{Fixed Cost} + \text{Hourly Cost}$
$$= \$25 + (\$6)(4)$$
$$= \$25 + \$24$$
$$= \$49$$

29) $x = $ the number of miles driven.
$\text{Total Cost} = \$133.40$
$\text{Fixed Cost} = 1(\$95.00) = \$95.00$
$\text{Cost of mileage} = \$0.60x$
$\text{Total Cost} = \text{Fixed Cost} + \text{Cost of mileage}$
$$133.40 = 95.00 + 0.60x$$
$$38.40 = 0.60x$$
$$64 = x$$
Charlie drove the truck 64 miles.

31) $x = $ the number of miles driven.
$\text{Total Cost} = \$57.60$
$\text{Fixed Cost} = 1(\$39.00) = \$39.00$
$\text{Cost of mileage} = \$0.15x$
$\text{Total Cost} = \text{Fixed Cost} + \text{Cost of mileage}$
$$57.60 = 39.00 + 0.15x$$
$$18.60 = 0.15x$$
$$124 = x$$
Luis drove the car 124 miles.

33) Let x = the distance riding the bike.

Then the distance walked $= \frac{1}{2}x + 1$.

$$\begin{pmatrix}\text{distance}\\\text{riding}\end{pmatrix} + \begin{pmatrix}\text{distance}\\\text{walking}\end{pmatrix} = 7$$

$$x \quad + \quad \frac{1}{2}x + 1 = 7$$

$$\frac{3}{2}x = 6$$

$$\frac{2}{3}\cdot\frac{3}{2}x = \frac{2}{3}\cdot 6$$

$$x = 4$$

$$\text{distance walking} = \frac{1}{2}x + 1$$

$$= \frac{1}{2}(4) + 1$$

$$= 3$$

Irina rode her bike 4 miles and walked her bike 3 miles.

35) x = first even integer

$x+2$ = second even integer

$x+4$ = third even integer

$$\frac{1}{6}\begin{pmatrix}\text{first even}\\\text{integer}\end{pmatrix} = \frac{1}{10}\left[\begin{pmatrix}\text{second even}\\\text{integer}\end{pmatrix} + \begin{pmatrix}\text{third even}\\\text{integer}\end{pmatrix}\right] - 3$$

$$\frac{1}{6}x \quad = \frac{1}{10}[\quad x+2 \quad + \quad x+4 \quad] - 3$$

$$\frac{1}{6}x = \frac{1}{10}[2x+6] - 3$$

$$30\left(\frac{1}{6}x\right) = 30\left(\frac{1}{10}[2x+6] - 3\right)$$

$$5x = 3[2x+6] - 90$$

$$5x = 6x + 18 - 90$$

$$5x = 6x - 72$$

$$-x = -72$$

$$x = 72$$

x = first even integer = 72

$x+2$ = second even integer = 74

$x+4$ = third even integer = 76

37) Let $x =$ the money that *Shrek* earned.
Then the money that *Harry Potter* earned $= x + 42$.
$(\textit{Shrek}\text{ money}) + (\textit{Harry Potter}\text{ money}) = 577.4$

$$x + x + 42 = 577.4$$
$$2x = 535.4$$
$$x = 267.7$$

$$\textit{Harry Potter}\text{ money} = x + 42$$
$$= 267.7 + 42$$
$$= 309.7$$

Shrek earned \$267.7 million, and *Harry Potter* earned \$309.7 million.

39) $x =$ the number of miles driven.
Total Cost $= \$54.75$
Fixed Cost $= 1(\$39.95) = \$39.95$
Cost of mileage $= \$0.10x$
Total Cost $=$ Fixed Cost $+$ Cost of mileage

$$54.75 = 39.95 + 0.10x$$
$$14.80 = 0.10x$$
$$148 = x$$

Jonas drove the car 148 miles.

41) Let $x =$ the number of albums sold by Avril (in millions).
Then the number of albums sold by Nelly (in millions) $= x + 0.8$,
and the number of albums soldy by Eminem (in millions) $= x + 3.5$

$$\begin{pmatrix}\text{Avril}\\ \text{albums}\end{pmatrix} + \begin{pmatrix}\text{Nelly}\\ \text{albums}\end{pmatrix} + \begin{pmatrix}\text{Eminem}\\ \text{albums}\end{pmatrix} = 16.6$$
$$x + x + 0.8 + x + 3.5 = 16.6$$
$$3x + 4.3 = 16.6$$
$$3x = 12.3$$
$$x = 4.1$$

$$\text{Nelly albums} = x + 0.8$$
$$= 4.1 + 0.8 = 4.8$$
$$\text{Eminem albums} = x + 3.5$$
$$= 4.1 + 3.5 = 7.6$$

Avril sold 4.1 million albums,
Nelly sold 4.9 million albums,
and Eminem sold 7.6 million albums.

43) Let $x =$ the length of the shortest piece.
Then the length of one piece $= x+6$,
and the length of the longest piece $= 2x$.

$$\begin{pmatrix}\text{shortest}\\ \text{piece}\end{pmatrix}+\begin{pmatrix}\text{one}\\ \text{piece}\end{pmatrix}+\begin{pmatrix}\text{longest}\\ \text{piece}\end{pmatrix}=58$$

$$x + x+6 + 2x = 58$$

$$4x = 52$$

$$x = 13$$

$$\text{one piece} = x+6 = 13+6 = 19$$

$$\text{longest piece} = 2x = 2(13) = 26$$

The cable will be cut into 13 ft, 19 ft, and 26 ft pieces.

Section 3.4: Exercises

1) Amount of Discount = (Rate of Discount)(Original Price)
$= (0.15) \cdot (75.00)$
$= 10.50$
Sale Price = Original Price − Amount of Discount
$= 75.00 - 10.50$
$= 63.75$
The sale price is \$63.75.

3) Amount of Discount = (Rate of Discount)(Original Price)
$= (0.30) \cdot (16.50)$
$= 4.95$
Sale Price = Original Price − Amount of Discount
$= 16.50 - 4.95$
$= 11.55$
The sale price is \$11.55.

5) Amount of Discount = (Rate of Discount)(Original Price)
$= (0.60) \cdot (29.00)$
$= 17.40$
Sale Price = Original Price − Amount of Discount
$= 29.00 - 17.40$
$= 11.60$
The sale price is \$11.60.

7) Let x = the original price of the camera.
Sale Price = Original Price – Amount of Discount

$$119 = x - 0.15x$$
$$119 = 0.85x$$
$$140 = x$$

The original price was $140.00.

9) Let x = the original price of the calendar.
Sale Price = Original Price – Amount of Discount

$$4.38 = x - 0.60x$$
$$4.38 = 0.40x$$
$$10.95 = x$$

The original price was $10.95.

11) Let x = the original price of the coffe maker.
Sale Price = Original Price – Amount of Discount

$$22.75 = x - 0.30x$$
$$22.75 = 0.70x$$
$$32.50 = x$$

The original price was $32.50.

13) Let x = the number of countries in the 1980 Olympics.
Countries in 1984 = Countries in 1980 + Amount of Increase

$$140 = x + x(0.75)$$
$$140 = 1.75x$$
$$80 = x$$

80 countries participated in the 1980 Olympics.

15) Let x = the number of Starbucks in 1994.
Starbucks in 2004 = Starbucks in 1994 + Amount of Increase

$$7{,}569 = x + x(16.81)$$
$$7{,}569 = 17.81x$$
$$424.99 \approx x$$

There were 425 Starbucks stores in 1994.

17) $I = PRT$
$P = 800,\ R = 0.04,\ T = 1$
$I = (800)(0.04)(1) = 32$
Jenna earned \$32 in interest.

19) $I = PRT$
$P = 6500,\ R = 0.06,\ T = 1$
$I = (6500)(0.06)(1)$
$= 390$
$\text{Total} = P + I$
$= 6500 + 390 = 6890$
There will be \$6890 in the account.

21) $I = PRT$
$\text{Total Interest Earned} = (4000)(0.065)(1) + (1500)(0.08)(1)$
$= 260 + 120 = 380$
Rachel earned a total of \$380 in interest from the two accounts.

23) $x =$ amount Amir invested in the 6% account.
$15{,}000 - x =$ amount Amir invested in the 7% account.
Total Interest Earned = Interest from 6% account + Interest from 7% account

$$960 = x(0.06)(1) + (15{,}000 - x)(0.07)(1)$$
$$100(960) = 100\left[x(0.06)(1) + (15{,}000 - x)(0.07)(1)\right]$$
$$96{,}000 = 6x + 7(15{,}000 - x)$$
$$96{,}000 = 6x + 105{,}000 - 7x$$
$$-9000 = -x$$
$$9000 = x$$

amount invested in $7\% = 15{,}000 - x$
$= 15{,}000 - 9000 = 6000$
Amir invested \$9000 in the 6% account and \$6000 in the 7% account.

25) $x =$ amount Enrique invested in the 6% account.
$x + 200 =$ amount Enrique invested in the 5% account.
Total Interest Earned = Interest from 6% account + Interest from 5% account

$$164 = x(0.06)(1) + (x + 200)(0.05)(1)$$
$$100(164) = 100\left[x(0.06)(1) + (x + 200)(0.05)(1)\right]$$
$$16{,}400 = 6x + 5(x + 200)$$
$$16{,}400 = 6x + 5x + 1000$$
$$15{,}400 = 11x$$
$$1400 = x$$

amount invested in $5\% = x + 200$
$= 1400 + 200 = 1600$
Enrique invested \$1400 in the 6% account and \$1600 in the 5% account.

27) $x =$ amount Clarissa invested in the 9.5% account.
$7000 - x =$ amount Clarissa invested in the 7% account.
Total Interest Earned = Interest from 9.5% account + Interest from 7% account

$$560 = x(0.095)(1) + (7000-x)(0.07)(1)$$
$$1000(560) = 1000\left[x(0.095)(1)+(7000-x)(0.07)(1)\right]$$
$$560,000 = 95x + 70(7000-x)$$
$$560,000 = 95x + 490,000 - 70x$$
$$70,000 = 25x$$
$$2800 = x$$

amount invested in 7% $= 7000 - x$
$$= 7000 - 2800 = 4200$$

Clarissa invested \$2800 in the 9.5% account and \$4200 in the 7% account.

29) Let $x =$ Cheryl's Cost.
Price in store = Cheryl's Cost + Amount of Increase

$$14.00 = x + x(0.60)$$
$$14.00 = 1.60x$$
$$8.75 = x$$

Each stuffed animal cost Cheryl \$8.75.

31) Let $x =$ people collecting unemployment benefits in September 2005.
Benefits 2006 = Benefits 2005 – Amount of Decrease

$$8330 = x - x(0.02)$$
$$8330 = 0.98x$$
$$8500 = x$$

8500 people were getting unemployment benefits in September 2005.

33) $x =$ amount Tamara invested in the 3% CD account.
$2x =$ amount Tamara invested in the 4% IRA account.
$x + 1000 =$ amount Tamara invested in the 5% mutual fund account.
Total Interest Earned = Interest from CD + Interest from IRA + Interest from mutual fund

$$290 = x(0.03)(1) + 2x(0.04)(1) + (x+1000)(0.05)(1)$$
$$100(290) = 100\left[x(0.03)(1)+2x(0.04)(1)+(x+1000)(0.05)(1)\right]$$
$$29,000 = 3x + 8x + 5(x+1000)$$
$$29,000 = 3x + 8x + 5x + 5000$$
$$29,000 = 16x + 5000$$
$$24,000 = 16x$$
$$1500 = x$$

amount invested CD $= 2x$ amount invested mutual fund $= x + 1000$

$$= x(1500) \qquad = 1500 + 1000$$

$$= 3000 \qquad = 2500$$

Tamara invested \$1500 in the CD, \$3000 in the IRA and \$2500 in the mutual fund.

35) Let $x =$ the original price of the cell phone.

Sale Price = Original Price − Amount of Discount

$$63.20 = x - 0.20x$$

$$63.20 = 0.80x$$

$$79.00 = x$$

The original price was \$79.00.

37) Let $x =$ Zoe's previous salary.

Current Salary = Previous Salaray + Amount of Increase

$$40{,}144 = x + x(0.04)$$

$$40{,}144 = 1.04x$$

$$38{,}600 = x$$

Zoe's salary was \$38,600 last year.

39) $x =$ amount Jerry invested in the 5% account.

$20{,}000 - x =$ amount Jerry invested in the 9% account.

Total Interest Earned = Interest from 5% account + Interest from 9% account

$$1560 = x(0.05)(1) + (20{,}000 - x)(0.09)(1)$$

$$100(1560) = 100\left[x(0.05)(1) + (20{,}000 - x)(0.09)(1)\right]$$

$$156{,}000 = 5x + 9(20{,}000 - x)$$

$$156{,}000 = 5x + 180{,}000 - 9x$$

$$-24{,}000 = -4x$$

$$6000 = x$$

amount invested in 9% $= 20{,}000 - x$

$$= 20{,}000 - 6{,}000 = 14{,}000$$

Jerry invested \$6000 in the 5% account and \$14,000 in the 9% account.

Section 3.5: Exercises

1) $l =$ length of pool
$V = 1700,\ w = 17,\ h = 4$
$V = lwh$
$1700 = l(17)(4)$
$1700 = 68l$
$25 = l$
The length of the pool is 25 ft.

3) $w =$ width of the printer area
$A = 48,\ l = 8$
$A = lw$
$48 = 8w$
$6 = w$
The width of the printed area is 6 in.

5) $A =$ area of Big Ben's face
$r = 11.5$
$A = \pi r^2$
$A = \pi(11.5)^2$
$A = 132.25\pi$
$A \approx 132.25(3.14)$
$A \approx 415 \text{ ft}^2$
The area is about 415 ft^2.

7) $w =$ width of the lane
$A = lw$
$228 = 19w$
$12 = w$
The width of the lane is 12 ft.

9) $h =$ the height of the can $V = 24\pi,\ r = 2$
$V = \pi r^2 h$
$24\pi = \pi(2)^2 h$
$24\pi = 4\pi h$
$6 = h$
The height of the can is 6 in.

11) $m\angle A = x^\circ \; m\angle B = (2x)^\circ$

$$m\angle A + m\angle B + m\angle C = 180$$
$$x + 2x + 102 = 180$$
$$3x = 78$$
$$x = 26$$
$$m\angle A = x^\circ = 26^\circ$$
$$m\angle B = (2x)^\circ = (2(26))^\circ = 52^\circ$$

13) $m\angle A = (2x+15)^\circ \; m\angle B = (6x-11)^\circ \; m\angle C = (8x)^\circ$

$$m\angle A + m\angle B + m\angle C = 180$$
$$2x+15 + 6x-11 + 8x = 180$$
$$16x+4 = 180$$
$$16x = 176$$
$$x = 11$$
$$m\angle A = (2x+15)^\circ = \left[2(11)+15\right]^\circ = 37^\circ$$
$$m\angle B = (6x-11)^\circ = \left[6(11)-11\right]^\circ = 55^\circ$$
$$m\angle C = (8x)^\circ = \left[8(11)\right]^\circ = 88^\circ$$

15) $(9x+5)^\circ = (10x-2)^\circ$

$$9x+5 = 10x-2$$
$$5 = x-2$$
$$7 = x$$
$$(9x+5)^\circ = \left[9(7)+5\right]^\circ = 68^\circ$$
$$(10x-2)^\circ = \left[10(7)-2\right]^\circ = 68^\circ$$

The labeled angles measure 68°.

17) $(9x-75)^\circ = (6x)^\circ$

$$9x-75 = 6x$$
$$-75 = -3x$$
$$25 = x$$
$$(9x-75)^\circ = \left[9(25)-75\right]^\circ = 150^\circ$$
$$(6x)^\circ = \left[6(25)\right]^\circ = 150^\circ$$

The labeled angles measure 150°.

19) $(10x+13)^\circ+(4x-1)^\circ=180^\circ$

$$10x+13 \quad + \quad 4x-1 \;=180$$
$$14x+12=180$$
$$14x=168$$
$$x=12$$
$$(10x+13)^\circ=[10(12)+13]^\circ=133^\circ$$
$$(4x-1)^\circ=[4(12)-1]^\circ=47^\circ$$

21) $(3x+19)^\circ+(5x+1)^\circ=180^\circ$

$$3x+19 \quad + \quad 5x+1 \;=180$$
$$8x+20=180$$
$$8x=160$$
$$x=20$$
$$(3x+19)^\circ=[3(20)+19]^\circ=79^\circ$$
$$(5x+1)^\circ=[5(20)+1]^\circ=101^\circ$$

23) $180-x$

25) Let $x=$ the measure of the angle.

$180-x=$ measure of the supplement

$10(\text{angle})=7+\;\text{supplement}$

$$10x \quad =7+\;(180-x)$$
$$10x=187-x$$
$$11x=187$$
$$x=17$$

The measure of the angle is 17°.

27) Let $x=$ the measure of the angle.

$180-x=$ measure of the supplement

$90-x=$ measure of the complement

$$4(\text{complement}) = 2(\text{supplement}) - 40$$
$$4(90 - x) = 2(180 - x) \quad -40$$
$$360 - 4x = 360 - 2x - 40$$
$$360 - 4x = 320 - 2x$$
$$360 = 320 + 2x$$
$$40 = 2x$$
$$20 = x$$

The measure of the angle, complement, and supplement are 20°, 70°, and 160°.

29) Let $x =$ the measure of the angle.
$180 - x =$ measure of the supplement
$90 - x =$ measure of the complement

$$\text{angle} + 3(\text{complement}) = 55 + \text{supplement}$$
$$x \quad + \quad 3(90 - x) = 55 + \ 180 - x$$
$$x + 270 - 3x = 235 - x$$
$$-2x + 270 = 235 - x$$
$$270 = 235 + x$$
$$35 = x$$

The measure of the angle is 35°.

31) Let $x =$ the measure of the angle.
$180 - x =$ measure of the supplement
$90 - x =$ measure of the complement

$$3(\text{angle}) + 2(\text{supplement}) = 400$$
$$3x \quad + \quad 2(180 - x) = 400$$
$$3x + 360 - 2x = 400$$
$$x + 360 = 400$$
$$x = 40$$

The measure of the angle is 40°.

33) $I = Prt$

$$240 = (3000)(0.04)t$$
$$240 = 120t$$
$$2 = t$$

35) $V = lwh$
$96 = (8)w(3)$
$96 = 24w$
$4 = w$

37) $P = 2l + 2w$
$50 = 2l + 2(7)$
$50 = 2l + 14$
$36 = 2l$
$18 = l$

39) $V = \frac{1}{3}\pi r^2 h$
$54\pi = \frac{1}{3}\pi(9)^2 h$
$54\pi = \frac{1}{3}\pi 81h$
$54\pi = 27\pi h$
$2 = h$

41) $S = 2\pi r^2 + 2\pi rh$
$120\pi = 2\pi(5)^2 + 2\pi(5)h$
$120\pi = 2\pi(25) + 2\pi(5)h$
$120\pi = 50\pi + 10\pi h$
$70\pi = 10\pi h$
$7 = h$

43) $A = \frac{1}{2}h(b_1 + b_2)$
$790 = \frac{1}{2}h(29 + 50)$
$790 = \frac{1}{2}h(79)$
$790 = \frac{79}{2}h$
$20 = h$

45) a) $x + 12 = 35$
$x + 12 - 12 = 35 - 12$
$x = 23$

b) $x + n = p$
$x + n - n = p - n$
$x = p - n$

c) $x + q = v$
$x + q - q = v - q$
$x = v - q$

47) a) $5n = 30$
$\frac{5n}{5} = \frac{30}{5}$
$n = 6$

b) $yn = c$
$\frac{yn}{y} = \frac{c}{y}$
$n = \frac{c}{y}$

c) $wn = d$
$\frac{wn}{w} = \frac{d}{w}$
$n = \frac{d}{w}$

49) a) $\frac{c}{3} = 7$
$3 \cdot \frac{c}{3} = 3 \cdot 7$
$c = 21$

b) $\frac{c}{u} = r$
$r \cdot \frac{c}{r} = r \cdot u$
$c = ur$

c)
$$\frac{c}{x}=t$$
$$x\cdot\frac{c}{t}=x\cdot t$$
$$c=xt$$

51) a)
$$8d-7=17$$
$$8d-7+7=17+7$$
$$8d=24$$
$$\frac{8d}{8}=\frac{24}{8}$$
$$d=3$$

b)
$$kd-a=z$$
$$kd-a+a=z+a$$
$$kd=z+a$$
$$\frac{kd}{k}=\frac{z+a}{k}$$
$$d=\frac{z+a}{k}$$

53) a)
$$6z+19=4$$
$$6z+19-19=4-19$$
$$6z=-15$$
$$\frac{6z}{6}=\frac{-15}{6}$$
$$z=-\frac{5}{2}$$

b)
$$yz+t=w$$
$$yz+t-t=w-t$$
$$yz=w-t$$
$$\frac{yz}{y}=\frac{w-t}{y}$$
$$z=\frac{w-t}{y}$$

55)
$$F=ma$$
$$\frac{F}{a}=\frac{ma}{a}$$
$$\frac{F}{a}=m$$

57)
$$n=\frac{c}{v}$$
$$n\cdot v=\frac{c}{v}\cdot v$$
$$nv=c$$

59)
$$E=\sigma T^4$$
$$\frac{E}{T^4}=\frac{\sigma T^4}{T^4}$$
$$\frac{E}{T^4}=\sigma$$

61)
$$V=\frac{1}{3}\pi r^2 h$$
$$3\cdot V=3\cdot\frac{1}{3}\pi r^2 h$$
$$3V=\pi r^2 h$$
$$\frac{3V}{\pi r^2}=\frac{\pi r^2 h}{\pi r^2}$$
$$\frac{3V}{\pi r^2}=h$$

63)
$$R=\frac{E}{I}$$
$$I\cdot R=I\cdot\frac{E}{I}$$
$$IR=E$$

65)
$$I=PRT$$
$$I=PRT$$
$$\frac{I}{PT}=\frac{PRT}{PT}$$
$$\frac{I}{PT}=R$$

67)
$$P=2l+2w$$
$$P-2w=2l+2w-2w$$
$$P-2w=2l$$
$$\frac{P-2w}{2}=\frac{2l}{2}$$
$$\frac{P-2w}{2}=l$$

69)
$$H=\frac{D^2N}{2.5}$$
$$2.5\cdot H=2.5\cdot\frac{D^2N}{2.5}$$
$$2.5H=D^2N$$
$$\frac{2.5H}{D^2}=\frac{D^2N}{D^2}$$
$$\frac{2.5H}{D^2}=N$$

71)
$$A=\frac{1}{2}h(b_1+b_2)$$
$$2\cdot A=2\cdot\frac{1}{2}h(b_1+b_2)$$
$$2A=h(b_1+b_2)$$
$$\frac{2A}{h}=\frac{h(b_1+b_2)}{h}$$
$$\frac{2A}{h}=b_1+b_2$$
$$\frac{2A}{h}-b_1=b_1-b_1+b_2$$
$$\frac{2A}{h}-b_1=b_2 \text{ or } b_2=\frac{2A-hb_1}{h}$$

73)
$$S=\frac{\pi}{4}(4h^2+c^2)$$
$$\frac{4}{\pi}\cdot S=\frac{4}{\pi}\cdot\frac{\pi}{4}(4h^2+c^2)$$
$$\frac{4S}{\pi}=4h^2+c^2$$
$$\frac{4S}{\pi}-c^2=4h^2+c^2-c^2$$
$$\frac{4S}{\pi}-c^2=4h^2$$
$$\frac{4S}{4\pi}-\frac{c^2}{4}=\frac{4h^2}{4}$$
$$\frac{S}{\pi}-\frac{c^2}{4}=h^2$$

75) a)
$$P=2l+2w$$
$$P-2l=2l-2l+2w$$
$$P-2l=2w$$
$$\frac{P-2l}{2}=\frac{2w}{2}$$
$$\frac{P-2l}{2}=w$$

b)
$$w=\frac{P-2l}{2}$$
$$w=\frac{28-2(11)}{2}$$
$$w=\frac{28-22}{2}$$
$$w=\frac{6}{2}$$
$$w=3\text{ cm}$$

77) a)
$$C = \frac{5}{9}(F - 32)$$
$$\frac{9}{5} \cdot C = \frac{9}{5} \cdot \frac{5}{9}(F - 32)$$
$$\frac{9}{5}C = F - 32$$
$$\frac{9}{5}C + 32 = F - 32 + 32$$
$$\frac{9}{5}C + 32 = F$$

b)
$$F = \frac{9}{5}C + 32$$
$$F = \frac{9}{5}(25) + 32$$
$$F = 45 + 32 = 77^\circ$$

Section 3.6: Exercises

1) $\frac{15}{25} = \frac{3}{5}$

3) $\frac{60}{45} = \frac{4}{3}$

5) A ratio is a quotient of two quantities. A proportion is a statement that two ratios are equal.

7) False
$$2 \cdot 45 = 15 \cdot 8$$
$$90 \neq 120$$

9) True
$$42 \cdot 11 = 77 \cdot 6$$
$$462 = 462$$

11)
$$\frac{m}{9} = \frac{10}{45}$$
$$45m = 90$$
$$m = 2$$

13)
$$\frac{120}{50} = \frac{x}{4}$$
$$480 = 50x$$
$$\frac{48}{5} = x$$

15)
$$\frac{2a + 3}{8} = \frac{2}{16}$$
$$16(2a + 3) = 16$$
$$32a + 48 = 16$$
$$32a = -32$$
$$a = -1$$

17)
$$\frac{n - 4}{5} = \frac{5n - 2}{10}$$
$$10(n - 4) = 5(5n - 2)$$
$$10n - 40 = 25n - 10$$
$$-40 = 15n - 10$$
$$-30 = 15n$$
$$-2 = n$$

19) Let x = bulls-eyes expected.
$$\frac{9}{15} = \frac{x}{50}$$
$$450 = 15x$$
$$30 = x$$
David can expect to hit the bulls-eye 30 times out of 50 tries.

21) Let $x =$ caffeine in an 18 oz. serving

$$\frac{12}{55} = \frac{18}{x}$$
$$12x = 990$$
$$x = 82.5$$

There are 82.5 mg of caffeine in an 18 oz serving of Mountain Dew.

23) Let $x =$ divorces expected.

$$\frac{4.8}{1000} = \frac{x}{35,000}$$
$$168,000 = 1000x$$
$$168 = x$$

There would be 168 divorces in a town of 35,000 people.

25) $$\frac{7}{5} = \frac{14}{x}$$
$$7x = 70$$
$$x = 10$$

27) $$\frac{12}{20} = \frac{x}{\frac{65}{3}}$$
$$260 = 20x$$
$$13 = x$$

29) $$\frac{x}{28} = \frac{45}{20}$$
$$20x = 1260$$
$$63 = x$$

31) a) $(\$0.10) \cdot 8 = \$0.80$

b) $10¢ \cdot 8 = 80¢$

33) a) $(\$0.01) \cdot 217 = \$2.17$

b) $(1¢) \cdot 217 = 217¢$

35) a) $(\$0.25) \cdot 9 + (\$0.10) \cdot 7$
$= \$2.25 + \$0.70$
$= \$2.95$

b) $(25¢) \cdot 9 + (10¢) \cdot 7$
$= 225¢ + 70¢$
$= 295¢$

37) a) $0.25 \cdot q = 0.25q$

b) $25 \cdot q = 25q$

39) a) $0.10 \cdot d = 0.10d$

b) $10 \cdot d = 10d$

41) a) $0.01 \cdot p + 0.25 \cdot q = 0.01p + 0.25q$

b) $1 \cdot p + 25 \cdot q = p + 25q$

43) $x =$ number of nickels; $8 + x =$ number of quarters

Value of Nickels + Value of Quarters = Total Value

$$0.05x + 0.25(x+8) = 4.70$$

$$100(0.05x + 0.25(x+8)) = 100(4.70)$$

$$5x + 25(x+8) = 470$$

$$5x + 25x + 200 = 470$$

$$30x = 270$$

$$x = 9$$

quarters $= x + 8$

$= 9 + 8 = 17$

There are 9 nickels and 17 quarters.

45) $x =$ number of \$1 bills; $25 - x =$ number of \$5 bills

Value of \$1 bills + Value of \$5 bills = Total Value

$$1x + 5(25-x) = 69.00$$

$$x + 125 - 5x = 69$$

$$-4x = -56$$

$$x = 14$$

\$5 bills $= 25 - x$

$= 25 - 14 = 11$

There are 11-\$5 bills and 14-\$1 bills.

47) $x =$ number of adult tickets; $\frac{x}{2} =$ number of children's ticket

Rev. from adult tickets + Rev. from children's tickets = Total Revenue

$$9x + 7\left(\frac{x}{2}\right) = 475.00$$

$$2\left(9x + \frac{7}{2}x\right) = 2(475)$$

$$18x + 7x = 950$$

$$25x = 950$$

$$x = 38$$

children's tickets $= \frac{x}{2}$

$= \frac{38}{2}$

$= 19$

There were 38 adult tickets and 19 children's tickets sold.

49) $\text{oz of alcohol} = (0.05)(40)$
$= 2$
There are 2 oz of alcohol in the 40 oz solution.

51) $\text{ml of acid} = (0.10)(60) + (0.04)(40)$
$= 6 + 1.6$
$= 7.6$
There are 7.6 ml of acid in the mixture.

53) x = number of oz of 4% acid solution
$24 - x$ = number of oz of 10% acid solution

Solution	Concentration	Number of oz of solution	Number of oz of acid in the solution
4%	0.04	x	$0.04x$
10%	0.10	$24 - x$	$0.10(24 - x)$
6%	0.06	24	$0.06(24)$

$$0.04x + 0.10(24 - x) = 0.06(24)$$
$$100(0.04x + 0.10(24 - x)) = 100(0.06(24))$$
$$4x + 10(24 - x) = 6(24)$$
$$4x + 240 - 10x = 144$$
$$-6x = -96$$
$$x = 16$$

number of oz of 10% solution $= 24 - x$
$= 24 - 16$
$= 8$

Mix 16 oz of the 4% solution and 8 oz of the 10% solution.

55) x = number of liters of 40% antifreeze solution
$x + 5$ = number of liters of 60% antifreeze solution

Solution	Concentration	Number of liters of solution	Number of liters of antifreeze in the solution
40%	0.4	x	$0.4x$
70%	0.7	5	$0.7(5)$
60%	0.6	$x + 5$	$0.6(x + 5)$

$$0.4x + 0.7(5) = 0.6(x+5)$$
$$10(0.4x + 0.7(5)) = 10(0.6(x+5))$$
$$4x + 7(5) = 6(x+5)$$
$$4x + 35 = 6x + 30$$
$$5 = 2x$$
$$2\frac{1}{2} = x$$

Add $2\frac{1}{2}$ liters of the 40% antifreeze solution.

57) $x =$ number of lbs of Aztec Coffee
$5 - x =$ number of lbs of Cinnamon Coffee

Coffee	Price per Pound	Number of lbs of coffee	Value
Aztec	\$6.00	x	$6x$
Cinnamon	\$8.00	$5-x$	$8(5-x)$
Winterfest	\$7.20	5	$7.2(5)$

$$6x + 8(5-x) = 7.2(5)$$
$$6x + 40 - 8x = 36$$
$$-2x = -4$$
$$x = 2$$

number of lbs of Cinnamon Coffee $= 5 - x$
$= 5 - 2 = 3$

Mix 2 lbs of the Aztec and 3 lbs of the Cinnamon coffee.

59) $t =$ the amount of time traveling until they are 6 miles apart
Truck's distance $+ 6 =$ Car's distance

	d	= r	· t
Truck	35t	35	t
Car	45t	45	t

$$35t + 6 = 45t$$
$$6 = 10t$$
$$\frac{3}{5} = t$$
$$\frac{3}{5} \cdot 60 = 36 \text{ min}$$

61) t = the amount of time traveling until they are 105 miles apart

	d =	r ·	t
Ajay	30t	30	t
Rohan	40t	40	t

$$\text{Ajay's distance} + \text{Rohan's distance} = 105$$
$$30t + 40t = 105$$
$$70t = 105$$
$$t = 1\frac{1}{2}$$

They will be 105 miles apart at 4:30 pm.

63) r = the rate of the car

	d =	r ·	t
Van	30	30	1
Car	$\frac{1}{2}r$	r	$\frac{1}{2}$

$$\text{Van's distance} + \text{Car's distance} = 54$$
$$30 + \frac{1}{2}r = 54$$
$$\frac{1}{2}r = 24$$
$$r = 48 \text{ mph}$$

65) x = number of quarters; $x+7$ = number of dimes

$$\text{Value of Quarters} + \text{Value of Dimes} = \text{Total Value}$$
$$0.25x + 0.10(x+7) = 6.30$$
$$100(0.25x + 0.10(x+7)) = 100(6.30)$$
$$25x + 10(x+7) = 630$$
$$25x + 10x + 70 = 630$$
$$35x = 560$$
$$x = 16$$

$$\text{dimes} = x + 7$$
$$= 16 + 7$$
$$= 3$$

There are 16 quarters and 23 dimes.

67) Let x = the number of calories in 3 cups.

$$\frac{\frac{1}{2}}{260} = \frac{3}{x}$$

$$\frac{1}{2}x = 780$$

$$x = 1560$$

There are 1560 calories in 3 cups of Cherry Garcia.

69) r = the speed of the small plane

$2r$ = the speed of the jet

	d	= r	· t
Small Plane	$\frac{1}{2}r$	r	$\frac{1}{2}$
Jet	r	2r	$\frac{1}{2}$

Small Plane's distance + 100 = Jet's distance

$$\frac{1}{2}r \quad + \quad 100 = r$$

$$100 = \frac{1}{2}r$$

$$200 = r$$

the speed of the jet $= 2r = 2(200) = 400$

The jet is traveling at 400 mph, and the small plane is traveling at 200 mph.

71) x = the cc of a 0.05% steroid solution

$20 - x$ = the cc of a 0.03% steroid solution

Solution	Concentration	Number of cc of solution	Number of cc of steroid in solution
8%	0.08	x	$0.08x$
3%	0.03	$20 - x$	$0.03(20 - x)$
5%	0.05	20	$0.05(20)$

$$0.08x + 0.03(20 - x) = 0.05(20)$$

$$100(0.08x + 0.03(20 - x)) = 100(0.05(20))$$

$$8x + 3(20 - x) = 5(20)$$

$$8x + 60 - 3x = 100$$

$$5x = 40$$

$$x = 8$$

number of cc of 0.03% steroid solution $= 20 - x = 20 - 8 = 12$

The pharmacist should use 8 cc of the 0.08% solution and 12 cc of the 0.03% solution.

73) $x =$ the number of gallons pure acid
$x+6 =$ the number of gallons of a 20% solution

Solution	Concentration	Number of gallons of solution	Number of gallons of acid in solution
100%	1.00	x	$1.00x$
4%	0.04	6	$0.04(6)$
20%	0.20	$x+6$	$0.20(x+6)$

$$1.00x+0.04(6)=0.20(x+6)$$
$$100(1.00x+0.04(6))=100(0.20(x+6))$$
$$100x+4(6)=20(x+6)$$
$$100x+24=20x+120$$
$$80x=96$$
$$x=1\frac{1}{5}$$

Add $1\frac{1}{5}$ gallons of pure acid.

Section 3.7: Exercises

1) 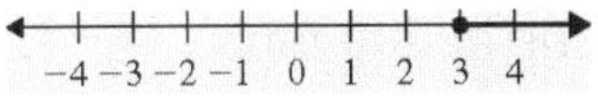

a) $\{x \mid x \geq 3\}$

b) $[3, \infty)$

3) 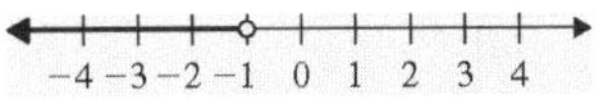

a) $\{c \mid c < -1\}$

b) $(-\infty, -1)$

5) 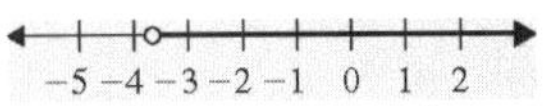

a) $\left\{w \mid w > -\frac{11}{3}\right\}$

b) $\left(-\frac{11}{3}, \infty\right)$

7) 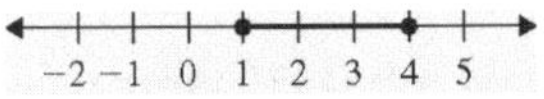

a) $\{n \mid 1 \leq n \leq 4\}$

b) $[1, 4]$

9) 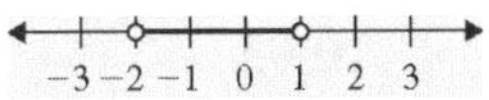

a) $\{a \mid -2 < a < 1\}$

b) $(-2, 1)$

11) 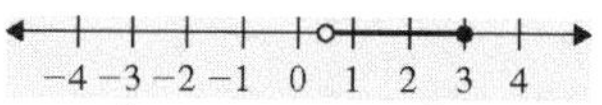

a) $\left\{z \mid \frac{1}{2} < z \le 3\right\}$

b) $\left(\frac{1}{2}, 3\right]$

13) You use parentheses when there is a $<$ or $>$ symbol or when you use ∞ or $-\infty$.

15) $n - 8 \le -3$

$n - 8 + 8 \le -3 + 8$

$n \le 5$

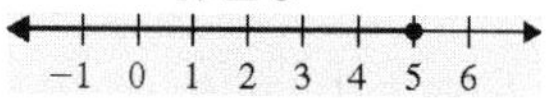

a) $\{n \mid n \le 5\}$

b) $(-\infty, 5]$

17) $y + 5 \ge 1$

$y + 5 - 5 \ge 1 - 5$

$y \ge -4$

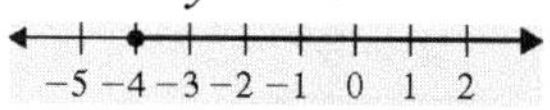

a) $\{y \mid y \ge -4\}$

b) $[-4, \infty)$

19) $3c > 12$

$\frac{3c}{3} > \frac{12}{3}$

$c > 4$

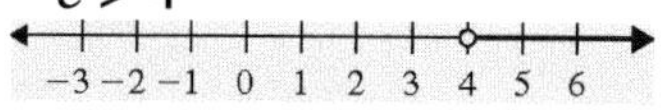

a) $\{c \mid c > 4\}$

b) $(4, \infty)$

21) $15k < -55$

$\frac{15k}{15} < \frac{-55}{15}$

$k < -\frac{11}{3}$

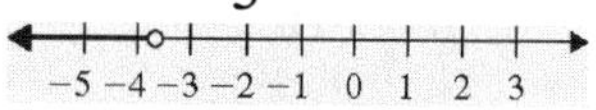

a) $\left\{k \mid k < -\frac{11}{3}\right\}$

b) $\left(-\infty, -\frac{11}{3}\right)$

23) $-4b \le 32$

$\frac{-4b}{-4} \ge \frac{32}{-4}$

$b \ge -8$

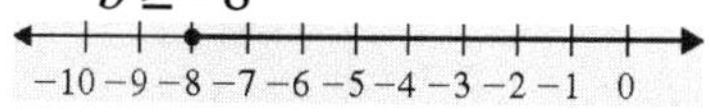

a) $\{b \mid b \ge -8\}$

b) $[-8, \infty)$

25) $-14w > -42$

$\frac{-14w}{-14} < \frac{-42}{-14}$

$w < 3$

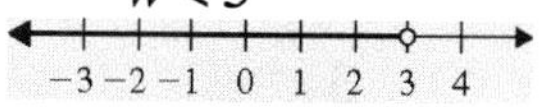

a) $\{w \mid w < 3\}$

b) $(-\infty, 3)$

27) $\frac{1}{3}x < -2$

$3 \cdot \frac{1}{3}x < 3 \cdot (-2)$

$x < -6$

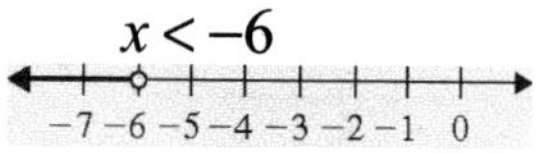

a) $\{x \mid x < -6\}$

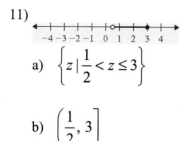

b) $(-\infty, -6)$

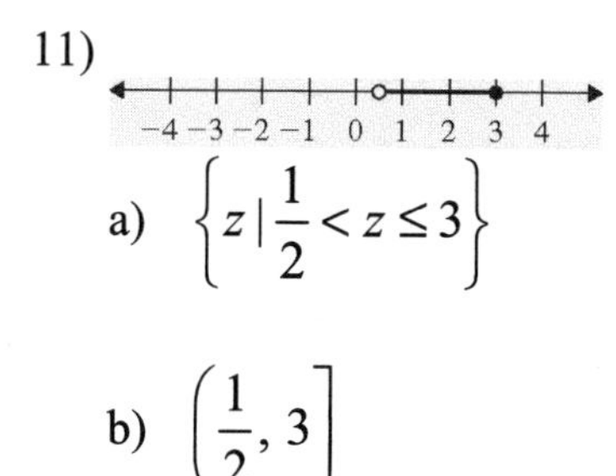

29)
$$-\frac{2}{5}p \geq 4$$
$$-\frac{5}{2}\cdot\left(-\frac{2}{5}p\right) \leq -\frac{5}{2}\cdot 4$$
$$p \leq -10$$

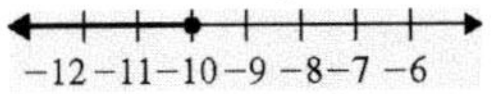

a) $\{p \mid p \leq -10\}$

b) $(-\infty, -10]$

31)
$$8z + 19 > 11$$
$$8z + 19 - 19 > 11 - 19$$
$$8z > -8$$
$$\frac{8z}{8} > \frac{-8}{8}$$
$$z > -1$$

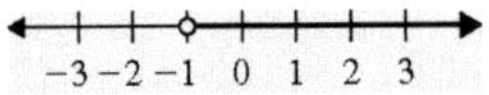

$(-1, \infty)$

33)
$$12 - 7t \geq 15$$
$$12 - 12 - 7t \geq 15 - 12$$
$$-7t \geq 3$$
$$\frac{-7t}{-7} \leq \frac{3}{-7}$$
$$t \leq -\frac{3}{7}$$

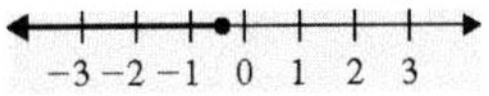

$\left(-\infty, -\frac{3}{7}\right]$

35)
$$-23 - w < -20$$
$$-23 + 23 - w < -20 + 23$$
$$-w < 3$$
$$w > -3$$

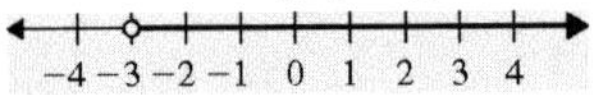

$(-3, \infty)$

37)
$$7a + 4(5 - a) \leq 4 - 5a$$
$$7a + 20 - 4a \leq 4 - 5a$$
$$3a + 20 \leq 4 - 5a$$
$$8a \leq -16$$
$$a \leq -2$$

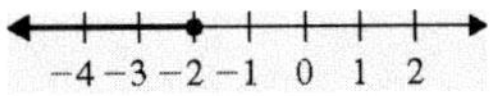

$(-\infty, -2]$

39)
$$9c + 17 > 14c - 3$$
$$20 > 5c$$
$$4 > c$$

$(-\infty, 4)$

41)
$$\frac{8}{3}(2k+1) > \frac{1}{6}k + \frac{8}{3}$$
$$6\left(\frac{8}{3}(2k+1)\right) > 6\left(\frac{1}{6}k + \frac{8}{3}\right)$$
$$16(2k+1) > k + 16$$
$$32k + 16 > k + 16$$
$$31k > 0$$
$$k > 0$$

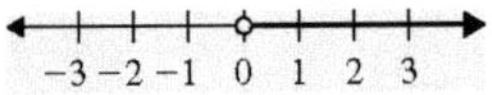

$(0, \infty)$

43)
$$0.04x+0.12(10-x)\geq 0.08(10)$$
$$100\big(0.04x+0.12(10-x)\big)\geq 100\big(0.08(10)\big)$$
$$4x+12(10-x)\geq 8(10)$$
$$4x+120-12x\geq 80$$
$$-8x\geq -40$$
$$x\leq 5$$

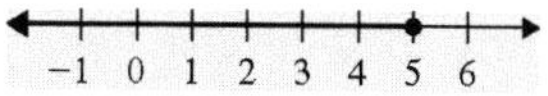

$(-\infty, 5]$

45)
$$-8\leq a-5\leq -4$$
$$-8+5\leq a-5+5\leq -4+5$$
$$-3\leq a\leq 1$$

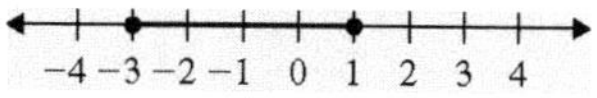

$[-3, 1]$

47) $9<6n<18$
$$\frac{9}{6}<\frac{6n}{6}<\frac{18}{6}$$
$$\frac{3}{2}<n<3$$

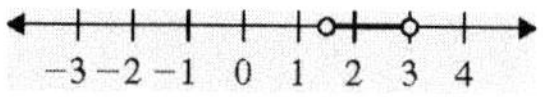

$\left(\frac{3}{2}, 3\right)$

49)
$$-19\leq 7p+9\leq 2$$
$$-19-9\leq 7p+9-9\leq 2-9$$
$$-28\leq 7p\leq -7$$
$$\frac{-28}{7}\leq\frac{7p}{7}\leq\frac{-7}{7}$$
$$-4\leq p\leq -1$$

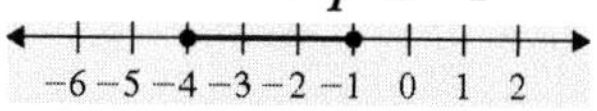

$[-4, -1]$

51) $$-6 \le 4c - 13 < -1$$
$$-6 + 13 \le 4c - 13 + 13 < -1 + 13$$
$$7 \le 4c < 12$$
$$\frac{7}{4} \le \frac{4c}{4} < \frac{12}{4}$$
$$\frac{7}{4} \le c < 3$$

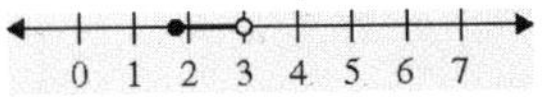

$$\left[\frac{7}{4}, 3\right)$$

53) $$2 < \frac{3}{4}u + 8 < 11$$
$$2 - 8 < \frac{3}{4}u + 8 - 8 < 11 - 8$$
$$-6 < \frac{3}{4}u < 3$$
$$\frac{4}{3} \cdot (-6) < \frac{4}{3} \cdot \frac{3}{4}u < \frac{4}{3} \cdot 3$$
$$-8 < u < 4$$

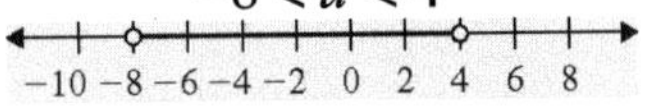

$$(-8, 4)$$

55) $$-\frac{1}{2} \le \frac{5d + 2}{6} \le 0$$
$$6 \cdot \left(-\frac{1}{2}\right) \le 6 \cdot \frac{5d + 2}{6} \le 6 \cdot 0$$
$$-3 \le 5d + 2 \le 0$$
$$-3 - 2 \le 5d + 2 - 2 \le 0 - 2$$
$$-5 \le 5d \le -2$$
$$\frac{-5}{5} \le \frac{5d}{5} \le \frac{-2}{5}$$
$$-1 \le d \le -\frac{2}{5}$$

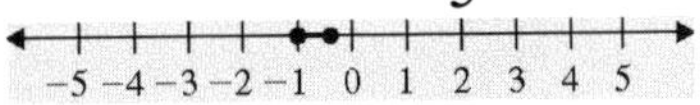

$$\left[-1, -\frac{2}{5}\right]$$

57) $-13 \le 14 - 9h < 5$

$$-13 - 14 \le 14 - 14 - 9h < 5 - 14$$
$$-27 \le -9h < -9$$
$$\frac{-27}{-9} \ge \frac{-9h}{-9} > \frac{-9}{-9}$$
$$3 \ge h > 1$$

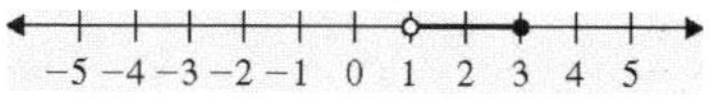

$(1, 3]$

59) $0 \le 4 - 3w \le 7$

$$0 - 4 \le 4 - 4 - 3w \le 7 - 4$$
$$-4 \le -3w \le 3$$
$$\frac{-4}{-3} \ge \frac{-3w}{-3} \ge \frac{3}{-3}$$
$$\frac{4}{3} \ge w \ge -1$$

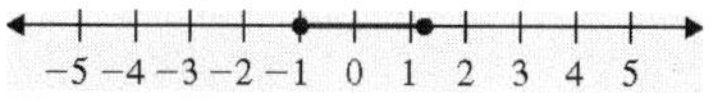

$\left[-1, \frac{4}{3}\right]$

61) Let x = the number of 15 min intervals after the first 75 min.

$15x + 75 = \text{Total Time}$

$$\begin{pmatrix}\text{Total}\\\text{Cost}\end{pmatrix} = \begin{pmatrix}\text{Cost of First}\\\text{75 min}\end{pmatrix} + \begin{pmatrix}\text{Cost of}\\\text{Add. min}\end{pmatrix}$$
$$= \quad 19 \quad + \quad 5x$$

$$19 + 5x \le 50$$
$$5x \le 31$$
$$x \le 6.2$$

Oscar can afford at most 6 intervals.

$$\text{Total Time} = 15x + 75$$
$$= 15(6) + 75$$
$$= 165 \text{ min or}$$

2 hours and 45 min

63) Let x = the number of $\frac{1}{5}$ mi. $\frac{1}{5}x$ = Total Mileage

$$\begin{pmatrix}\text{Total}\\\text{Cost}\end{pmatrix} = \begin{pmatrix}\text{Initial}\\\text{Cost}\end{pmatrix} + \begin{pmatrix}\text{Cost per}\\\text{Mile}\end{pmatrix}$$

$$= \quad 2 \quad + \quad 0.25x$$

$$2 + 0.25x \le 12$$
$$0.25x \le 10$$
$$x \le 40$$

$$\text{Total Mileage} = \frac{1}{5}x$$
$$= \frac{1}{5}(40) = 8 \text{ miles}$$

65) Let x = the grade she needs to make

$87 + 94 + x$ = Total of three scores

$$\frac{87 + 94 + x}{3} \ge 90$$
$$3 \cdot \frac{181 + x}{3} \ge 3 \cdot 90$$
$$181 + x \ge 270$$
$$x \ge 89$$

Melinda must make an 89 or higher.

Section 3.8: Exercises

1) $A \cap B$ means "A intersect B." $A \cap B$ is the set of all numbers which are in set A and set B.

3) $\{8\}$

5) $\{2, 4, 5, 6, 7, 8, 9, 10\}$

7) $\varnothing$

9) $\{1, 2, 3, 4, 5, 6, 8, 10\}$

11)

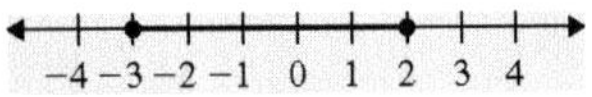

$[-3, 2]$

13)

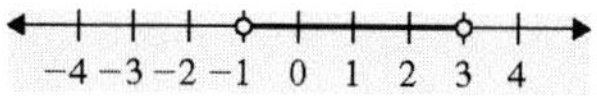

$(-1, 3)$

15)

$[3, \infty)$

17)

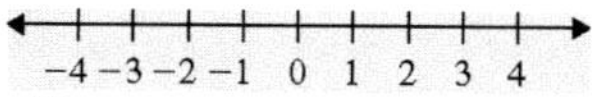

$\varnothing$

19)

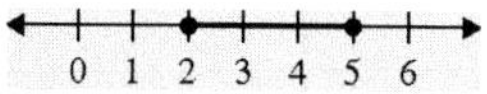

$[2, 5]$

21) $b-7>-9$ and $8b<24$

$b>-2$ and $b<3$

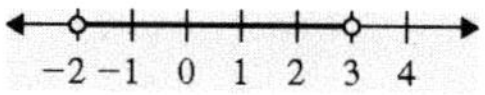

$(-2, 3)$

23) $5w+9\le 29$ and $\frac{1}{3}w-8>-9$

$5w\le 20$ and $\frac{1}{3}w>-1$

$w\le 4$ and $w>-3$

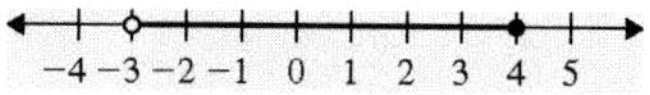

$(-3, 4]$

25) $2m+15\ge 19$ and $m+6<5$

$2m\ge 4$ and $m<-1$

$m\ge 2$ and $m<-1$

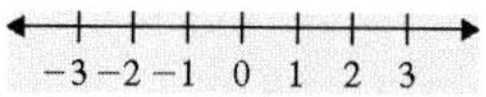

$\varnothing$

27) $r-10>-10$ and $3r-1>8$

$r>0$ and $3r>9$

$r>0$ and $r>3$

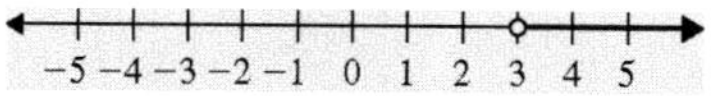

$(3, \infty)$

29) $9-n\le 13$ and $n-8\le -7$

$-n\le 4$ and $n\le 1$

$n\ge -4$ and $n\le 1$

$[-4, 1]$

31)

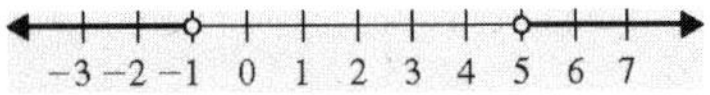

$(-\infty, -1)\cup(5, \infty)$

33)

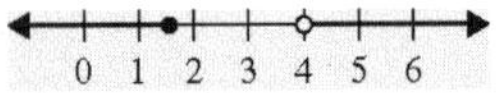

$\left(-\infty, \dfrac{5}{3}\right]\cup(4, \infty)$

35)

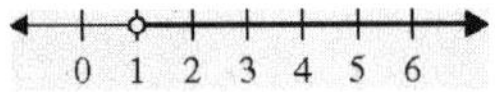

$(1, \infty)$

37)

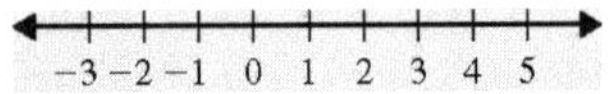

$(-\infty, \infty)$

39)

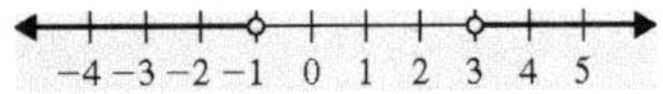

$(-\infty, -1)\cup(3, \infty)$

41) $6m\le 21$ or $m-5>1$

$m\le \dfrac{7}{2}$ or $m>6$

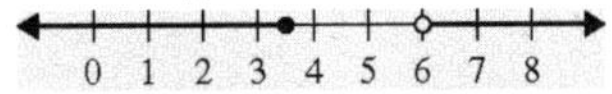

$\left(-\infty, \dfrac{7}{2}\right]\cup(6, \infty)$

43) $3t+4>-11$ or $t+19>17$

$3t>-15$ or $t>-2$

$t>-5$ or $t>-2$

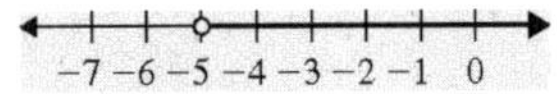

$(-5, \infty)$

45) $-2v-5\le 1$ or $\frac{7}{3}v<-14$

$-2v\le 6$ or $v<-6$

$v\ge -3$ or $v<-6$

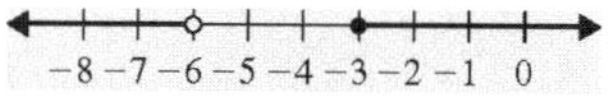

$(-\infty, -6)\cup[-3, \infty)$

47) $c+3\ge 6$ or $\frac{4}{5}c\le 10$

$c\ge 3$ or $c\le \frac{25}{2}$

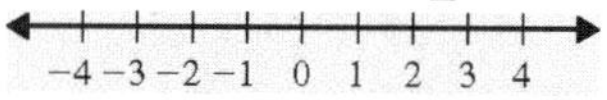

$(-\infty, \infty)$

49) $7-6n\ge 19$ or $n+14<11$

$-6n\ge 12$ or $n<-3$

$n\le -2$ or $n<-3$

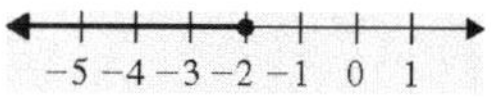

$(-\infty, -2]$

Chapter 3 Review

1) Yes.

$$2n+13=10$$
$$2\left(-\frac{3}{2}\right)+13=10$$
$$-3+13=10$$
$$10=10$$

3) $-9z=30$

$$\frac{-9z}{-9}=\frac{30}{-9}$$
$$z=-\frac{10}{3}$$

The solution set is $\left\{-\frac{10}{3}\right\}$.

5) $$\begin{aligned} 21 &= k+2 \\ 21-2 &= k+2-2 \\ 19 &= k \end{aligned}$$
The solution set is $\{19\}$.

7) $$\begin{aligned} -\frac{4}{9}w &= -\frac{10}{7} \\ -\frac{9}{4}\cdot\left(-\frac{4}{9}w\right) &= -\frac{9}{4}\cdot\left(-\frac{10}{7}\right) \\ w &= \frac{45}{14} \end{aligned}$$
The solution set is $\left\{\frac{45}{14}\right\}$.

9) $$\begin{aligned} 21 &= 0.6q \\ 10(21) &= 10(0.6q) \\ 210 &= 6q \\ \frac{210}{6} &= \frac{6q}{6} \\ 35 &= q \end{aligned}$$
The solution set is $\{35\}$.

11) $$\begin{aligned} 8b-7 &= 57 \\ 8b-7+7 &= 57+7 \\ 8b &= 64 \\ \frac{8b}{8} &= \frac{64}{8} \\ b &= 8 \end{aligned}$$
The solution set is $\{8\}$.

13) $$\begin{aligned} 6 &= 15+\frac{9}{2}v \\ 6-15 &= 15-15+\frac{9}{2}v \\ \frac{2}{9}\cdot(-9) &= \frac{2}{9}\cdot\frac{9}{2}v \\ -2 &= v \end{aligned}$$
The solution set is $\{-2\}$.

15) $$\frac{2}{7}-\frac{3}{4}k=-\frac{17}{14}$$

$$28\left(\frac{2}{7}-\frac{3}{4}k\right)=28\left(-\frac{17}{14}\right)$$

$$8-21k=-34$$

$$14-8-21k=-34-8$$

$$-21k=-42$$

$$k=2$$

The solution set is $\{2\}$.

17) $$4p+9+2(p-12)=15$$

$$4p+9+2p-24=15$$

$$6p-15=15$$

$$6p-15+15=15+15$$

$$6p=30$$

$$p=5$$

The solution set is $\{5\}$.

19) $$11x+13=2x-5$$

$$11x-2x+13=2x-2x-5$$

$$9x+13=-5$$

$$9x+13-13=-5-13$$

$$9x=-18$$

$$x=-2$$

The solution set is $\{-2\}$.

21) $$6-5(4d-3)=7(3-4d)+8d$$

$$6-20d+15=21-28d+8d$$

$$21-20d=21-20d$$

$$21-20d+20d=21-20d+20d$$

$$21=21$$

The solution set is $\{\text{all real numbers}\}$.

23)
$$0.05m+0.11(6-m)=0.08(6)$$
$$100(0.05m+0.11(6-m))=100(0.08(6))$$
$$5m+11(6-m)=8(6)$$
$$5m+66-11m=48$$
$$-6m+66=48$$
$$-6m=-18$$
$$m=3$$
The solution set is $\{3\}$.

25) Let $x=$ a number
$$x-12=5$$
$$x-12+12=5+12$$
$$x=17$$
The number is 17.

27) Let $x=$ a number
$$x+9=2x-1$$
$$x+9=2x-1$$
$$9=x-1$$
$$10=x$$
The number is 10.

29) $26-c$

31) Let $x=$ the number of Clarkson CDs sold. Then the number of Aiken CDs sold $=x+316{,}000$.
$$\begin{pmatrix}\text{Clarkson}\\ \text{CDs}\end{pmatrix}+\begin{pmatrix}\text{Aiken}\\ \text{CDs}\end{pmatrix}=910{,}000$$
$$x+x+316{,}000=910{,}000$$
$$2x=594{,}000$$
$$x=297{,}000$$
$$\text{Aiken CDs}=x+316{,}000$$
$$=297{,}000+316{,}000$$
$$=613{,}000$$
Clarkson sold 297,000 copies and Aiken sold 613,000 copies.

33) Let x = the length of the gravel portion.
Then the length of the paved portion $= 3x$.
gravel portion + paved portion = 500

$$x \quad + \quad 3x \quad = 500$$
$$4x = 500$$
$$x = 125$$

The gravel portion of the road is 125 ft long.

35) New Revenue = Old Revenue − Amount of Decrease

$$= \quad 8200 \quad - \quad (0.18)\cdot 8200$$
$$= 8200 - 1476$$
$$= 6724$$

Imelda's revenue was \$6724 during the first month of construction.

37) Let x = the value of the market in 1998.
Value in 2002 = Value in 1998 + Amount of Increase

$$88 = \quad x \quad + \quad x(3.42)$$
$$88 = 4.42x$$
$$19.9 \approx x$$

The pet market was valued at \$19.9 million in 1998.

39) h = height of triangle
$b = 9,\ A = 54$

$$A = \frac{1}{2}bh$$
$$54 = \frac{1}{2}(9)h$$
$$108 = 9h$$
$$12 = h$$

The height of the triangle is 12 cm.

41) $(2x-1)^\circ + (3x-19)^\circ = 180^\circ$

$$2x - 1 \quad + \quad 3x - 19 = 180$$
$$5x - 20 = 180$$
$$5x = 200$$
$$x = 40$$
$$(2x-1)^\circ = (2(40)-1)^\circ = 79^\circ$$
$$(3x-19)^\circ = (3(40)-19)^\circ = 101^\circ$$

43) Let $x =$ the measure of the angle.

$180 - x =$ measure of the supplement.

$$3(\text{angle}) = 12 + \text{ supplement}$$

$$3x = 12 + (180 - x)$$

$$3x = 192 - x$$

$$4x = 192$$

$$x = 48$$

The measure of the angle is 48°.

45) $pV = nRT$

$$\frac{pV}{nT} = \frac{nRT}{nT}$$

$$\frac{pV}{nT} = R$$

47) $\frac{k}{12} = \frac{15}{9}$

$$9k = 180$$

$$k = 20 \qquad \{20\}$$

49) Let $x =$ girls expected to use alcohol.

$$\frac{1}{5} = \frac{x}{85}$$

$$85 = 5x$$

$$17 = x$$

17 out of the 85 girls would be expected to use alcohol.

51) $x =$ number of dimes; $91 - x =$ number of quarters

Value of Dimes + Value of Quarters = Total Value

$$0.10x + 0.25(91 - x) = 14.05$$

$$100(0.10x + 0.25(91 - x)) = 100(14.05)$$

$$10x + 25(91 - x) = 1405$$

$$10x + 2275 - 25x = 1405$$

$$-15x = -870$$

$$x = 58$$

$\text{quarters} = 91 - x$

$= 91 - 58 = 33$

There are 58 dimes and 33quarters.

53) t = the amount of time traveling for Peter

$t-\frac{1}{4}$ = the amount of time traveling for Mitchell

	d	= r	· t
Peter	30t	30	t
Mitchell	$40\left(t-\frac{1}{4}\right)$	40	$t-\frac{1}{4}$

Peter's distance = Mitchell's distance

$$30t = 40\left(t-\frac{1}{4}\right)$$

$$30t = 40t - 10$$

$$-10t = -10$$

$$t = 1$$

Mitchell's Time $= t-\frac{1}{4} = 1-\frac{1}{4} = \frac{3}{4}$

$\frac{3}{4}\text{hr} = 45\,\text{min}$

It will take Mitchell 45 minutes to catch Peter.

55)
$$-10y+7>32$$

$$-10y+7-7>32-7$$

$$-10y>25$$

$$\frac{-10y}{-10}<\frac{25}{-10}$$

$$y<-\frac{5}{2}$$

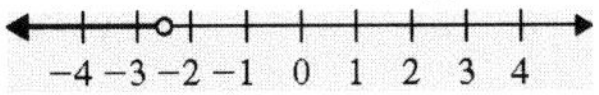

$\left(-\infty, -\frac{5}{2}\right)$

57)
$$-15<4p-7\le 5$$

$$-15+7<4p-7+7\le 5+7$$

$$-8<4p\le 12$$

$$\frac{-8}{4}<\frac{4p}{4}\le\frac{12}{4}$$

$$-2<p\le 3$$

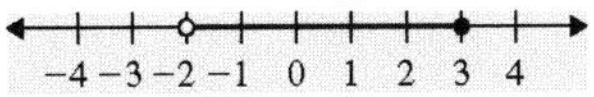

$(-2, 3]$

59) Let x = the grade she needs to make
$95+91+86+x=$ Total of four scores

$$\frac{95+91+86+x}{4} \geq 90$$
$$4 \cdot \frac{272+x}{4} \geq 90 \cdot 4$$
$$272+x \geq 360$$
$$x \geq 88$$

Renee must make an 88 or higher.

61) $a+6 \leq 9$ and $7a-2 \geq 5$
$a \leq 3$ and $7a \geq 7$
$a \leq 3$ and $a \geq 1$

$[1, 3]$

63) $8-y<9$ or $\frac{1}{10}y > \frac{3}{5}$
$-y<1$ or $y>6$
$y>-1$ or $y>6$

$(-1, \infty)$

Chapter 3 Test

1) $-12q = 20$
$$\frac{-12q}{-12} = \frac{20}{-12}$$
$$q = -\frac{5}{3} \qquad \left\{-\frac{5}{3}\right\}$$

3) $5-3(2c+7)=4c-1-5c$
$$5-6c-21=4c-1-5c$$
$$-16-6c=-c-1$$
$$-5c=15$$
$$c=-3 \qquad \{-3\}$$

5) $6(4t-7)=3(8t+5)$
$$24t-42=24t+15$$
$$-42 \neq 15$$
$\varnothing$

7) Let x = a number.
$$2x-9=33$$
$$2x=42$$
$$x=21$$
The number is 21.

9) $x =$ quarts of regular oil
$5 - x =$ quarts of synthetic oil

Oil	Price per Quart	Number of lbs of Nuts	Value
regular	\$1.20	x	$1.20x$
synthetic	\$3.40	$5-x$	$3.40(5-x)$
mixture	\$1.86	5	$1.86(5)$

$$1.20x + 3.40(5-x) = 1.86(5)$$
$$100(1.20x + 3.40(5-x)) = 100(1.86(5))$$
$$120x + 340(5-x) = 186(5)$$
$$120x + 1700 - 340x = 930$$
$$-220x + 1700 = 930$$
$$-220x = -770$$
$$x = 3.5$$

quarts of synthetic oil $= 5 - x$
$$= 5 - 3.5 = 1.5$$

Bob should mix 3.5 qt of regular oil with 1.5 qt of synthetic oil.

11) Let $w =$ the width of the site.
$$P = 2l + 2w$$
$$460 = 2(160) + 2w$$
$$460 = 320 + 2w$$
$$140 = 2w$$
$$70 = w$$
The width of the site is 70 ft.

13)
$$S = 2\pi r^2 + 2\pi rh$$
$$S - 2\pi r^2 = 2\pi r^2 - 2\pi r^2 + 2\pi rh$$
$$S - 2\pi r^2 = 2\pi rh$$
$$\frac{S}{2\pi r} - \frac{2\pi r^2}{2\pi r} = \frac{2\pi rh}{2\pi r}$$
$$\frac{S - 2\pi r^2}{2\pi r} = h$$

15) $9-3x<5x+3$
$$6<8x$$
$$\frac{3}{4}<x$$
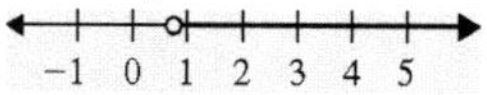

$$\left(\frac{3}{4},\ \infty\right)$$

17) $y-8\le -5$ and $2y\ge 0$
$$y\le 3 \text{ and } y\ge 0$$
$$[0,\ 3]$$

Cumulative Review: Chapters 1-3.

1) $\dfrac{5}{12}-\dfrac{7}{9}=\dfrac{15}{36}-\dfrac{28}{36}=-\dfrac{13}{36}$

3) $52-12\div 4+3\cdot 5=52-3+15$
$$=64$$

5) $0, 8, \sqrt{81}, -2$

7) $\dfrac{12k^{11}}{18k^{4}}=\dfrac{2}{3}k^{11-4}=\dfrac{2}{3}k^{7}$

9) $\left(-7m^{8}\right)\left(\dfrac{5}{21}m^{-6}\right)=-\dfrac{5}{3}m^{8-6}$
$$=-\frac{5}{3}m^{2}$$

11) $279{,}000{,}000$: $279{,}000{,}000.\times 10^{8}$
$$=2.79\times 10^{8}$$

13) $-14<6y+10<3$
$$-24<6y<-7$$
$$-4<y<-\frac{7}{6}$$
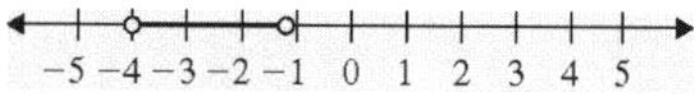

$$\left(-4,\ -\frac{7}{6}\right)$$

15)
$$A=P+PRT$$
$$A-P=P-P+PRT$$
$$\frac{A-P}{PT}=\frac{PRT}{PT}$$
$$\frac{A-P}{PT}=R$$

17) t = time riding

	d	= r	· t
#1	8t	8	t
#2	$10t$	10	t

distance #1 + distance #2 = 9
$$8t + 10t = 9$$
$$18t=9$$
$$t=\frac{1}{2}$$

It will take a $\frac{1}{2}$ hour to meet.

Section 4.1 Exercises

1) Answers may vary.

3) Yes
$$7x+2y=4$$
$$7(2)+2(-5)=4$$
$$14-10=4$$
$$4=4$$

5) Yes
$$-2x-y=13$$
$$-2(-8)-(3)=13$$
$$16-3=13$$
$$13=13$$

7) No
$$y=5x-6$$
$$(11)=5(3)-6$$
$$11=15-6$$
$$11\neq 9$$

9) Yes
$$y=-\frac{3}{2}x-19$$
$$(-4)=-\frac{3}{2}(-10)-19$$
$$-4=15-19$$
$$-4=-4$$

11) No
$$x=13$$
$$(5)=13$$
$$5\neq 13$$

13) $y=-x+6$
$$y=-(9)+6$$
$$y=-3$$

15)
$$2x-15y=13$$
$$2x-15\left(-\frac{4}{3}\right)=13$$
$$2x+20=13$$
$$2x=-7$$
$$x=-\frac{7}{2}$$

17) $y=-4$

19) $y=-3x+4$

$y=-3(0)+4$, $y=0+4$, $y=4$

$y=-3(2)+4$, $y=-6+4$, $y=-2$

$y=-3(1)+4$, $y=-3+4$, $y=1$

$y=-3(-1)+4$, $y=3+4$, $y=7$

x	y
0	4
1	1
2	−2
−1	7

21) $y=12x+10$

$y=12(0)+10$, $y=0+10$, $y=10$

$(-2)=12x+10$, $-12=12x$, $1=x$

$y=12\left(-\frac{1}{2}\right)+10$, $y=-6+10$, $y=4$

$(46)=12x+10$, $36=12x$, $3=x$

x	y
0	10
$-\frac{1}{2}$	4
-1	-2
3	46

23) $3x-y=8$

$3x-(0)=8$
$3x=8$
$x=\frac{8}{3}$

$3(0)-y=8$
$-y=8$
$y=-8$

$3x-(-7)=8$
$3x+7=8$
$3x=1$
$x=\frac{1}{3}$

$3(5)-y=8$
$15-y=8$
$-y=-7$
$y=7$

x	y
$\frac{8}{3}$	0
0	-8
5	7
$\frac{1}{3}$	-7

25) $x=5$

x	y
5	0
5	4
5	-1
5	-8

27) Answers may vary.

29) A: $(5, 1)$ quadrant I
B: $(2, -3)$ quadrant IV
C: $(-2, 4)$ quadrant II
D: $(-3, -4)$ quadrant III
E: $(3, 0)$ no quadrant
F: $(0, -2)$ no quadrant

31-34)

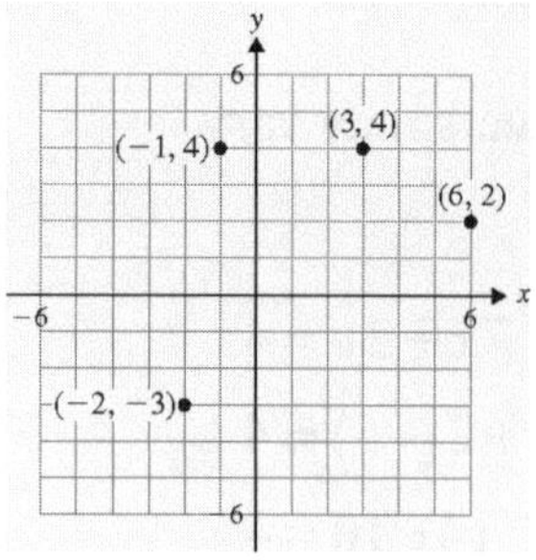

35-38)

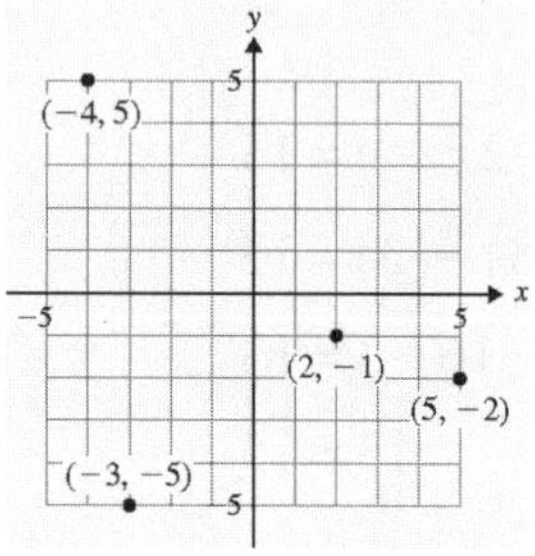

39-42)

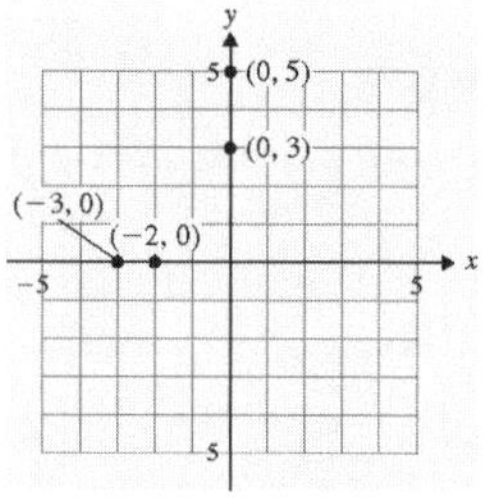

43-46)

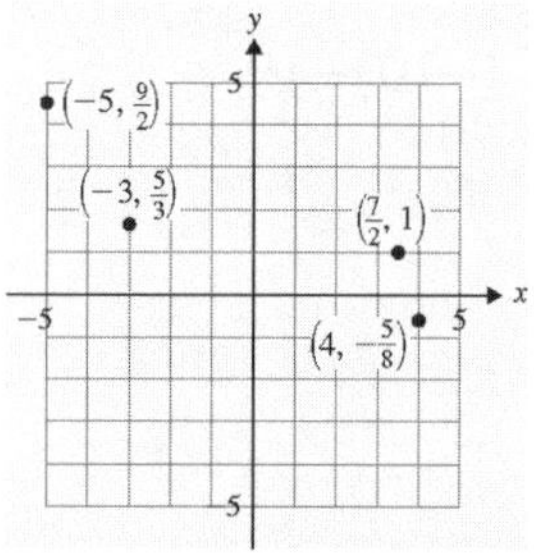

47-48)

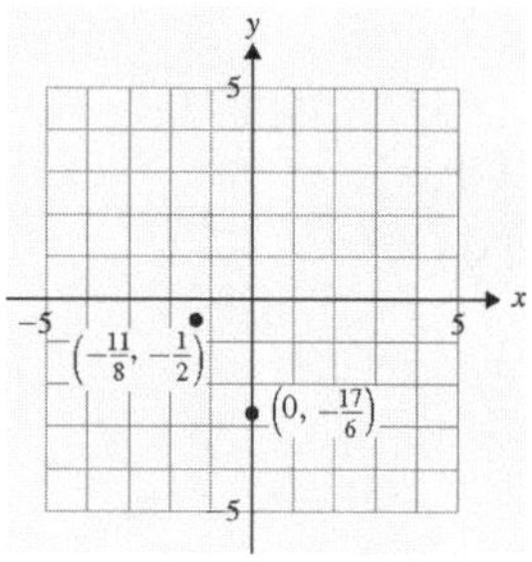

49) $y=-4x+3$

$$\begin{aligned} y&=-4(0)+3 \\ y&=0+3 \\ y&=3 \end{aligned} \qquad \begin{aligned} (0)&=-4y+3 \\ -3&=-4y \\ \frac{3}{4}&=y \end{aligned}$$

$$\begin{aligned} y&=-4(2)+3 \\ y&=-8+3 \\ y&=-5 \end{aligned} \qquad \begin{aligned} (7)&=-4y+3 \\ 4&=-4y \\ -1&=y \end{aligned}$$

x	y
0	3
$\frac{3}{4}$	0
2	−5
−1	7

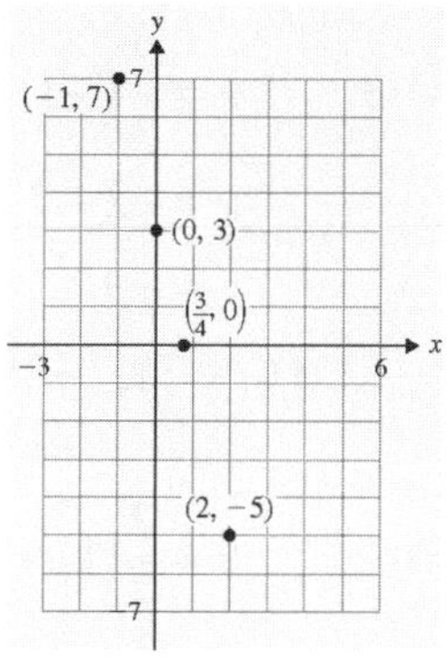

51) $y=x$

$$\begin{aligned} (0)&=x \\ 0&=x \end{aligned} \qquad \begin{aligned} y&=(-2) \\ y&=-2 \end{aligned}$$

$$\begin{aligned} (4)&=x \\ 4&=x \end{aligned} \qquad \begin{aligned} y&=(-3) \\ y&=-3 \end{aligned}$$

x	y
0	0
4	4
−2	−2
−3	−3

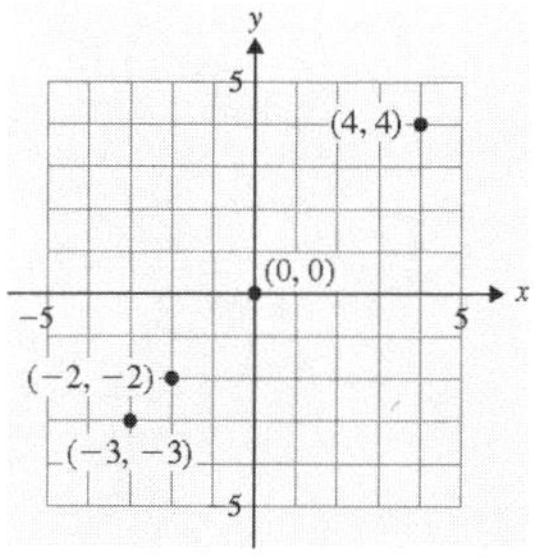

53) $3x-2y=6$

$$\begin{aligned} 3(0)-2y&=6 \\ -2y&=6 \\ y&=-3 \end{aligned} \qquad \begin{aligned} 3x-2(0)&=6 \\ 3x&=6 \\ x&=2 \end{aligned}$$

$$\begin{aligned} 3(-3)-2y&=6 \\ -9-2y&=6 \\ -2y&=15 \\ y&=-\frac{15}{2} \end{aligned} \qquad \begin{aligned} 3x-2(3)&=6 \\ 3x-6&=6 \\ 3x&=12 \\ x&=4 \end{aligned}$$

x	y
0	−3
2	0
4	3
−3	$-\frac{15}{2}$

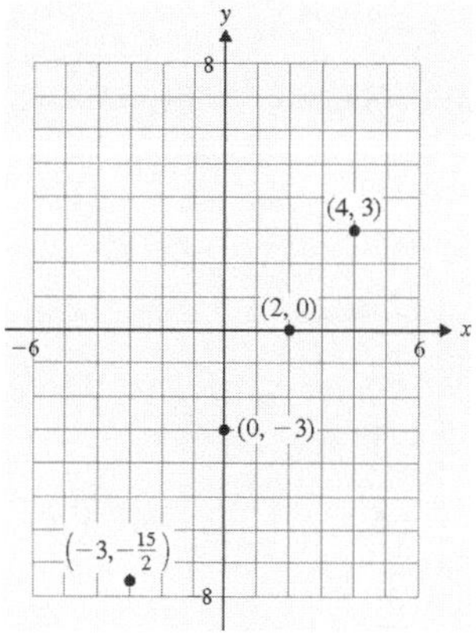

55) $x=-6$

x	y
−6	0
−6	−5
−6	2
−6	4

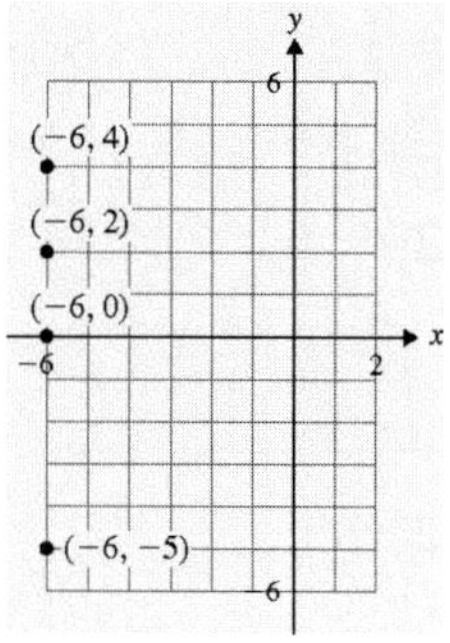

57) $y=-\frac{5}{4}x+2$

$y=-\frac{5}{4}(0)+2$
$y=0+2$
$y=2$

$y=-\frac{5}{4}(8)+2$
$y=-10+2$
$y=-8$

$y=-\frac{5}{4}(4)+2$
$y=-5+2$
$y=-3$

$y=-\frac{5}{4}(1)+2$
$y=-\frac{5}{4}+\frac{8}{4}$
$y=\frac{3}{4}$

x	y
0	2
4	−3
8	−8
1	$\frac{3}{4}$

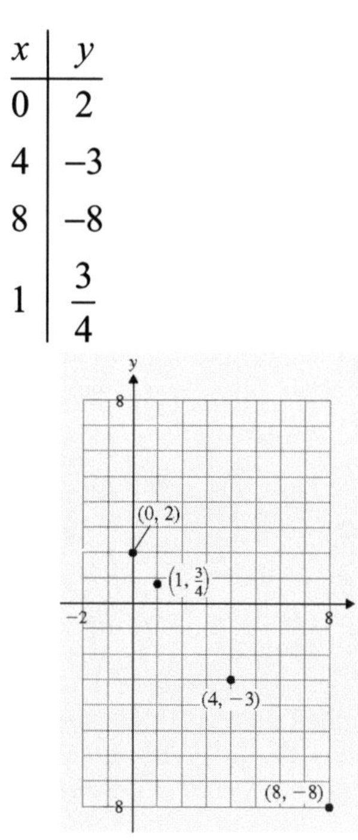

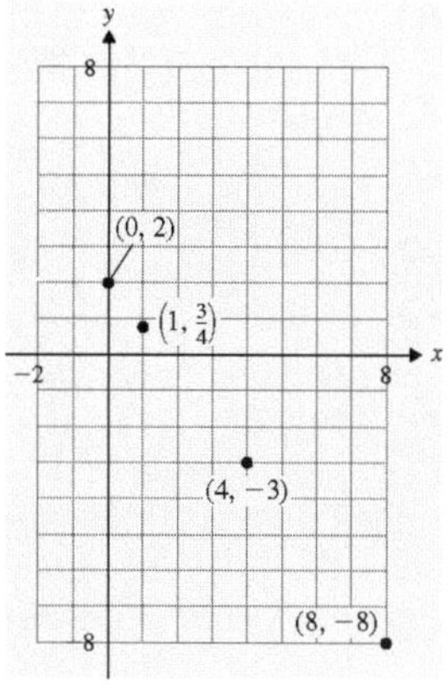

59) $y=\frac{2}{3}x-7$

a) $x=3$ $\quad y=\frac{2}{3}(3)-7$
$y=2-7$
$y=-5$

$x=6$ $\quad y=\frac{2}{3}(6)-7$
$y=4-7$
$y=-3$

$x=-3$ $\quad y=\frac{2}{3}(-3)-7$
$y=-2-7$
$y=-9$

$(3,-5),(6,-3),(-3,-9)$

b) $x=1$ $\quad y=\frac{2}{3}(1)-7$
$y=\frac{2}{3}-\frac{21}{3}$
$y=-\frac{19}{3}$

$x=5$ $\quad y=\frac{2}{3}(5)-7$
$y=\frac{10}{3}-\frac{21}{3}$
$y=-\frac{11}{3}$

$x=-2$ $\quad y=\frac{2}{3}(-2)-7$
$y=-\frac{4}{3}-\frac{21}{3}$
$y=-\frac{25}{3}$

$\left(1,-\frac{19}{3}\right),\left(5,-\frac{11}{3}\right),\left(-2,-\frac{25}{3}\right)$

c) The x-values in part a) are multiples of the denominator of $\frac{2}{3}$. So when you multiply $\frac{2}{3}$ by a multiple of 3 the fraction is eliminated.

61) positive

63) positive

65) negative

67) zero

69) a) $(16, 1300), (17, 1300), (18, 1600), (19, 1500)$

b)

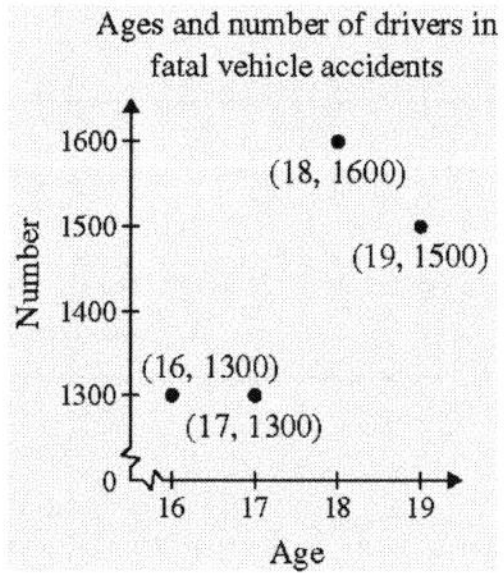

c) There were 1600 18-year-old drivers involved in fatal motor accidents in 2002.

71) a) $y = 20x + 100$

$$y = 20(1) + 100 \quad y = 20(4) + 100$$
$$y = 20 + 100 \quad y = 80 + 100$$
$$y = 120 \quad y = 180$$

$$y = 20(3) + 100 \quad y = 20(6) + 100$$
$$y = 60 + 100 \quad y = 120 + 100$$
$$y = 160 \quad y = 220$$

x	y
1	120
3	160
4	180
6	220

$(1, 120), (3, 160), (4, 180), (6, 220)$

b)

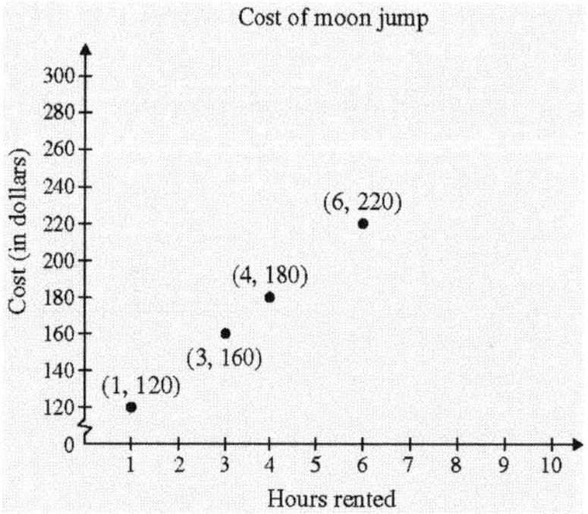

c) The cost of renting the moon jump for 4 hours is \$180.

d) Yes, they lie on a straight line.

e)
$$y = 20x + 100$$
$$280 = 20x + 100$$
$$180 = 20x$$
$$9 = x$$
A customer could rent the moon jump for 9 hours.

Section 4.2 Exercises

1) Every linear equation in two variables has an infinite number of solutions.

3) $y=3x-1$

$y=3(0)-1 \qquad y=3(2)-1$

$y=0-1 \qquad y=6-1$

$y=-1 \qquad y=5$

$y=3(1)-1 \qquad y=3(-1)-1$

$y=3-1 \qquad y=-3-1$

$y=2 \qquad y=-4$

x	y
0	−1
1	2
2	5
−1	−4

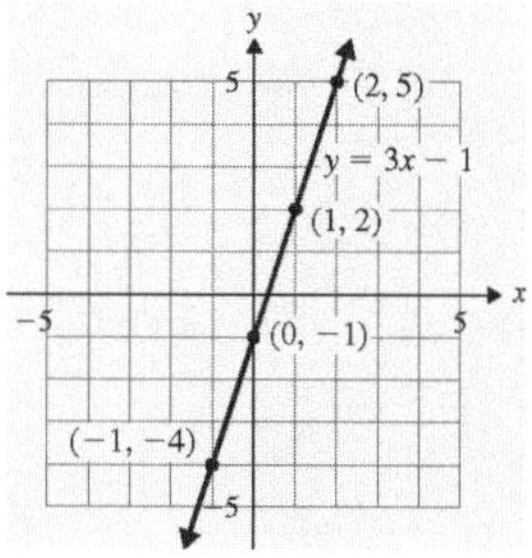

5) $y=-\dfrac{2}{3}x+4$

$y=-\dfrac{2}{3}(0)+4 \qquad y=-\dfrac{2}{3}(3)+4$

$y=0+4 \qquad y=-2+4$

$y=4 \qquad y=2$

$y=-\dfrac{2}{3}(-3)+4 \qquad y=-\dfrac{2}{3}(6)+4$

$y=2+4 \qquad y=-4+4$

$y=6 \qquad y=0$

x	y
0	4
−3	6
3	2
6	0

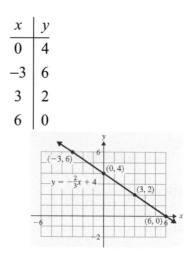

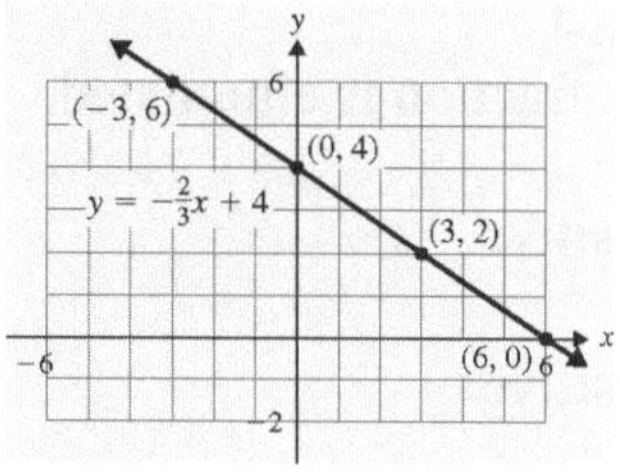

7) $-3x+6y=9$

$-3(0)+6y=9 \qquad -3x+6(4)=9$

$6y=9 \qquad -3x+24=9$

$y=\dfrac{3}{2} \qquad -3x=-15$

$x=5$

$-3x+6(0)=9 \qquad -3(-1)+6y=9$

$-3x=9 \qquad 3+6y=9$

$x=-3 \qquad 6y=6$

$y=1$

x	y
0	$\dfrac{3}{2}$
−3	0
5	4
−1	1

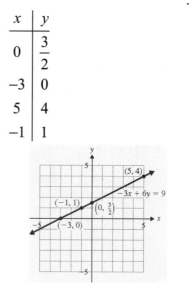

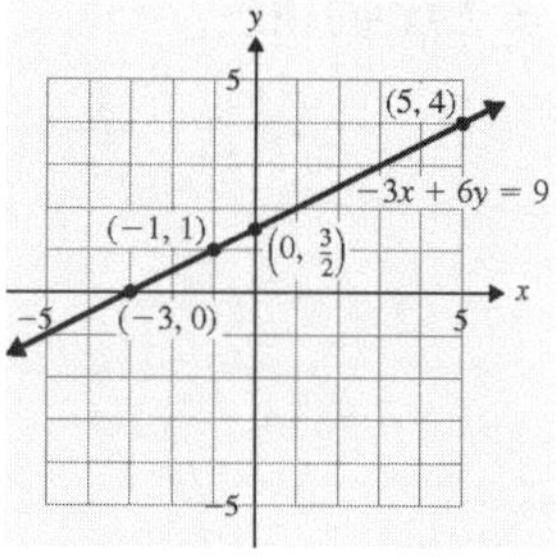

9) $y+4=0$

$y=-4$

x	y
0	-4
-3	-4
-1	-4
2	-4

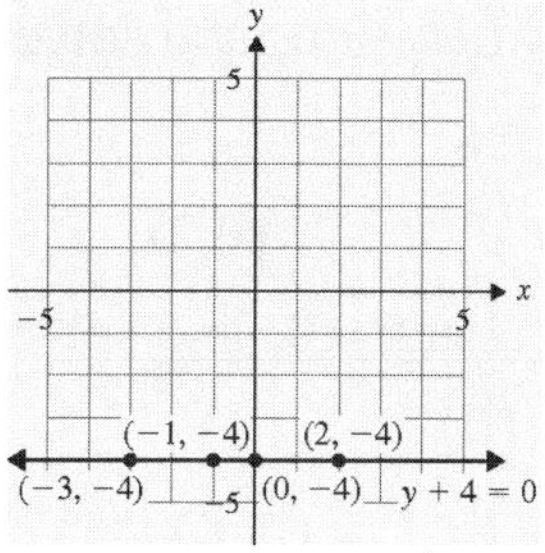

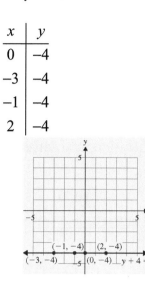

11) Let $x=0$, and solve for y.

13) $y=-2x+6$

x-int: Let $y=0$, and solve for x.

$(0)=-2x+6$

$-6=-2x$

$3=x \qquad (3, 0)$

y-int: Let $x=0$, and solve for y.

$y=-2(0)+6$

$y=0+6$

$y=6 \qquad (0, 6)$

Let $x=1$.

$y=-2(1)+6$

$y=-2+6$

$y=4 \qquad (1, 4)$

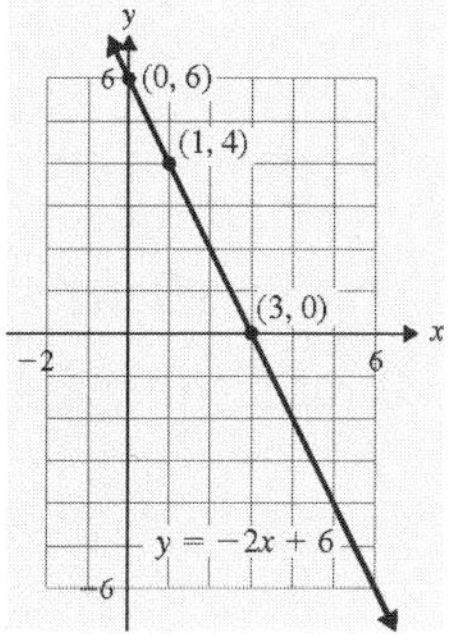

15) $3x-4y=12$

x-int: Let $y=0$, and solve for x.

$3x-4(0)=12$

$3x-0=12$

$3x=12$

$x=4 \qquad (4, 0)$

y-int: Let $x=0$, and solve for y.

$3(0)-4y=12$

$0-4y=12$

$-4y=12$

$y=-3 \qquad (0, -3)$

Let $x=2$.

$3(2)-4y=12$

$6-4y=12$

$-4y=6$

$y=-\frac{3}{2} \qquad \left(2, -\frac{3}{2}\right)$

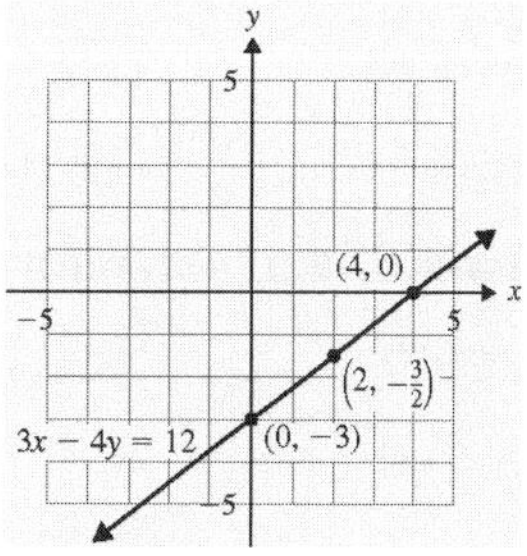

17) $x=\frac{1}{4}y-1$

x-int: Let $y=0$, and solve for x.

$x=\frac{1}{4}(0)-1$

$x=0-1$

$x=-1 \qquad (-1, 0)$

y-int: Let $x=0$, and solve for y.

$(0)=\frac{1}{4}y-1$

$1=\frac{1}{4}y$

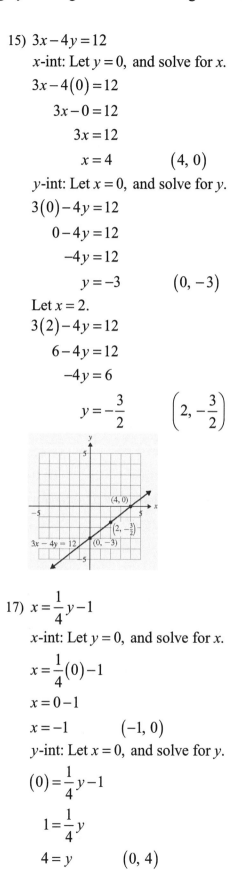

$4=y \qquad (0, 4)$

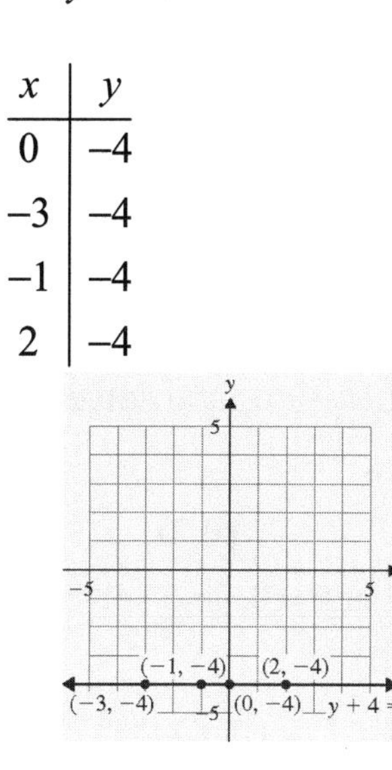

Let $y=8$.

$x=\frac{1}{4}(8)-1$

$x=2-1$

$x=1 \qquad (1, 8)$

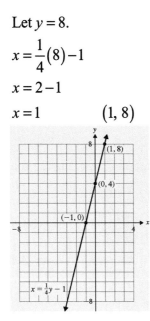

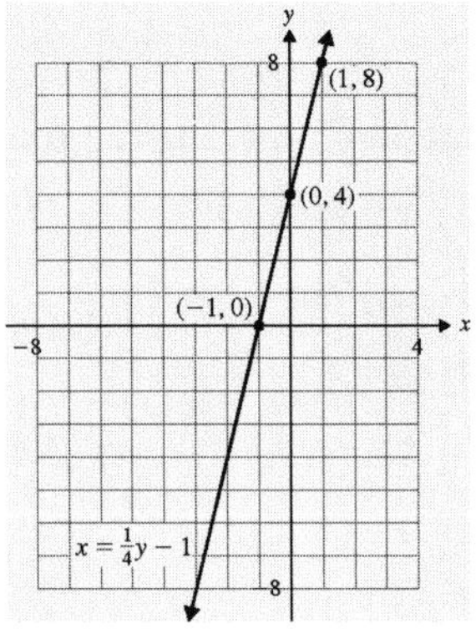

19) $2x+3y=-6$

x-int: Let $y=0$, and solve for x.

$2x+3(0)=-6$

$2x+0=-6$

$2x=-6$

$x=-3 \qquad (-3, 0)$

y-int: Let $x=0$, and solve for y.

$2(0)+3y=-6$

$0+3y=-6$

$3y=-6$

$y=-2 \qquad (0, -2)$

Let $x=1$.

$2(1)+3y=-6$

$2+3y=-6$

$3y=-8$

$y=-\frac{8}{3} \qquad \left(1, -\frac{8}{3}\right)$

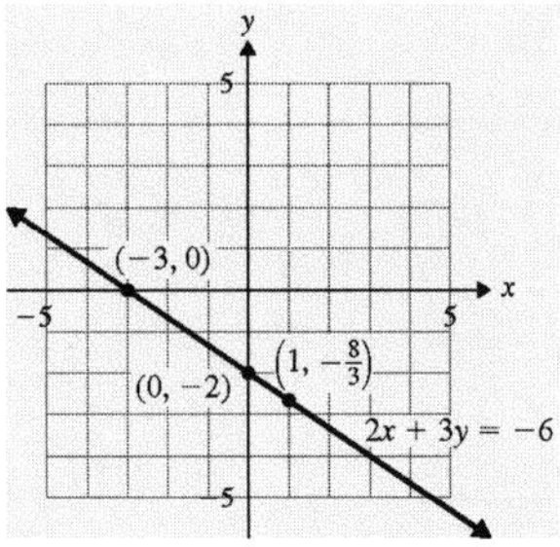

21) $y=-x$

x-int: Let $y=0$, and solve for x.

$(0)=-x$

$0=x \qquad (0, 0)$

y-int: Let $x=0$, and solve for y.

$y=-(0)$

$y=0 \qquad (0, 0)$

Let $x=1$.

$y=-(1)$

$y=-1 \qquad (1, -1)$

Let $x=-1$.

$y=-(-1)$

$y=1 \qquad (-1, 1)$

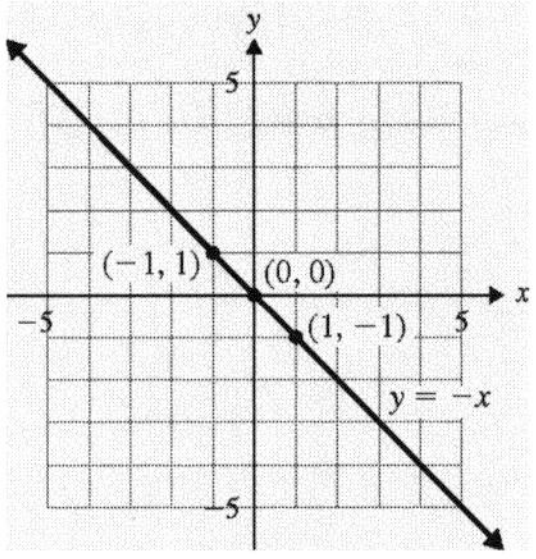

23) $5y-2x=0$

x-int: Let $y=0$, and solve for x.

$5(0)-2x=0$

$0-2x=0$

$-2x=0$

$x=0 \qquad (0, 0)$

y-int: Let $x=0$, and solve for y.

$5y-2(0)=0$

$5y-0=0$

$5y=0$

$y=0 \qquad (0,\ 0)$

Let $x=5$.

$5y-2(5)=0$

$5y-10=0$

$5y=10$

$y=2 \qquad (5,\ 2)$

Let $x=-5$.

$5y-2(-5)=0$

$5y+10=0$

$5y=-10$

$y=-2 \qquad (-5,\ -2)$

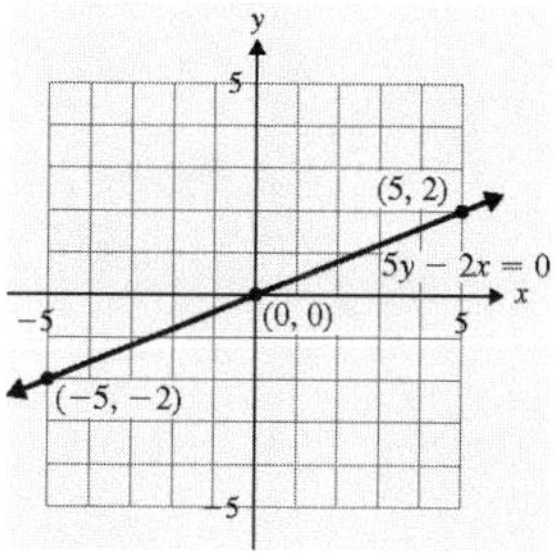

25) $x=5$

$(5,\ 0),\ (5,\ 2),\ (5,\ -1)$

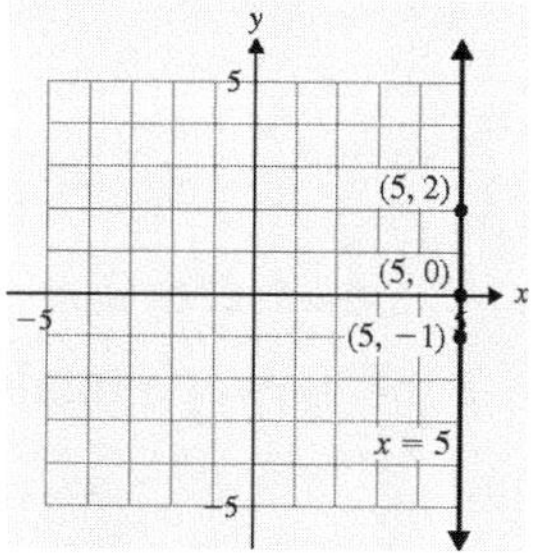

27) $y=0$

$(0,\ 0),\ (1,\ 0),\ (-2,\ 0)$

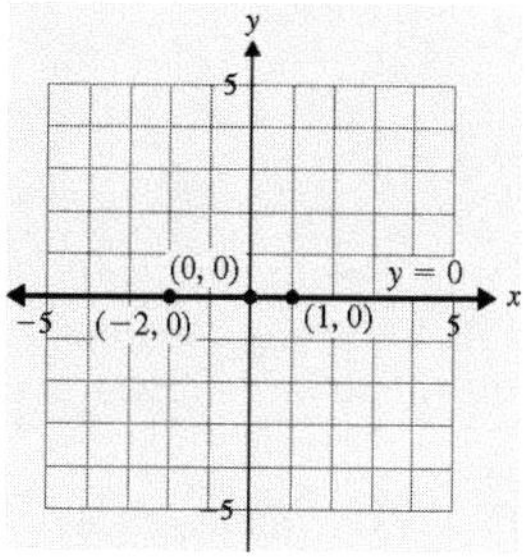

29) $y+3=0$

$y=-3$

$(0,\ -3),\ (1,\ -3),\ (-3,\ -3)$

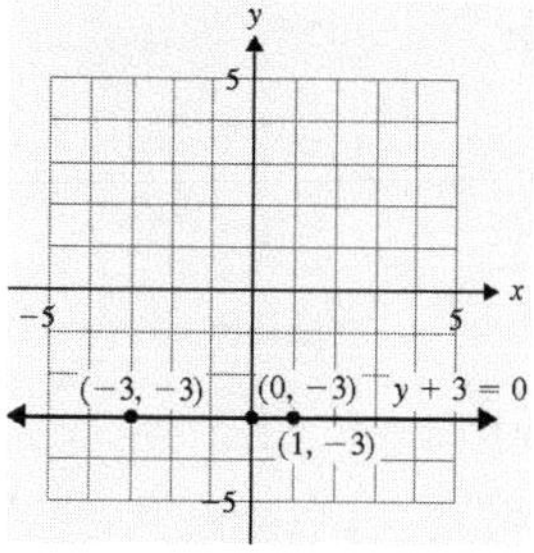

31) $x+3y=8$

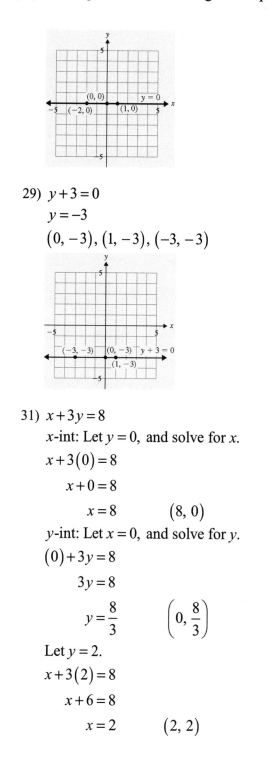

x-int: Let $y=0$, and solve for x.

$x+3(0)=8$

$x+0=8$

$x=8 \qquad (8,\ 0)$

y-int: Let $x=0$, and solve for y.

$(0)+3y=8$

$3y=8$

$y=\frac{8}{3} \qquad \left(0,\ \frac{8}{3}\right)$

Let $y=2$.

$x+3(2)=8$

$x+6=8$

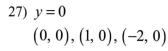

$x=2 \qquad (2,\ 2)$

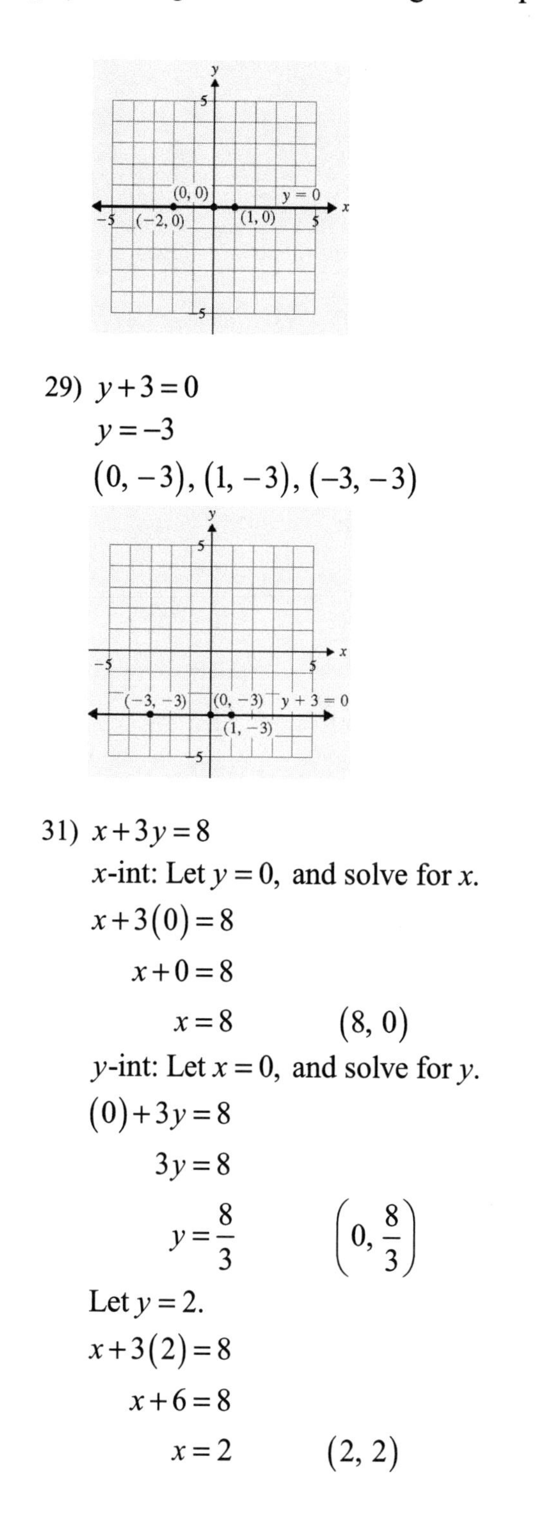

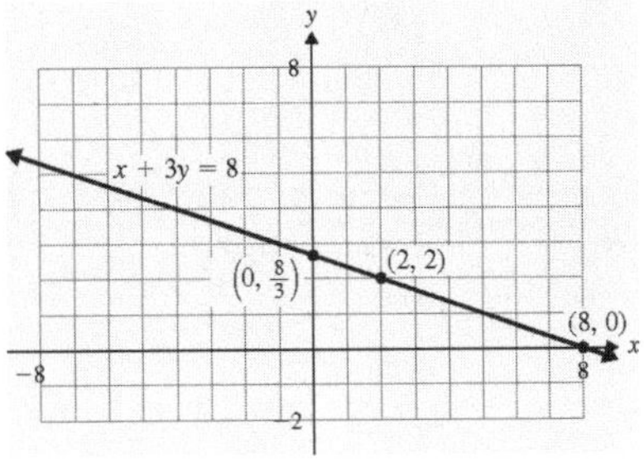

33) $y=1.5x$

a) $x=0$ $\quad y=1.5(0)$
$y=0$

$x=10$ $\quad y=1.5(10)$
$y=15$

$x=20$ $\quad y=1.5(20)$
$y=30$

$x=60$ $\quad y=1.5(60)$
$y=90$

x	y
0	0
10	15
20	30
60	90

$(0, 0), (10, 15),$
$(20, 30), (60, 90)$

b) $(0, 0)$: Before engineers began working $(x=0 \text{ days})$, the tower did not move toward vertical $(y=0)$.
$(10, 15)$: After 10 days of working, the Tower was moved 15 mm toward vertical.
$(20, 30)$: After 20 days of working, the Tower was moved 30 mm toward vertical.
$(60, 90)$: After 60 days of working, the Tower was moved 90 mm toward vertical.

c)

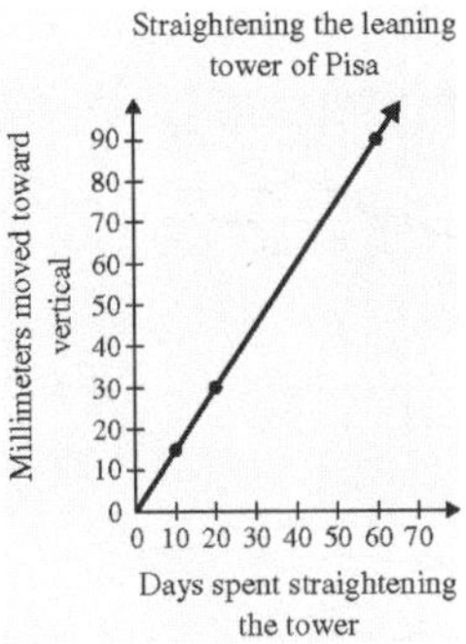

d) $y=1.5x$
$450=1.5x$
$300=x$
It took the engineers 300 days to move the Tower a total of 450 mm.

35) a) Answers may vary.

b) $y=-0.001x+29.86$
$x=0$
$y=-0.001(0)+29.86$
$y=0+29.86$
$y=29.86 \text{ in}$

$x = 1000$

$y = -0.001(1000) + 29.86$

$y = -1 + 29.86$

$y = 28.86$ in

$x = 3500$

$y = -0.001(3500) + 29.86$

$y = -3.5 + 29.86$

$y = 26.36$ in

$x = 5000$

$y = -0.001(5000) + 29.86$

$y = -5 + 29.86$

$y = 24.86$ in

c)

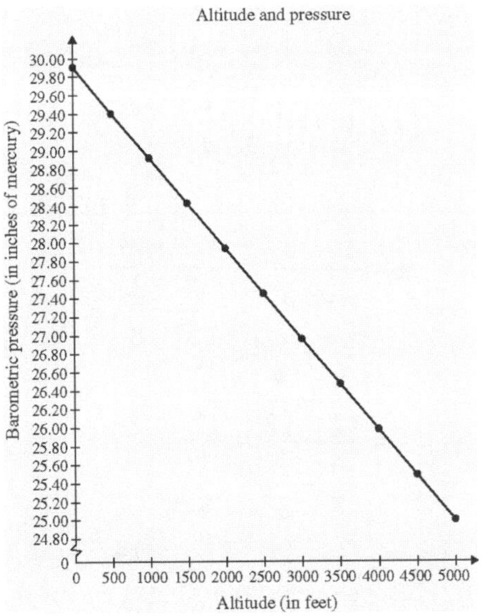

d) No, because the problem states that the equation applies to altitudes 0 ft-5000 ft.

Section 4.3 Exercises

1) The slope of a line is the ratio of vertical change to horizontal change.
It is $\dfrac{\text{change in } y}{\text{change in } x}$ or $\dfrac{\text{rise}}{\text{run}}$ or $\dfrac{y_2 - y_1}{x_2 - x_1}$
where (x_1, y_1) and (x_2, y_2) are points on the line.

3) It slants downward from left to right.

5) undefined

7) a) Vertical change: 3 units
Horizontal change: 4 units

$$\text{Slope} = \frac{3}{4}$$

b) $(x_1, y_1) = (1, -1)$
$(x_2, y_2) = (5, 2)$

$$m = \frac{y_2 - y_1}{x_2 - x_1} = \frac{2-(-1)}{5-1} = \frac{3}{4}$$

9) a) Vertical change: -2 units
Horizontal change: 6 units

$$\text{Slope} = \frac{-2}{6} = -\frac{1}{3}$$

b) $(x_1, y_1) = (-3, -2)$
$(x_2, y_2) = (3, -4)$

$$m = \frac{y_2 - y_1}{x_2 - x_1} = \frac{-4-(-2)}{3-(-3)} = \frac{-2}{6}$$

$$= -\frac{1}{3}$$

11) a) Vertical change: -5 units
Horizontal change: 1 unit

$$\text{Slope} = \frac{-5}{1} = -5$$

b) $(x_1, y_1) = (-1, 4)$
$(x_2, y_2) = (0, -1)$

$$m = \frac{y_2 - y_1}{x_2 - x_1} = \frac{-1-4}{0-(-1)} = \frac{-5}{1} = -5$$

13) a) Vertical change: 0 units
Horizontal change: 4 units

$$\text{Slope} = \frac{0}{4} = 0$$

b) $(x_1, y_1) = (-1, -5)$
$(x_2, y_2) = (3, -5)$

$$m = \frac{y_2 - y_1}{x_2 - x_1} = \frac{-5-(-5)}{3-(-1)} = \frac{0}{4} = 0$$

15)

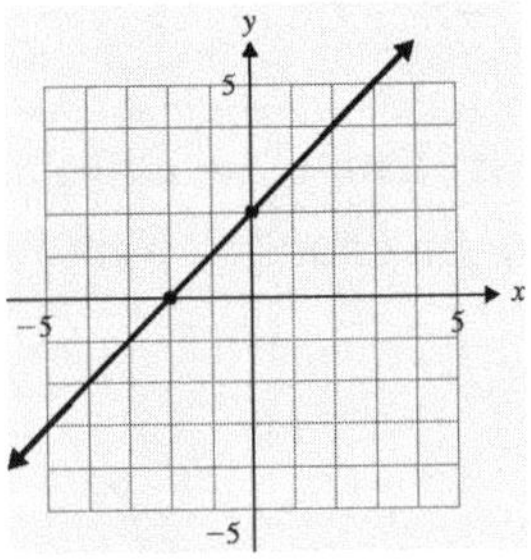

17) $(x_1, y_1) = (3, 2)$
$(x_2, y_2) = (9, 5)$

$$m = \frac{y_2 - y_1}{x_2 - x_1} = \frac{5-2}{9-3} = \frac{3}{6} = \frac{1}{2}$$

19) $(x_1, y_1) = (-2, 8)$
$(x_2, y_2) = (2, 4)$

$$m = \frac{y_2 - y_1}{x_2 - x_1} = \frac{4-8}{2-(-2)} = \frac{-4}{4} = -1$$

21) $(x_1, y_1) = (9, 2)$
$(x_2, y_2) = (0, 4)$

$$m = \frac{y_2 - y_1}{x_2 - x_1} = \frac{4-2}{0-9} = \frac{2}{-9} = -\frac{2}{9}$$

23) $(x_1, y_1) = (3, 5)$
$(x_2, y_2) = (-1, 5)$

$$m = \frac{y_2 - y_1}{x_2 - x_1} = \frac{5-5}{-1-3} = \frac{0}{-4} = 0$$

25) $(x_1, y_1) = (3, 2)$
$(x_2, y_2) = (3, -1)$

$$m = \frac{y_2 - y_1}{x_2 - x_1} = \frac{-1-2}{3-3} = \frac{-3}{0}$$

Slope is undefined.

27) $(x_1, y_1) = \left(\frac{3}{8}, -\frac{1}{3}\right)$
$(x_2, y_2) = \left(\frac{1}{2}, \frac{1}{4}\right)$

$$m = \frac{y_2 - y_1}{x_2 - x_1} = \frac{\frac{1}{4} - \left(-\frac{1}{3}\right)}{\frac{1}{2} - \frac{3}{8}}$$

$$= \frac{\frac{3}{12} + \frac{4}{12}}{\frac{4}{8} - \frac{3}{8}} = \frac{\frac{7}{12}}{\frac{1}{8}}$$

$$= \frac{7}{12} \div \frac{1}{8} = \frac{7}{12} \cdot 8 = \frac{14}{3}$$

29) $(x_1, y_1) = (-1.7, -1.2)$
$(x_2, y_2) = (2.8, -10.2)$

$$m = \frac{y_2 - y_1}{x_2 - x_1} = \frac{-10.2-(-1.2)}{2.8-(-1.7)}$$

$$= \frac{-9}{4.5} = -2$$

31) a) $m = \frac{\text{rise}}{\text{run}} = \frac{10}{12} = \frac{5}{6}$

b) $m = \frac{\text{rise}}{\text{run}} = \frac{8}{12} = \frac{2}{3}$

c) convert the ft to in: 2 ft = 24 in.

$m = \frac{\text{rise}}{\text{run}} = \frac{8}{24} = \frac{1}{3}$;

$\frac{1}{3} = \frac{x}{12}$

$3x = 12$

$x = 4$; 4-12 pitch

33) a) $22,000

b) The line slants downward from left to right, therefore it has a negative slope.

c) The value of the car is decreasing over time.

d) $m = \frac{\text{rise}}{\text{run}} = \frac{-2000}{1} = -2000$

The value of the car is decreasing by $2000 per year.

35)

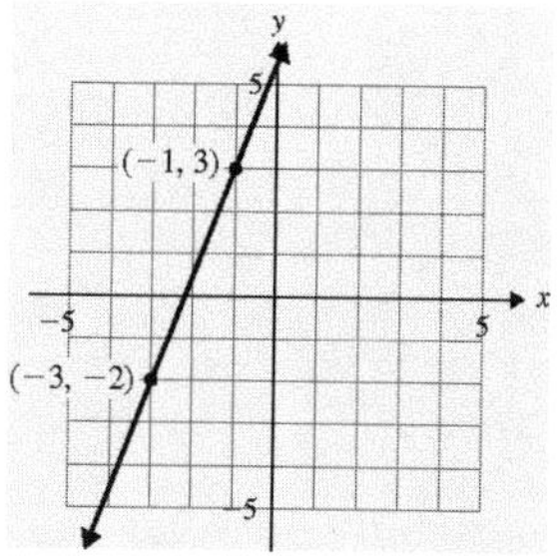

37)

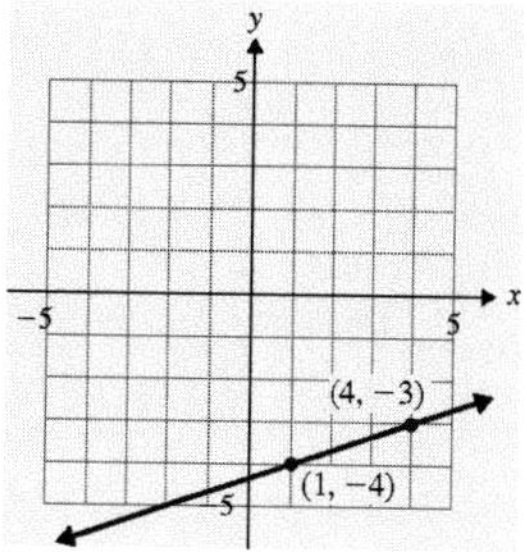

39)

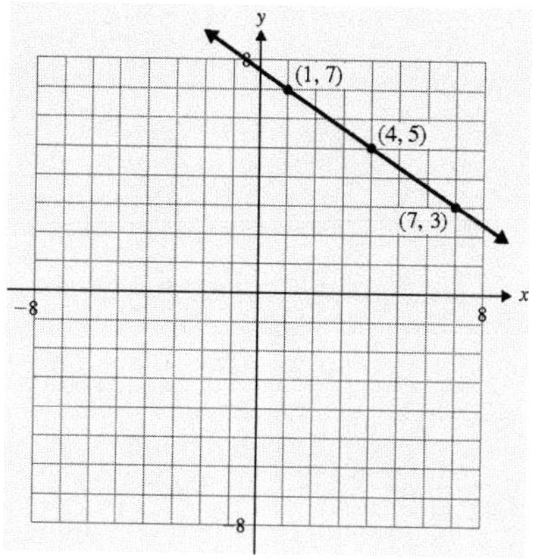

41)

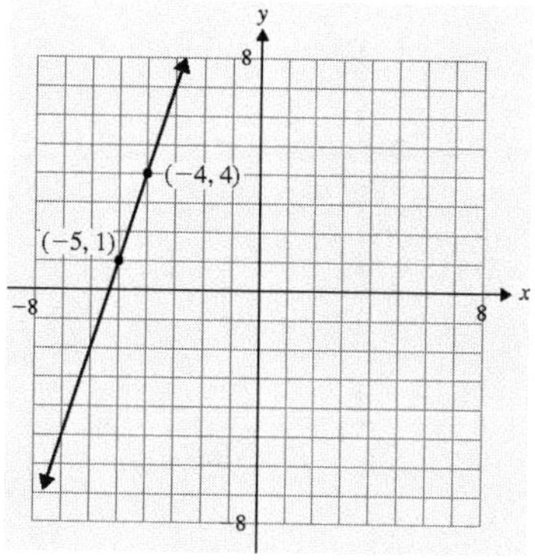

43)

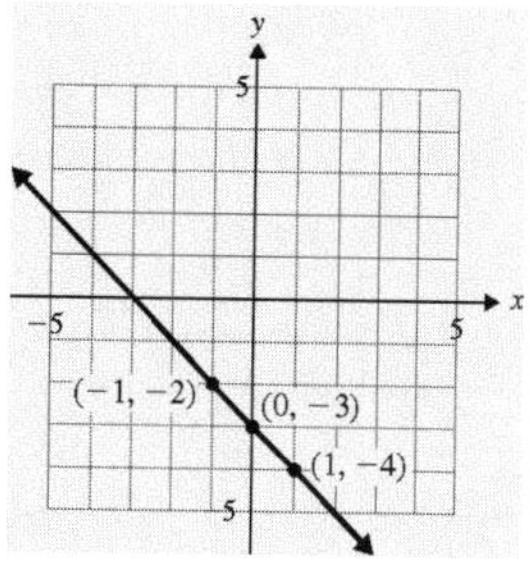

45)

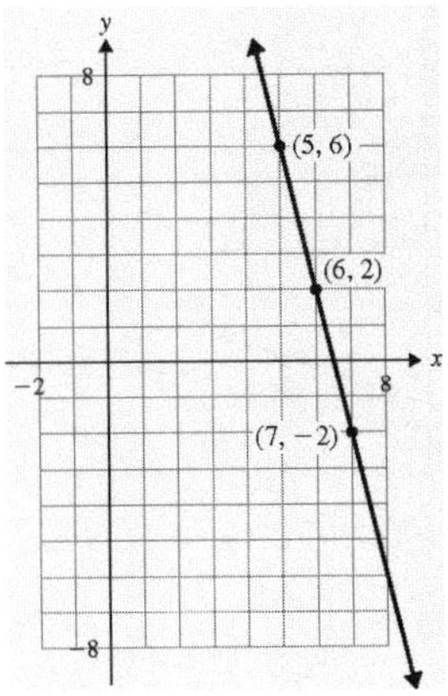

47)

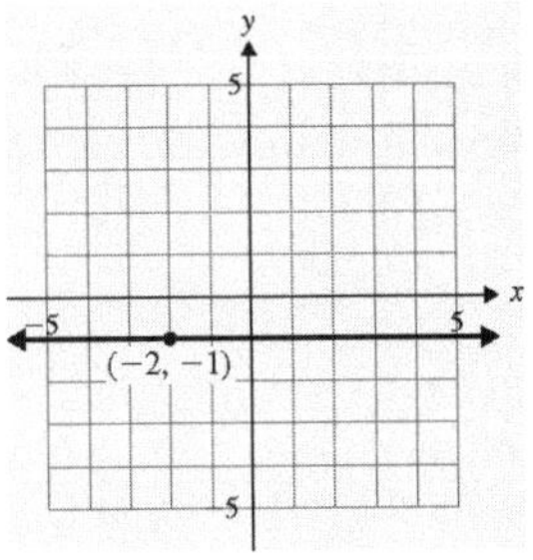

49)

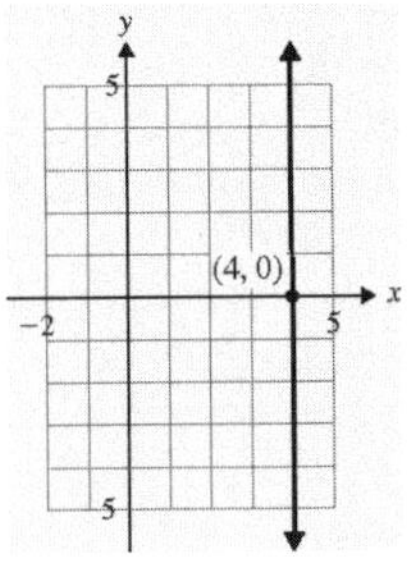

51)

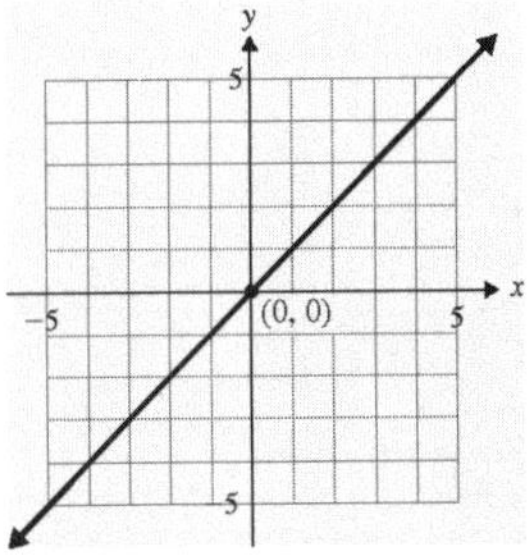

Section 4.4 Exercises

1) The slope is m, and the y-intercept is $(0,\ b)$.

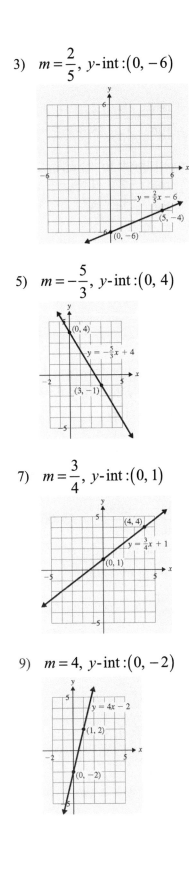

3) $m=\frac{2}{5},\ y\text{-int}:(0,\ -6)$

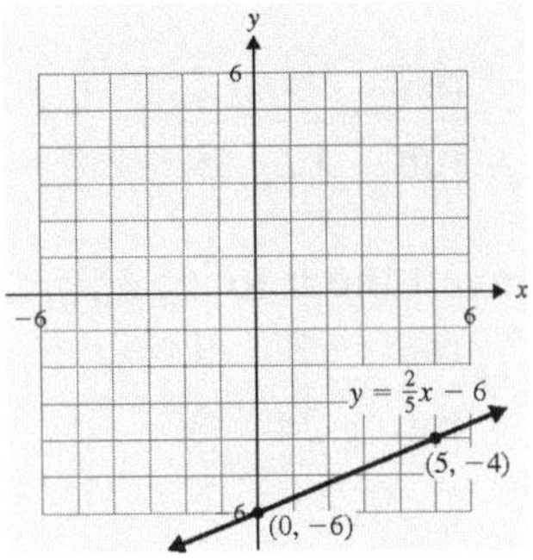

5) $m=-\frac{5}{3},\ y\text{-int}:(0,\ 4)$

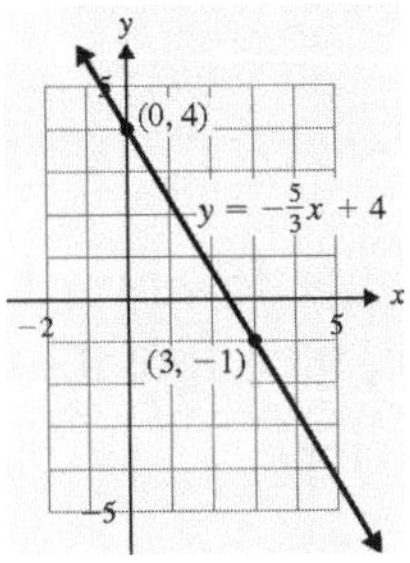

7) $m=\frac{3}{4},\ y\text{-int}:(0,\ 1)$

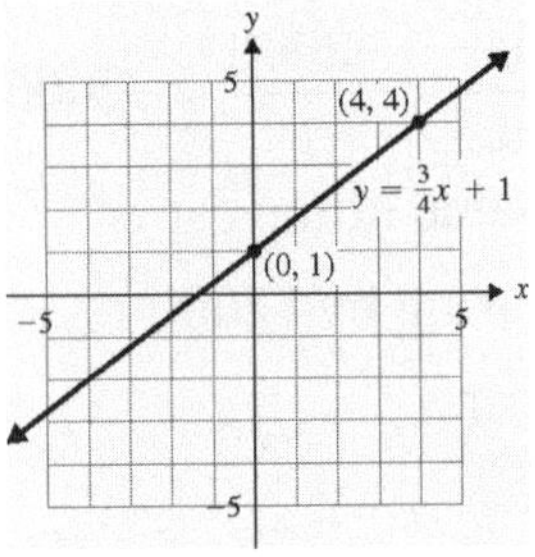

9) $m=4,\ y\text{-int}:(0,\ -2)$

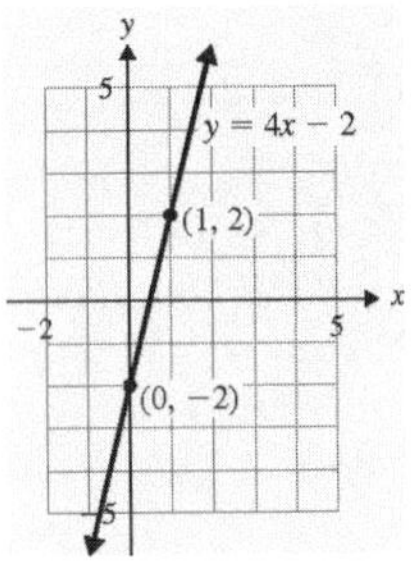

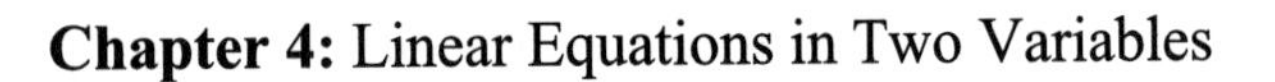

11) $m=-1,\ y\text{-int}:(0,\ 5)$

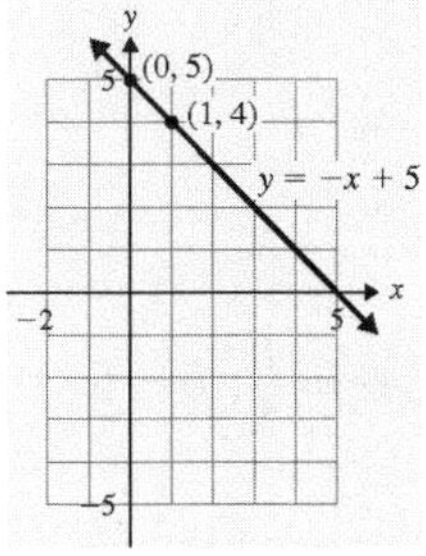

13) $m=\dfrac{3}{2},\ y\text{-int}:\left(0,\ \dfrac{1}{2}\right)$

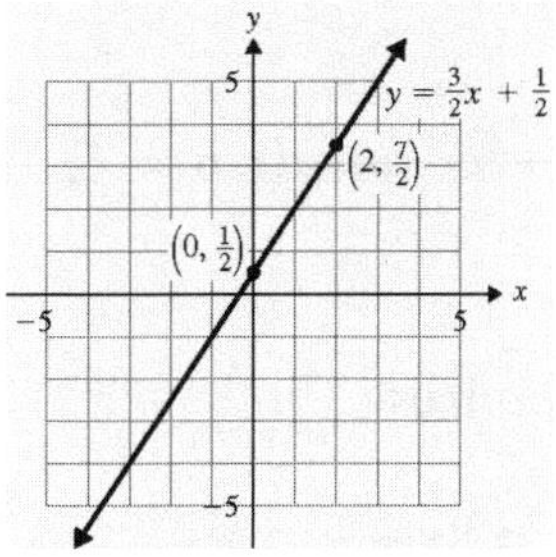

15) $m=0,\ y\text{-int}:(0,\ -2)$

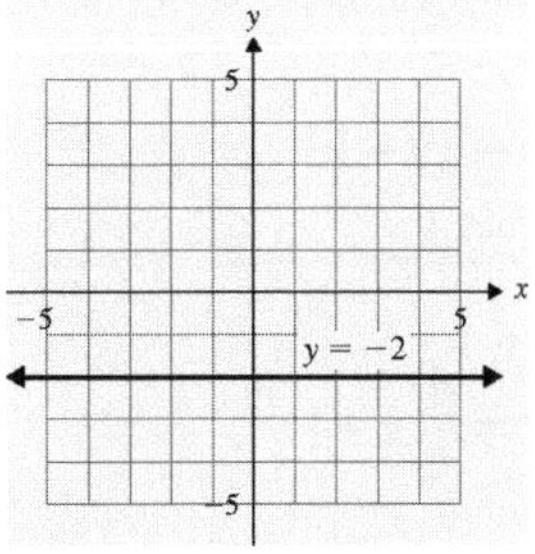

17) $x+3y=-6$

$$3y=-x-6$$

$$y=-\frac{1}{3}x-2$$

$m=-\dfrac{1}{3},\ y\text{-int}:(0,\ -2)$

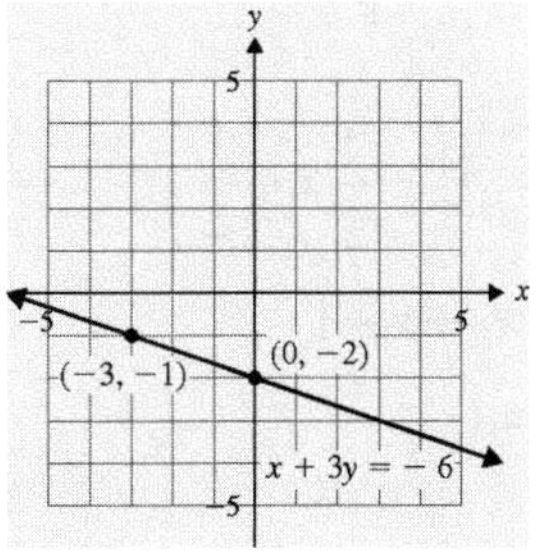

19) $12x-8y=32$

$$-8y=-12x+32$$

$$y=\frac{3}{2}x-4$$

$m=\dfrac{3}{2},\ y\text{-int}:(0,\ -4)$

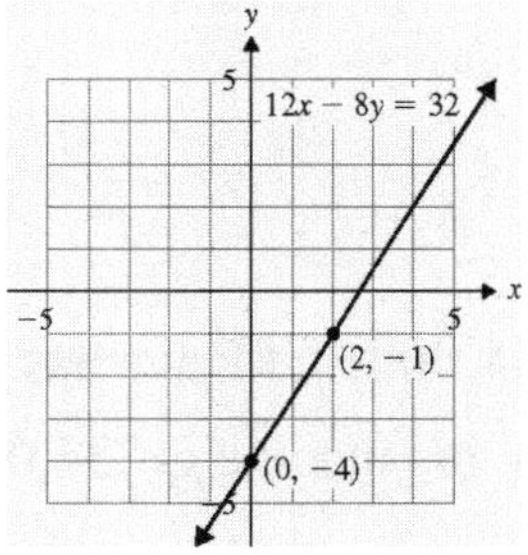

21) $x+9=2$

$$x=-7$$

The slope is undefined, and no y-intercept

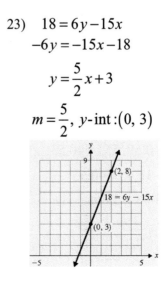

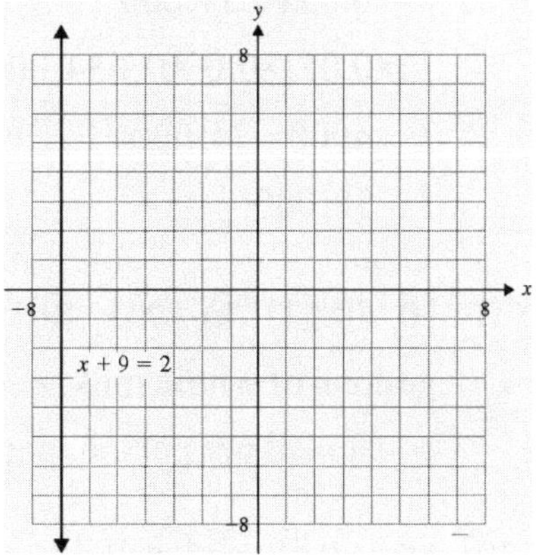

23) $18=6y-15x$

$$-6y=-15x-18$$

$$y=\frac{5}{2}x+3$$

$m=\dfrac{5}{2},\ y\text{-int}:(0,\ 3)$

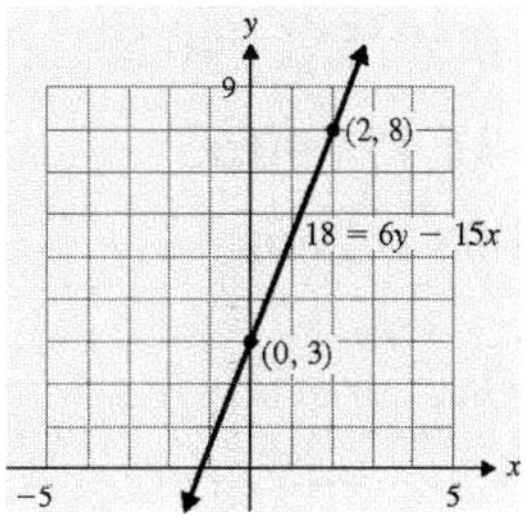

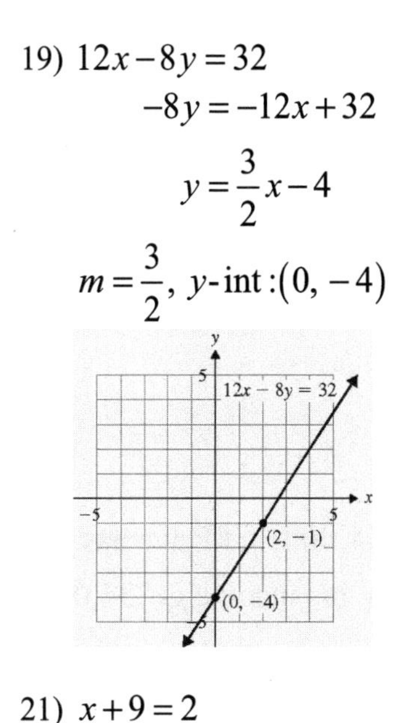

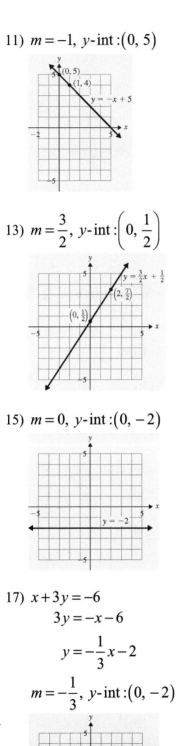

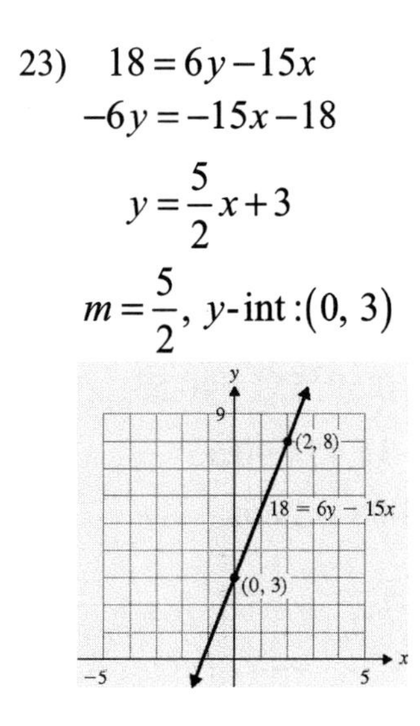

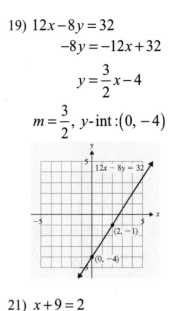

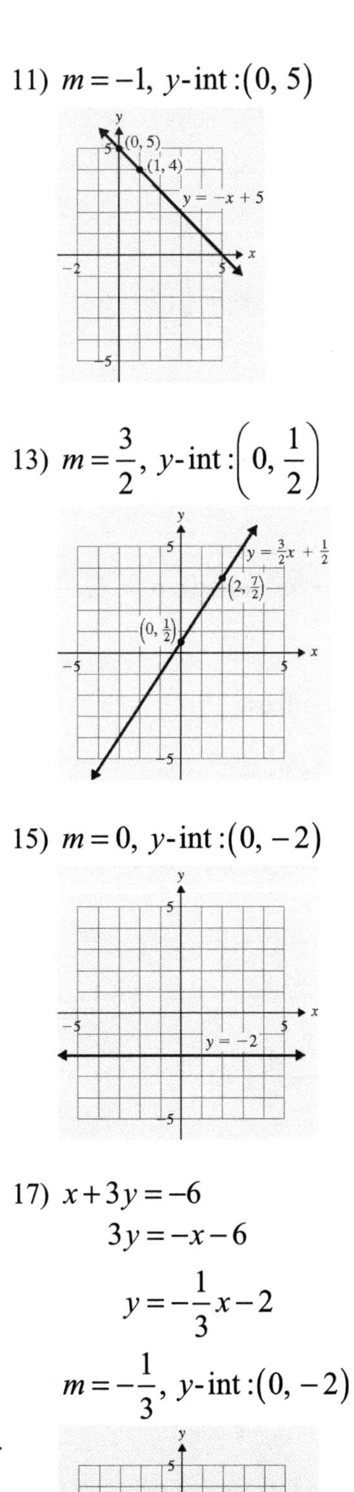

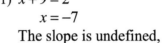

25) $y=0$

$m=0$, $y\text{-int}:(0, 0)$

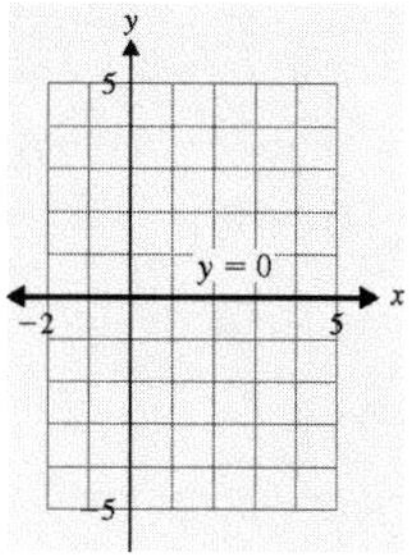

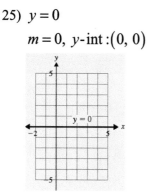

27) a) $(0, 34,000)$; If Dave has \$0 in sales, his income is \$34,000.

b) $m=0.05$; Dave earns \$0.05 for every \$1 in sales.

c) \$38,000
$$I=0.05x+34,000$$
$$I=0.05(80,000)+34,000$$
$$I=4000+34,000$$
$$I=38,000$$

29) a) $(0, 40.53)$; In 1945, 40.53 gallons of whole milk were consumed per person, per year.

b) $m=-0.59$; Since 1945, Americans have been consuming 0.59 fewer gallons of milk each year.

c) Estimate: 7.5 gallons
$$y=-0.59x+40.53$$
$$y=-0.59(55)+40.53$$
$$y=-32.45+40.53$$
$$y=8.08 \text{ gallons}$$

31) $y=mx+b \quad m=-4;\ b=7$
$$y=-4x+7$$

33) $y=mx+b \quad m=\frac{8}{5};\ b=-6$
$$y=\frac{8}{5}x-6$$

35) $y=mx+b \quad m=\frac{1}{3};\ b=5$
$$y=\frac{1}{3}x+5$$

37) $y=mx+b \quad m=-1;\ b=0$
$$y=-x$$

39) $y=mx+b \quad m=0;\ b=-2$
$$y=-2$$

Section 4.5 Exercises

1)
$$y=8x+3$$
$$-8x+y=3$$
$$8x-y=-3$$

3)
$$x=-2y-11$$
$$x+2y=-11$$

5)
$$y=\frac{4}{5}x-1$$
$$5\cdot y=5\left(\frac{4}{5}x-1\right)$$
$$5y=4x-5$$
$$-4x+5y=-5$$
$$4x-5y=5$$

7)
$$y=-\frac{3}{2}x+\frac{1}{4}$$
$$4\cdot y=4\left(-\frac{3}{2}x+\frac{1}{4}\right)$$
$$4y=-6x+1$$
$$6x+4y=1$$

9) Use $y = mx + b$ and substitute the slope and y-intercept values into the equation.

11) $y = mx + b \quad m = -7;\ b = 2$
$y = -7x + 2$

13) $y = mx + b \quad m = 1;\ b = -3$
$y = x - 3$
$-x + y = -3$
$x - y = 3$

15) $y = mx + b \quad m = -\frac{1}{3};\ b = -4$
$y = -\frac{1}{3}x - 4$
$3 \cdot y = 3\left(-\frac{1}{3}x - 4\right)$
$3y = -x - 12$
$x + 3y = -12$

17) $y = mx + b \quad m = 1;\ b = 0$
$y = x$

19) a) $y - y_1 = m(x - x_1)$

b) Use the point-slope formula and substitute the point on the line and the slope into the equation.

21) $(x_1, y_1) = (1, 6);\ m = 5$
$y - y_1 = m(x - x_1)$
$y - 6 = 5(x - 1)$
$y - 6 = 5x - 5$
$y = 5x + 1$

23) $(x_1, y_1) = (-9, 4);\ m = -1$
$y - y_1 = m(x - x_1)$
$y - 4 = -1(x - (-9))$
$y - 4 = -1(x + 9)$
$y - 4 = -x - 9$
$y = -x - 5$

25) $(x_1, y_1) = (-2, -1);\ m = 4$
$y - y_1 = m(x - x_1)$
$y - (-1) = 4(x - (-2))$
$y + 1 = 4x + 8$
$-4x + y = 7$
$4x - y = -7$

27) $(x_1, y_1) = (12, 3);\ m = -\frac{2}{3}$
$y - y_1 = m(x - x_1)$
$y - 3 = -\frac{2}{3}(x - 12)$
$y - 3 = -\frac{2}{3}x + 8$
$3(y - 3) = 3\left(-\frac{2}{3}x + 8\right)$
$3y - 9 = -2x + 24$
$2x + 3y = 33$

29) $(x_1, y_1) = (-4, -5);\ m = \frac{1}{6}$
$y - y_1 = m(x - x_1)$
$y - (-5) = \frac{1}{6}(x - (-4))$
$y + 5 = \frac{1}{6}x + \frac{2}{3}$
$y = \frac{1}{6}x - \frac{13}{3}$

31) $(x_1, y_1)=(6,0);\ m=-\frac{5}{9}$

$$y-y_1=m(x-x_1)$$
$$y-0=-\frac{5}{9}(x-6)$$
$$y=-\frac{5}{9}x+\frac{10}{3}$$
$$9\cdot y=9\left(-\frac{5}{9}x+\frac{10}{3}\right)$$
$$9y=-5x+30$$
$$5x+9y=30$$

33) Use the points to find the slope of the line, and then use the slope and either one of the points in the point-slope formula.

35) $m=\frac{8-4}{7-3}=\frac{4}{4}=1$

$$(x_1, y_1)=(3, 4)$$
$$y-y_1=m(x-x_1)$$
$$y-4=1(x-3)$$
$$y-4=x-3$$
$$y=x+1$$

37) $m=\frac{3-4}{1-(-2)}=\frac{-1}{3}=-\frac{1}{3}$

$$(x_1, y_1)=(-2, 4)$$
$$y-y_1=m(x-x_1)$$
$$y-4=-\frac{1}{3}(x-(-2))$$
$$y-4=-\frac{1}{3}(x+2)$$
$$y-4=-\frac{1}{3}x-\frac{2}{3}$$
$$y=-\frac{1}{3}x+\frac{10}{3}$$

39) $m=\frac{-2-(-5)}{3-(-1)}=\frac{3}{4}$

$$(x_1, y_1)=(-1, -5)$$
$$y-y_1=m(x-x_1)$$
$$y-(-5)=\frac{3}{4}(x-(-1))$$
$$y+5=\frac{3}{4}x+\frac{3}{4}$$
$$4(y+5)=4\left(\frac{3}{4}x+\frac{3}{4}\right)$$
$$4y+20=3x+3$$
$$-3x+4y=-17$$
$$3x-4y=17$$

41) $m=\frac{-3-1}{6-4}=\frac{-4}{2}=-2$

$$(x_1, y_1)=(4, 1)$$
$$y-y_1=m(x-x_1)$$
$$y-1=-2(x-4)$$
$$y-1=-2x+8$$
$$2x+y=9$$

43) $m=\frac{7.2-4.2}{3.1-2.5}=\frac{3}{0.6}=5.0$

$$(x_1, y_1)=(2.5, 4.2)$$
$$y-y_1=m(x-x_1)$$
$$y-4.2=5.0(x-2.5)$$
$$y-4.2=5.0x-12.5$$
$$y=5.0x-8.3$$

45) $m=\frac{-1-0}{2-(-6)}=\frac{-1}{8}=-\frac{1}{8}$

$$(x_1, y_1)=(-6, 0)$$

$$y - y_1 = m(x - x_1)$$
$$y - 0 = -\frac{1}{8}(x - (-6))$$
$$y = -\frac{1}{8}(x + 6)$$
$$-8y = x + 6$$
$$-x - 8y = 6$$
$$x + 8y = -6$$

47) $m = \dfrac{-1-(-4)}{2-0} = \dfrac{3}{2}$

$y = mx + b \quad m = \dfrac{3}{2};\ b = -4$

$y = \dfrac{3}{2}x - 4$

49) $m = \dfrac{-1-3}{-1-(-5)} = \dfrac{-4}{4} = -1$

$(x_1, y_1) = (-5, 3)$
$$y - y_1 = m(x - x_1)$$
$$y - 3 = -1(x - (-5))$$
$$y - 3 = -1(x + 5)$$
$$y - 3 = -x - 5$$
$$y = -x - 2$$

51) Horizontal Line
$(c, d) = (0, 5)$
$y = d$
$y = 5$

53) $(x_1, y_1) = (-4, -1);\ m = 4$
$$y - y_1 = m(x - x_1)$$
$$y - (-1) = 4(x - (-4))$$
$$y + 1 = 4x + 16$$
$$y = 4x + 15$$

55) $y = mx + b \quad m = \dfrac{8}{3};\ b = -9$

$y = \dfrac{8}{3}x - 9$

57) $m = \dfrac{1-(-2)}{-5-(-1)} = \dfrac{3}{-4} = -\dfrac{3}{4}$

$(x_1, y_1) = (-1, -2)$
$$y - y_1 = m(x - x_1)$$
$$y - (-2) = -\frac{3}{4}(x - (-1))$$
$$y - (-2) = -\frac{3}{4}(x + 1)$$
$$y + 2 = -\frac{3}{4}x - \frac{3}{4}$$
$$y = -\frac{3}{4}x - \frac{11}{4}$$

59) Vertical Line
$(c, d) = (3, 5)$
$x = c$
$x = 3$

61) Horizontal Line
$(c, d) = (0, -8)$
$y = d$
$y = -8$

63) $(x_1, y_1) = (4, -5);\ m = -\dfrac{1}{2}$
$$y - y_1 = m(x - x_1)$$
$$y - (-5) = -\frac{1}{2}(x - 4)$$
$$y + 5 = -\frac{1}{2}x + 2$$
$$y = -\frac{1}{2}x - 3$$

65) $m = \frac{0-2}{6-0} = \frac{-2}{6} = -\frac{1}{3}$

$y = mx + b \quad m = -\frac{1}{3};\ b = 2$

$y = -\frac{1}{3}x + 2$

67) $y = mx + b \quad m = 0;\ b = 0$

$y = 1x + 0$

$y = x$

69) a) $m = 8700$

The year 2001 corresponds to $x = 3$.

Then 1,284,000 corresponds to $y = 1{,}284{,}000$.

A point on the line is $(3,\ 1284000) = (x_1,\ y_1)$.

$$y - y_1 = m(x - x_1)$$
$$y - 1{,}284{,}000 = 8700(x - 3)$$
$$y - 1{,}284{,}000 = 8700x - 26{,}100$$
$$y = 8700x + 1{,}257{,}900$$

b) The population of Maine is increasing by 8700 people per year.

c) $y = 8700x + 1{,}257{,}900$

In 1998, $x = 0$

$y = 8700(0) + 1{,}257{,}900$

$y = 1{,}257{,}900$ people

In 2002, $x = 4$

$y = 8700(4) + 1{,}257{,}900$

$y = 64{,}800 + 1{,}257{,}900$

$y = 1{,}292{,}700$ people

d)
$$y = 8700x + 1{,}257{,}900$$
$$1{,}431{,}900 = 8700x + 1{,}257{,}900$$
$$174{,}000 = 8700x$$
$$20 = x$$

Twenty years after 1998, the population would reach 1,431,900. This will be the year 2018.

71) a) $1997:(0, 124);\quad 2002:(5, 92)$

$m=\dfrac{92-124}{5-0}=\dfrac{-32}{5}=-6.4;$

$b=124$

$y=mx+b$

$y=-6.4x+124$

b) The number of farms with milk cows is decreasing by 6.4 thousand or 6400 per year.

c) $y=-6.4x+124$

In 2004, $x=7$

$y=-6.4(7)+124$

$y=-44.8+124$

$y=79.2$ thousand or 79,200

73) a) $2000:(0, 6479);$

$2003:(3, 43436)$

$m=\dfrac{43,435-6479}{3-0}=\dfrac{36,956}{3}$

$\approx 12,318.7;\quad b=6479$

$y=mx+b$

$y=12,318.7x+6479$

b) The number of registered hybrid vehicles is increasing by 12,318.7 per year.

c) $y=12,318.7x+6479$

In 2002, $x=2$

$y=12,318.7(2)+6479$

$y=24,637.4+6479$

$y=31,116.4$

This is slightly lower than the actual value.

d) $y=12,318.7x+6479$

In 2010, $x=10$

$y=12,318.7(10)+6479$

$y=123,187+6479$

$y=129,666$

Section 4.6 Exercises

1) Perpendicular lines intersect at 90° angles and have slopes which are negative reciprocals of each other.

3) $y=-8x-6 \qquad y=\dfrac{1}{8}x+3$

$m=-8 \qquad m=\dfrac{1}{8}$

perpendicular

5) $y=\dfrac{2}{9}x+4 \qquad 4x-18y=9$

$-18y=-4x+9$

$y=\dfrac{2}{9}x-\dfrac{1}{2}$

$m=\dfrac{2}{9} \qquad m=\dfrac{2}{9}$

parallel

7) $-3x+2y=-10 \qquad 3x-4y=-2$

$2y=3x-10 \qquad -4y=-3x-2$

$y=\dfrac{3}{2}x-5 \qquad y=\dfrac{3}{4}x+\dfrac{1}{2}$

$m=\dfrac{3}{2} \qquad m=\dfrac{3}{4}$

neither

9) $y=x \qquad x+y=7$

$y=-x+7$

$m=1 \qquad m=-1$

perpendicular

11) $4y-9x=2$ $\quad$ $4x-9y=9$

$4y=9x+2$ $\quad$ $-9y=-4x+9$

$y=\frac{9}{4}x+\frac{1}{2}$ $\quad$ $y=\frac{4}{9}-1$

$m=\frac{9}{4}$ $\quad$ $m=\frac{4}{9}$

neither

13) $4x-3y=18$ $\quad$ $-8x+6y=5$

$-3y=-4x+18$ $\quad$ $6y=8x+5$

$y=\frac{4}{3}x-6$ $\quad$ $y=\frac{4}{3}x+\frac{5}{6}$

$m=\frac{4}{3}$ $\quad$ $m=\frac{4}{3}$

parallel

15) Vertical lines are parallel.

17) perpendicular

19) $L_1: m=\frac{-13-2}{6-1}=\frac{-15}{5}=-3$

$y-2=-3(x-1)$

$y-2=-3x+3$

$y=-3x+5$

$L_2: m=\frac{-10-5}{3-(-2)}=\frac{-15}{5}=-3$

$y-5=-3(x-(-2))$

$y-5=-3x-6$

$y=-3x+1$

parallel

21) $L_1: m=\frac{8-(-7)}{2-(-1)}=\frac{15}{3}=5$

$L_2: m=\frac{4-2}{0-10}=\frac{2}{-10}=-\frac{1}{5}$

perpendicular

23) $L_1: m=\frac{3-(-1)}{7-5}=\frac{4}{2}=2$

$L_2: m=\frac{5-0}{4-(-6)}=\frac{5}{10}=\frac{1}{2}$

neither

25) $L_1: m=\frac{2-2}{0-(-3)}=\frac{0}{3}=0$

$y=2$

$L_2: m=\frac{-1-(-1)}{-2-1}=\frac{0}{-3}=0$

$y=-1$

parallel

27) $L_1: m=\frac{-1-4}{-5-(-5)}=\frac{-5}{0}$ undefined

$x=5$

$L_2: m=\frac{2-2}{-3-(-1)}=\frac{0}{-2}=0$

$y=2$

perpendicular

29) $y=mx+b$ $\quad$ $m=4;\ b=2$

$y=4x+2$

31) $(x_1, y_1)=(4, 5);\ m=\frac{1}{2}$

$y-y_1=m(x-x_1)$

$y-5=\frac{1}{2}(x-4)$

$2(y-5)=2\left(\frac{1}{2}x-2\right)$

$2y-10=x-4$

$-x+2y=6$

$x-2y=-6$

33) Determine the slope.

$$4x+3y=-6$$
$$3y=-4x-6$$
$$y=-\frac{4}{3}x-2$$

$$(x_1, y_1)=(-9, 4);\ m=-\frac{4}{3}$$
$$y-y_1=m(x-x_1)$$
$$y-4=-\frac{4}{3}(x-(-9))$$
$$y-4=-\frac{4}{3}(x+9)$$
$$3(y-4)=3\left(-\frac{4}{3}x-12\right)$$
$$3y-12=-4x-36$$
$$4x+3y=-24$$

35) Determine the slope.

$$x+5y=10$$
$$5y=-x+10$$
$$y=-\frac{1}{5}x+2$$

$$(x_1, y_1)=(15, 7);\ m=-\frac{1}{5}$$
$$y-y_1=m(x-x_1)$$
$$y-7=-\frac{1}{5}(x-15)$$
$$y-7=-\frac{1}{5}x+3$$
$$y=-\frac{1}{5}x+10$$

37) $(x_1, y_1)=(6, -3);\ m_{\text{perp}}=-\frac{3}{2}$

$$y-y_1=m(x-x_1)$$
$$y-(-3)=-\frac{3}{2}(x-6)$$
$$y+3=-\frac{3}{2}x+9$$
$$y=-\frac{3}{2}x+6$$

39) $(x_1, y_1)=(10, 0);\ m_{\text{perp}}=\frac{1}{5}$

$$y-y_1=m(x-x_1)$$
$$y-0=\frac{1}{5}(x-10)$$
$$5\cdot y=5\left(\frac{1}{5}x-2\right)$$
$$5y=x-10$$
$$-x+5y=-10$$
$$x-5y=10$$

41) Determine the slope.

$$x+y=9$$
$$y=-x+9$$
$$m=-1$$
$$(x_1, y_1)=(-5, -5);\ m_{\text{perp}}=1$$
$$y-y_1=m(x-x_1)$$
$$y-(-5)=1(x-(-5))$$
$$y+5=x+5$$
$$y=x$$

43) Determine the slope.

$$24x-15y=10$$
$$-15y=-24x+10$$
$$\frac{-15y}{-15}=\frac{-24}{-15}x+\frac{10}{-15}$$
$$y=\frac{8}{5}x-\frac{2}{3} \qquad m=\frac{8}{5}$$

$(x_1, y_1) = (16, -7);\ m_{perp} = -\frac{5}{8}$

$y - y_1 = m(x - x_1)$

$y - (-7) = -\frac{5}{8}(x - 16)$

$8(y + 7) = 8\left(-\frac{5}{8}x + 10\right)$

$8y + 56 = -5x + 80$

$5x + 8y = 24$

45) Determine the slope.

$2x - 6y = -3$

$-6y = -2x - 3$

$y = \frac{1}{3}x + \frac{1}{2}$

$m = \frac{1}{3}$

$(x_1, y_1) = (2, 2);\ m_{perp} = -3$

$y - y_1 = m(x - x_1)$

$y - 2 = -3(x - 2)$

$y - 2 = -3x + 6$

$y = -3x + 8$

47) $(x_1, y_1) = (1, -3);\ m = 2$

$y - y_1 = m(x - x_1)$

$y - (-3) = 2(x - 1)$

$y + 3 = 2x - 2$

$y = 2x - 5$

49) $(c, d) = (-1, -5);\ m$ is undefined

m_{perp} is undefined

$x = c$

$x = -1$

51) $(c, d) = (2, 1);\ m_{perp}$ undefined

$x = c$

$x = 2$

53) Determine the slope.

$21x - 6y = 2$

$-6y = -21x + 2$

$\frac{-6y}{-6} = \frac{-21}{-6}x + \frac{2}{-6}$

$y = \frac{7}{2}x - \frac{1}{3} \qquad m = \frac{7}{2}$

$(x_1, y_1) = (4, -1);\ m_{perp} = -\frac{2}{7}$

$y - y_1 = m(x - x_1)$

$y - (-1) = -\frac{2}{7}(x - 4)$

$y + 1 = -\frac{2}{7}x + \frac{8}{7}$

$y = -\frac{2}{7}x + \frac{1}{7}$

55) $(c, d) = \left(-3, -\frac{5}{2}\right);\ m = 0$

$y = d$

$y = -\frac{5}{2}$

Section 4.7 Exercises

1) a) any set of ordered pairs

b) Answers may very.

c) Answers may very.

3) Domain: $\{-8, -2, 1, 5\}$

Range: $\{-3, 4, 6, 13\}$

Function

5) Domain: $\{1, 9, 25\}$

Range: $\{-3, -1, 1, 5, 7\}$

Not a function

7) Domain: $\{-2, 1, 2, 5\}$
Range: $\{6, 9, 30\}$
Function

9) Domain: $\{-1, 2, 5, 8\}$
Range: $\{-7, -3, 12, 19\}$
Not a function

11) Domain: $(-\infty, \infty)$
Range: $(-\infty, \infty)$
Function

13) Domain: $(-\infty, 4]$
Range: $(-\infty, \infty)$
Not a function

15) Domain: $(-\infty, \infty)$
Range: $(-\infty, 6]$
Function

17) yes

19) yes

21) no

23) no

25) yes

27) Domain: $(-\infty, \infty)$; Function

29) Domain: $(-\infty, \infty)$; Function

31) Domain: $[0, \infty)$; Not a function

33) $x \neq 0$
Domain: $(-\infty, 0) \cup (0, \infty)$;
Function

35) $x + 4 = 0$
$x = -4$
Domain: $(-\infty, -4) \cup (-4, \infty)$;
Function

37) $x - 5 = 0$
$x = 5$
Domain: $(-\infty, 5) \cup (5, \infty)$;
Function

39) $x + 8 = 0$
$x = -8$
Domain: $(-\infty, -8) \cup (-8, \infty)$;
Function

41) $5x - 3 = 0$
$5x = 3$
$x = \frac{3}{5}$
Domain: $\left(-\infty, \frac{3}{5}\right) \cup \left(\frac{3}{5}, \infty\right)$;
Function

43) $3x + 4 = 0$
$3x = -4$
$x = -\frac{4}{3}$
Domain: $\left(-\infty, -\frac{4}{3}\right) \cup \left(-\frac{4}{3}, \infty\right)$;
Function

45) $4x = 0$
$x = 0$
Domain: $(-\infty, 0) \cup (0, \infty)$;
Function

47) $2x+1=0$

$2x=-1$

$x=-\frac{1}{2}$

Domain: $\left(-\infty, -\frac{1}{2}\right)\cup\left(-\frac{1}{2}, \infty\right)$;

Function

49) $9-3x=0$

$-3x=-9$

$x=3$

Domain: $(-\infty, 3)\cup(3, \infty)$;

Function

Section 4.8 Exercises

1) y is a function, and y is a function of x.

3) a) $y=5(3)-8$

$y=15-8$

$y=7$

b) $f(3)=5(3)-8$

$f(3)=15-8$

$f(3)=7$

5) $f(5)=-4(5)+7$

$f(5)=-20+7$

$f(5)=-13$

7) $f(0)=-4(0)+7$

$f(0)=0+7$

$f(0)=7$

9) $g(4)=(4)^2+9(4)-2$

$g(4)=16+36-2$

$g(4)=50$

11) $g(-1)=(-1)^2+9(-1)-2$

$g(-1)=1-9-2$

$g(-1)=-10$

13) $g\left(-\frac{1}{2}\right)=\left(-\frac{1}{2}\right)^2+9\left(-\frac{1}{2}\right)-2$

$g\left(-\frac{1}{2}\right)=\frac{1}{4}-\frac{9}{2}-2=\frac{1}{4}-\frac{18}{4}-\frac{8}{4}$

$g\left(-\frac{1}{2}\right)=-\frac{25}{4}$

15) $f(-1)=10,\ f(4)=-5$

17) $f(-1)=6,\ f(4)=2$

19) $10=-3x-2$

$12=-3x$

$-4=x$

21) $5=\frac{2}{3}x+1$

$4=\frac{2}{3}x$

$6=x$

23) a) $f(c)=-7(c)+2=-7c+2$

b) $f(t)=-7(t)+2=-7t+2$

c) $f(a+4)=-7(a+4)+2$

$f(a+4)=-7a-28+2$

$f(a+4)=-7a-26$

d) $f(z-9) = -7(z-9)+2$

$f(z-9) = -7z+63+2$

$f(z-9) = -7z+65$

e) $g(k) = (k)^2 - 5(k)+12$

$g(k) = k^2 - 5k + 12$

f) $g(m) = (m)^2 - 5(m)+12$

$g(m) = m^2 - 5m + 12$

25) $f(x) = x-4$

$f(0) = (0)-4$

$f(0) = -4$

$f(1) = (1)-4$

$f(1) = -3$

x	$f(x)$
0	−4
1	−3
3	−1

$f(3) = (3)-4$

$f(3) = -1$

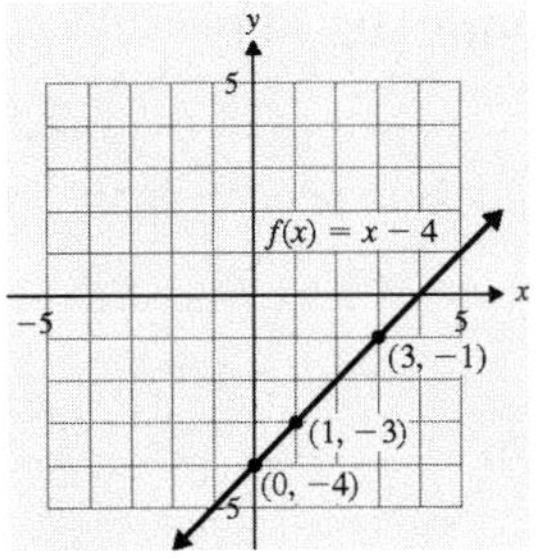

27) $f(x) = \frac{2}{3}x+2$

$f(-3) = \frac{2}{3}(-3)+2$

$f(-3) = -2+2 = 0$

$f(0) = \frac{2}{3}(0)+2$

$f(0) = 0+2 = 2$

x	$f(x)$
−3	0
0	2
3	4

$f(3) = \frac{2}{3}(3)+2$

$f(3) = 2+2 = 4$

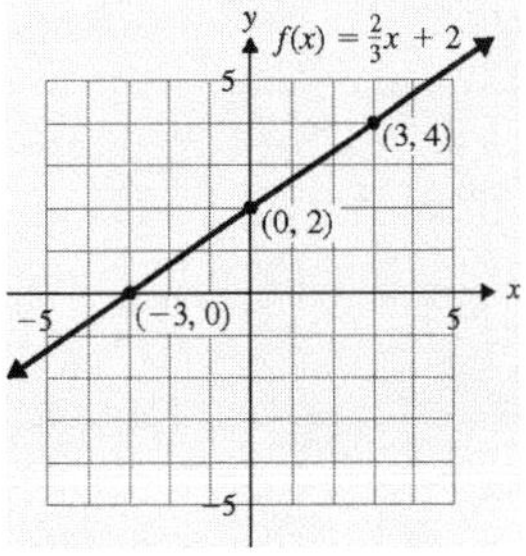

29) $h(x) = -3$

$h(-2) = -3$

$h(0) = -3$

$h(1) = -3$

x	$h(x)$
−2	−3
0	−3
1	−3

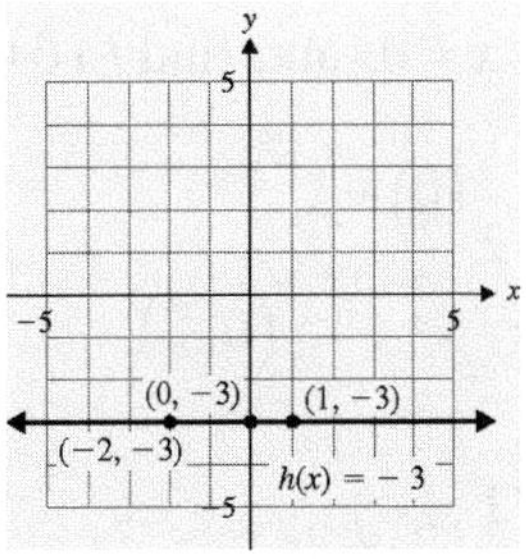

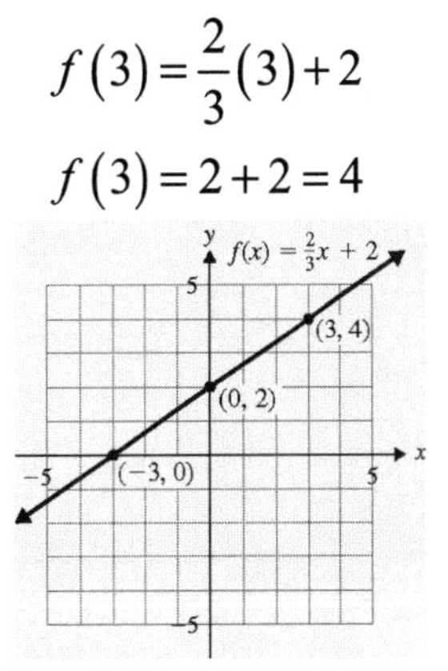

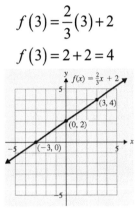

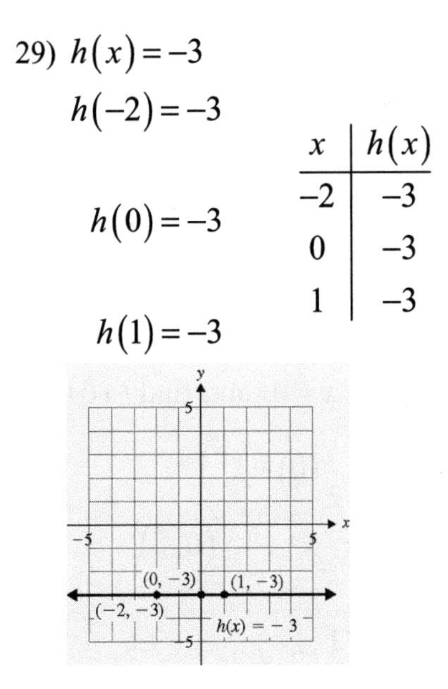

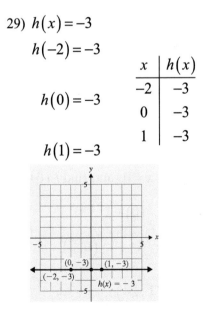

31) $g(x)=3x+3$

x-int: Let $g(x)=0,$ and solve for x.

$0=3x+3$

$-3=3x$

$-1=x \qquad (-1,\ 0)$

y-int: Let $x=0,$ and find $g(0)$.

$g(0)=3(0)+3$

$g(0)=3 \qquad (0,\ 3)$

Let $x=-2$.

$g(-2)=3(-2)+3$

$g(-2)=-6+3=-3 \qquad (-2,\ -3)$

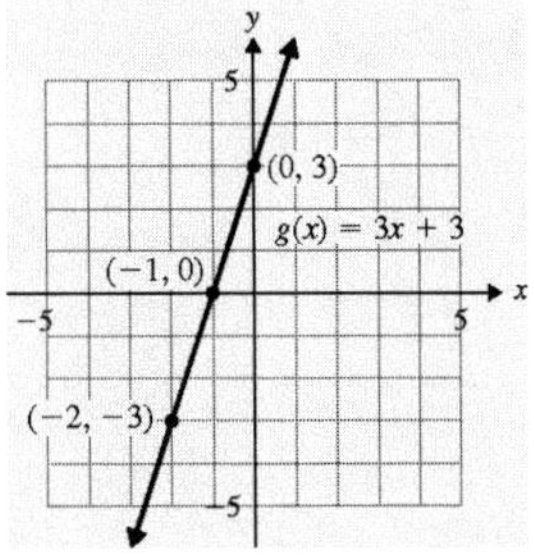

33) $f(x)=-\dfrac{1}{2}x+2$

x-int: Let $f(x)=0,$ and solve for x.

$0=-\dfrac{1}{2}x+2$

$-2=-\dfrac{1}{2}x$

$4=x \qquad (4,\ 0)$

y-int: Let $x=0,$ and find $f(0)$.

$f(0)=-\dfrac{1}{2}(0)+2$

$f(0)=2 \qquad (0,\ 2)$

Let $x=2$.

$f(2)=-\dfrac{1}{2}(2)+2$

$f(2)=-1+2=1 \qquad (2,\ 1)$

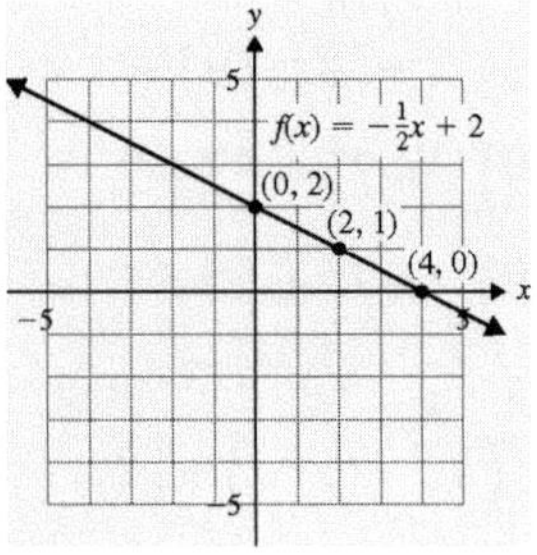

35) $h(x)=x$

x-int: Let $h(x)=0,$ and solve for x.

$0=x \qquad (0,\ 0)$ - also $y-$int

Let $x=-2$.

$h(-2)=(-2)=-2 \qquad (-2,\ -2)$

Let $x=1$.

$h(1)=(1)=1 \qquad (1,\ 1)$

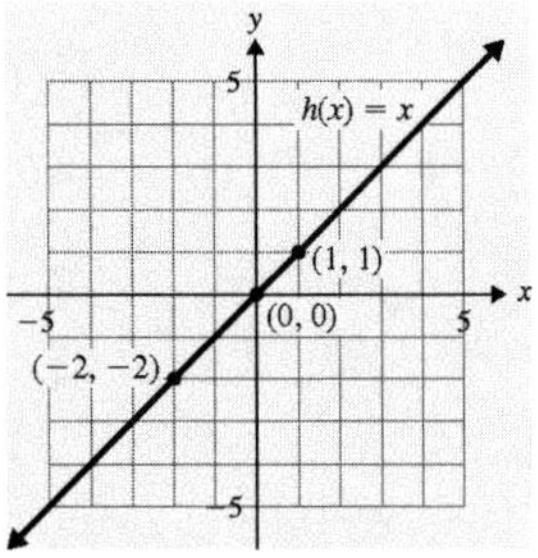

37) $m=-4,\ y$-int:$(0,\ -1)$

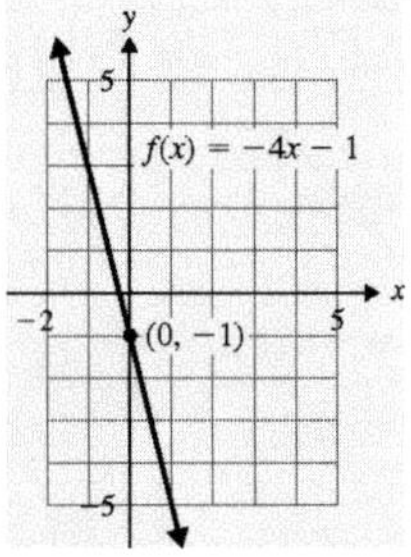

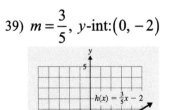

39) $m=\dfrac{3}{5},\ y$-int:$(0,\ -2)$

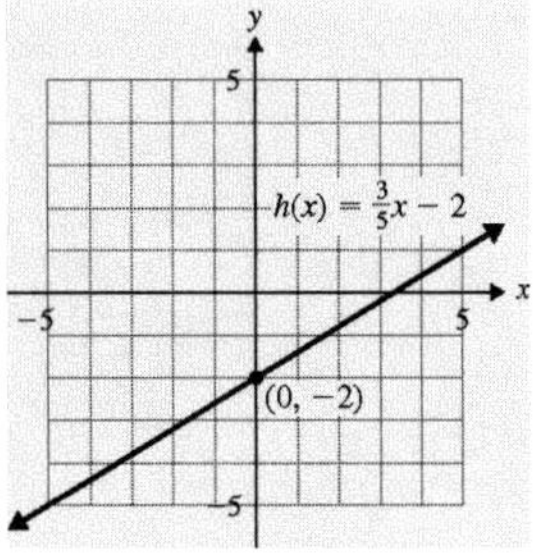

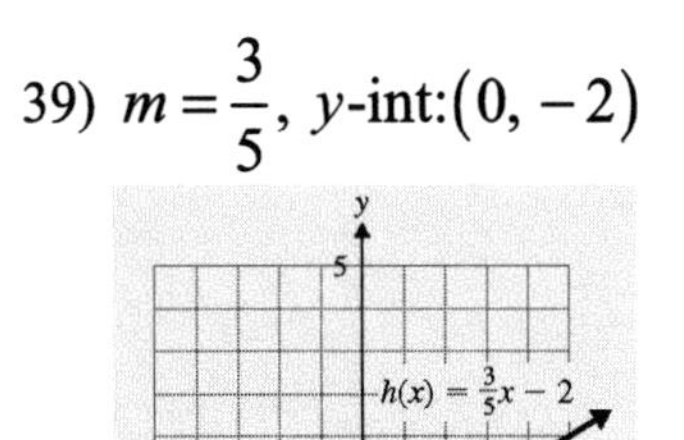

41) $m = 2$, y-int: $\left(0, \dfrac{1}{2}\right)$

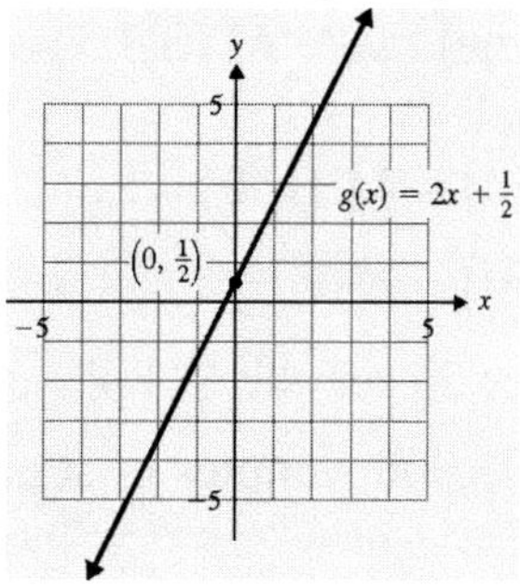

43) $m = -\dfrac{5}{2}$, y-int: $(0, 4)$

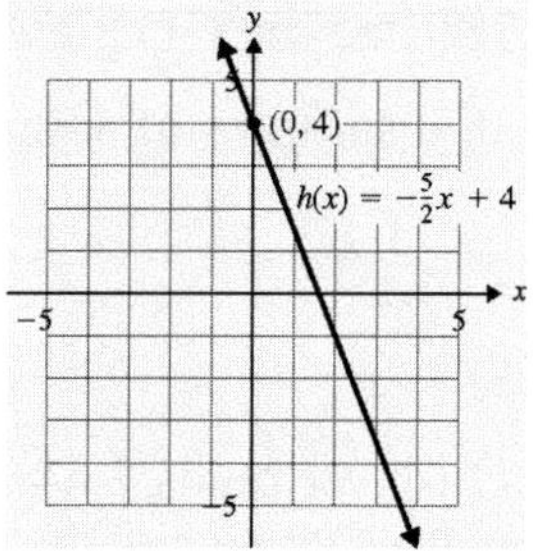

45)

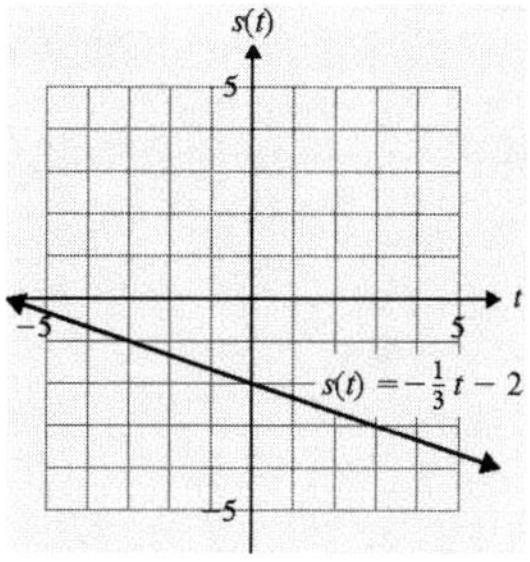

47)

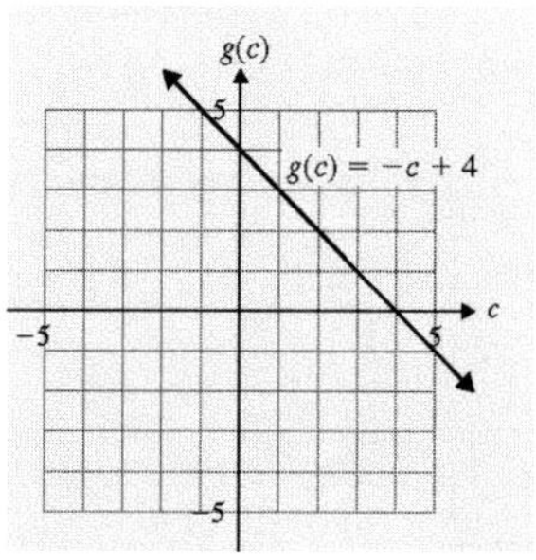

49)

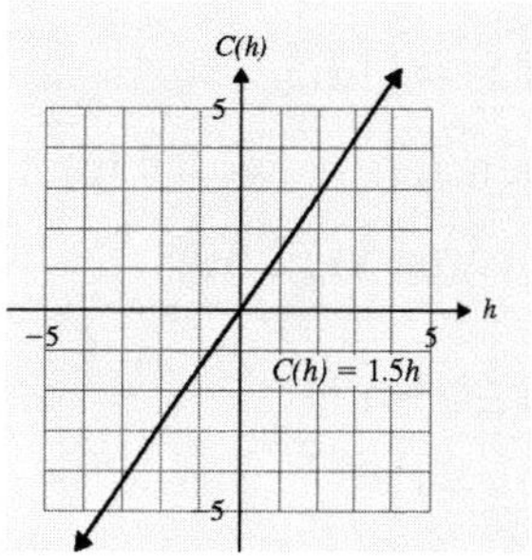

51) a) $D(2) = 54(2)$

$D(2) = 108$

The truck can travel 108 miles in 2 hours.

b) $D(4) = 54(4)$

$D(4) = 216$

The truck can travel 216 miles in 4 hours.

c) $135 = 54t$

$2.5 = t$

In 2.5 hours the truck can travel 135 miles.

d)

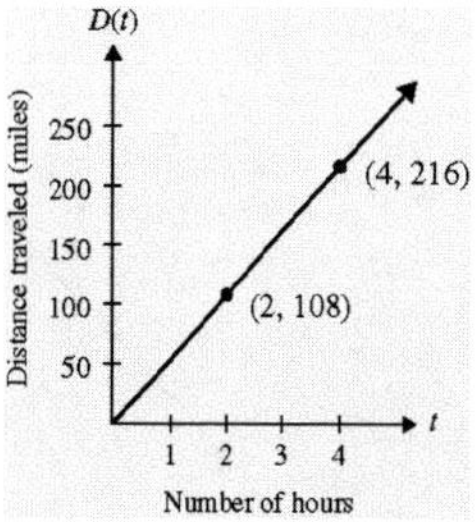

53) a) $E(10) = 7.50(10)$

$E(10) = 75.0$

When Jenelle works for 10 hr, she earns \$75.00.

b) $E(15)=7.50(15)$

$E(15)=112.5$

When Jenelle works for 15 hr, she earns \$112.50.

c) $210=7.50t$

$28=t$

For Jenelle to earn \$210.00, she must work 28 hr.

55) a) $D(12)=21.13(12)$

$D(12)=253.56$

253.56 MB can be recorded in 12 seconds.

b) $D(60)=21.13(60)$

$D(60)=1267.8$

1267.8 MB can be recorded in 1 minute.

c) $422.6=21.13t$

$20=t$

In 20 seconds 422.6 MB can be recorded.

d)

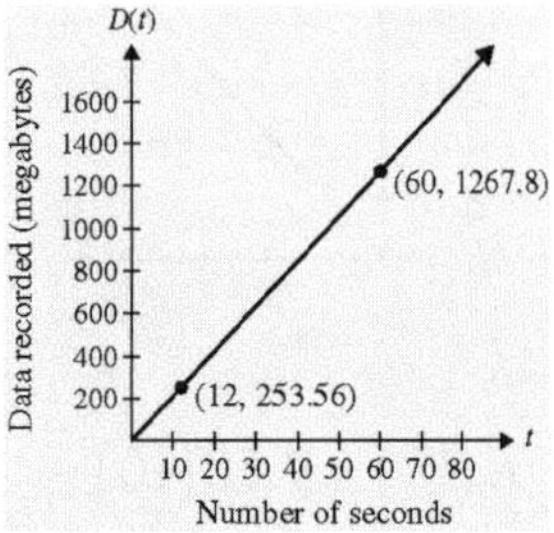

57) a) $F(2)=30(2)$

$F(2)=60$

60,000 fingerprints can be compared in 2 seconds.

b) $105=30s$

$3.5=s$

It would take AFIS 3.5 seconds.

59) a) $S(50)=44.1(50)$

$S(50)=2205$

After 50 sec the CD player reads 2,205,000 samples of sound.

b) $S(180)=44.1(180)$

$S(108)=7938$

After 180 sec the CD player reads 7,938,000 samples of sound.

c) $2646=44.1t$

$60=t$

The CD player reads 2,646,000 samples of sound in 60 sec (or 1 minute).

Chapter 4 Review

1) Yes

$$4x-y=9$$
$$4(1)-(-5)=9$$
$$4+5=9$$
$$9=9$$

3) Yes

$$y=\frac{5}{4}x+\frac{1}{2}$$
$$(3)=\frac{5}{4}(2)+\frac{1}{2}$$
$$3=\frac{5}{2}+\frac{1}{2}$$
$$3=3$$

5) $y=-6x+10$
$y=-6(-3)+10$
$y=18+10$
$y=28$
$(-3,28)$

7) $y=-8$
$(5,-8)$

9) $y=x-11$

$y=(0)-11$ $\quad y=(-1)-11$
$y=-11$ $\quad y=-12$

$y=(3)-11$ $\quad y=(-5)-11$
$y=-8$ $\quad y=-16$

x	y
0	−11
3	−8
−1	−12
−5	−16

11)

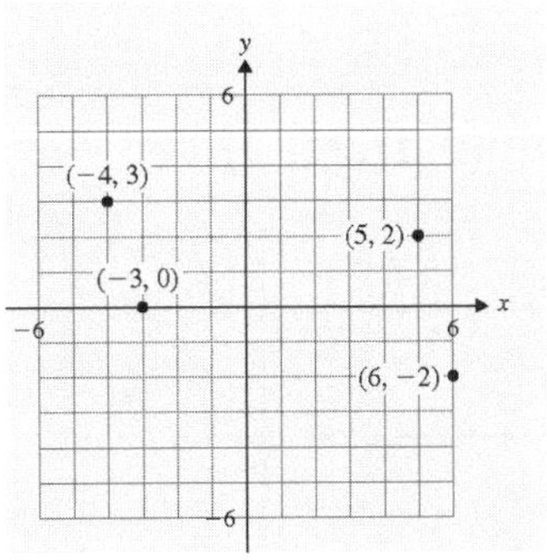

13) a) $y=0.10x$

$y=0.10(1)$ $\quad y=0.10(7)$
$y=0.10$ $\quad y=0.70$

$y=0.10(2)$ $\quad y=0.10(10)$
$y=0.20$ $\quad y=1.00$

x	y
1	0.10
2	0.20
7	0.70
10	1.00

$(1, 0.10), (2, 0.20),$
$(7, 0.70), (10, 1.00)$

b)

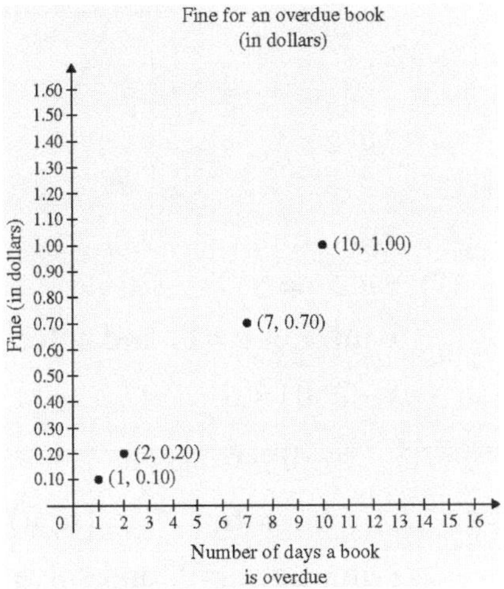

c) If a book is 14 days overdue, the fine is \$1.40.

15) $y=-2x+3$

$y=-2(0)+3$ $\quad y=-2(2)+3$
$y=0+3$ $\quad y=-4+3$
$y=3$ $\quad y=-1$

$y=-2(1)+3$ $\quad y=-2(-2)+3$
$y=-2+3$ $\quad y=4+3$
$y=1$ $\quad y=7$

x	y
0	3
1	1
2	−1
−2	7

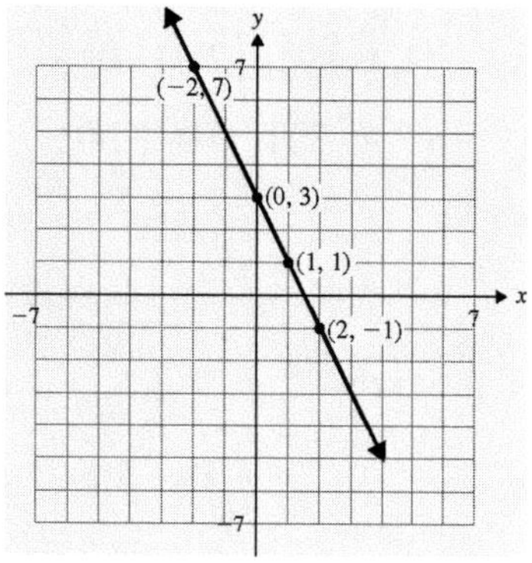

17) $x-2y=6$

x-int: Let $y=0$, and solve for x.

$x-2(0)=6$

$x-0=6$

$x=6 \qquad (6, 0)$

y-int: Let $x=0$, and solve for y.

$(0)-2y=6$

$-2y=6$

$y=-3 \qquad (0, -3)$

Let $x=2$.

$(2)-2y=6$

$-2y=4$

$y=-2 \qquad (2, -2)$

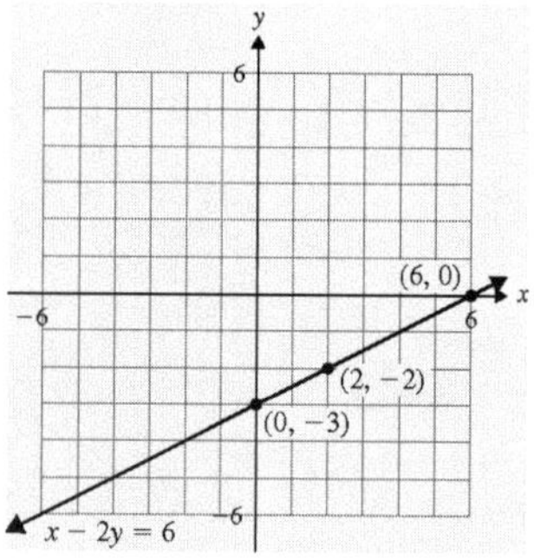

19) $y=-\dfrac{1}{6}x+4$

x-int: Let $y=0$, and solve for x.

$(0)=-\dfrac{1}{6}x+4$

$-4=-\dfrac{1}{6}x$

$24=x \qquad (24, 0)$

y-int: Let $x=0$, and solve for y.

$y=-\dfrac{1}{6}(0)+4$

$y=0+4$

$y=4 \qquad (0, 4)$

Let $x=12$.

$y=-\dfrac{1}{6}(12)+4$

$y=-2+4$

$y=2 \qquad (12, 2)$

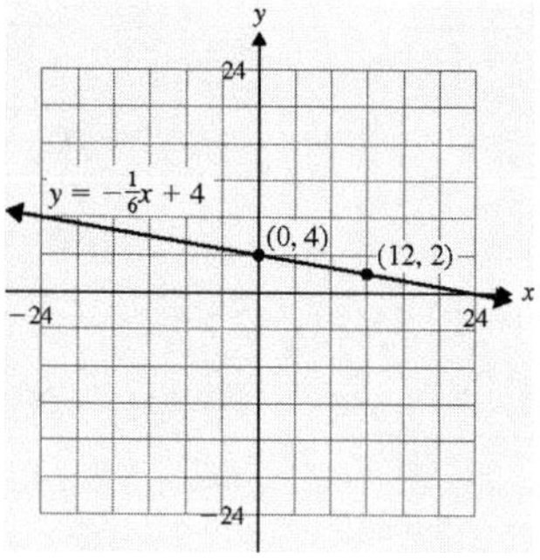

21) $x=5$

$(5, 0), (5, 1), (5, 2)$

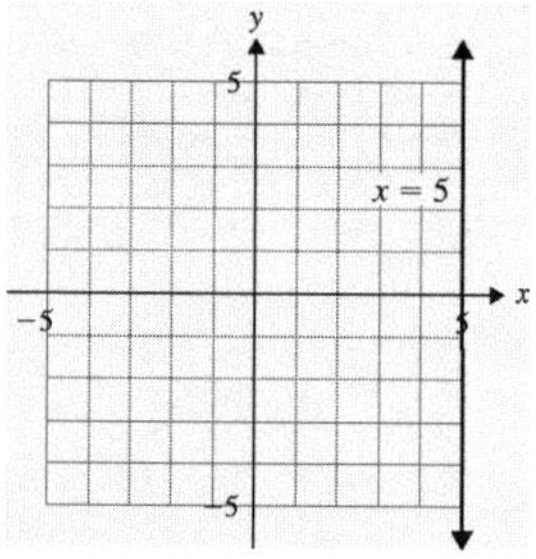

23) $m=\dfrac{\text{rise}}{\text{run}}=\dfrac{-2}{5}=-\dfrac{2}{5}$

25) $(x_1, y_1)=(1, 7)$

$(x_2, y_2)=(-4, 2)$

$m=\dfrac{y_2-y_1}{x_2-x_1}=\dfrac{2-7}{-4-1}=\dfrac{-5}{-5}=1$

27) $(x_1, y_1)=(-2, 5)$

$(x_2, y_2)=(3, -8)$

$$m=\frac{y_2-y_1}{x_2-x_1}=\frac{-8-5}{3-(-2)}=\frac{-13}{5}=-\frac{13}{5}$$

29) $(x_1, y_1)=\left(\frac{3}{2}, -1\right)$

$(x_2, y_2)=\left(-\frac{5}{2}, 7\right)$

$$m=\frac{y_2-y_1}{x_2-x_1}=\frac{7-(-1)}{-\frac{5}{2}-\frac{3}{2}}=\frac{8}{-\frac{8}{2}}=-2$$

31) $(x_1, y_1)=(9, 0)$

$(x_2, y_2)=(9, 4)$

$$m=\frac{y_2-y_1}{x_2-x_1}=\frac{4-0}{9-9}=\frac{4}{0}$$

Slope is undefined.

33) a) In 2002, one share of the stock was worth $32.

b) The slope is positive, so the value of the stock is increasing over time.

c) $m=\frac{6}{2}=3$. The value of the stock is increasing by $3.00 per year.

35)

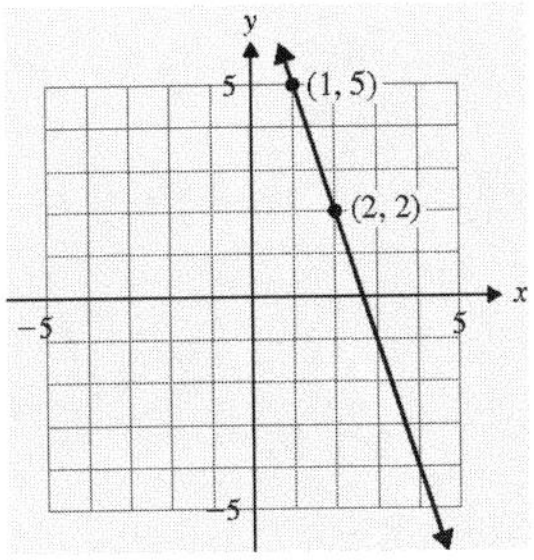

37)

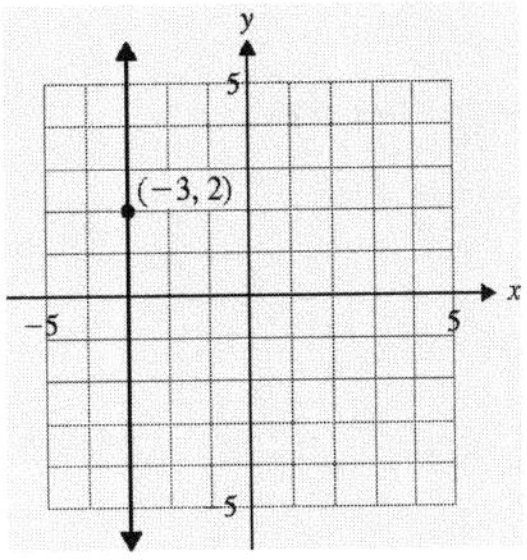

39) $m=1$, y-int :$(0, -3)$

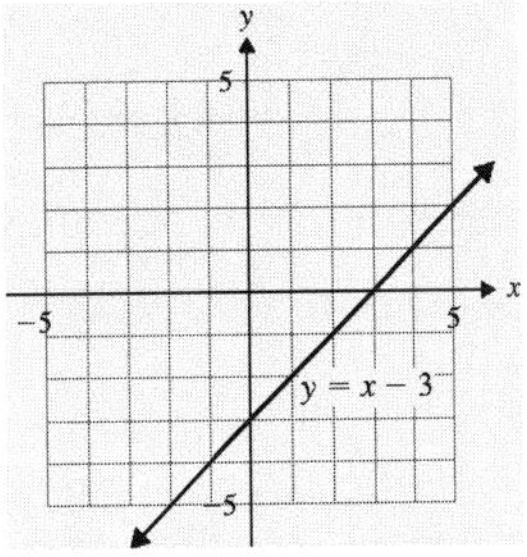

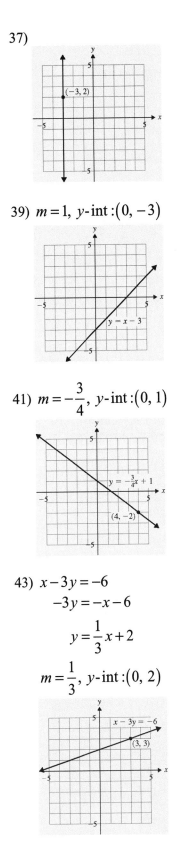

41) $m=-\frac{3}{4}$, y-int :$(0, 1)$

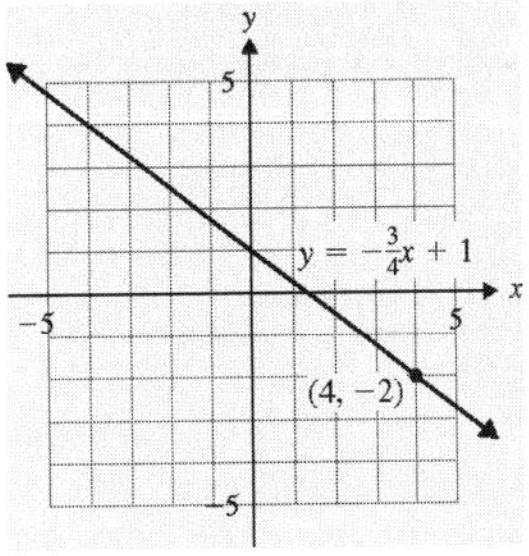

43) $x-3y=-6$

$-3y=-x-6$

$y=\frac{1}{3}x+2$

$m=\frac{1}{3}$, y-int :$(0, 2)$

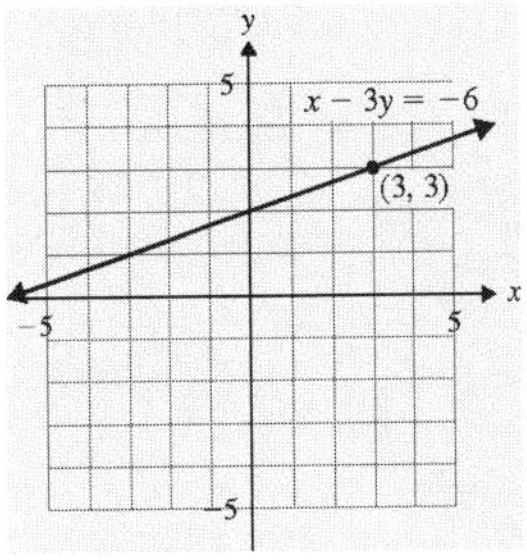

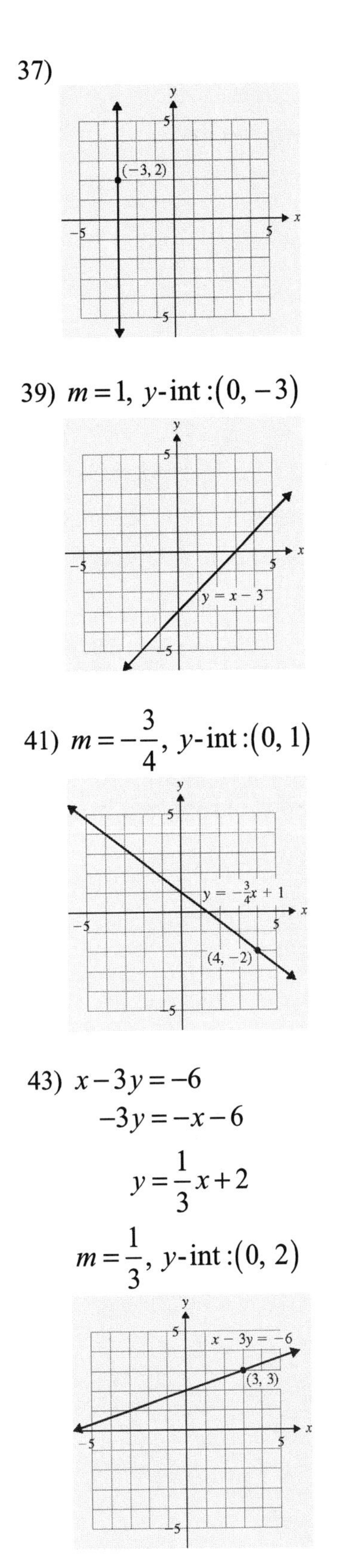

45) $x+y=0$

$y=-x$

$m=-1,\ y\text{-int}:(0,\ 0)$

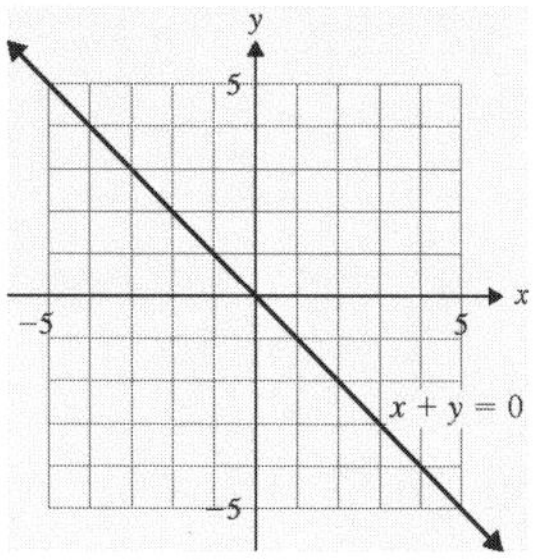

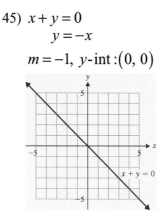

47) a) $(0,\ 5920.1)$; In 1998 the amount of money spent for personal consumption was \$5920.1 billion.

b) It has been increasing by \$371.5 billion per year.

c) Estimate: \$7400 billion.

$y=371.5x+5920.1$

$y=371.5(4)+5920.1$

$y=1486+5920.1$

$y=7406.1$

\$7406.1 billion

49) $(x_1,\ y_1)=(2,\ 5);\ m=7$

$y-y_1=m(x-x_1)$

$y-5=7(x-2)$

$y-5=7x-14$

$y=7x-9$

51) $y=mx+b \quad m=-\frac{4}{9};\ b=2$

$y=-\frac{4}{9}x+2$

53) $m=\frac{-2-(-6)}{-9-3}=\frac{4}{-12}=-\frac{1}{3}$

$(x_1,\ y_1)=(3,\ -6)$

$y-y_1=m(x-x_1)$

$y-(-6)=-\frac{1}{3}(x-3)$

$y+6=-\frac{1}{3}x+1$

$y=-\frac{1}{3}x-5$

55) Horizontal Line

$(c,\ d)=(1,\ 9)$

$y=d$

$y=9$

57) $m=\frac{7-2}{8-(-2)}=\frac{5}{10}=\frac{1}{2}$

$(x_1,\ y_1)=(-2,\ 2)$

$y-y_1=m(x-x_1)$

$y-2=\frac{1}{2}(x-(-2))$

$2(y-2)=2\left(\frac{1}{2}x+1\right)$

$2y-4=x+2$

$-x+2y=6$

$x-2y=-6$

59) $(x_1,\ y_1)=\left(\frac{4}{3},\ 1\right);\ m=-3$

$y-y_1=m(x-x_1)$

$y-1=-3\left(x-\frac{4}{3}\right)$

$y-1=-3x+4$

$3x+y=5$

61) $y=mx+b \quad m=6;\ b=0$

$y=(6)x-0$

$-6x+y=0$

$6x-y=0$

63) $m=\frac{-5-1}{-7-1}=\frac{-6}{-8}=\frac{3}{4}$

$(x_1, y_1)=(1, 1)$

$$y-y_1=m(x-x_1)$$
$$y-1=\frac{3}{4}(x-1)$$
$$4(y-1)=4\left(\frac{3}{4}x-\frac{3}{4}\right)$$
$$4y-4=3x-3$$
$$-3x+4y=1$$
$$3x-4y=-1$$

65) a) $2001:(944.2);\ 2004:(1502.9)$

$$m=\frac{1502.9-944.2}{3-0}=\frac{558.7}{3}$$
$$m\approx 186.2;\quad b=944.2$$
$$y=mx+b$$
$$y=186.2x+944.2$$

b) The number of worldwide wireless subscribers is increasing by 186.2 million per year.

c) $y=186.2x+944.2$

In 2003, $x=2$

$$y=186.2(2)+944.2$$
$$y=372.4+944.2$$

$y=1316.6$ million

This is slightly less than the number given on the chart.

67) $9x-4y=-1 \qquad -27x+12y=2$

$$-4y=-9x-1 \qquad 12y=27x+2$$
$$y=\frac{9}{4}x+\frac{1}{4} \qquad y=\frac{9}{4}x+\frac{1}{6}$$
$$m=\frac{9}{4} \qquad m=\frac{9}{4}$$

parallel

69) perpendicular

71) $y=6x-7 \qquad 4x+y=9$

$$y=-4x+9$$
$$m=6 \qquad m=-4$$

neither

73) $(x_1, y_1)=(-2, -4);\ m=5$

$$y-y_1=m(x-x_1)$$
$$y-(-4)=5(x-(-2))$$
$$y+4=5x+10$$
$$y=5x+6$$

75) Determine the slope.

$$2x+y=5$$
$$y=-2x+5$$
$$(x_1, y_1)=(1, -9);\ m=-2$$
$$y-y_1=m(x-x_1)$$
$$y-(-9)=-2(x-1)$$
$$y+9=-2x+2$$
$$2x+y=-7$$

77) Determine the slope.

$$5x-3y=7$$
$$-3y=-5x+7$$
$$y=\frac{5}{3}x-\frac{7}{3}$$
$$(x_1, y_1)=(4, 8);\ m=\frac{5}{3}$$
$$y-y_1=m(x-x_1)$$
$$y-8=\frac{5}{3}(x-4)$$
$$y-8=\frac{5}{3}x-\frac{20}{3}$$
$$y=\frac{5}{3}x+\frac{4}{3}$$

79) $(x_1, y_1) = (6, 5);\ m_{\text{perp}} = 2$

$$y - y_1 = m(x - x_1)$$
$$y - 5 = 2(x - 6)$$
$$y - 5 = 2x - 12$$
$$y = 2x - 7$$

81) Determine the slope.

$$2x - 11y = 11$$
$$-11y = -2x + 11$$
$$y = \frac{2}{11}x - 1$$
$$(x_1, y_1) = (2, -7);\ m_{\text{perp}} = -\frac{11}{2}$$
$$y - y_1 = m(x - x_1)$$
$$y - (-7) = -\frac{11}{2}(x - 2)$$
$$y + 7 = -\frac{11}{2}x + 11$$
$$y = -\frac{11}{2}x + 4$$

83) Determine the slope.

$$4x - y = -3$$
$$-y = -4x - 3$$
$$y = 4x + 3$$
$$(x_1, y_1) = (-8, -3);\ m_{\text{perp}} = -\frac{1}{4}$$
$$y - y_1 = m(x - x_1)$$
$$y - (-3) = -\frac{1}{4}(x - (-8))$$
$$4(y + 3) = 4\left(-\frac{1}{4}x - 2\right)$$
$$4y + 12 = -x - 8$$
$$x + 4y = -20$$

85) $x = 2$

87) $x = -1$

89) Domain: $\{-4, 0, 2, 5\}$
Range: $\{-9, 3, 9, 18\}$
Function

91) Domain: $\{-6, 2, 5\}$
Range: $\{0, 1, 8, 18\}$
Not a function

93) Domain: $(-\infty, \infty)$
Range: $(-\infty, \infty)$
Function

95) Domain: $(-\infty, \infty)$;
Function

97) $x \neq 0$
Domain: $(-\infty, 0) \cup (0, \infty)$;
Function

99) Domain: $(-\infty, \infty)$; Function

101) $f(3) = -14;\ f(-2) = -5$

103) $f(3) = -2;\ f(-2) = 1$

105) a) $f(4) = 5(4) - 12 = 20 - 12 = 8$

b) $f(-3) = 5(-3) - 12$
$f(-3) = -15 - 12 = -27$

c) $g(3) = (3)^2 + 6(3) + 5$
$g(3) = 9 + 18 + 5 = 32$

d) $g(0) = (0)^2 + 6(0) + 5$
$g(0) = 0 + 5 = 5$

e) $f(a) = 5(a) - 12 = 5a - 12$

f) $g(t)=(t)^2+6(t)+5$

$g(t)=t^2+6t+5$

g) $f(k+8)=5(k+8)-12$

$f(k+8)=5k+40-12$

$f(k+8)=5k+28$

h) $f(c-2)=5(c-2)-12$

$f(c-2)=5c-10-12$

$f(c-2)=5c-22$

107) $15=4x-9$

$24=4x$

$6=x$

109) $f(x)=-2x+6$

$g(0)=-2(0)+6$

$g(0)=6$

$g(2)=-2(2)+6$

$g(2)=-4+6=2$

x	$f(x)$
0	6
2	2
3	0

$g(3)=-2(3)+6$

$g(3)=-6+6=0$

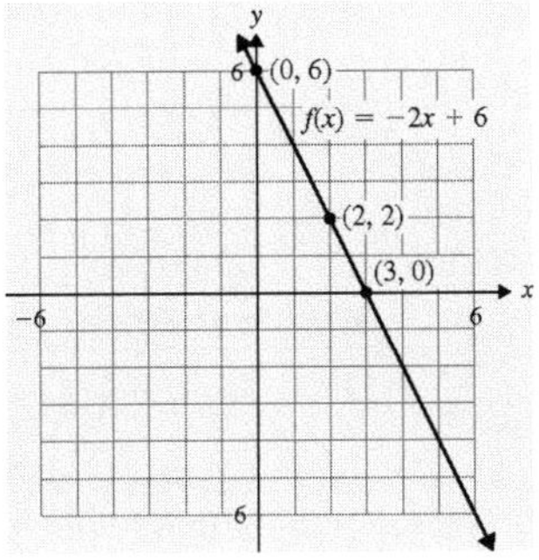

111)

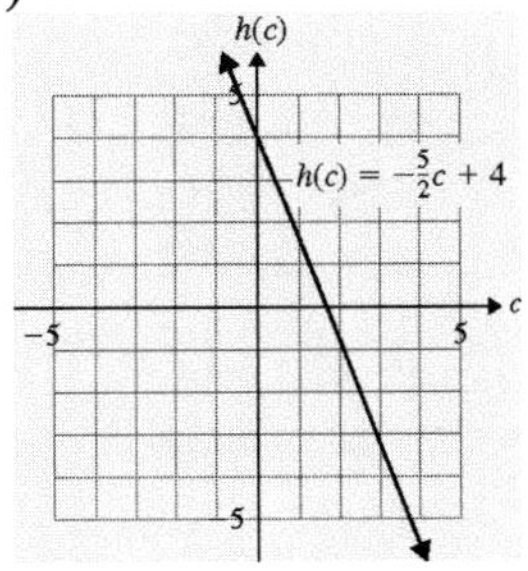

113) a) $f(2)=480(2)$

$f(2)=960$ MB

$f(6)=480(6)$

$f(6)=2880$ MB

b) $1200=480t$

$2.5=t$ 2.5 sec

Chapter 4 Test

1) Yes

$5x+3y=6$

$5(9)+3(-13)=6$

$45-39=6$

$6=6$

3) negative; positive

5)

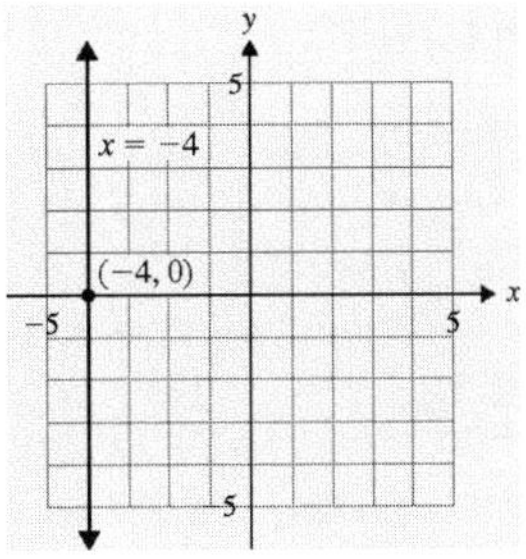

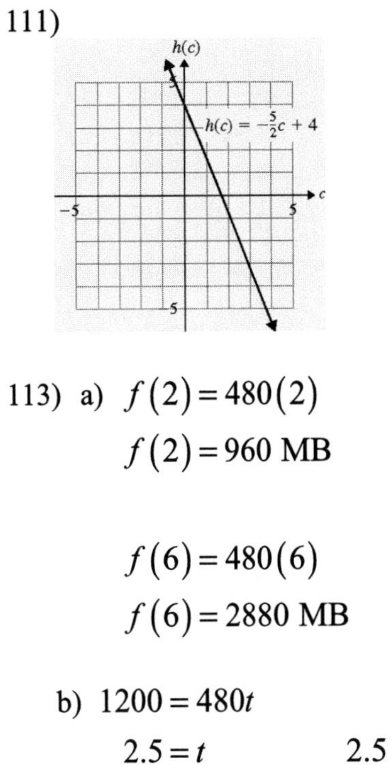

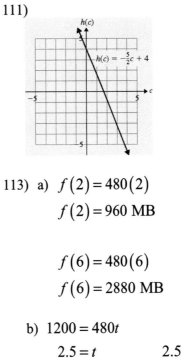

7)

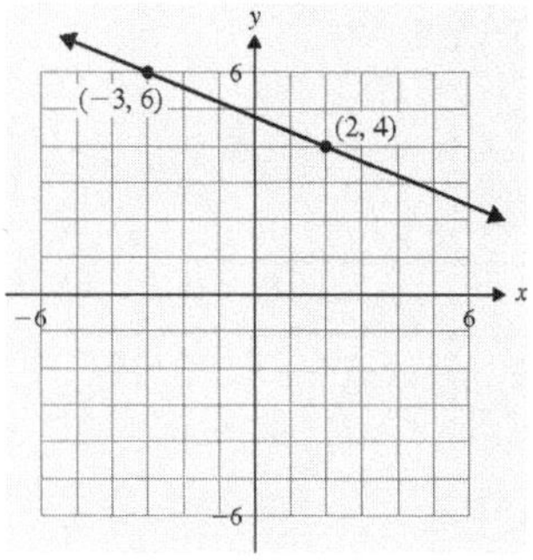

9) $6x-2y=8$

$$-2y=-6x+8$$
$$y=3x-4$$

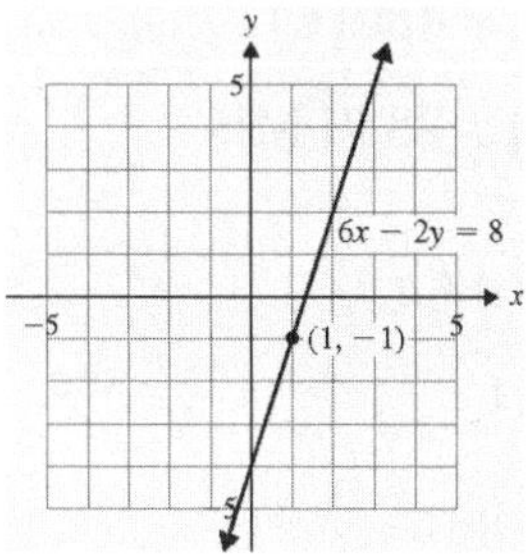

11) $(x_1,\ y_1)=(3,\ 8);\ m=\dfrac{1}{2}$

$$y-y_1=m(x-x_1)$$
$$y-8=\frac{1}{2}(x-3)$$
$$2(y-8)=2\left(\frac{1}{2}x-\frac{3}{2}\right)$$
$$2y-16=x-3$$
$$-x+2y=13$$
$$x-2y=-13$$

13) a) $(x_1,\ y_1)=(12,\ -5);$

$$m_{\text{perp}}=\frac{1}{3}$$

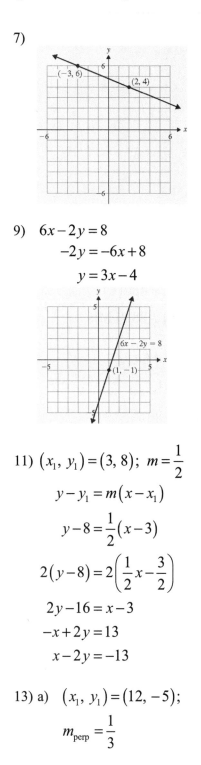

$$y-y_1=m(x-x_1)$$
$$y-(-5)=\frac{1}{3}(x-12)$$
$$y+5=\frac{1}{3}x-4$$
$$y=\frac{1}{3}x-9$$

b) Determine the slope.

$$5x-2y=2$$
$$-2y=-5x+2$$
$$y=\frac{5}{2}x-1$$
$$(x_1,\ y_1)=(8,\ 14);\ m=\frac{5}{2}$$
$$y-y_1=m(x-x_1)$$
$$y-14=\frac{5}{2}(x-8)$$
$$y-14=\frac{5}{2}x-20$$
$$y=\frac{5}{2}x-6$$

15) Domain: $\{-2, 1,\ 3, 8\}$

Range: $\{-5, -1,\ 1, 4\}$

Function

17) a) $(-\infty, \infty)$ b) yes

19) $f(2)=-3$

21) $f(6)=-4(6)+2=-24+2=-22$

23) $g(t)=(t)^2-3(t)+7=t^2-3t+7$

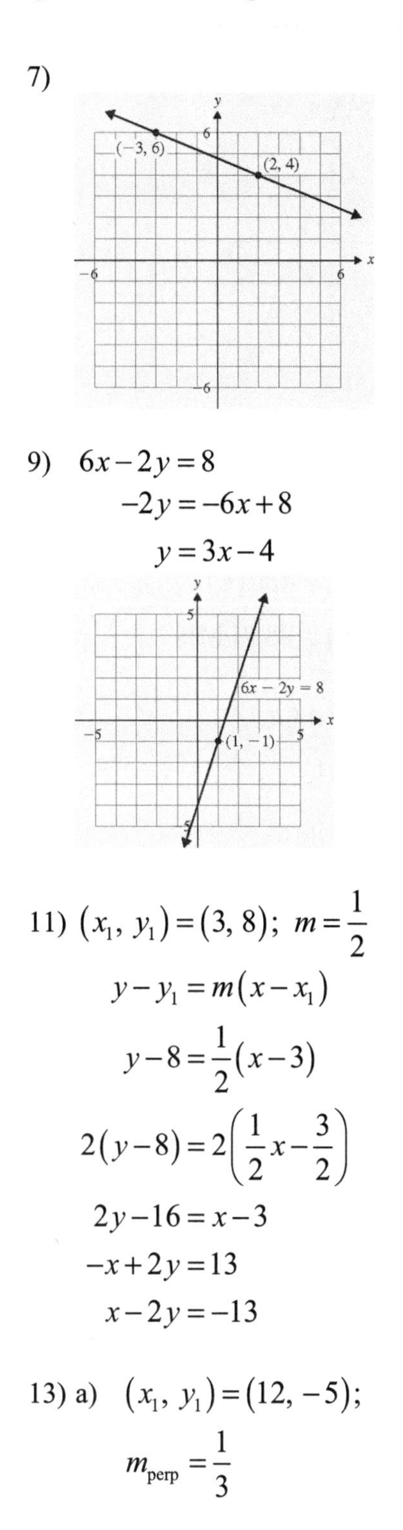

25)

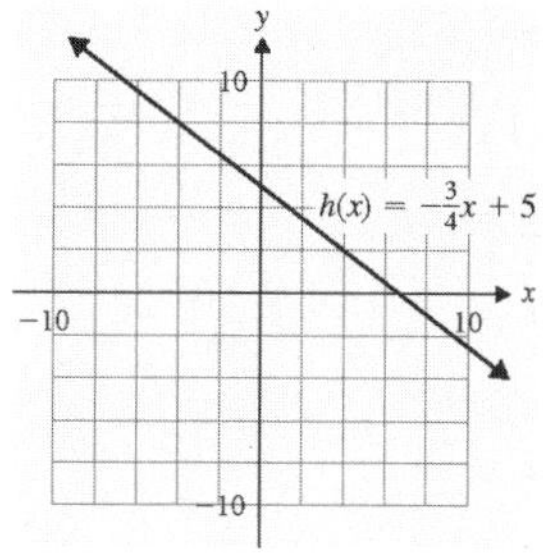

Cumulative Review: Chapters 1-4

1) $\dfrac{252}{840} = \dfrac{252 \div 84}{840 \div 84} = \dfrac{3}{10}$

3) $-2^6 = -64$

5) $3 - \dfrac{2}{5} = \dfrac{15}{5} - \dfrac{2}{5} = \dfrac{13}{5}$

$(3t^4)(-7t^{10}) = -21t^{14}$

9) $12 - 5(2n+9) = 3n + 2(n+6)$

$$\begin{aligned} 12 - 10n - 45 &= 3n + 2n + 12 \\ -33 - 10n &= 5n + 12 \\ -33 &= 15n + 12 \\ -45 &= -15n \\ -3 &= n \end{aligned}$$

The solution set is $\{-3\}$.

11) Let x = the number of calories in CM.

CD = CM − Amount of Decrease

$270 = x - x(0.10)$

$270 = 0.90x$

$300 = x$

There are 300 calories in Chunky Monkey ice cream.

13) Let x = the age of the daughter.
Then $3x - 7$ = Lynette's age.

$$\begin{pmatrix}\text{Daughter's}\\ \text{Age}\end{pmatrix} + \begin{pmatrix}\text{Lynette's}\\ \text{Age}\end{pmatrix} = 57$$

$$\begin{aligned} x + 3x - 7 &= 57 \\ 4x - 7 &= 57 \\ 4x &= 64 \\ x &= 16 \end{aligned}$$

$$\begin{aligned} \text{Lynette's Age} &= 3x - 7 \\ &= 3(16) - 7 \\ &= 48 - 7 \\ &= 41 \end{aligned}$$

$$\begin{aligned} \text{Daughter's Age} &= x \\ &= 16 \end{aligned}$$

15)

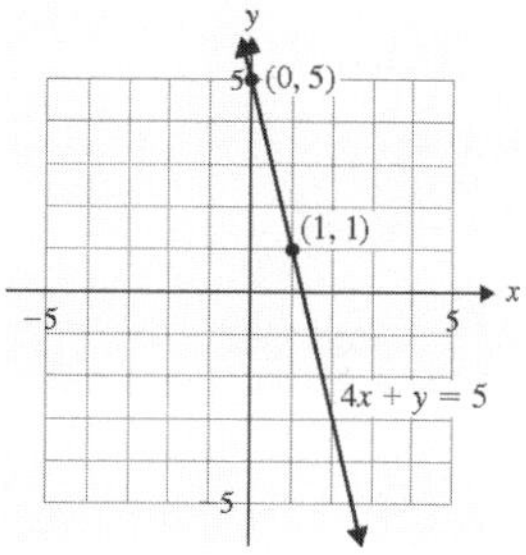

17) $(x_1, y_1) = (4, -12);\ m_{\text{perp}} = -3$

$$\begin{aligned} y - y_1 &= m(x - x_1) \\ y - (-12) &= -3(x - 4) \\ y + 12 &= -3x + 12 \\ y &= -3x \end{aligned}$$

19) $f(-5) = 8(-5) + 3 = -40 + 3 = -37$

21) $f(t+2) = 8(t+2) + 3$

$f(t+2) = 8t + 16 + 3 = 8t + 19$

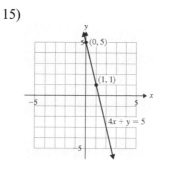

Section 5.1 Exercises

1) Yes

$$2x-3y=-15$$
$$2(3)-3(7)=-15$$
$$6-21=-15$$
$$-15=-15$$

$$-x+y=4$$
$$-(3)+7=4$$
$$4=4$$

3) No

$$3x+2y=4$$
$$3(-2)+2(5)=4$$
$$-6+10=4$$
$$4=4$$

$$4x-y=-3$$
$$4(-2)-(5)=-3$$
$$-8-5=-3$$
$$-13\neq-3$$

5) Yes

$$10x+7y=-13$$
$$10\left(\frac{3}{2}\right)+7(-4)=-13$$
$$15-28=-13$$
$$-13=-13$$

$$-6x-5y=11$$
$$-6\left(\frac{3}{2}\right)-5(-4)=11$$
$$-9+20=11$$
$$11=11$$

7) No

$$y=5x-7$$
$$-2=5(-1)-7$$
$$-2=-5-7$$
$$-2\neq-12$$

$$3x+9=y$$
$$3(-1)+9=-2$$
$$-3+9=-2$$
$$6\neq-2$$

9) The lines are parallel.

11) (3, 1)

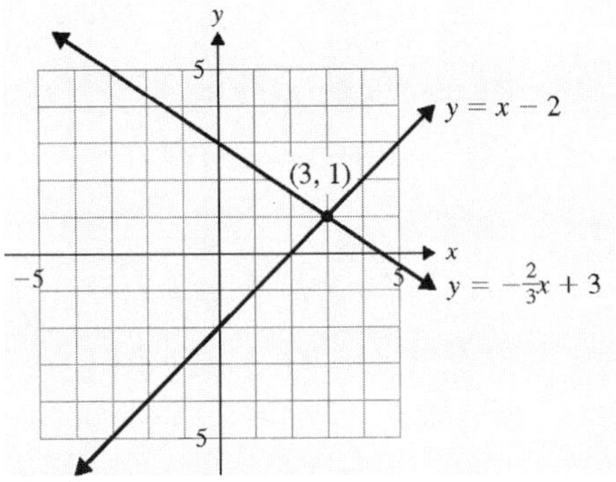

13) (−1, −1)

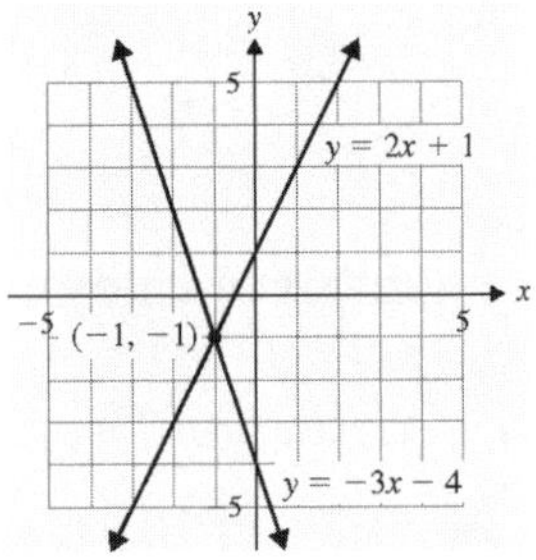

15) $(4, -5)$

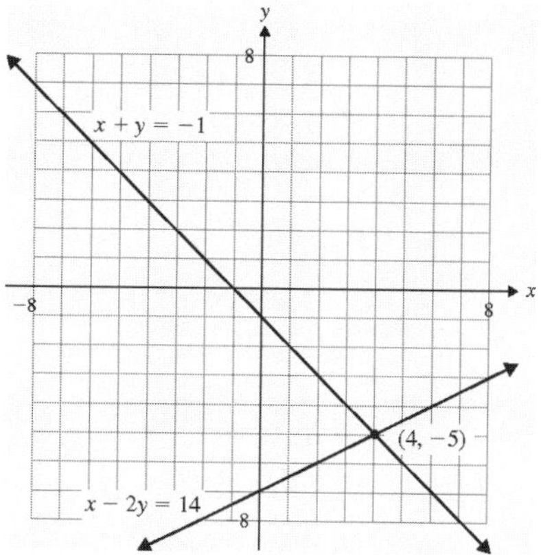

17) $\varnothing$; inconsistent system

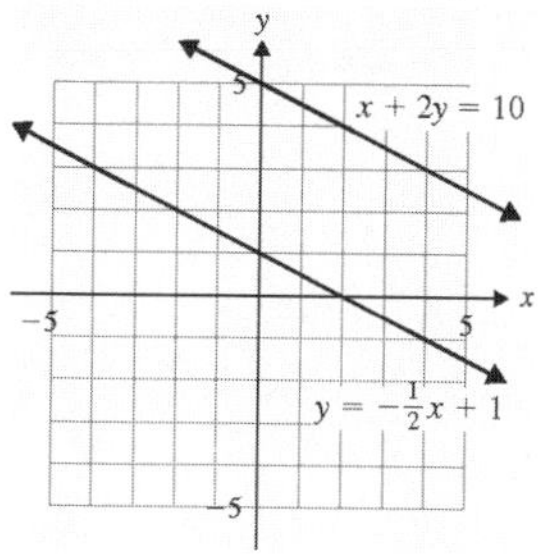

19) infinite number of solutions of the form$\{(x, y) \mid 6x - 3y = 12\}$; dependent system

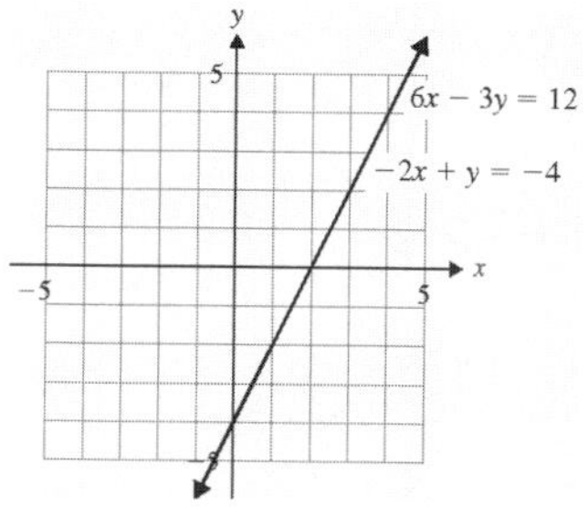

21) $\varnothing$; inconsistent system

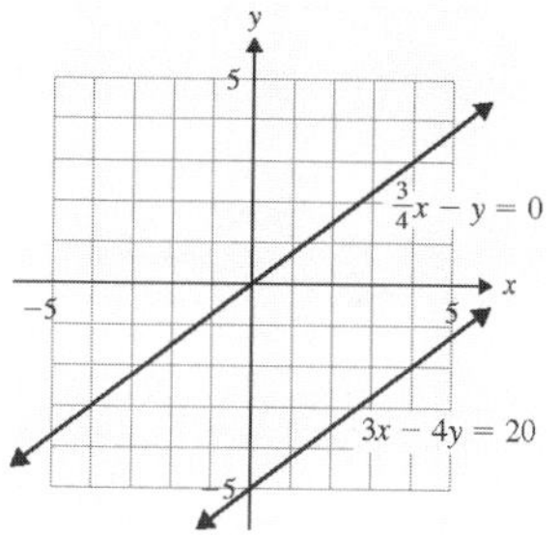

23) $(4, 5)$

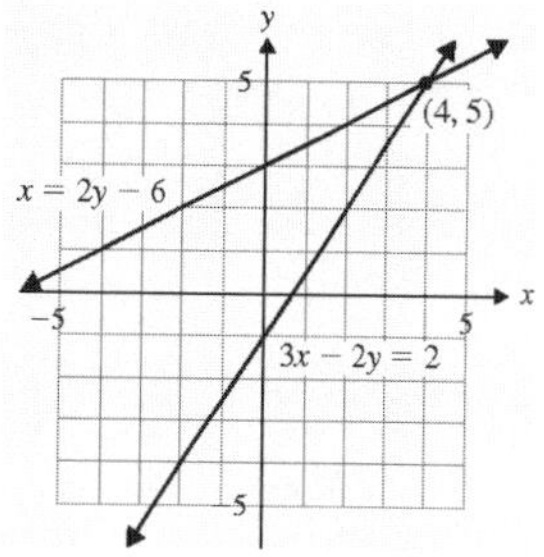

25) infinite number of solutions of the form$\{(x, y) \mid y = -3x + 1\}$; dependent system

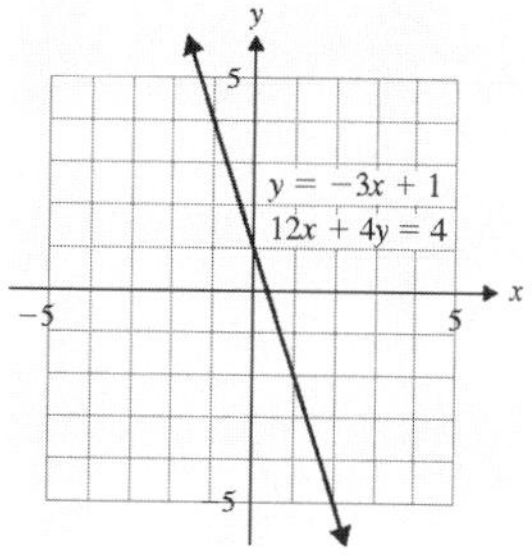

27) $(-1, -3)$

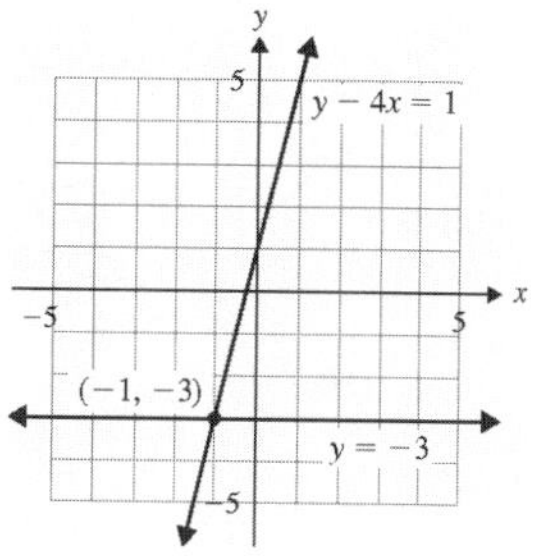

29) $(-1, 1)$

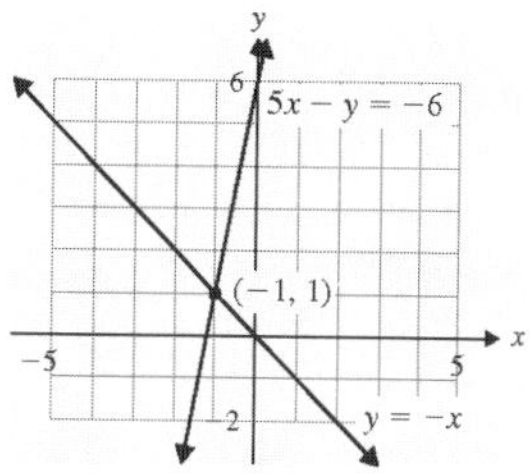

31) $\varnothing$; inconsistent system

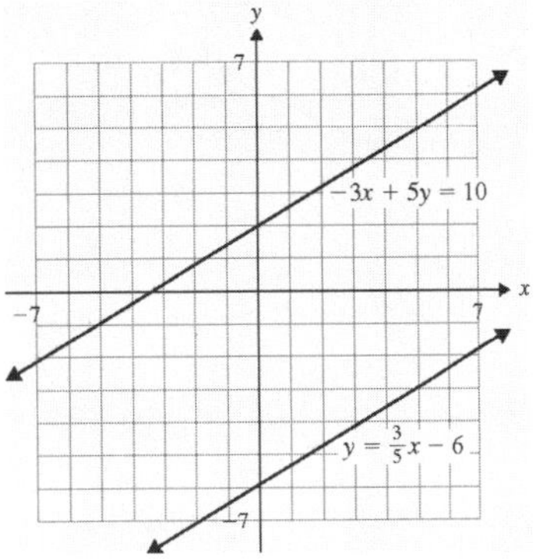

33) Answers may vary.

$$x+y=6$$
$$x-y=4$$

35) Answers may vary.

$$x+y=-5$$
$$x-y=3$$

37) Answers may vary.

$$2x+y=2$$
$$2x-y=-4$$

39) D. $(5,-3)$ is in quadrant IV.

41) C. $(0,-5)$ is on the y-axis not the x-axis.

43) The slopes are different.

45) $y=\frac{3}{2}x+\frac{7}{2}$

$$-9x+6y=21$$
$$6y=9x+21$$
$$\frac{6y}{6}=\frac{9}{6}x+\frac{21}{6}$$
$$y=\frac{3}{2}x+\frac{7}{2}$$

The lines are the same, so there are an infinite number of solutions

47) $5x-2y=-11$

$$-2y=-5x-11$$
$$y=\frac{5}{2}x+\frac{11}{2}$$

$$x+6y=18$$
$$6y=-x+18$$
$$y=-\frac{1}{6}x+3$$

The slopes are different, so there will be one solution.

49) $x+y=10$

$$y=-x+10$$

$$-9x-9y=2$$
$$-9y=9x+2$$
$$y=-x-\frac{2}{9}$$

The lines are parallel, so there is no solution.

51) $5x-y=-2$

$$-y=-5x-2$$
$$y=5x+2$$

$$x+6y=2$$
$$\frac{6y}{6}=\frac{-x}{6}+\frac{2}{6}$$
$$y=-\frac{1}{6}x+\frac{1}{3}$$

The slopes are different, so there will be one solution.

53) $y=-2$

$$y=3$$

The lines are parallel, so there is no solution.

55) a) There are more snowboarders than ice/figure skaters after the year 2001.
b) 2001; 5.3 million

c) 1999 – 2001

d) 1999 – 2001

57) 4

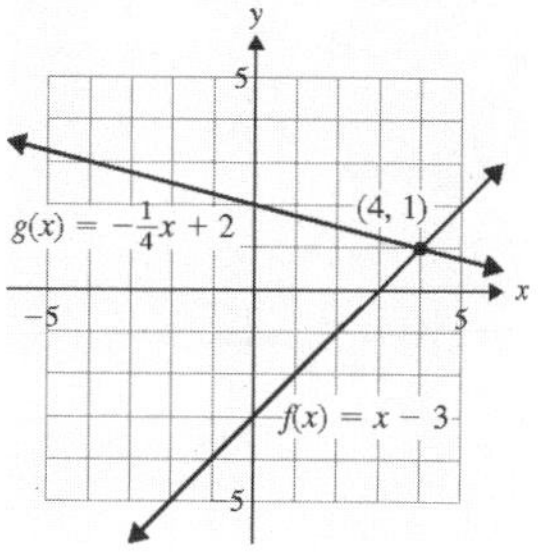

59) −1

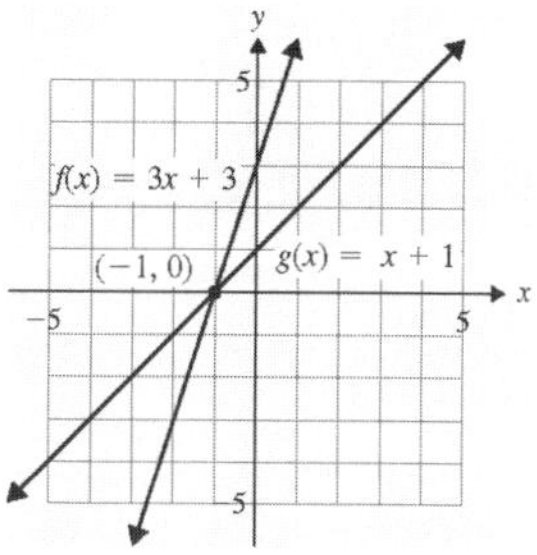

61) 0

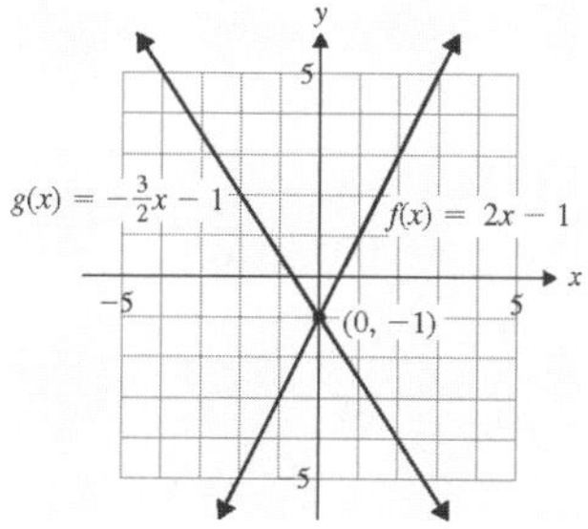

63) $(3, -2)$

65) $(-5, -1)$

67) $(-0.5, -1.25)$

Section 5.2 Exercises

1) It is the only variable with a coefficient of 1.

3) Substitute $y = 3x + 4$ into

$$-6x + y = -2$$
$$-6x + (3x + 4) = -2$$
$$-3x + 4 = -2$$
$$-3x = -6$$
$$x = 2$$

Substitute $x = 2$ into

$$y = 3x + 4$$
$$y = 3(2) + 4$$
$$y = 6 + 4$$
$$y = 10 \qquad (2, 10)$$

5) Substitute $x = y + 6$ into

$$2x - y = 5$$
$$2(y + 6) - y = 5$$
$$2y + 12 - y = 5$$
$$y + 12 = 5$$
$$y = -7$$

Substitute $y = -7$ into

$$x = y + 6$$
$$x = -7 + 6$$
$$x = -1 \qquad (-1, -7)$$

7) $x - 6y = 4$

$$x = 6y + 4$$

Substitute $x = 6y + 4$ into

$$2x + 5y = 8$$
$$2(6y + 4) + 5y = 8$$
$$12y + 8 + 5y = 8$$
$$17y + 8 = 8$$
$$17y = 0$$
$$y = 0$$

Substitute $y=0$ into
$x=6y+4$
$x=6(0)+4$
$x=4$ $(4,\ 0)$

9) Substitute $x=6-15y$ into
$2x+30y=9$
$2(6-15y)+30y=9$
$12-30y+30y=9$
$12\neq 9$
$\varnothing$

11) $6x+y=-6$
$y=-6x-6$
Substitute $y=-6x-6$ into
$-12x-2y=12$
$-12x-2(-6x-6)=12$
$-12x+12x+12=12$
$12=12$
infinite number of solutions of the form $\{(x,\ y)\,|\,6x+y=-6\}$

13) $10x+y=-5$
$y=-10x-5$
Substitute $y=-10x-5$ into
$-5x+2y=10$
$-5x+2(-10x-5)=10$
$-5x-20x-10=10$
$-25x-10=10$
$-25x=20$
$x=-\frac{4}{5}$

Substitute $x=-\frac{4}{5}$ into
$y=-10x-5$
$y=-10\left(-\frac{4}{5}\right)-5$
$y=8-5$
$y=3$ $\left(-\frac{4}{5},\ 3\right)$

15) $x+2y=6$
$x=-2y+6$
Substitute $x=-2y+6$ into
$x+20y=-12$
$(-2y+6)+20y=-12$
$18y+6=-12$
$18y=-18$
$y=-1$
Substitute $y=-1$ into
$x=-2y+6$
$x=-2(-1)+6$
$x=2+6$
$x=8$ $(8,\ -1)$

17) $6y-x=0$
$-x=-6y$
$x=6y$
Substitute $x=6y$ into
$2x-9y=-2$
$2(6y)-9y=-2$
$12y-9y=-2$
$3y=-2$
$y=-\frac{2}{3}$

Substitute $y=-\frac{2}{3}$ into

$x=6y$

$x=6\left(-\frac{2}{3}\right)$

$x=-4$ $\left(-4,-\frac{2}{3}\right)$

19) $2x-y=3$

$-y=-2x+3$

$y=2x-3$

Substitute $y=2x-3$ into

$9y-18x=5$

$9(2x-3)-18x=5$

$18x-27-18x=5$

$-27\neq 5$

$\varnothing$

21) $2x-y=6$

$-y=-2x+6$

$y=2x-6$

Substitute $y=2x-6$ into

$3y=-18-x$

$3(2x-6)=-18-x$

$6x-18=-18-x$

$7x=0$

$x=0$

Substitute $x=0$ into

$y=2x-6$

$y=2(0)-6$

$y=0-6$

$y=-6$ $(0,-6)$

23) $2x-5y=-4$

$2x=5y-4$

$x=\frac{5}{2}y-2$

Substitute $x=\frac{5}{2}y-2$ into

$8x-9y=6$

$8\left(\frac{5}{2}y-2\right)-9y=6$

$20y-16-9y=6$

$11y-16=6$

$11y=22$

$y=2$

Substitute $y=2$ into

$x=\frac{5}{2}y-2$

$x=\frac{5}{2}(2)-2$

$x=5-2$

$x=3$ $(3, 2)$

25) $4x+6y=-12$

$4x=-6y-12$

$x=-\frac{3}{2}y-3$

Substitute $x=-\frac{3}{2}y-3$ into

$9y-2x=22$

$9y-2\left(-\frac{3}{2}y-3\right)=22$

$9y+3y+6=22$

$12y+6=22$

$12y=16$

$y=\frac{16}{12}$

$y=\frac{4}{3}$

Substitute $y=\frac{4}{3}$ into

$x=\frac{3}{2}y-3$

$x=-\frac{3}{2}\left(\frac{4}{3}\right)-3$

$x=-2-3$

$x=-5$ $\left(-5,\frac{4}{3}\right)$

27) $4y-10x=-8$

$4y=10x-8$

$y=\frac{10}{4}x-2$

$y=\frac{5}{2}x-2$

Substitute $y=\frac{5}{2}x-2$ into

$15x-6y=12$

$15x-6\left(\frac{5}{2}x-2\right)=12$

$15x-15x+12=12$

$12=12$

infinite number of solutions of the form $\{(x,\ y)\,|\,4y-10x=-8\}$

29) Multiply the equation by the LCD of the fractions to eliminate the fractions.

31) $\frac{1}{4}x-\frac{1}{2}y=1$

$4\left(\frac{1}{4}x-\frac{1}{2}y\right)=4\cdot 1$

$x-2y=4$

$x=2y+4$

$\frac{2}{3}x+\frac{1}{6}y=\frac{25}{6}$

$6\left(\frac{2}{3}x+\frac{1}{6}y\right)=6\cdot\frac{25}{6}$

$4x+y=25$

Substitute $x=2y+4$ into

$4x+y=25$

$4(2y+4)+y=25$

$8y+16+y=25$

$9y+16=25$

$9y=9$

$y=1$

Substitute $y=1$ into

$x=2y+4$

$x=2(1)+4$

$x=2+4$

$x=6$ $(6,\ 1)$

33) $\frac{x}{10}-\frac{y}{2}=\frac{13}{10}$

$10\left(\frac{x}{10}-\frac{y}{2}\right)=10\cdot\frac{13}{10}$

$x-5y=13$

$x=5y+13$

$\frac{1}{3}x+\frac{5}{4}y=-\frac{3}{2}$

$12\left(\frac{1}{3}x+\frac{5}{4}y\right)=12\left(-\frac{3}{2}\right)$

$4x+15y=-18$

Substitute $x=5y+13$ into

$4x+15y=-18$

$4(5y+13)+15y=-18$

$20y+52+15y=-18$

$35y+52=-18$

$35y=-70$

$y=-2$

Substitute $y=-2$ into

$x=5y+13$

$x=5(-2)+13$

$x=-10+13$

$x=3$ $\qquad (3,-2)$

35) $\frac{3}{4}x+\frac{5}{2}y=5$

$4\left(\frac{3}{4}x+\frac{5}{2}y\right)=4\cdot 5$

$3x+10y=20$

$\frac{3}{2}x-\frac{1}{6}y=-\frac{1}{3}$

$6\left(\frac{3}{2}x-\frac{1}{6}y\right)=6\left(-\frac{1}{3}\right)$

$9x-y=-2$

$-y=-9x-2$

$y=9x+2$

Substitute $y=9x+2$ into

$3x+10y=20$

$3x+10(9x+2)=20$

$3x+90x+20=20$

$93x+20=20$

$93x=0$

$x=0$

Substitute $x=0$ into

$y=9x+2$

$y=9(0)+2$

$y=0+2$

$y=2$ $\qquad (0,2)$

37) $\frac{5}{3}x-\frac{4}{3}y=-\frac{4}{3}$

$3\left(\frac{5}{3}x-\frac{4}{3}y\right)=3\left(-\frac{4}{3}\right)$

$5x-4y=-4$

Substitute $y=2x+4$ into

$5x-4y=-4$

$5x-4(2x+4)=-4$

$5x-8x-16=-4$

$-3x-16=-4$

$-3x=12$

$x=-4$

Substitute $x=-4$ into

$y=2x+4$

$y=2(-4)+4$

$y=-8+4$

$y=-4$ $\qquad (-4,-4)$

39) $0.2x-0.1y=0.1$

$10(0.2x-0.1y)=10\cdot 0.1$

$2x-y=1$

$0.01x+0.04y=0.23$

$100(0.01x+0.04y)=100\cdot 0.23$

$x+4y=23$

$x=-4y+23$

Substitute $x=-4y+23$ into

$2x-y=1$

$2(-4y+23)-y=1$

$-8y+46-y=1$

$-9y+46=1$

$-9y=-45$

$y=5$

Substitute $y=5$ into

$x=-4y+23$

$x=-4(5)+23$

$x=-20+23$

$x=3$ $\qquad (3,5)$

41) $0.1x + 0.5y = 0.4$

$10(0.1x + 0.5y) = 10(0.4)$

$x + 5y = 4$

$x = -5y + 4$

$-0.03x + 0.01y = 0.2$

$100(-0.03x + 0.01y) = 100 \cdot 0.2$

$-3x + y = 20$

Substitute $x = -5y + 4$ into

$-3x + y = 20$

$-3(-5y + 4) + y = 20$

$15y - 12 + y = 20$

$16y - 12 = 20$

$16y = 32$

$y = 2$

Substitute $y = 2$ into

$x = -5y + 4$

$x = -5(2) + 4$

$x = -10 + 4$

$x = -6$ $(-6, 2)$

43) $0.3x - 0.1y = 5$

$10(0.3x - 0.1y) = 10 \cdot 5$

$3x - y = 50$

$3x - 50 = y$

$0.15x + 0.1y = 4$

$100(0.15x + 0.1y) = 100 \cdot 4$

$15x + 10y = 400$

Substitute $y = 3x - 50$ into

$15x + 10y = 400$

$15x + 10(3x - 50) = 400$

$15x + 30x - 500 = 400$

$45x - 500 = 400$

$45x = 900$

$x = 20$

Substitute $x = 20$ into

$y = 3x - 50$

$y = 3(20) - 50$

$y = 60 - 50$

$y = 10$ $(20, 10)$

45) $5(2x - 3) + y - 6x = -24$

$10x - 15 + y - 6x = -24$

$4x - 15 + y = -24$

$4x + y = -9$

$y = -4x - 9$

$8y - 3(2y + 3) + x = -6$

$8y - 6y - 9 + x = -6$

$2y + x = 3$

Substitute $y = -4x - 9$ into

$2y + x = 3$

$2(-4x - 9) + x = 3$

$-8x - 18 + x = 3$

$-7x - 18 = 3$

$-7x = 21$

$x = -3$

Substitute $x = -3$ into

$y = -4x - 9$

$y = -4(-3) - 9$

$y = 12 - 9$

$y = 3$ $(-3, 3)$

47) $7x + 3(y - 2) = 7y + 6x - 1$

$7x + 3y - 6 = 7y + 6x - 1$

$x - 4y = 5$

$x = 4y + 5$

$18 + 2(x - y) = 4(x + 2) - 5y$

$18 + 2x - 2y = 4x + 8 - 5y$

$-2x + 3y = -10$

Substitute $x = 4y + 5$ into

$$-2x+3y=-10$$
$$-2(4y+5)+3y=-10$$
$$-8y-10+3y=-10$$
$$-5y-10=-10$$
$$-5y=0$$
$$y=0$$

Substitute $y=0$ into

$$x=4y+5$$
$$x=4(0)+5$$
$$x=0+5$$
$$x=5 \qquad (5,\ 0)$$

49) $9y-4(2y+3)=-2(4x+1)$

$$9y-8y-12=-8x-2$$
$$y-12=-8x-2$$
$$8x+y=10$$
$$y=-8x+10$$

$$16-5(2x+3)=2(4-y)$$
$$16-10x-15=8-2y$$
$$1-10x=8-2y$$
$$2y-10x=7$$

Substitute $y=-8x+10$ into

$$2y-10x=7$$
$$2(-8x+10)-10x=7$$
$$-16x+20-10x=7$$
$$-26x+20=7$$
$$-26x=-13$$
$$x=\frac{1}{2}$$

Substitute $x=\frac{1}{2}$ into

$$y=-8x+10$$
$$y=-8\left(\frac{1}{2}\right)+10$$
$$y=-4+10$$
$$y=6 \qquad \left(\frac{1}{2},\ 6\right)$$

51) a) Rent-for-Less: $y=0.40x$

$$y=0.40(60)$$
$$y=\$24$$

Frugal: $y=0.30x+12$

$$y=0.30(60)+12$$
$$y=18+12$$
$$y=\$30$$

b) Rent-for-Less: $y=0.40x$

$$y=0.40(160)$$
$$y=\$64$$

Frugal: $y=0.30x+12$

$$y=0.30(160)+12$$
$$y=48+12$$
$$y=\$60$$

c) Substitute $y=0.40x$ into

$$y=0.30x+12$$
$$0.40x=0.30x+12$$
$$0.10x=12$$
$$x=120$$

Substitute $x=120$ into

$$y=0.40x$$
$$y=0.40(120)$$
$$y=48 \qquad (120,\ 48)$$

If the car is driven 120 miles, the cost would be the same from each company: $48.

d) If a car is driven less than 120 miles, it is cheaper to rent from Rent-for-Less. If a car is driven more than 120 miles, it is cheaper to rent from Frugal Rentals. If a car is driven exactly 120 miles, the cost is the same from each company.

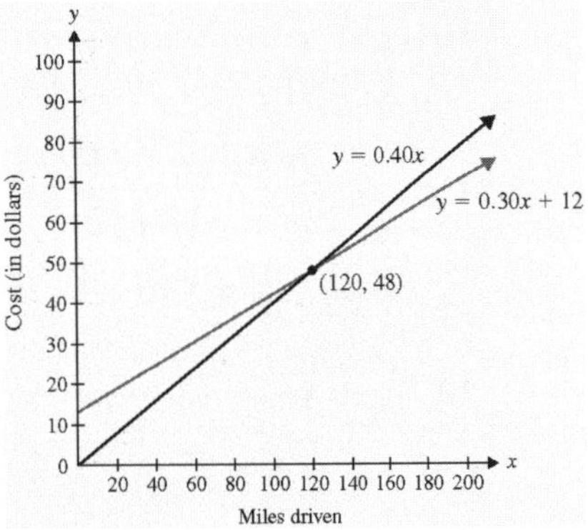

Section 5.3 Exercises

1) Add the equations.

$$\begin{array}{r} x-3y=1 \\ +\ -x+y=-3 \\ \hline -2y=-2 \end{array}$$

$$y=1$$

Substitute $y=1$ into

$$x-3y=1$$
$$x-3(1)=1$$
$$x-3=1$$
$$x=4 \qquad (4, 1)$$

3) Add the equations.

$$\begin{array}{r} 3x+5y=-10 \\ +\ 7x-5y=10 \\ \hline 10x=0 \end{array}$$

$$x=0$$

Substitute $x=0$ into

$$3x+5y=-10$$
$$3(0)+5y=-10$$
$$0+5y=-10$$
$$5y=10$$
$$y=-2 \qquad (0, -2)$$

5) $3x+2y=-1$

$$-3(3x+2y)=-3(-1)$$
$$-9x-6y=3$$

Add the equations.

$$\begin{array}{r} 7x+6y=3 \\ +\ -9x-6y=3 \\ \hline -2x=6 \end{array}$$

$$x=-3$$

Substitute $x=-3$ into

$$7x+6y=3$$
$$7(-3)+6y=3$$
$$-21+6y=3$$
$$6y=24$$
$$y=4 \qquad (-3, 4)$$

7) $3x-y=4$

$$2(3x-y)=2(4)$$
$$6x-2y=8$$

Add the equations.

$$\begin{array}{r} 6x-2y=8 \\ +\ -6x+2y=-8 \\ \hline 0=0 \end{array}$$

infinite number of solutions of the form $\{(x, y)\,|\,3x-y=4\}$

9) $3x+2y=-9$

$-2(3x+2y)=-2(-9)$

$-6x-4y=18$

$2x-7y=19$

$3(2x-7y)=3(19)$

$6x-21y=57$

Add the equations.

$$\begin{array}{r} -6x-4y=18 \\ +\ 6x-21y=57 \\ \hline -25y=75 \end{array}$$

$y=-3$

Substitute $y=-3$ into

$3x+2y=-9$

$3x+2(-3)=-9$

$3x-6=-9$

$3x=-3$

$x=-1$ $(-1, -3)$

11) $x=12-4y$

$x+4y=12$

$-2(x+4y)=-2(12)$

$-2x-8y=-24$

$2x-7=9y$

$2x-9y=7$

Add the equations.

$$\begin{array}{r} -2x-8y=-24 \\ +\ 2x-9y=7 \\ \hline -17y=-17 \end{array}$$

$y=1$

Substitute $y=1$ into

$x=12-4y$

$x=12-4(1)$

$x=12-4$

$x=8$ $(8, 1)$

13) $2x-9=8y$

$2x-8y=9$

$5(2x-8y)=5(9)$

$10x-40y=45$

$20y-5x=6$

$-5x+20y=6$

$2(-5x+20y)=2(6)$

$-10x+40y=12$

Add the equations.

$$\begin{array}{r} 10x-40y=45 \\ +\ -10x+40y=12 \\ \hline 0\neq 57 \end{array}$$

$\varnothing$

15) $y=6x-10$

$-6x+y=-10$

$-5(-6x+y)=-5(-10)$

$30x-5y=50$

Add the equations.

$$\begin{array}{r} 30x-5y=50 \\ +\ -4x+5y=-11 \\ \hline 26x=39 \end{array}$$

$x=\frac{39}{26}$

$x=\frac{3}{2}$

Substitute $x=\frac{3}{2}$ into

$y=6x-10$

$y=6\left(\frac{3}{2}\right)-10$

$y=9-10$

$y=-1$ $\left(\frac{3}{2}, -1\right)$

17) $3x+4y=9$
$3(3x+4y)=3(9)$
$9x+12y=27$

$5x+6y=16$
$-2(5x+6y)=-2(16)$
$-10x-12y=-32$
Add the equations.
$$\begin{array}{r} 9x+12y=27 \\ +\ -10x-12y=-32 \\ \hline -x=-5 \\ x=5 \end{array}$$
Substitute $x=5$ into
$3x+4y=9$
$3(5)+4y=9$
$15+4y=9$
$4y=-6$
$y=\frac{-6}{4}$
$y=-\frac{3}{2}$ $\left(5, -\frac{3}{2}\right)$

19) $-2x-11=16y$
$-2x-16y=11$

$x=3-8y$
$x+8y=3$
$2(x+8y)=2(3)$
$2x+16y=6$
Add the equations.
$$\begin{array}{r} -2x-16y=11 \\ +\ 2x+16y=6 \\ \hline 0\neq 17 \end{array}$$
$\varnothing$

21) $7x+2y=12$
$2(7x+2y)=2(12)$
$14x+4y=24$

$24-14x=4y$
$-14x-4y=-24$
Add the equations.
$$\begin{array}{r} 14x+4y=24 \\ +\ -14x-4y=-24 \\ \hline 0=0 \end{array}$$
infinite number of solutions of the form $\{(x, y) \mid 7x+2y=12\}$

23) $9x-7y=-14$
$3(9x-7y)=3(-14)$
$27x-21y=-42$

$4x+3y=6$
$7(4x+3y)=7(6)$
$28x+21y=42$
Add the equations.
$$\begin{array}{r} 27x-21y=-42 \\ +\ 28x+21y=42 \\ \hline 55x=0 \\ x=0 \end{array}$$
Substitute $x=0$ into
$4x+3y=6$
$4(0)+3y=6$
$0+3y=6$
$3y=6$
$y=2$ $(0, 2)$

25) Eliminate the decimals. Multiply the first equation by 10, and multiply the second equation by 100.

27) $\frac{x}{4}+\frac{y}{2}=-1$
$4\left(\frac{x}{4}+\frac{y}{2}\right)=4(-1)$
$x+2y=-4$

$$-9(x+2y)=-9(-4)$$
$$-9x-18y=36$$

$$\frac{3}{8}x+\frac{5}{3}y=-\frac{7}{12}$$
$$24\left(\frac{3}{8}x+\frac{5}{3}y\right)=24\left(-\frac{7}{12}\right)$$
$$9x+40y=-14$$

Add the equations.

$$\begin{array}{r} -9x-18y=36 \\ +\ 9x+40y=-14 \\ \hline 22y=22 \end{array}$$
$$y=1$$

Substitute $y=1$ into

$$x+2y=-4$$
$$x+2(1)=-4$$
$$x+2=-4$$
$$x=-6 \qquad (-6, 1)$$

29) $$\frac{x}{2}-\frac{y}{5}=\frac{1}{10}$$
$$10\left(\frac{x}{2}-\frac{y}{5}\right)=10\left(\frac{1}{10}\right)$$
$$5x-2y=1$$
$$3(5x-2y)=3(1)$$
$$15x-6y=3$$

$$\frac{x}{3}+\frac{y}{4}=\frac{5}{6}$$
$$12\left(\frac{x}{3}+\frac{y}{4}\right)=12\left(\frac{5}{6}\right)$$
$$4x+3y=10$$
$$2(4x+3y)=2(10)$$
$$8x+6y=20$$

Add the equations.

$$\begin{array}{r} 15x-6y=3 \\ +\ 8x+6y=20 \\ \hline 23x=23 \end{array}$$
$$x=1$$

Substitute $x=1$ into

$$5x-2y=1$$
$$5(1)-2y=1$$
$$5-2y=1$$
$$-2y=-4$$
$$y=2 \qquad (1, 2)$$

31) $$x+\frac{3}{2}y=13$$
$$2\left(x+\frac{3}{2}y\right)=2(13)$$
$$2x+3y=26$$

$$-\frac{1}{8}x+\frac{1}{4}y=\frac{1}{8}$$
$$8\left(-\frac{1}{8}x+\frac{1}{4}y\right)=8\left(\frac{1}{8}\right)$$
$$-x+2y=1$$
$$2(-x+2y)=2(1)$$
$$-2x+4y=2$$

Add the equations.

$$\begin{array}{r} 2x+3y=26 \\ +\ -2x+4y=2 \\ \hline 7y=28 \end{array}$$
$$y=4$$

Substitute $y=4$ into

$$-x+2y=1$$
$$-x+2(4)=1$$
$$-x+8=1$$
$$-x=-7$$
$$x=7 \qquad (7, 4)$$

33) $0.1x+2y=-0.8$

$10(0.1x+2y)=10(-0.8)$

$x+20y=-8$

$-3(x+20y)=-3(-8)$

$-3x-60y=24$

$0.03x+0.10y=0.26$

$100(0.03x+0.10y)=100(0.26)$

$3x+10y=26$

Add the equations.

$$\begin{array}{r} -3x-60y=24 \\ +\ \ 3x+10y=26 \\ \hline -50y=50 \end{array}$$

$y=-1$

Substitute $y=-1$ into

$x+20y=-8$

$x+20(-1)=-8$

$x-20=-8$

$x=12$ $(12, -1)$

35) $0.02x+0.07y=-0.24$

$100(0.02x+0.07y)=100(-0.24)$

$2x+7y=-24$

$2(2x+7y)=2(-24)$

$4x+14y=-48$

$0.05y-0.04x=0.10$

$-0.04x+0.05y=0.10$

$100(-0.04x+0.05y)=100(0.10)$

$-4x+5y=10$

Add the equations.

$$\begin{array}{r} 4x+14y=-48 \\ +\ \ -4x+5y=10 \\ \hline 19y=-38 \end{array}$$

$y=-2$

Substitute $y=-2$ into

$2x+7y=-24$

$2x+7(-2)=-24$

$2x-14=-24$

$2x=-10$

$x=-5$ $(-5, -2)$

37) $2(y-6)=3y+4(x-5)$

$2y-12=3y+4x-20$

$-4x-y=-8$

$2(-4x-y)=2(-8)$

$-8x-2y=-16$

$2(4x+3)-5=2(1-y)+5x$

$8x+6-5=2-2y+5x$

$8x+1=2-2y+5x$

$3x+2y=1$

Add the equations.

$$\begin{array}{r} -8x-2y=-16 \\ +\ \ 3x+2y=1 \\ \hline -5x=-15 \end{array}$$

$x=3$

Substitute $x=3$ into

$-4x-y=-8$

$-4(3)-y=-8$

$-12-y=-8$

$-y=4$

$y=-4$ $(3, -4)$

39) $20+3(2y-3)=4(2y-1)-9x$

$20+6y-9=8y-4-9x$

$11+6y=8y-4-9x$

$9x-2y=-15$

$$5(3x-4)+8y=3x+7(y-1)$$
$$15x-20+8y=3x+7y-7$$
$$12x+y=13$$
$$2(12x+y)=2(13)$$
$$24x+2y=26$$

Add the equations.

$$\begin{array}{r} 9x-2y=-15 \\ +\ 24x+2y=26 \\ \hline 33x=11 \end{array}$$
$$x=\frac{11}{33}$$
$$x=\frac{1}{3}$$

Substitute $x=\frac{1}{3}$ into

$$12x+y=13$$
$$12\left(\frac{1}{3}\right)+y=13$$
$$4+y=13$$
$$y=9 \qquad \left(\frac{1}{3}, 9\right)$$

41) $6(x-3)+x-4y=1+2(x-9)$

$$6x-18+x-4y=1+2x-18$$
$$7x-18-4y=-17+2x$$
$$5x-4y=1$$
$$2(5x-4y)=2(1)$$
$$10x-8y=2$$

$$4(2y-3)+10x=5(x+1)-4$$
$$8y-12+10x=5x+5-4$$
$$8y-12+10x=5x+1$$
$$5x+8y=13$$

Add the equations.

$$\begin{array}{r} 10x-8y=2 \\ +\ 5x+8y=13 \\ \hline 15x=15 \end{array}$$
$$x=1$$

Substitute $x=1$ into

$$5x-4y=1$$
$$5(1)-4y=1$$
$$5-4y=1$$
$$-4y=-4$$
$$y=1 \qquad (1, 1)$$

43) Eliminate y.

$$4x+5y=-6$$
$$-8(4x+5y)=-8(-6)$$
$$-32x-40y=48$$

$$3x+8y=15$$
$$5(3x+8y)=5(15)$$
$$15x+40y=75$$

Add the equations.

$$\begin{array}{r} -32x-40y=48 \\ +\ 15x+40y=75 \\ \hline -17x=123 \end{array}$$
$$x=-\frac{123}{17}$$

Eliminate x.

$$4x+5y=-6$$
$$-3(4x+5y)=-3(-6)$$
$$-12x-15y=18$$
$$3x+8y=15$$
$$4(3x+8y)=4(15)$$
$$12x+32y=60$$

Add the equations.

$$-12x-15y=18$$
$$+\ 12x+32y=60$$
$$17y=78$$
$$y=\frac{78}{17}$$
$$\left(-\frac{123}{17},\frac{78}{17}\right)$$

45) Eliminate y.

$$2x-7y=-10$$
$$2(2x-7y)=2(-10)$$
$$4x-14y=-20$$

$$6x+2y=15$$
$$7(6x+2y)=7(15)$$
$$42x+14y=105$$

Add the equations.

$$4x-14y=-20$$
$$+\ 42x+14y=105$$
$$46x=85$$
$$x=\frac{85}{46}$$

Eliminate x.

$$2x-7y=-10$$
$$-3(2x-7y)=-3(-10)$$
$$-6x+21y=30$$

Add the equations.

$$-6x+21y=30$$
$$+\ \ 6x+2y=15$$
$$23y=45$$
$$y=\frac{45}{23}$$
$$\left(\frac{85}{46},\frac{45}{23}\right)$$

47) Eliminate y.

$$8x-4y=-21$$
$$3(8x-4y)=3(-21)$$
$$24x-12y=-63$$

$$-5x+6y=12$$
$$2(-5x+6y)=2(12)$$
$$-10x+12y=24$$

Add the equations.

$$24x-12y=-63$$
$$+\ -10x+12y=24$$
$$14x=-39$$
$$x=-\frac{39}{14}$$

Eliminate x.

$$8x-4y=-21$$
$$5(8x-4y)=5(-21)$$
$$40x-20y=-105$$

$$-5x+6y=12$$
$$8(-5x+6y)=8(12)$$
$$-40x+48y=96$$

Add the equations.

$$40x-20y=-105$$
$$+\ -40x+48y=96$$
$$28y=-9$$
$$y=-\frac{9}{28}$$
$$\left(-\frac{39}{14},-\frac{9}{28}\right)$$

49) $x+y=8$

$$-1(x+y)=-1(8)$$
$$-x-y=-8$$

Add the equations.

$$-x-y=-8$$
$$+\ x+y=c$$
$$0=c-8$$

a) 8

b) c can be any real number except 8

51) $2x-3y=5$

$-2(2x-3y)=-2(5)$

$-4x+6y=-10$

Add the equations.

$$\begin{array}{r} -4x+6y=-10 \\ + \quad ax-6y=10 \\ \hline ax-4x=0 \end{array}$$

a) 4

b) a can be any real number except 4

53) Substitute $(2,-1)$ into $-x+by=-7$ and solve for b.

$-(2)+b(-1)=-7$

$-2-b=-7$

$-b=-5$

$b=5$

Mid-Chapter Summary

1) Elimination method; none of the coefficients is 1 or -1

$2x-3y=-8$

$-4(2x-3y)=-4(-8)$

$-8x+12y=32$

Add the equations.

$$\begin{array}{r} 8x-5y=10 \\ + \; -8x+12y=32 \\ \hline 7y=42 \end{array}$$

$y=6$

Substitute $y=6$ into

$2x-3y=-8$

$2x-3(6)=-8$

$2x-18=-8$

$2x=10$

$x=5$ $\qquad (5, 6)$

3) Since the coefficient of x in the first equation is 1, you can solve for x and use substitution. Or, multiply the first equation by -3 and use the elimination method. Either method will work well.

$x+6y=-10$

$x=-6y-10$

Substitute $x=-6y-10$ into

$3x-8y=-4$

$3(-6y-10)-8y=-4$

$-18y-30-8y=-4$

$-26y-30=-4$

$-26y=26$

$y=-1$

Substitute $y=-1$ into

$x=-6y-10$

$x=-6(-1)-10$

$x=6-10$

$x=-4$ $\qquad (-4, -1)$

5) Substitution; the second equation is solved for x and does not contain any fractions

Substitute $x=y+8$ into

$y-4x=-11$

$y-4(y+8)=-11$

$y-4y-32=-11$

$-3y-32=-11$

$-3y=21$

$y=-7$

Substitute $y=-7$ into

$x=y+8$

$x=(-7)+8$

$x=1 \qquad (1, -7)$

7) Substitute $y=2x+1$ into

$$9x-2y=8$$
$$9x-2(2x+1)=8$$
$$9x-4x-2=8$$
$$5x-2=8$$
$$5x=10$$
$$x=2$$

Substitute $x=2$ into

$y=2x+1$

$y=2(2)+1$

$y=4+1$

$y=5 \qquad (2, 5)$

9) $8y-x=-11$

$-x+8y=-11$

Add the equations.

$$\begin{array}{r} -x+8y=-11 \\ +\quad x+10y=2 \\ \hline 18y=-9 \end{array}$$

$$y=\frac{-9}{18}$$
$$y=-\frac{1}{2}$$

Substitute $y=-\frac{1}{2}$ into

$$8y-x=-11$$
$$8\left(-\frac{1}{2}\right)-x=-11$$
$$-4-x=-11$$
$$-x=-7$$
$$x=7 \qquad \left(7, -\frac{1}{2}\right)$$

11) $10x+4y=7$

$$3(10x+4y)=3(7)$$
$$30x+12y=21$$

$$15x+6y=-2$$
$$-2(15x+6y)=-2(-2)$$
$$-30x-12y=4$$

Add the equations.

$$\begin{array}{r} 30x+12y=21 \\ +\ -30x-12y=4 \\ \hline 0\neq 25 \end{array}$$

$\varnothing$

13) $\frac{1}{3}x+\frac{3}{2}y=-\frac{1}{2}$

$$6\left(\frac{1}{3}x+\frac{3}{2}y\right)=6\left(-\frac{1}{2}\right)$$
$$2x+9y=-3$$

Add the equations.

$$\begin{array}{r} -2x-7y=5 \\ +\quad 2x+9y=-3 \\ \hline 2y=2 \end{array}$$

$$y=1$$

Substitute $y=1$ into

$$-2x-7y=5$$
$$-2x-7(1)=5$$
$$-2x-7=5$$
$$-2x=12$$
$$x=-6 \qquad (-6, 1)$$

15) Substitute $y=-6$ into

$$5x+2y=3$$
$$5x+2(-6)=3$$
$$5x-12=3$$
$$5x=15$$
$$x=3 \qquad (3, -6)$$

17) $5y-4x=8$

$-4x+5y=8$

$-3(-4x+5y)=-3(8)$

$12x-15y=-24$

$10x+3y=11$

$5(10x+3y)=5(11)$

$50x+15y=55$

Add the equations.

$$\begin{array}{r} 12x-15y=-24 \\ +\ \ 50x+15y=55 \\ \hline 62x=31 \end{array}$$

$x=\frac{31}{62}$

$x=\frac{1}{2}$

Substitute $x=\frac{1}{2}$ into

$-4x+5y=8$

$-4\left(\frac{1}{2}\right)+5y=8$

$-2+5y=8$

$5y=10$

$y=2$ $\left(\frac{1}{2}, 2\right)$

19) Substitute $y=-6x+5$ into

$12x+2y=10$

$12x+2(-6x+5)=10$

$12x-12x+10=10$

$10=10$

infinite number of solutions of the form $\{(x, y) \mid y=-6x+5\}$

21) $0.01x+0.02y=0.28$

$100(0.01x+0.02y)=100(0.28)$

$x+2y=28$

$x=-2y+28$

$0.04x-0.03y=0.13$

$100(0.04x-0.03y)=100(0.13)$

$4x-3y=13$

Substitute $x=-2y+28$ into

$4x-3y=13$

$4(-2y+28)-3y=13$

$-8y+112-3y=13$

$-11y+112=13$

$-11y=-99$

$y=9$

Substitute $y=9$ into

$x=-2y+28$

$x=-2(9)+28$

$x=-18+28$

$x=10$ $(10, 9)$

23) $6(2x-3)=y+4(x-3)$

$12x-18=y+4x-12$

$8x-6=y$

$5(3x+4)+4y=11-3y+27x$

$15x+20+4y=11-3y+27x$

$-12x+7y=-9$

Substitute $y=8x-6$ into

$-12x+7y=-9$

$-12x+7(8x-6)=-9$

$-12x+56x-42=-9$

$44x=33$

$x=\frac{33}{44}$

$x=\frac{3}{4}$

Substitute $x=\frac{3}{4}$ into

$$y=8x-6$$
$$y=8\left(\frac{3}{4}\right)-6$$
$$y=6-6$$
$$y=0 \qquad \left(\frac{3}{4},\,0\right)$$

25)
$$y=\frac{5}{6}x+\frac{10}{3}$$
$$-\frac{5}{6}x+y=\frac{10}{3}$$
$$6\left(-\frac{5}{6}x+y\right)=6\left(\frac{10}{3}\right)$$
$$-5x+6y=20$$
$$3(-5x+6y)=3(20)$$
$$-15x+18y=60$$

$$\frac{1}{2}x+\frac{9}{8}y=-2$$
$$8\left(\frac{1}{2}x+\frac{9}{8}y\right)=8(-2)$$
$$4x+9y=-16$$
$$-2(4x+9y)=-2(-16)$$
$$-8x-18y=32$$

Add the equations.

$$\begin{array}{r} -15x+18y=60 \\ +\quad -8x-18y=32 \\ \hline -23x=92 \end{array}$$
$$x=-4$$

Substitute $x=-4$ into

$$-5x+6y=20$$
$$-5(-4)+6y=20$$
$$20+6y=20$$
$$6y=0$$
$$y=0 \qquad (-4,\,0)$$

27) $(2,\,2)$

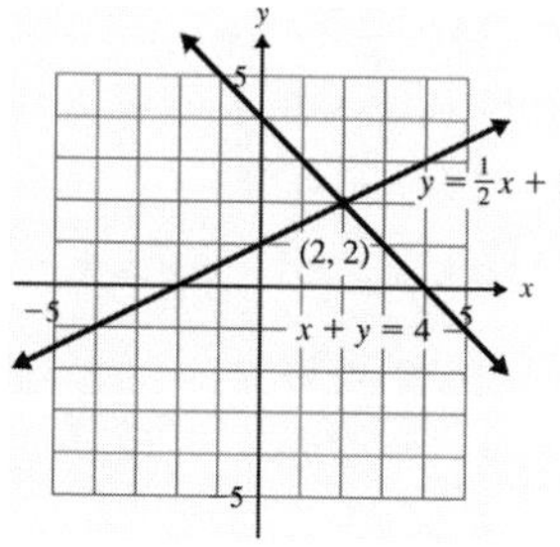

29) $(-2,\,-1)$

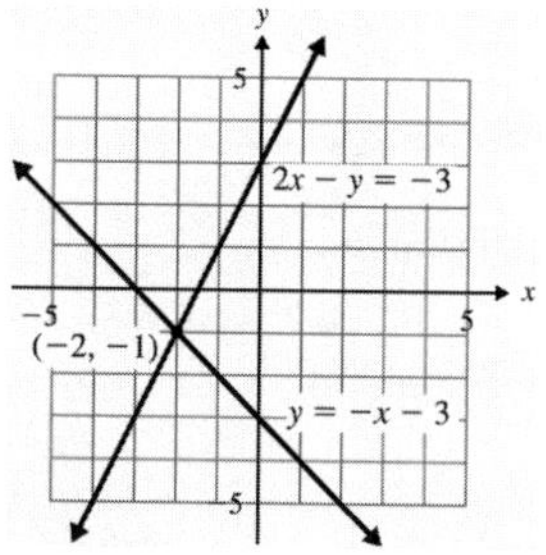

31) $(5,\,-3)$

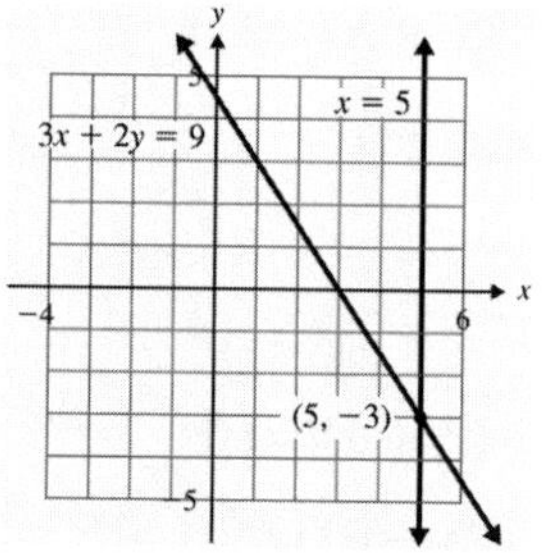

Section 5.4 Exercises

1) Let $x =$ one number
$y =$ other number
The sum of the numbers is 36
$$x + y = 36$$
One number is two more than the other number
$$x = 2 + y$$
The system of equations is $x + y = 36$
$x = 2 + y$

Substitute $x = 2 + y$ into $x + y = 36$ and solve.
$$(2 + y) + y = 36$$
$$2 + 2y = 36$$
$$2y = 34$$
$$y = 17$$
Substitute $y = 17$ into $x = 2 + y$
$$x = 2 + 17$$
$$x = 19$$
The numbers are 19 and 17.

3) Let $x =$ number of *Aviator* Awards, and
let $y =$ number of *Neverland* Awards.
Aviator $= 4 +$ *Neverland*
$$x = 4 + y$$
Aviator + *Neverland* $= 18$
$$x + y = 18$$
Use substitution.
$$(4 + y) + y = 18$$
$$4 + 2y = 18$$
$$2y = 14$$
$$y = 7$$
$$x = 4 + y;\ y = 7$$
$$x = 7 + 4$$
$$x = 11$$
The Aviator, 11; *Finding Neverland*, 7

5) Let $x =$ number of IHOPs, and
let $y =$ number of Waffle Houses.
IHOP + Waffle = 2626
$$x + y = 2626$$
IHOP = Waffle − 314
$$x = y - 314$$

Use substitution.

$(y-314)+y=2626$

$2y-314=2626$

$2y=2940$

$y=1470$

$x=y-314;\ y=1470$

$x=1470-314$

$x=1156$

IHOP: 1156; Waffle House: 1470

7) Let $x=$ number of people in 1939, and let $y=$ number of people in 2004.

people in 1939 + 38,968 = people in 2004

$x+38{,}968=y$

people in 1939 + people in 2004 = 49,968

$x+y=49{,}968$

Use substitution.

$x+(x+38{,}968)=49{,}968$

$2x+38{,}968=49{,}968$

$2x=11{,}000$

$x=5500$

$y=x+38{,}968;\ x=5500$

$y=5500+38{,}968$

$y=44{,}468$

1939: 5500; 2004: 44,468

9) Let $x=$ pounds of chicken, and let $y=$ pounds of beef.

chicken = beef − 6.3

$x=y-6.3$

chicken + beef = 120.5

$x+y=120.5$

Use substitution.

$y+(y-6.3)=120.5$

$2y-6.3=120.5$

$2y=126.8$

$y=63.4$

$x = y - 6.3;\ y = 63.4$

$x = 63.4 - 6.3$

$x = 57.1$

beef: 63.4 lb; chicken: 57.1 lb

11) Let $w =$ the width, and

let $l =$ the length.

length $= 2 \cdot$ width

$l \quad = 2 \cdot \quad w$

$2l + 2w = 78$

Use substitution.

$2(2w) + 2w = 78$

$4w + 2w = 78$

$6w = 78$

$w = 13$

$l = 2w;\ w = 13$

$l = 2(13)$

$l = 26$

length: 26 in; width: 13 in

13) Let $w =$ the width, and

let $h =$ the height.

$2h + 2w = 220$

width = height − 50

$w \quad = \quad h \quad - 50$

Use substitution.

$2h + 2(h - 50) = 220$

$2h + 2h - 100 = 220$

$4h - 100 = 220$

$4h = 320$

$h = 80$

$w = h - 50;\ h = 80$

$w = 80 - 50$

$w = 30$

height: 80 in; width: 30 in

15) Let $w =$ the width, and

let $l =$ the length.

$2l + 2w = 28$

length $= 4 +$ width

$l \quad = 4 + \quad w$

Use substitution.

$2(4 + w) + 2w = 28$

$8 + 2w + 2w = 28$

$8 + 4w = 28$

$4w = 20$

$w = 5$

$l = 4 + w;\ w = 5$

$l = 4 + 5$

$l = 9$

length: 9 cm; width: 5 cm

17) $x^\circ = \frac{2}{3} y^\circ$

supplementary angles

$x^\circ + y^\circ = 180^\circ$

Use substitution.

$\frac{2}{3} y + y = 180$

$\frac{5}{3} y = 180$

$y = 108$

$x = \frac{2}{3} y;\ y = 108$

$x = \frac{2}{3}(108)$

$x = 72$

$m\angle x = 72^\circ;\ m\angle y = 108^\circ$

19) Let $x =$ cost of a *Marc Anthony* ticket and let $y =$ cost of a *Santana* ticket.

Jennifer's Purchase:

$5 \cdot \textit{Marc Anthony} + 2 \cdot \textit{Santana} = 563$

$5x \quad + \quad 2y \quad = 563$

Carlos's Purchase:

$3 \cdot \textit{Marc Anthony} + 6 \cdot \textit{Santana} = 657$

$3x \quad + \quad 6y \quad = 657$

$$5x+2y=563$$
$$-3(5x+2y)=-3(563)$$
$$-15x-6y=-1689$$

Add the equations.

$$\begin{array}{r} -15x-6y=-1689 \\ +\ \ 3x+6y=\ \ 657 \\ \hline -12x=-1032 \end{array}$$
$$x=86$$

Substitute $x=86$ into

$$5x+2y=563$$
$$5(86)+2y=563$$
$$430+2y=563$$
$$2y=133$$
$$y=66.5$$

Marc Anthony: \$86; *Santana*: \$66.50

21) Let x = cost of a two-item meal
let y = cost of a three-item meal

$$3\cdot\text{two-item}+1\cdot\text{three-item}=21.96$$
$$3x \quad + \quad y \quad =21.96$$
$$2\cdot\text{two-item}+2\cdot\text{three-item}=23.16$$
$$2x \quad + \quad 2y \quad =23.16$$

$$3x+y=21.96$$
$$y=21.96-3x$$

Use substitution.

$$2x+2(21.96-3x)=23.16$$
$$2x+43.92-6x=23.16$$
$$-4x+43.92=23.16$$
$$-4x=-20.76$$
$$x=5.19$$
$$y=21.96-3x;\ x=5.19$$
$$y=21.96-3(5.19)$$
$$y=21.96-15.57$$
$$y=6.39$$

two-item: \$5.19; three-item: \$6.39

23) Let x = cost of a key chain, and
let y = cost of a postcard.

$$3\cdot\text{key chain}+5\cdot\text{postcard}=10.00$$
$$3x \quad + \quad 5y \quad =10.00$$
$$2\cdot\text{key chain}+3\cdot\text{postcard}=6.50$$
$$2x \quad + \quad 3y \quad =6.50$$

$$3x+5y=10.00$$
$$-2(3x+5y)=-2(10.00)$$
$$-6x-10y=-20.00$$
$$2x+3y=6.50$$
$$3(2x+3y)=3(6.50)$$
$$6x+9y=19.50$$

Add the equations.

$$\begin{array}{r} -6x-10y=-20.00 \\ +\ \ 6x+\ 9y=\ \ 19.50 \\ \hline -y=-0.50 \end{array}$$
$$y=0.50$$

Substitute $y=0.50$ into

$$3x+5y=10.00$$
$$3x+5(0.50)=10.00$$
$$3x+2.50=10.00$$
$$3x=7.50$$
$$x=2.50$$

key chain : \$2.50; postcard: \$0.50

25) Let x = cost of a cantaloupe
let y = cost of a watermelon

$$3\cdot\text{cantaloupe}+1\cdot\text{watermelon}=7.50$$
$$3x \quad + \quad y \quad =7.50$$
$$2\cdot\text{cantaloupe}+2\cdot\text{watermelon}=9.00$$
$$2x \quad + \quad 2y \quad =9.00$$

$$3x+y=7.50$$
$$y=7.50-3x$$

Use substitution.

$$2x+2(7.50-3x)=9.00$$
$$2x+15-6x=9.00$$
$$15.00-4x=9.00$$
$$-4x=-6.00$$
$$x=1.50$$

$y=7.50-3x;\ x=1.50$

$y=7.50-3(1.50)$

$y=7.50-4.50$

$y=3.00$

cantaloupe: \$1.50;

watermelon: \$3.00

27) Let x = cost of a hamburger
let y = cost of a small fry
$6\cdot\text{hamburger}+1\cdot\text{small fry}=3.91$
$$6x \quad + \quad y \quad =3.91$$
$8\cdot\text{hamburger}+2\cdot\text{small fry}=5.94$
$$8x \quad + \quad 2y \quad =5.94$$

$6x+y=3.91$
$y=3.91-6x$
Use substitution.
$$8x+2(3.91-6x)=5.94$$
$$8x+7.82-12x=5.94$$
$$-4x+7.82=5.94$$
$$-4x=-1.88$$
$$x=0.47$$
$y=3.91-6x;\ x=0.47$

$y=3.91-6(0.47)$

$y=3.91-2.82$

$y=1.09$

hamburger: \$0.47; small fry: \$1.09

29) x = number of ounces of 9% solution
y = number of ounces of 17% solution

Solution	Concentration	Number of ounces of solution	Number of ounces of alcohol in the solution
9%	0.09	x	$0.09x$
17%	0.17	y	$0.17y$
15%	0.15	12	$0.15(12)$

$x+y=12$ $\qquad$ $0.09x+0.17y=0.15(12)$

$y=12-x$ $\qquad$ $100(0.09x+0.17y)=100[0.15(12)]$

$9x+17y=15(12)$

Use substitution.

$9x+17(12-x)=15(12)$

$9x+204-17x=180$ $\qquad$ $y=12-x;\ x=3$

$-8x+204=180$ $\qquad$ $y=12-3$

$-8x=-24$ $\qquad$ $y=9$

$x=3$

9% : 3 oz; 17% : 9 oz

31) $x=$ number of pounds of peanuts

$y=$ number of pounds of cashews

Nuts	Price per Pound	Number of lbs of Nuts	Value
cashews	\$1.80	x	$1.80x$
pistachios	\$4.50	y	$4.50y$
mixture	\$2.61	10	$2.61(10)$

$x+y=10$ $\qquad$ $1.80x+4.50y=2.61(10)$

$y=10-x$ $\qquad$ $100(1.80x+4.50y)=100[2.61(10)]$

$180x+450y=261(10)$

Use substitution.

$180x+450(10-x)=261(10)$

$180x+4500-450x=2610$ $\qquad$ $y=10-x;\ x=7$

$-270x+4500=2610$ $\qquad$ $y=10-7$

$-270x=-1890$ $\qquad$ $y=3$

$x=7$

peanuts : 7 lb; cashews : 3 lb

33) $x =$ amount Sally invested in the 3% account.
$y =$ amount Sally invested in the 5% account.

$x + y = 4000$

$y = 4000 - x$

Total Interest Earned = Interest from 3% account + Interest from 5% account

$$144 \quad = \quad x(0.03)(1) \quad + \quad y(0.05)(1)$$

$$100(144) = 100\left[x(0.03)(1) + y(0.05)(1)\right]$$

$$14,400 = 3x + 5y$$

Use substitution.

$14,400 = 3x + 5(4000 - x)$

$14,400 = 3x + 20,000 - 5x$ $y = 4000 - x;\ x = 2800$

$14,400 = -2x + 20,000$ $y = 4000 - 2800$

$-5600 = -2x$ $y = 1200$

$2800 = x$

Sally invested \$2800 in the 3% account and \$1200 in the 5% account.

35) $q =$ number of quarters; $d =$ number of dimes

$q + d = 110$

$q = 110 - d$

Value of Quarters + Value of Dimes = Total Value

$$0.25q \quad + \quad 0.10d \quad = \quad 18.80$$

$$100(0.25q + 0.10d) = 100(18.80)$$

$$25q + 10d = 1880$$

Use substituion.

$25(110 - d) + 10d = 1880$

$2750 - 25d + 10d = 1880$ $q = 110 - d;\ d = 58$

$2750 - 15d = 1880$ $q = 110 - 58$

$-15d = -870$ $q = 52$

$d = 58$

There are 52 quarters and 58 dimes.

37) $x =$ number of liters of 100% solution
$y =$ number of liters of 10% solution

Solution	Concentration	Number of liters of solution	Number of liters of acid in the solution
100%	1.00	x	$1.00x$
10%	0.10	y	$0.10y$
40%	0.40	12	$0.40(12)$

$x+y=12$

$y=12-x$

$1.00x+0.10y=0.40(12)$

$10(1.00x+0.10y)=10[0.40(12)]$

$10x+y=4(12)$

Use substitution.

$10x+(12-x)=4(12)$

$9x+12=48$

$9x=36$

$x=4$

$y=12-x;\ x=4$

$y=12-4$

$y=8$

4 liters of pure acid; 8 liters of 10% solution

39) Car's distance + Truck's distance = 330

	d	r	t
Car	$3x$	x	3
Truck	$3y$	y	3

$y=x-10$ $\qquad 3x+3y=220$

Use substitution.

$3x+3(x-10)=330$

$3x+3x-30=330$

$6x-30=330$

$6x=360$

$x=60$

$y=x-10;\ x=60$

$y=60-10$

$y=50$

Car: 60 mph; Truck: 50 mph

41) Walk time = Bike time

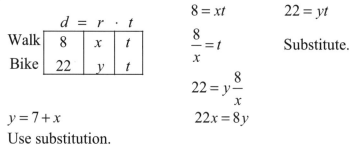

	d	r	t
Walk	8	x	t
Bike	22	y	t

$8=xt$ $\qquad 22=yt$

$\frac{8}{x}=t$ $\qquad$ Substitute.

$22=y\frac{8}{x}$

$22x=8y$

$y=7+x$

Use substitution.

$$22x = 8(7 + x)$$
$$22x = 56 + 8x$$
$$14x = 56$$
$$x = 4$$

$$y = 7 + x;\ x = 4$$
$$y = 7 + 4$$
$$y = 11$$

walking: 4 mph; biking: 11 mph

43) Nick's distance + Scott's distance = 13

	d	$=\ r$	$\cdot\ t$
Nick	$0.5x$	x	0.5
Scott	$0.5y$	y	0.5

$y = x - 2$ $\qquad$ $0.5x + 0.5y = 13$

Use substitution.

$$0.5x + 0.5(x - 2) = 13$$
$$0.5x + 0.5x - 1 = 13$$
$$x = 14$$

$$y = x - 2;\ x = 14$$
$$y = 14 - 2$$
$$y = 12$$

Nick: 14 mph; Scott: 12 mph

Section 5.5 Exercises

1)
$$4x + 3y - 7z = -6$$
$$4(-2) + 3(3) - 7(1) = -6$$
$$-8 + 9 - 7 = -6$$
$$-6 = -6$$
$$x - 2y + 5z = -3$$
$$(-2) - 2(3) + 5(1) = -3$$
$$-2 - 6 + 5 = -3$$
$$-3 = -3$$
$$-x + y + 2z = 7$$
$$-(-2) + (3) + 2(1) = 7$$
$$2 + 3 + 2 = 7$$
$$7 = 7$$

The ordered triple is a solution of the system.

3)
$$-x + y - 2z = 2$$
$$-(0) + (6) - 2(2) = 2$$
$$6 - 4 = 2$$
$$2 = 2$$
$$3x - y + 5z = 4$$
$$3(0) - (6) + 5(2) = 4$$
$$0 - 6 + 10 = 4$$
$$4 = 4$$
$$2x + 3y - z = 7$$
$$2(0) + 3(6) - (2) = 7$$
$$0 + 18 - 2 = 7$$
$$16 \neq 7$$

The ordered triple is not a solution of the system.

5) I $x+3y+z=3$

II $4x-2y+3z=7$

III $-2x+y-z=-1$

Add $I\cdot(-3)$ and II

$$\begin{array}{r} -3x-9y-3z=-9 \\ +\underline{\quad 4x-2y+3z=7} \\ \text{A} \quad x-11y \quad =-2 \end{array}$$

Add I and III

$$\begin{array}{r} x+3y+z=3 \\ +\underline{\quad -2x+y-z=-1} \\ \text{B} \quad -x+4y \quad =2 \end{array}$$

Add A and B

$$\begin{array}{r} x-11y=-2 \\ +\underline{\quad -x+4y=2} \\ -7y=0 \\ y=0 \end{array}$$

Substitute $y=0$ into B

$$\begin{aligned} -x+4(0)&=2 \\ -x&=2 \\ x&=-2 \end{aligned}$$

Substitute $x=-2$ and $y=0$ into I

$$\begin{aligned} (-2)+3(0)+z&=3 \\ -2+0+z&=3 \\ z&=5 \end{aligned}$$

$(-2,0,5)$

7) I $5x+3y-z=-2$

II $-2x+3y+2z=3$

III $x+6y+z=-1$

Add $I\cdot(-1)$ and II

$$\begin{array}{r} -5x-3y+z=2 \\ +\underline{\quad -2x+3y+2z=3} \\ \text{A} \quad -7x+3z=5 \end{array}$$

Add $I\cdot(-2)$ and III

$$\begin{array}{r} -10x-6y+2z=4 \\ +\underline{\quad x+6y+z=-1} \\ \text{B} \quad -9x+3z=3 \end{array}$$

Add A and $B\cdot(-1)$

$$\begin{array}{r} -7x+3z=5 \\ +\underline{\quad 9x-3z=-3} \\ 2x=2 \\ x=1 \end{array}$$

Substitute $x=1$ into B

$$\begin{aligned} -9(1)+3z&=3 \\ -9+3z&=3 \\ 3z&=12 \\ z&=4 \end{aligned}$$

Substitute $x=1$ and $z=4$ into III

$$\begin{aligned} (1)+6y+(4)&=-1 \\ 6y+5&=-1 \\ 6y&=-6 \\ y&=-1 \end{aligned}$$

$(1,-1,4)$

9) I $3a+5b-3c=-4$

II $a-3b+c=6$

III $-4a+6b+2c=-6$

Add I and $II\cdot(-3)$

$$\begin{array}{r} 3a+5b-3c=-4 \\ +\underline{\quad -3a+9b-3c=-18} \\ \text{A} \quad 14b-6c=-22 \end{array}$$

Add $II\cdot(4)$ and III

$$\begin{array}{r} 4a-12b+4c=24 \\ +\underline{\quad -4a+6b+2c=-6} \\ \text{B} \quad -6b+6c=18 \end{array}$$

Add A and B

$$\begin{array}{r} 14b-6c=-22 \\ \underline{-6b+6c=18} \\ 8b=-4 \\ b=-\frac{1}{2} \end{array}$$

Substitute $b = -\frac{1}{2}$ into B

$$-6\left(-\frac{1}{2}\right) + 6c = 18$$
$$3 + 6c = 18$$
$$6c = 15$$
$$c = \frac{15}{6} = \frac{5}{2}$$

Substitute $b = -\frac{1}{2}$ and $c = \frac{5}{2}$ into III

$$-4a + 6\left(-\frac{1}{2}\right) + 2\left(\frac{5}{2}\right) = -6$$
$$-4a - 3 + 5 = -6$$
$$-4a = -8$$
$$a = 2$$

$$\left(2, -\frac{1}{2}, \frac{5}{2}\right)$$

11) I $a - 5b + c = -4$

II $3a + 2b - 4c = -3$

III $6a + 4b - 8c = 9$

Add $\text{I} \cdot (-3)$ and II

$$\begin{array}{rl} & -3a + 15b - 3c = 12 \\ + & \underline{3a + 2b - 4c = -3} \\ \text{A} & 17b - 7c = 9 \end{array}$$

Add $\text{I} \cdot (-6)$ and III

$$\begin{array}{rl} & -6a + 30b - 6c = 24 \\ + & \underline{6a + 4b - 8c = 9} \\ \text{B} & 34b - 14c = 33 \end{array}$$

Add $\text{A} \cdot (-2)$ and B

$$\begin{array}{rl} & -34b + 14c = -18 \\ + & \underline{34b - 14c = 33} \\ & 0 \neq 15 \end{array}$$

$\varnothing$

13) I $-15x - 3y + 9z = 3$

II $5x + y - 3z = -1$

III $10x + 2y - 6z = -2$

Equation $\text{II} = \text{I} \cdot (-3)$

Equation $\text{II} = \text{III} \cdot (2)$

$\{(x, y, z) \mid 5x + y - 3z = -1\}$

15) I $-3a + 12b - 9c = -3$

II $5a - 20b + 15c = 5$

III $-a + 4b - 3c = -1$

Equation $\text{III} = \text{I}/3$

Equation $\text{III} = \text{II}/5$

$\{(a, b, c) \mid -a + 4b - 3c = -1\}$

17) I $5x - 2y + z = -5$

II $x - y - 2z = 7$

III $4y + 3z = 5$

Add I and $\text{II} \cdot (-5)$

$$\begin{array}{rl} & 5x - 2y + z = -5 \\ + & \underline{-5x + 5y + 10z = -35} \\ \text{A} & 3y + 11z = -40 \end{array}$$

Add $\text{A} \cdot (4)$ and $\text{III} \cdot (-3)$

$$\begin{array}{rl} & 12y + 44z = -160 \\ + & \underline{-12y - 9z = -15} \\ & 35z = -175 \\ & z = -5 \end{array}$$

Substitute $z = -5$ into III

$$4y + 3(-5) = 5$$
$$4y - 15 = 5$$
$$4y = 20$$
$$y = 5$$

Substitute $y = 5$ and $z = -5$ into II

$$x - (5) - 2(-5) = 7$$
$$x - 5 + 10 = 7$$
$$x + 5 = 7$$
$$x = 2$$

$(2,5,-5)$

19) I $\quad a+15b=5$

II $\quad 4a+10b+c=-6$

III $\quad -2a-5b-2c=-3$

Add II and III$\cdot(2)$

$$\begin{array}{r} 4a+10b+\ c=-6 \\ +\ \underline{-4a-10b-4c=-6} \\ -3c=-12 \\ c=4 \end{array}$$

Add I$\cdot$(2) and III

$$\begin{array}{r} 2a+30b\qquad =10 \\ +\ \underline{-2a-5b-2c=-3} \\ \text{A}\qquad 25b-2c=7 \end{array}$$

Substitute $c=4$ into A

$$25b-2(4)=7$$
$$25b-8=7$$
$$25b=15$$
$$b=\frac{15}{25}=\frac{3}{5}$$

Substitute $b=\frac{3}{5}$ and $c=4$ into II

$$4a+10\left(\frac{3}{5}\right)+(4)=-6$$
$$4a+6+4=-6$$
$$4a=-16$$
$$a=-4$$

$\left(-4,\frac{3}{5},4\right)$

21) I $\quad x+2y+3z=4$

II $\quad -3x+y=-7$

III $\quad 4y+3z=-10$

Add II and III

$$\begin{array}{r} -3x+y\qquad =-7 \\ +\ \underline{\qquad 4y+3z=-10} \\ \text{A}\quad -3x+5y+3z=-17 \end{array}$$

Add I and A$\cdot(-1)$

$$\begin{array}{r} x+2y+3z=4 \\ +\ \underline{3x-5y-3z=17} \\ \text{B}\qquad 4x-3y=21 \end{array}$$

Add II$\cdot$(3) and B

$$\begin{array}{r} -9x+3y=-21 \\ +\ \underline{4x-3y=21} \\ -5x=0 \\ x=0 \end{array}$$

Substitute $x=0$ into II

$$-3(0)+y=-7$$
$$y=-7$$

Substitute $y=-7$ into III

$$4(-7)+3z=-10$$
$$-28+3z=-10$$
$$3z=18$$
$$z=6$$

$(0,-7,6)$

23) I $\quad -5x+z=-3$

II $\quad 4x-y=-1$

III $\quad 3y-7z=1$

Solve I for z.

$z=5x-3$

Substitute $z=5x-3$ into III

$$3y-7(5x-3)=1$$
$$3y-35x+21=1$$

A $\quad -35x+3y=-20$

Add II$\cdot$(3) and A

$$\begin{array}{r} 12x-3y=-3 \\ +\ \underline{-35x+3y=-20} \\ -23x=-23 \\ x=1 \end{array}$$

Substitute $x=1$ into I

$$-5(1)+z=-3$$
$$-5+z=-3$$
$$z=2$$

Substitute $x=1$ into II

$4(1)-y=-1$
$4-y=-1$
$-y=-5$
$y=5$
$(1,5,2)$

25) I $4a+2b=-11$
II $-8a-3c=-7$
III $b+2c=1$
Solve III for b.
$b=1-2c$
Substitute $b=1-2c$ into I
$4a+2(1-2c)=-11$
$4a+2-4c=-11$
A $4a-4c=-13$
Add II and A $\cdot(2)$
$-8a-3c=-7$
$+\ \underline{8a-8c=-26}$
$-11c=-33$
$c=3$
Substitute $c=3$ into II
$-8a-3(3)=-7$
$-8a-9=-7$
$-8a=2$
$a=-\frac{2}{8}=-\frac{1}{4}$
Substitute $c=3$ into III
$b+2(3)=1$
$b+6=1$
$b=-5$
$\left(-\frac{1}{4},-5,3\right)$

27) I $6x+3y-3z=-1$
II $10x+5y-5z=4$
III $x-3y+4z=6$
Add I and III.
$6x+3y-3z=-1$
$+\ \underline{x-3y+4z=6}$
A $7x+z=5$
Add II $\cdot(3)$ and III $\cdot(5)$
$30x+15y-15z=12$
$+\ \underline{5x-15y+20z=30}$
B $35x+5z=42$
Add A $\cdot(-5)$ and B.
$-35x-5z=-25$
$+\ \underline{35x+5z=42}$
$0\neq 17$
$\varnothing$

29) I $7x+8y-z=16$
II $-\frac{1}{2}x-2y+\frac{3}{2}z=1$
III $\frac{4}{3}x+4y-3z=-\frac{2}{3}$
Add I and II $\cdot(14)$
$7x+8y-z=16$
$+\ \underline{-7x-28y+21z=14}$
A $-20y+20z=30$
Add II $\cdot(8)$ and III $\cdot(3)$
$-4x-16y+12z=8$
$+\ \underline{4x+12y-9z=-2}$
B $-4y+3z=6$
Add A and B $\cdot(-5)$
$-20y+20z=30$
$+\ \underline{20y-15z=-30}$
$5z=0$
$z=0$
Substitute $z=0$ into A
$-4y+3(0)=6$
$-4y=6$
$y=-\frac{6}{4}=-\frac{3}{2}$

Substitute $y=-\frac{3}{2}$ and $z=0$ into I

$7x+8\left(-\frac{3}{2}\right)-0=16$

$7x-12=16$

$7x=28$

$x=4$

$\left(4,-\frac{3}{2},0\right)$

31) I $2a-3b=-4$

II $3b-c=8$

III $-5a+4c=-4$

Solve II for c.

$c=3b-8$

Substitute $c=3b-8$ into III

$-5a+4(3b-8)=-4$

$-5a+12b-32=-4$

A $-5a+12b=28$

Add $\text{I}\cdot(4)$ and A

$8a-12b=-16$

$+\ \underline{-5a+12b=28}$

$3a=12$

$a=4$

Substitute $a=4$ into I

$2(4)-3b=-4$

$8-3b=-4$

$-3b=-12$

$b=4$

Substitute $a=4$ into III

$-5(4)+4c=-4$

$-20+4c=-4$

$4c=16$

$c=4$

$(4,4,4)$

33) I $-4x+6y+3z=3$

II $-\frac{2}{3}x+y+\frac{1}{2}z=\frac{1}{2}$

III $12x-18y-9z=-9$

Equation I = II / 6

Equation I = III $\cdot(-3)$

$\{(x,y,z)\,|\,-4x+6y+3z=3\}$

35) I $a+b+9c=-3$

II $-5a-2b+3c=10$

III $4a+3b+6c=-15$

Add $\text{I}\cdot(5)$ and II

$5a+5b+45c=-15$

$+\ \underline{-5a-2b+3c=10}$

A $3b+48c=-5$

Add $\text{I}\cdot(-4)$ and III

$-4a-4b-36c=12$

$+\ \underline{4a+3b+6c=-15}$

B $-b-30c=-3$

Add A and $\text{B}\cdot(3)$

$3b+48c=-5$

$+\ \underline{-3b-90c=-9}$

$-42c=-14$

$c=\frac{-14}{-42}=\frac{1}{3}$

Substitute $c=\frac{1}{3}$ into B

$-b-30\left(\frac{1}{3}\right)=-3$

$-b-10=-3$

$-b=7$

$b=-7$

Substitute $b=-7$ and $c=\frac{1}{3}$ into I

$a+(-7)+9\left(\frac{1}{3}\right)=-3$

$a-7+3=-3$

$a-4=-3$

$a=1$

$\left(1,-7,\frac{1}{3}\right)$

37) I $2x-y+4z=-1$

II $x+3y+z=-5$

III $-3x+2y=7$

Add I and II $\cdot(-4)$

$$\begin{array}{rr} & 2x-\ y\ +4z=-1 \\ + & \underline{-4x-12y-4z=20} \\ \text{A} & -2x-13y=19 \end{array}$$

Add III $\cdot$ (2) and A $\cdot(-3)$

$$\begin{array}{rr} & -6x+4y=\ 14 \\ + & \underline{6x+39y=-57} \\ & 43y=-43 \\ & y=-1 \end{array}$$

Substitute $y=-1$ into III

$-3x+2(-1)=7$

$-3x-2=7$

$-3x=9$

$x=-3$

Substitute $x=-3$ and $y=-1$ into II

$(-3)+3(-1)+z=-5$

$-3-3+z=-5$

$-6+z=-5$

$z=1$

$(-3,-1,1)$

39) $x=$ cost of a hot dog

$y=$ cost of an order of fries

$z=$ cost of a large soda

Use the information from the problem to get the following system of equations.

M $2x+2y+z=9.00$

L $2x+y+2z=9.50$

C $3x+2y+z=11.00$

Add M and L $\cdot(-2)$

$$\begin{array}{rr} & 2x+2y+z=9.00 \\ + & \underline{-4x-2y-4z=-19.00} \\ \text{A} & -2x-3z=-10.00 \end{array}$$

Add L $\cdot(-2)$ and C

$$\begin{array}{rr} & -4x-2y-4z=-19.00 \\ + & \underline{3x+2y+z=11.00} \\ \text{B} & -x-3z=-8.00 \end{array}$$

Add A and B $\cdot(-1)$

$$\begin{array}{rr} & -2x-3z=-10.00 \\ + & \underline{x+3z=\ 8.00} \\ & -x=-2.00 \\ & x=2.00 \end{array}$$

Substitute $x=2.00$ into A

$-2(2.00)-3z=-10.00$

$-4.00-3z=-10.00$

$-3z=-6.00$

$z=2.00$

Substitute $x=2.00$ and $y=2.00$ into L

$2(2.00)+y+2(2.00)=9.50$

$4.00+y+4.00=9.50$

$y+8.00=9.50$

$y=1.50$

hot dog: \$2.00, fries: \$1.50, soda: \$2.00

41) c = grams of protein in a Clif Bar.
b = grams of protein in a Balance Bar.
p = grams of protein in a PowerBar.

Use the information from the problem to get the following system of equations.

I $\quad c = b - 4$

II $\quad p = b + 9$

III $\quad c + b + p = 50$

Substitute I and II into III

$$(b-4)+b+(b+9)=50$$
$$3b+5=50$$
$$3b=45$$
$$b=15$$

Substitute $b=15$ into I

$c=15-4$

$c=11$

Substitute $b=15$ into II

$p=15+9$

$p=24$

Clif Bar: 11g, Balance Bar: 15g, PowerBar: 24g

43) k = revenue collected by the Knicks.
l = revenue collected by the Lakers.
b = revenue collected by the Bulls.
Use the information from the problem to get the following system of equations.

I $\quad k + l + b = 428$

II $\quad l = b + 30$

III $\quad k = l + 11$

Solve III for l

$l = k - 11$

Substitute $l = k - 11$ into II

$k - 11 = b + 30$

A $\quad k = b + 41$

Substitute II and A into I

$$(b+41)+(b+30)+b=428$$
$$3b+71=428$$
$$3b=357$$
$$b=119$$

Substitute $b=119$ into II

$l=119+30$

$l=149$

Substitute $l=149$ into III

$k=149+11$

$k=160$

Knicks: $160 million, Lakers: $149 million, Bulls: $119 million

45) v = cost of a value date
r = cost of a regular date
p = cost of a prime date
Use the information from the problem to get the following system of equations.

I $\quad 4v+4r+3p=286$

II $\quad 4v+3r+2p=220$

III $\quad 3v+3r+p=167$

Add I and III $\cdot(-3)$

$$\begin{array}{rr} & 4v+4r+3p=286 \\ + & \underline{-9v-9r-3p=-501} \\ \text{A} & -5v-5r=-215 \end{array}$$

Add II and III $\cdot(-2)$

$$\begin{array}{rr} & 4v+3r+2p=220 \\ + & \underline{-6v-6r-2p=-334} \\ \text{B} & -2v-3r=-114 \end{array}$$

Add A $\cdot$ (2) and B $\cdot(-5)$

$$\begin{array}{r} -10v-10r=-430 \\ \underline{10v+15r=570} \\ 5r=140 \\ r=28 \end{array}$$

Substitute $r=28$ into B

$$-2v-3(28)=-114$$
$$-2v-84=-114$$
$$-2v=-30$$
$$v=15$$

Substitute $v=15$ and $p=28$ into III

$$3(15)+3(28)+p=167$$
$$45+84+p=167$$
$$p+129=167$$
$$p=38$$

Value: \$15, Regular: \$28,

Prime: \$38

47) $l=$ measure of the largest angle.

$m=$ measure of the middle angle.

$s=$ measure of the smallest angle.

Use the information from the problem to get the following system of equations.

I $\quad l=2m$

II $\quad s=m-28$

III $\quad l+m+s=180$

Substitute I and II into III

$$(2m)+m+(m-28)=180$$
$$4m-28=180$$
$$4m=208$$
$$m=52$$

Substitute $m=52$ into I

$$l=2(52)=104$$

Substitute $m=52$ into II

$$s=52-28=24$$

$104^\circ,\ 52^\circ,\ 24^\circ$

49) $l=$ measure of the largest angle.

$m=$ measure of the middle angle.

$s=$ measure of the smallest angle.

Use the information from the problem to get the following system of equations.

I $\quad s=l-44$

II $\quad s+m=l+20$

III $\quad l+m+s=180$

Add II$\cdot(-1)$ and III

$$l-m-s=-20$$
$$+\quad \underline{l+m+s=180}$$
$$2l=160$$
$$l=80$$

Substitute $l=80$ into I

$$s=80-44=36$$

Substitute $l=80$ and $s=36$ into III

$$80+m+36=180$$
$$m+116=180$$
$$m=64$$

$80^\circ,\ 64^\circ,\ 36^\circ$

51) $l=$ length of the longest side

$m=$ length of the medium side

$s=$ length of the shortest side

Use the information from the problem to get the following system of equations.

I $\quad l+m+s=29$

II $\quad l=s+5$

III $\quad m+s=l+5$

Add I and III$\cdot(-1)$

$$l+m+s=29$$
$$+\quad \underline{l-m-s=-5}$$
$$2l=24$$
$$l=12$$

Substitute $l=12$ into II

$$12=s+5$$
$$7=s$$

Substitute $l=12$ and $s=7$ into I

$$12+m+7=29$$
$$m+19=29$$
$$m=10$$

12 cm, 10 cm, 7 cm

Chapter 5 Review

1) No

$$-2x+y=3$$
$$-2(-4)+(-5)=3$$
$$8-5=3$$
$$3=3$$

$$3x-y=-17$$
$$3(-4)-(-5)=-17$$
$$-12+5=-17$$
$$-7\neq-17$$

3) Yes

$$3x+4y=0$$
$$3\left(\frac{2}{3}\right)+4\left(-\frac{1}{2}\right)=0$$
$$2-2=-6$$
$$0=0$$

$$9x+2y=5$$
$$9\left(\frac{2}{3}\right)+2\left(-\frac{1}{2}\right)=5$$
$$6-1=5$$
$$5=5$$

5) (1, 1)

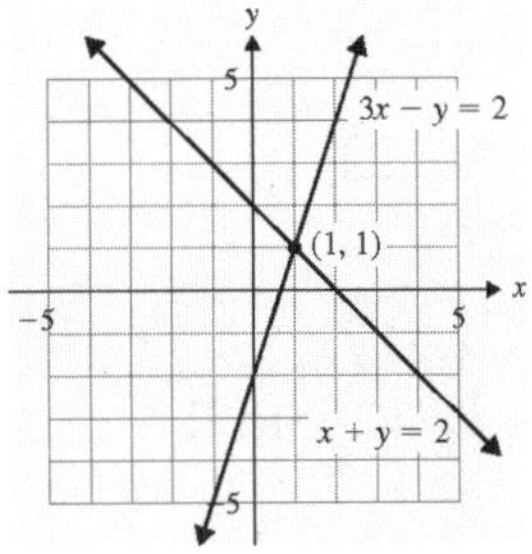

7) (−2, 3)

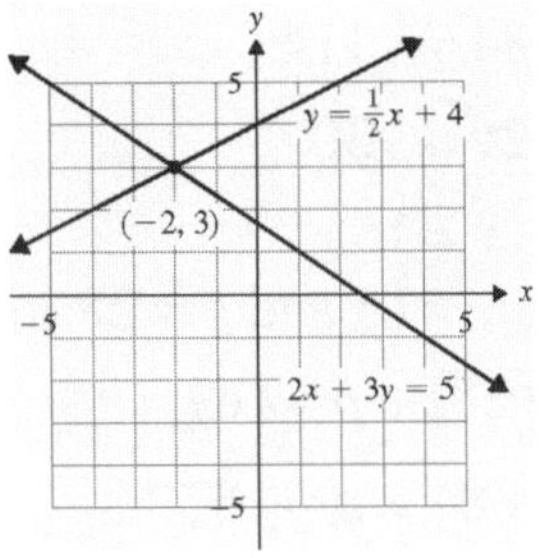

9) $x+7y=-3$

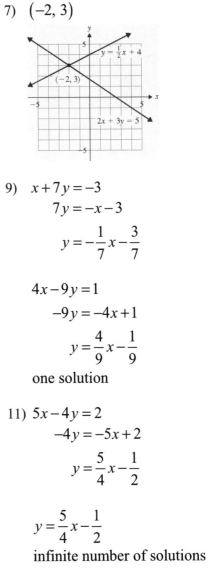

$$7y=-x-3$$
$$y=-\frac{1}{7}x-\frac{3}{7}$$

$$4x-9y=1$$
$$-9y=-4x+1$$
$$y=\frac{4}{9}x-\frac{1}{9}$$

one solution

11) $5x-4y=2$

$$-4y=-5x+2$$
$$y=\frac{5}{4}x-\frac{1}{2}$$

$$y=\frac{5}{4}x-\frac{1}{2}$$

infinite number of solutions

13) a) during the 4th quarter of 2003

b) During the 3rd quarter of 2003 approximately 89.1% of flights left on time.

15) Substitute $y=\frac{5}{6}x-2$ into

$$6y-5x=-12$$
$$6\left(\frac{5}{6}x-2\right)-5x=-12$$
$$5x-12-5x=-12$$
$$-12=-12$$

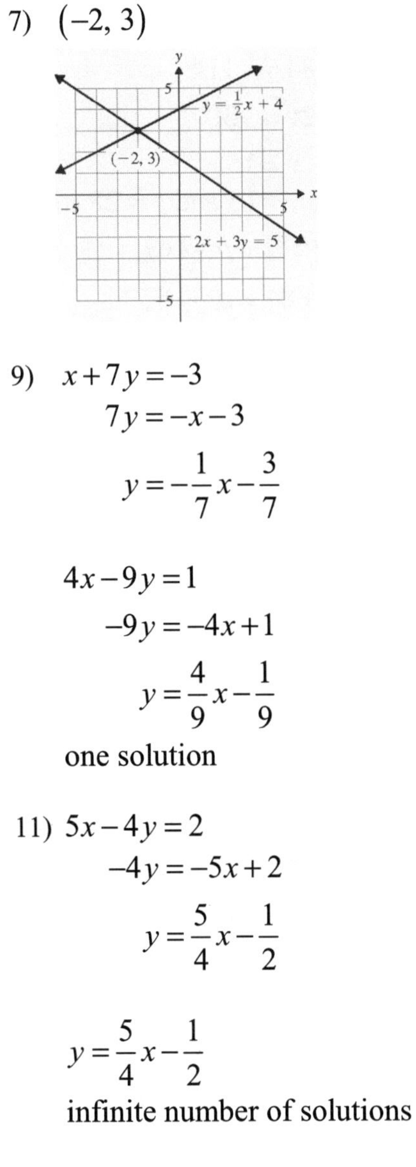

infinite number of solutions of the form $\left\{(x, y) \mid y=\frac{5}{6}x-2\right\}$

17) $6x-y=-3$

$6x+3=y$

Substitute $y=6x+3$ into

$$15x+2y=15$$
$$15x+2(6x+3)=15$$
$$15x+12x+6=15$$
$$27x+6=15$$
$$27x=9$$
$$x=\frac{9}{27}$$
$$x=\frac{1}{3}$$

Substitute $x=\frac{1}{3}$ into

$$y=6x+3$$
$$y=6\left(\frac{1}{3}\right)+3$$
$$y=2+3$$
$$y=5 \qquad \left(\frac{1}{3}, 5\right)$$

19) Substitute $x=20-8y$ into

$$x+8y=2$$
$$(20-8y)+8y=2$$
$$20 \neq 2$$

$\varnothing$

21) $2(2x-3)=y+3$

$$4x-6=y+3$$
$$4x-9=y$$
$$8+5(y-6)=8(x+3)-9x-y$$
$$8+5y-30=8x+24-9x-y$$
$$5y-22=-x+24-y$$
$$x+6y=46$$

Substitute $y=4x-9$ into

$$x+6y=46$$
$$x+6(4x-9)=46$$
$$x+24x-54=46$$
$$25x-54=46$$
$$25x=100$$
$$x=4$$

Substitute $x=4$ into

$$y=4x-9$$
$$y=4(4)-9$$
$$y=16-9$$
$$y=7 \qquad (4, 7)$$

23) Add the equations.

$$\begin{array}{r} x-y=-8 \\ +\ 3x+y=-12 \\ \hline 4x=-20 \end{array}$$
$$x=-5$$

Substitute $x=-5$ into

$$x-y=-8$$
$$(-5)-y=-8$$
$$-y=-3$$
$$y=3 \qquad (-5, 3)$$

25) $4x-5y=-16$

$$3(4x-5y)=3(-16)$$
$$12x-15y=-48$$

$$-3x+4y=13$$
$$4(-3x+4y)=4(13)$$
$$-12x+16y=52$$

Add the equations.

$$\begin{array}{r} 12x-15y=-48 \\ +\ -12x+16y=\ \ 52 \\ \hline y=4 \end{array}$$

Substitute $y=4$ into

$$4x-5y=-16$$
$$4x-5(4)=-16$$
$$4x-20=-16$$
$$4x=4$$
$$x=1 \qquad (1,\ 4)$$

27) $0.12x+0.01y=0.06$

$$100(0.12x+0.01y)=100(0.06)$$
$$12x+y=6$$
$$-2(12x+y)=-2(6)$$
$$-24x-2y=-12$$

$$0.5x+0.2y=-0.7$$
$$10(0.5x+0.2y)=10(-0.7)$$
$$5x+2y=-7$$

Add the equations.

$$\begin{array}{r} -24x-2y=-12 \\ +\ \ \ 5x+2y=-7 \\ \hline -19x=-19 \end{array}$$
$$x=1$$

Substitute $x=1$ into

$$12x+y=6$$
$$12(1)+y=6$$
$$12+y=6$$
$$y=-6 \qquad (1,\ -6)$$

29) $\frac{3}{4}x-y=\frac{1}{2}$

$$4\left(\frac{3}{4}x-y\right)=4\left(\frac{1}{2}\right)$$
$$3x-4y=2$$
$$2(3x-4y)=2(2)$$
$$6x-8y=4$$

$$-\frac{x}{3}+\frac{y}{2}=-\frac{1}{6}$$
$$6\left(-\frac{x}{3}+\frac{y}{2}\right)=6\left(-\frac{1}{6}\right)$$
$$-2x+3y=-1$$
$$3(-2x+3y)=3(-1)$$
$$-6x+9y=-3$$

Add the equations.

$$\begin{array}{r} 6x-8y=\ \ 4 \\ +\ -6x+9y=-3 \\ \hline y=1 \end{array}$$

Substitute $y=1$ into

$$3x-4y=2$$
$$3x-4(1)=2$$
$$3x-4=2$$
$$3x=6$$
$$x=2 \qquad (2,\ 1)$$

31) Eliminate y.

$$3x-y=13$$
$$7(3x-y)=7(13)$$
$$21x-7y=91$$

Add the equations.

$$\begin{array}{r} 2x+7y=-8 \\ +\ \ 21x-7y=91 \\ \hline 23x=83 \end{array}$$
$$x=\frac{83}{23}$$

Eliminate x.

$$2x+7y=-8$$
$$3(2x+7y)=3(-8)$$
$$6x+21y=-24$$
$$3x-y=13$$
$$-2(3x-y)=-2(13)$$
$$-6x+2y=-26$$

Add the equations.

$$\begin{array}{r} 6x+21y=-24 \\ +\ \ -6x+2y=-26 \\ \hline 23y=-50 \end{array}$$

$$y=-\frac{50}{23}$$

$$\left(\frac{83}{23},-\frac{50}{23}\right)$$

33) Let x = number of dogs, and let y = number of cats.
dogs = $2\cdot$ cats
$x = 2y$
dogs + cats = 51
$x + y = 51$
Use substitution.

$(2y)+y=51$ $x=2y;\ y=17$
$3y=51$ $x=2(17)$
$y=17$ $x=34$

34 dogs; 17 cats

35) Let x = the cost of a hot dog, and let y = the cost of a soda.
$4\cdot$ hot dog + $2\cdot$ soda = 26.50
$4x + 2y = 26.50$
$3\cdot$ hot dog + $4\cdot$ soda = 28.00
$3x + 4y = 28.00$

$$4x+2y=26.50$$
$$-2(4x+2y)=-2(26.50)$$
$$-8x-4y=-53.00$$

Add the equations.

$$\begin{array}{r} -8x-4y=-53.00 \\ +\ \ 3x+4y=\ \ 28.00 \\ \hline -5x=-25.00 \end{array}$$
$$x=5.00$$

Substitute $x=5.00$ into

$$4x+2y=26.50$$
$$4(5.00)+2y=26.50$$
$$20.00+2y=26.50$$
$$2y=6.50$$
$$y=3.25$$

hot dog : \$5.00; soda: \$3.25

37) Let w = the width, and let l = the length.
$P=2l+2w=66$
length = 3 + width
$l = 3 + w$
Use substitution.

$$2(3+w)+2w=66$$
$$6+2w+2w=66$$
$$6+4w=66$$
$$4w=60$$
$$w=15$$

$l=3+w;\ w=15$
$l=3+15$
$l=18$

width: 15 in; length: 18 in

39) $x =$ number of pounds of gummi bears
$y =$ number of pounds of jelly beans

Candy	Price per Pound	Number of lbs of Nuts	Value
bears	\$2.40	x	$2.40x$
beans	\$1.60	y	$1.60y$
mixture	\$1.92	10	$1.92(10)$

$x+y=10 \qquad 2.40x+1.60y=1.92(10)$

$y=10-x \qquad 100(2.40x+1.60y)=100[1.92(10)]$

$240x+160y=192(10)$

Use substitution.

$240x+160(10-x)=192(10)$

$240x+1600-160x=1920 \qquad y=10-x;\ x=4$

$80x+2600=1920 \qquad y=10-4$

$80x=320 \qquad y=6$

$x=4$

gummi bears : 4 lb; jelly beans : 6 lb

41) Car's distance + Bus' distance = 270

	d	$= r$	$\cdot\ t$
Car	$2.5x$	x	2.5
Bus	$2.5y$	y	2.5

$x=y+12$

$2.5x+2.5y=270$

Use substitution.

$02.5(y+12)+2.5y=270$

$2.5y+30+2.5y=270$

$5y+30=270$

$5y=240$

$y=48$

$x=y+12;\ y=48$

$x=48+12$

$x=60$

car: 60 mph; bus: 48 mph

43) $x-6y+4z=13$

$(-3)-6(-2)+4(1)=13$

$-3+12+4=13$

$13=13$

$5x+y+7z=8$

$5(-3)+(-2)+7(1)=8$

$-15-2+7=8$

$-10\neq 8$

$2x+3y-z=-5$

$2(-3)+3(-2)-(1)=-5$

$-6-6-1=-5$

$-13\neq -5$

The ordered triple is not a solution of the system.

45) I $2x-5y-2z=3$

II $x+2y+z=5$

III $-3x-y+2z=0$

Add I and II $\cdot(-2)$

$2x-5y-2z=3$

$+\ \underline{-2x-4y-2z=-10}$

A $-9y-4z=-7$

Add II $\cdot$ (3) and III

$3x+6y+3z=15$

$+\ \underline{-3x-y+2z=0}$

B $5y+5z=15$

Add A and B $\cdot(9/5)$

$-9y-4z=-7$

$\underline{9y+9z=27}$

$5z=20$

$z=4$

Substitute $z=4$ into B

$5y+5(4)=15$

$5y+20=15$

$5y=-5$

$y=-1$

Substitute $y=-1$ and $z=4$ into II

$x+2(-1)+(4)=5$

$x-2+4=5$

$x=3$

$(3,-1,4)$

47) I $5a-b+2c=-6$

II $-2a-3b+4c=-2$

III $a+6b-2c=10$

Add I $\cdot(-2)$ and II

$-10a+2b-4c=12$

$+\ \underline{-2a-3b+4c=-2}$

A $-12a-b=10$

Add I and III

$5a-b+2c=-6$

$+\ \underline{a+6b-2c=10}$

B $6a+5b=4$

Add A $\cdot$ (5) and B

$-60a-5b=50$

$+\ \underline{6a+5b=4}$

$-54a=54$

$a=-1$

Substitute $a=-1$ into A

$-12(-1)-b=10$

$12-b=10$

$-b=-2$

$b=2$

Substitute $a=-1$ and $b=2$ into I

$5(-1)-(2)+2c=-6$

$-5-2+2c=-6$

$-7+2c=-6$

$2c=1$

$c=\frac{1}{2}$

$\left(-1,\ 2,\ \frac{1}{2}\right)$

49) I $4x-9y+8z=2$

II $x+3y=5$

III $6y+10z=-1$

Add I and $\text{II}\cdot(-4)$

$$\begin{array}{rl} & 4x-9y+8z=2 \\ + & \underline{-4x-12y \quad =-20} \\ \text{A} & -21y+8z=-18 \end{array}$$

Add $\text{A}\cdot(-5)$ and III

$$\begin{array}{rl} & 105y-40z=90 \\ + & \underline{24y+40z=-4} \\ & 129y=86 \end{array}$$

$$y=\frac{86}{129}=\frac{2}{3}$$

Substitute $y=\frac{2}{3}$ into III

$$6\left(\frac{2}{3}\right)+10z=-1$$

$$4+10z=-1$$

$$10z=-5$$

$$z=-\frac{5}{10}=-\frac{1}{2}$$

Substitute $y=\frac{2}{3}$ into II

$$x+3\left(\frac{2}{3}\right)=5$$

$$x+2=5$$

$$x=3$$

$$\left(3, \frac{2}{3}, -\frac{1}{2}\right)$$

51) I $x+3y-z=0$

II $11x-4y+3z=8$

III $5x+15y-5z=1$

Add $\text{I}\cdot(-5)$ and III

$$\begin{array}{rl} & -5x-15y+5z=0 \\ + & \underline{5x+15y-5z=1} \\ & 0\neq 1 \end{array}$$

$\varnothing$

53) I $12a-8b+4c=8$

II $3a-2b+c=2$

III $-6a+4b-2c=-4$

Equation $\text{I}=\text{II}\cdot 4$

Equation $\text{I}=\text{III}\cdot(-2)$

$\{(a,b,c) \mid 3a-2b+c=2\}$

55) I $5y+2z=6$

II $-x+2y=-1$

III $4x-z=1$

Add $\text{II}\cdot(4)$ and III

$$\begin{array}{rl} & -4x+8y \quad =-4 \\ & \underline{4x \quad -z=1} \\ \text{A} & 8y-z=-3 \end{array}$$

Add $\text{A}\cdot(-1)$ and III

$$\begin{array}{rl} & 5y+2z=6 \\ + & \underline{16y-2z=-6} \\ & 21y=0 \end{array}$$

$$y=0$$

Substitute $y=0$ into I

$$5(0)+2z=6$$

$$2z=6$$

$$z=3$$

Substitute $y=0$ into II

$$-x+2(0)=-1$$

$$-x=-1$$

$$x=1$$

(1, 0, 3)

57) I $8x+z=7$

II $3y+2z=-4$

III $4x-y=5$

Add $\text{I}\cdot(-2)$ and II

$$\begin{array}{rl} & -16x \quad -2z = -14 \\ + & \underline{\quad 3y + 2z = -4} \\ \text{A} & -16x + 3y = -18 \end{array}$$

Add III $\cdot$ (3) and A

$$\begin{array}{rl} & 12x - 3y = -15 \\ + & \underline{-16x + 3y = -18} \\ & -4x = -3 \end{array}$$

$$x = \frac{3}{4}$$

Substitute $x = \frac{3}{4}$ into I

$$8\left(\frac{3}{4}\right) + z = 7$$

$$6 + z = 7$$

$$z = 1$$

Substitute $x = \frac{3}{4}$ into III

$$4\left(\frac{3}{4}\right) - y = 5$$

$$3 - y = 5$$

$$-y = 2$$

$$y = -2$$

$$\left(\frac{3}{4}, -2, 1\right)$$

59) $p =$ mg of sodium in a Powerade.
$r =$ mg of sodium in a Propel.
$g =$ mg of sodium in a Gatorade.
Use the information from the problem to get the following system of equations.

I $\quad p = r + 20$

II $\quad g = 2p$

III $\quad p + r + g = 200$

Subtract r in I to get equation A:

A $\quad p - r = 20$

Add A and III

$$\begin{array}{rl} & p - r \quad = 20 \\ & \underline{p + r + g = 200} \\ \text{B} & 2p + g = 220 \end{array}$$

Substitute II into B

$$2p + (2p) = 220$$

$$4p = 220$$

$$p = 55$$

Substitute $p = 55$ into II

$$g = 2(55) = 110$$

Substitute $p = 55$ into I

$$(55) = r + 20$$

$$35 = r$$

Propel: 35 mg, Powerade: 55 mg, Gatorade Extreme: 110 mg

61) $v =$ Verizon subscribers.
$c =$ Cingular subscribers.
$t =$ T-Mobile subscribers.
Use the information from the problem to get the following system of equations.

I $\quad v + c + t = 64.3$

II $\quad v = c + 10.5$

III $\quad t = c - 11.9$

Substitute II and III into I

$$(c + 10.5) + c + (c - 11.9) = 64.3$$

$$3c - 1.4 = 64.3$$

$$3c = 65.7$$

$$c = 21.9$$

Substitute $c = 21.9$ into II

$$v = (21.9) + 10.5$$

$$v = 32.4$$

Substitute $c = 21.9$ into III

$$t = (21.9) - 11.9$$

$$t = 10.0$$

Verizon: 32.4 million,
Cingular: 21.9 million,
T-Mobile: 10.0 million

63) $x =$ cost of a cone.
$y =$ cost of a shake.
$z =$ cost of a sundae.
Use the information from the problem to get the following system of equations.

$$\text{I} \quad 2x+3y+z=13.50$$
$$\text{II} \quad 3x+y+2z=13.00$$
$$\text{III} \quad 4x+y+z=11.50$$

Add $\text{I}\cdot(-2)$ and II

$$-4x-6y-2z=-27.00$$
$$+\quad \underline{3x+1y+2z=13.00}$$
$$\text{A} \quad -x-5y=-14.00$$

Add I and $\text{III}\cdot(-1)$

$$2x+3y+1z=13.50$$
$$+\quad \underline{-4x-1y-1z=-11.50}$$
$$\text{B} \quad -2x+2y=2.00$$

Add $\text{A}\cdot(-2)$ and B

$$2x+10y=28.00$$
$$+\quad \underline{-2x+2y=2.00}$$
$$12y=30.00$$
$$y=2.50$$

Substitute $y=2.50$ into A

$$-x-5(2.50)=-14.00$$
$$-x-12.50=-14.00$$
$$-x=-1.50$$
$$x=1.50$$

Substitute $x=1.50$ and $y=2.50$ into I

$$2(1.50)+3(2.50)+1z=13.50$$
$$3.00+7.50+z=13.50$$
$$z=3.00$$

ice cream cone: \$1.50, shake: \$2.50, sundae: \$3.00

65) $l =$ measure of the largest angle.
$m =$ measure of the middle angle.
$s =$ measure of the smallest angle.
Use the information from the problem to get the following system of equations.

$$\text{I} \quad s=\frac{1}{3}m$$
$$\text{II} \quad l=s+70$$
$$\text{III} \quad l+m+s=180$$

Solve I for m

$$s=\frac{1}{3}m$$
$$\text{A} \quad 3s=m$$

Substitute A and II into III

$$(s+70)+(3s)+s=180$$
$$5s+70=180$$
$$5s=110$$
$$s=22$$

Substitute $s=22$ into II

$$l=(22)+70$$
$$l=92$$

Substitute $s=22$ into A

$$3(22)=m$$
$$66=m$$

$92^\circ,\ 66^\circ,\ 22^\circ$

Chapter 5 Test

1) Yes

$$8x+y=1$$
$$8\left(\frac{3}{4}\right)+(-5)=1$$
$$6-5=1$$
$$1=1$$

$$-12x-4y=11$$
$$-12\left(\frac{3}{4}\right)-4(-5)=11$$
$$-9+20=11$$
$$11=11$$

3) $\varnothing$

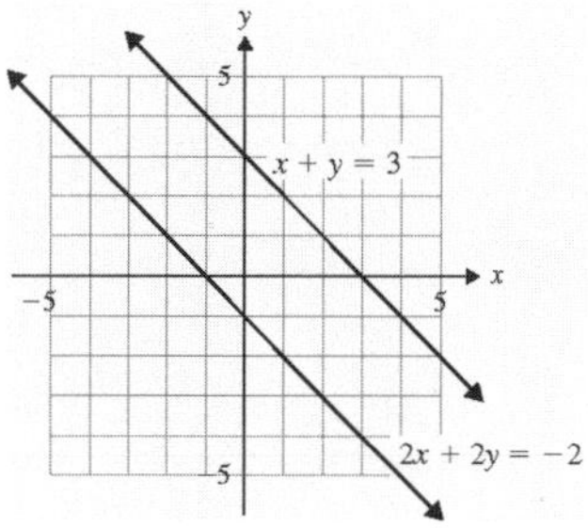

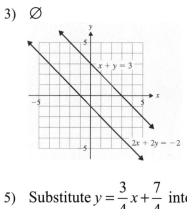

5) Substitute $y=\frac{3}{4}x+\frac{7}{4}$ into

$$-9x+12y=21$$
$$-9x+12\left(\frac{3}{4}x+\frac{7}{4}\right)=21$$
$$-9x+9x+21=21$$
$$21=21$$

infinite number of solutions of the form $\{(x,\ y)\,|\,-9x+12y=21\}$

7)
$$7x+8y=28$$
$$-3(7x+8y)=-3(28)$$
$$-21x-24y=-84$$
$$-5x+6y=-20$$
$$4(-5x+6y)=4(-20)$$
$$-20x+24y=-80$$

Add the equations.

$$\begin{array}{r} -21x-24y=-84 \\ +\ -20x+24y=-80 \\ \hline -41x=-164 \end{array}$$
$$x=4$$

Substitute $x=4$ into

$$-5x+6y=-20$$
$$-5(4)+6y=-20$$
$$-20+6y=-20$$
$$6y=0$$
$$y=0 \qquad (4,\ 0)$$

9) $x-8y=1$

$$x=8y+1$$

Substitute $x=8y+1$ into

$$-2x+9y=-9$$
$$-2(8y+1)+9y=-9$$
$$-16y-2+9y=-9$$
$$-7y-2=-9$$
$$-7y=-7$$
$$y=1$$

Substitute $y=1$ into

$$x=8y+1$$
$$x=8(1)+1$$
$$x=8+1$$
$$x=9 \qquad (9,\ 1)$$

11) $5(y+4)-9=-3(x-4)+4y$

$$5y+20-9=-3x+12+4y$$
$$5y+11=-3x+12+4y$$
$$3x+y=1$$
$$-3(3x+y)=-3(1)$$
$$-9x-3y=-3$$

$$13x-2(3x+2)=3(1-y)$$
$$13x-6x-4=3-3y$$
$$7x-4=3-3y$$
$$7x+3y=7$$

Add the equations.

$$\begin{array}{r} -9x-3y=-3 \\ +\ \ 7x+3y=\ \ 7 \\ \hline -2x=4 \end{array}$$
$$x=-2$$

Substitute $x=-2$ into

$$3x+y=1$$
$$3(-2)+y=1$$
$$-6+y=1$$
$$y=7 \qquad (-2, 7)$$

13) Let x = the cost of a box of screws
Let y = the cost of a box of nails

$$3\cdot\text{screws}+2\cdot\text{nails}=18$$
$$3x \quad + \quad 2y \quad =18$$
$$1\cdot\text{screws}+4\cdot\text{nails}=16$$
$$x \quad + \quad 4y \quad =16$$

$$x+4y=16$$
$$x=16-4y$$

Use substitution.

$$3(16-4y)+2y=18$$
$$48-12y+2y=18$$
$$48-10y=18$$
$$-10y=-30$$
$$y=3$$
$$x=16-4y;\ y=3$$
$$x=16-4(3)$$
$$x=16-12$$
$$x=4$$

screws: \$4; nails: \$3

5) $-8(3x^2-x-7)=-24x^2+8x+56$

7) $3c^2\cdot 5c^{-8}=15c^{2-8}=15c^{-6}=\dfrac{15}{c^6}$

9) $0.00008319=8.319\times 10^{-5}$

Cumulative Review: Chapters 1-5

1) $\dfrac{3}{10}-\dfrac{7}{15}=\dfrac{9}{30}-\dfrac{14}{30}=-\dfrac{5}{30}=-\dfrac{1}{6}$

3) $(5-8)^3+40\div 10-6$

$$=(-3)^3+40\div 10-6$$
$$=-27+40\div 10-6$$
$$=-27+4-6$$
$$=-29$$

11) $0.04(3p-2)-0.02p=0.1(p+3)$

$$100(0.04(3p-2)-0.02p)=100(0.1(p+3))$$
$$4(3p-2)-2p=10(p+3)$$
$$12p-8-2p=10p+30$$
$$10p-8=10p+30$$
$$-8\neq 30$$

$\varnothing$

13) Let $x =$ the number of plastic surgeries in 1997.

Surgeris in 2003 = Surgeries in 1997 + Amount of Increase

$$8,253,000 = x + x(2.93)$$
$$8,253,000 = 3.93x$$
$$2,100,000 = x$$

There were 2,100,000 plastic surgeries in 1997.

15)

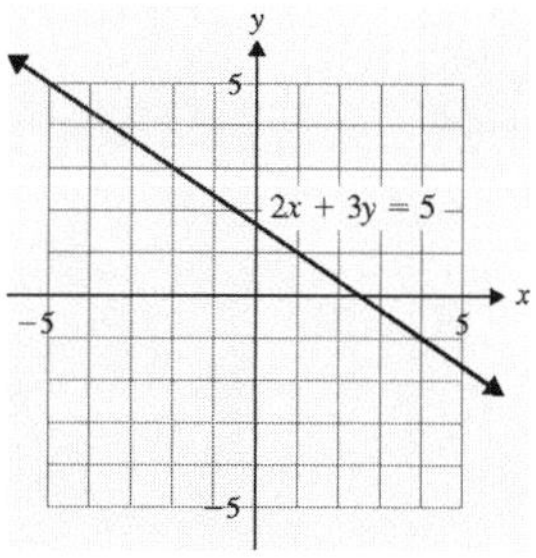

17) $m=\dfrac{-3-4}{1-(-7)}=\dfrac{-7}{8}=-\dfrac{7}{8}$

$(x_1, y_1)=(-7, 4)$

$$y-y_1=m(x-x_1)$$
$$y-4=-\frac{7}{8}(x-(-7))$$
$$y-4=-\frac{7}{8}(x+7)$$
$$y-4=-\frac{7}{8}x-\frac{49}{8}$$
$$y=-\frac{7}{8}x-\frac{17}{8}$$

19) $3x+4y=-11$

$$-3(3x+4y)=-3(-11)$$
$$-9x-12y=33$$

Add the equations.

$$\begin{array}{r} 9x+\ \ 7y=7 \\ +\ \ -9x-12y=33 \\ \hline -5y=40 \end{array}$$
$$y=-8$$

Substitute $y=-8$ into

$$3x+4y=-11$$
$$3x+4(-8)=-11$$
$$3x-32=-11$$
$$3x=21$$

$(7, -8)$

21) $\frac{5}{6}x-\frac{1}{2}y=\frac{2}{3}$

$6\left(\frac{5}{6}x-\frac{1}{2}y\right)=6\left(\frac{2}{3}\right)$

$5x-3y=4$

$-\frac{5}{4}x+\frac{3}{4}y=\frac{1}{2}$

$4\left(-\frac{5}{4}x+\frac{3}{4}y\right)=4\left(\frac{1}{2}\right)$

$-5x+3y=2$

Add the equations.

$$\begin{array}{r} 5x-3y=4 \\ +\quad -5x+3y=2 \\ \hline 0\neq 6 \end{array}$$

$\varnothing$

23) Let $a=$ the number of Aretha's awards.
Let $c=$ the number of Christina's awards.
Let $w=$ the number of Whitney's awards.

$a=w+10$

$c=\frac{1}{2}w$

$a+w+c+=25$

Use substitution.

$(w+10)+w+\left(\frac{1}{2}w\right)=25$

$\frac{5}{2}w+10=25$

$\frac{5}{2}w=15$

$w=6$

$a=w+10;\ w=6$

$a=6+10=16$

$c=\frac{1}{2}w;\ w=6$

$c=\frac{1}{2}(6)=3$

Aretha Franklin: 16;
Whitney Houston: 6;
Christina Aguilera: 3

Chapter 6: Polynomials

Section 6.1 Exercises

1) $2^2 \cdot 2^4 = 2^{2+4} = 2^6 = 64$

3) $\dfrac{(-4)^8}{(-4)^5} = (-4)^{8-5} = (-4)^3 = -64$

5) $6^{-1} = \dfrac{1}{6}$

7) $\left(\dfrac{1}{9}\right)^{-2} = 9^2 = 81$

9) $\left(\dfrac{3}{2}\right)^{-4} = \left(\dfrac{2}{3}\right)^4 = \dfrac{2^4}{3^4} = \dfrac{16}{81}$

11) $6^0 + \left(-\dfrac{1}{2}\right)^{-5} = 1 + (-2)^5$
$= 1 - 32$
$= -31$

13) $\dfrac{8^5}{8^7} = 8^{5-7} = 8^{-2} = \left(\dfrac{1}{8}\right)^2 = \dfrac{1}{64}$

15) $t^5 \cdot t^8 = t^{5+8} = t^{13}$

17) $(-8c^4)(2c^5) = -16c^{4+5} = -16c^9$

19) $(z^6)^4 = z^{6\cdot 4} = z^{24}$

21) $(5p^{10})^3 = 5^3 p^{10\cdot 3} = 125p^{30}$

23) $\left(-\dfrac{2}{3}a^7 b\right)^3 = \left(-\dfrac{2}{3}\right)^3 a^{7\cdot 3} b^3$
$= -\dfrac{8}{27}a^{21}b^3$

25) $\dfrac{f^{11}}{f^7} = f^{11-7} = f^4$

27) $\dfrac{35v^9}{5v^8} = 7v^{9-8} = 7v$

29) $\dfrac{9d^{10}}{54d^6} = \dfrac{d^{10-6}}{6} = \dfrac{d^4}{6}$

31) $\dfrac{x^3}{x^9} = x^{3-9} = x^{-6} = \dfrac{1}{x^6}$

33) $\dfrac{m^2}{m^3} = m^{2-3} = m^{-1} = \dfrac{1}{m}$

35) $\dfrac{45k^{-2}}{30k^2} = \dfrac{3}{2}k^{-2-2} = \dfrac{3}{2}k^{-4} = \dfrac{3}{2k^4}$

37) $5(2m^4n^7)^2 = 5(4m^8n^{14}) = 20m^8n^{14}$

39) $(6y^2)(2y^3)^2 = 6y^2 \cdot 4y^6 = 24y^8$

41) $\left(\dfrac{7a^4}{b^{-1}}\right)^{-2} = \left(\dfrac{b^{-1}}{7a^4}\right)^2 = \dfrac{b^{-2}}{49a^8} = \dfrac{1}{49a^8b^2}$

43) $\dfrac{a^{-12}b^7}{a^{-9}b^2} = a^{-12-(-9)}b^{7-2} = a^{-3}b^5 = \dfrac{b^5}{a^3}$

45) $(xy^{-3})^{-5} = x^{-5}y^{15} = \dfrac{y^{15}}{x^5}$

47) $\left(\dfrac{a^2b}{4c^2}\right)^{-3} = \left(\dfrac{4c^2}{a^2b}\right)^3 = \dfrac{64c^6}{a^6b^3}$

49) $\left(\frac{7h^{-1}k^9}{21h^{-5}k^5}\right)^{-2} = \left(\frac{21h^{-5}k^5}{7h^{-1}k^9}\right)^2$
$= \left(3h^{-5-(-1)}k^{5-9}\right)^2$
$= \left(3h^{-4}k^{-4}\right)^2$
$= 9h^{-8}k^{-8}$
$= \frac{9}{h^8k^8}$

51) $\left(\frac{15cd^{-4}}{5c^3d^{-10}}\right)^{-3} = \left(\frac{5c^3d^{-10}}{15cd^{-4}}\right)^3$
$= \left(\frac{1}{3}c^{3-1}d^{-10-(-4)}\right)^3$
$= \left(\frac{1}{3}c^2d^{-6}\right)^3$
$= \frac{1}{27}c^6d^{-18}$
$= \frac{c^6}{27d^{18}}$

53) $A = 5x \cdot 2x = 10x^2$ sq. units
$P = 2(5x) + 2(2x) = 10x + 4x$
$= 14x$ units

55) $A = \frac{1}{4}p \cdot \frac{3}{4}p = \frac{3}{16}p^2$ sq. units
$P = 2\left(\frac{1}{4}p\right) + 2\left(\frac{3}{4}p\right)$
$= \frac{1}{2}p + \frac{3}{2}p$
$= 2p$ units

57) $k^{4a} \cdot k^{2a} = k^{4a+2a} = k^{6a}$

59) $\left(g^{2x}\right)^4 = g^{2x \cdot 4} = g^{8x}$

61) $\frac{x^{7b}}{x^{4b}} = x^{7b-4b} = x^{3b}$

Section 6.2 Exercises

1) Yes. The coefficients are real numbers and the exponents are whole numbers.

3) No. One of the exponents is a negative number.

5) No. Two of the exponents are fractions.

7) binomial

9) trinomial

11) monomial

13) It is the same as the degree of the term in the polynomial with the highest degree.

15) Add the exponents on the variables.

17) Degree of poly = 3

Term	Coeff	Degree
$7y^3$	7	3
$10y^2$	10	2
$-y$	-1	1
2	2	0

19) Degree of poly = 5

Term	Coeff	Degree
$-9r^3s^2$	-9	5
$-r^2s^2$	-1	4
$\frac{1}{2}rs$	$\frac{1}{2}$	2
$6s$	6	1

21) a) $k^2+5k+8,\ k=2$

$(2)^2+5(2)+8=4+10+8=22$

b) $k^2+5k+8,\ k=-3$

$(-3)^2+5(-3)+8=9-15+8=2$

23) $2xy-7x+9,\ \ x=-4; y=3$

$2(-4)(3)-7(-4)+9=-24+28+9$

$=13$

25) $-x^2y+3xy+10y-1,\ \ x=-4; y=3$

$-(-4)^2(3)+3(-4)(3)+10(3)-1$

$=-16\cdot 3-36+30-1$

$=-48-36+30-1$

$=-55$

27) $\frac{2}{3}xy^2-3x+4y,\ \ x=-4; y=3$

$\frac{2}{3}(-4)(3)^2-3(-4)+4(3)$

$=\frac{2}{3}(-4)9+12+12$

$=-24+12+12$

$=0$

29) $(11w^2+2w-13)+(-6w^2+5w+7)$

$=5w^2+7w-6$

31) $(-p+16)+(-7p-9)=-8p+7$

33) $\left(-7a^4-\frac{3}{2}a+1\right)+\left(2a^4+9a^3-a^2-\frac{3}{8}\right)$

$=-5a^4+9a^3-a^2-\frac{3}{2}a+\frac{5}{8}$

35) $\left(\frac{11}{4}x^3-\frac{5}{6}\right)+\left(\frac{3}{8}x^3+\frac{11}{12}\right)$

$=\left(\frac{22}{8}x^3-\frac{10}{12}\right)+\left(\frac{3}{8}x^3+\frac{11}{12}\right)$

$=\frac{25}{8}x^3+\frac{1}{12}$

37) $(6.8k^3+3.5k^2-10k-3.3)$

$+(-4.2k^3+5.2k^2+2.7k-1.1)$

$=2.6k^3+8.7k^2-7.3k-4.4$

39)
$$\begin{array}{r} 12x-11 \\ +\ \underline{\ 5x+3} \\ 17x-8 \end{array}$$

41)
$$\begin{array}{r} 9r^2+16r+2 \\ +\ \underline{3r^2-10r+9} \\ 12r^2+6r+11 \end{array}$$

43)
$$\begin{array}{r} -2.6q^3-q^2+6.9q\ \ -1 \\ +\ \underline{4.1q^3\qquad\ -2.3q+16} \\ 1.5q^3-q^2+4.6q+15 \end{array}$$

45) $(8a^4-9a^2+17)-(15a^4+3a^2+3)$

$=(8a^4-9a^2+17)+(-15a^4-3a^2-3)$

$=-7a^4-12a^2+14$

47) $(j^2+18j+2)-(-7j^2+6j+2)$

$=(j^2+18j+2)+(+7j^2-6j-2)$

$=8j^2+12j$

49) $(19s^5 - 11s^2) - (10s^5 + 3s^4 - 8s^2 - 2)$
$= (19s^5 - 11s^2) + (-10s^5 - 3s^4 + 8s^2 + 2)$
$= 9s^5 - 3s^4 - 3s^2 + 2$

51) $(-3b^4 - 5b^2 + b + 2) - (-2b^4 + 10b^3 - 5b^2 - 18)$
$= (-3b^4 - 5b^2 + b + 2) + (2b^4 - 10b^3 + 5b^2 + 18)$
$= -b^4 - 10b^3 + b + 20$

53) $\left(-\frac{5}{7}r^2 + \frac{4}{9}r + \frac{2}{3}\right) - \left(-\frac{5}{14}r^2 - \frac{5}{9}r + \frac{11}{6}\right)$
$= \left(-\frac{10}{14}r^2 + \frac{4}{9}r + \frac{4}{6}\right) + \left(\frac{5}{14}r^2 + \frac{5}{9}r - \frac{11}{6}\right) = -\frac{5}{14}r^2 + r - \frac{7}{6}$

55)
$$\begin{array}{rr} & 17v + 3 \\ - & \underline{2v + 9} \end{array} \qquad \begin{array}{rr} & 17v + 3 \\ + & \underline{-2v - 9} \\ & 15v - 6 \end{array}$$

57)
$$\begin{array}{rr} & 2b^2 - 7b + 4 \\ - & \underline{3b^2 + 5b - 3} \end{array} \qquad \begin{array}{rr} & 2b^2 - 7b + 4 \\ + & \underline{-3b^2 - 5b + 3} \\ & -b^2 - 12b + 7 \end{array}$$

59)
$$\begin{array}{rr} & a^4 - 2a^3 + 6a^2 - 7a + 11 \\ - & \underline{-2a^4 + 9a^3 - a^2 \qquad\quad + 3} \end{array} \qquad \begin{array}{rr} & a^4 - 2a^3 + 6a^2 - 7a + 11 \\ + & \underline{2a^4 - 9a^3 + a^2 \qquad\quad - 3} \\ & 3a^4 - 11a^3 + 7a^2 - 7a + 8 \end{array}$$

61) Answers may vary.

63) No. If the coefficients of the like terms are opposite in sign, their sum will be zero. Example:
$(3x^2 + 4x + 5) + (2x^2 - 4x + 1) = 5x^2 + 6$

65) $(-3b^4 + 4b^2 - 6) + (2b^4 - 18b^2 + 4)$
$+ (b^4 + 5b^2 - 2)$
$= (-b^4 - 14b^2 - 2) + (b^4 + 5b^2 - 2) = -9b^2 - 4$

67) $\left(n^3-\frac{1}{2}n^2-4n+\frac{5}{8}\right)+\left(\frac{1}{4}n^3-n^2+7n-\frac{3}{4}\right)$

$=\left(\frac{4}{4}n^3-\frac{1}{2}n^2-4n+\frac{5}{8}\right)+\left(\frac{1}{4}n^3-\frac{2}{2}n^2+7n-\frac{6}{8}\right)=\left(\frac{5}{4}n^3-\frac{3}{2}n^2+3n-\frac{1}{8}\right)$

69) $\left(u^3+2u^2+1\right)-\left(4u^3-7u^2+u+9\right)+\left(8u^3-19u^2+2\right)$

$=\left(u^3+2u^2+1\right)+\left(-4u^3+7u^2-u-9\right)+\left(8u^3-19u^2+2\right)$

$=\left(-3u^3+9u^2-u-8\right)+\left(8u^3-19u^2+2\right)=5u^3-10u^2-u-6$

71) $\left(\frac{3}{8}k^2+k-\frac{1}{5}\right)-\left(2k^2+k-\frac{7}{10}\right)+\left(k^2-9k\right)$

$=\left(\frac{3}{8}k^2+k-\frac{2}{10}\right)+\left(-\frac{16}{8}k^2-k+\frac{7}{10}\right)+\left(\frac{8}{8}k^2-9k\right)$

$=\left(-\frac{13}{8}k^2+\frac{5}{10}\right)+\left(\frac{8}{8}k^2-9k\right)=-\frac{5}{8}k^2-9k+\frac{1}{2}$

73) $\left(2t^3-8t^2+t+10\right)-\left[\left(5t^3+3t^2-t+8\right)+\left(-6t^3-4t^2+3t+5\right)\right]$

$=\left(2t^3-8t^2+t+10\right)-\left[-t^3-t^2+2t+13\right]$

$=\left(2t^3-8t^2+t+10\right)+\left(t^3+t^2-2t-13\right)=3t^3-7t^2-t-3$

75) $\left(-12a^2+9\right)-\left(-9a^3+7a+6\right)+\left(12a^2-a+10\right)$

$=\left(-12a^2+9\right)+\left(9a^3-7a-6\right)+\left(12a^2-a+10\right)$

$=\left(9a^3-12a^2-7a+3\right)+\left(12a^2-a+10\right)=9a^3-8a+13$

77) $\left(4a+13b\right)-\left(a+5b\right)=\left(4a+13b\right)+\left(-a-5b\right)=3a+8b$

79) $\left(5m+\frac{5}{6}n+\frac{1}{2}\right)+\left(-6m+n-\frac{3}{4}\right)=\left(5m+\frac{5}{6}n+\frac{2}{4}\right)+\left(-6m+\frac{6}{6}n-\frac{3}{4}\right)$

$=-m+\frac{11}{6}n-\frac{1}{4}$

81) $\left(-12y^2z^2+5y^2z-25yz^2+16\right)+\left(17y^2z^2+2y^2z-15\right)=5y^2z^2+7y^2z-25yz^2+1$

83) $\left(8x^3y^2-7x^2y^2+7x^2y-3\right)+\left(2x^3y^2+x^2y-1\right)-\left(4x^2y^2+2x^2y+8\right)$
$=\left(10x^3y^2-7x^2y^2+8x^2y-4\right)+\left(-4x^2y^2-2x^2y-8\right)=10x^3y^2-11x^2y^2+6x^2y-12$

85) $\left(v^2-9\right)+\left(4v^2+3v+1\right)=5v^2+3v-8$

87) $\left(5g^2+3g+6\right)-\left(g^2-7g+16\right)=\left(5g^2+3g+6\right)+\left(-g^2+7g-16\right)=4g^2+10g-10$

89) $\left(2n^2+n+4\right)-\left[\left(4n^2+1\right)+\left(6n^2-10n+3\right)\right]=\left(2n^2+n+4\right)-\left[10n^2-10n+4\right]$
$=\left(2n^2+n+4\right)+\left[-10n^2+10n-4\right]$
$=-8n^2+11n$

91) $P=2(3x+8)+2(x-1)=6x+16+2x-2=8x+14$ units

93) $P=2\left(3w^2-2w+4\right)+2(w-7)=6w^2-4w+8+2w-14=6w^2-2w-6$ units

95) a) $f(-3)=5(-3)^2+7(-3)-8$
$f(-3)=5(9)+7(-3)-8$
$f(-3)=45-21-8$
$f(-3)=16$

b) $f(1)=5(1)^2+7(1)-8$
$f(1)=5(1)+7(1)-8$
$f(1)=5+7-8$
$f(1)=4$

97) a) $P(4)=(4)^3-3(4)^2+2(4)+5$
$P(4)=64-3(16)+2(4)+5$
$P(4)=64-48+8+5$
$P(4)=29$

b) $P(0)=(0)^3-3(0)^2+2(0)+5$
$P(0)=0-3(0)+2(0)+5$
$P(0)=0-0+0+5$
$P(0)=5$

99) $11=-4z+9$
$2=-4z$
$-\frac{2}{4}=z$
$-\frac{1}{2}=z$

101) $13=\frac{2}{5}z-3$
$16=\frac{2}{5}z$
$80=2z$
$40=z$

Section 6.3 Exercises

1) Answers may vary.

3) $(7k^4)(2k^2)=14k^6$

5) $(-4t)(6t^8)=-24t^9$

7) $\left(\frac{7}{10}d^9\right)\left(\frac{5}{2}d^2\right)=\frac{35}{20}d^{11}=\frac{7}{4}d^{11}$

9) $7y(4y-9)=(7y)(4y)+(7y)(-9)=28y^2-63y$

11) $-4b(9b+8)=(-4b)(9b)+(-4b)(8)=-36b^2-32b$

13) $6v^3(v^2-4v-2)=(6v^3)(v^2)+(6v^3)(-4v)+(6v^3)(-2)=6v^5-24v^4-12v^3$

15) $-3t^2(9t^3-6t^2-4t-7)=(-3t^2)(9t^3)+(-3t^2)(-6t^2)+(-3t^2)(-4t)+(-3t^2)(-7)$
$=-27t^5+18t^4+12t^3+21t^2$

17) $2x^3y(xy^2+8xy-11y+2)=(2x^3y)(xy^2)+(2x^3y)(8xy)+(2x^3y)(-11y)+(2x^3y)(2)$
$=2x^4y^3+16x^4y^2-22x^3y^2+4x^3y$

19) $-\frac{3}{4}t^4(20t^3+8t^2-5t)=\left(-\frac{3}{4}t^4\right)(20t^3)+\left(-\frac{3}{4}t^4\right)(8t^2)+\left(-\frac{3}{4}t^4\right)(-5t)$
$=-15t^7-6t^6+\frac{15}{4}t^5$

21) $2(10g^3+5g^2+4)-(2g^3-14g-20)=20g^3+10g^2+8-2g^3+14g+20$
$=18g^3+10g^2+14g+28$

23) $-(10r^3-14r+27)+3(3r^3-13r^2-15r+6)=-10r^3+14r-27+9r^3-39r^2-45r+18$
$=-r^3-39r^2-31r-9$

25) $7(a^3b^3+6a^3b^2+3)-9(6a^3b^3+9a^3b^2-12a^2b+8)$
$=7a^3b^3+42a^3b^2+21-54a^3b^3-81a^3b^2+108a^2b-72$
$=-47a^3b^3-39a^3b^2+108a^2b-51$

27) $(q+3)(5q^2-15q+9)=(q)(5q^2)+(q)(-15q)+(q)(9)+(3)(5q^2)+(3)(-15q)+(3)(9)$
$=5q^3-15q^2+9q+15q^2-45q+27$
$=5q^3-36q+27$

29) $(p-6)(2p^2+3p-5)=(p)(2p^2)+(p)(3p)+(p)(-5)+(-6)(2p^2)+(-6)(3p)+(-6)(-5)$
$=2p^3+3p^2-5p-12p^2-18p+30$
$=2p^3-9p^2-23p+30$

31) $(5y^3-y^2+8y+1)(3y-4)=(5y^3)(3y)+(5y^3)(-4)+(-y^2)(3y)+(-y^2)(-4)+(8y)(3y)$
$+(8y)(-4)+(1)(3y)+(1)(-4)$
$=15y^4-20y^3-3y^3+4y^2+24y^2-32y+3y-4$
$=15y^4-23y^3+28y^2-29y-4$

33) $\left(\frac{1}{2}k^2+3\right)(12k^2+5k-10)$
$=\left(\frac{1}{2}k^2\right)(12k^2)+\left(\frac{1}{2}k^2\right)(5k)+\left(\frac{1}{2}k^2\right)(-10)+(3)(12k^2)+(3)(5k)+(3)(-10)$
$=6k^4+\frac{5}{2}k^3-5k^2+36k^2+15k-30=6k^4+\frac{5}{2}k^3+31k^2+15k-30$

35) $(a^2-a+3)(a^2+4a-2)$
$=(a^2)(a^2)+(a^2)(4a)+(a^2)(-2)+(-a)(a^2)+(-a)(4a)$
$+(-a)(-2)+(3)(a^2)+(3)(4a)+(3)(-2)$
$=a^4+4a^3-2a^2-a^3-4a^2+2a+3a^2+12a-6=a^4+3a^3-3a^2+14a-6$

37) $(3v^2-v+2)(-8v^3+6v^2+5)$
$=(3v^2)(-8v^3)+(3v^2)(6v^2)+(3v^2)(5)+(-v)(-8v^3)+(-v)(6v^2)$
$+(-v)(5)+(2)(-8v^3)+(2)(6v^2)+(2)(5)$
$=-24v^5+18v^4+15v^2+8v^4-6v^3-5v-16v^3+12v^2+10$
$=-24v^5+26v^4-22v^3+27v^2-5v+10$

39) $(2x-3)(4x^2-5x+2)$
$=(2x)(4x^2)+(2x)(-5x)+(2x)(2)+(-3)(4x^2)+(-3)(-5x)+(-3)(2)$
$=8x^3-10x^2+4x-12x^2+15x-6=8x^3-22x^2+19x-6$

$$\begin{array}{r} 4x^2-5x+2 \\ \times \qquad 2x-3 \\ \hline -12x^2+15x-6 \\ +\ 8x^3-10x^2+4x \qquad \\ \hline 8x^3-22x^2+19x-6 \end{array}$$

41) First, Outer, Inner, Last

43) $(w+8)(w+7)=w^2+7w+8w+56$
$=w^2+15w+56$

45) $(k-5)(k+9)=k^2+9k-5k-45$
$=k^2+4k-45$

47) $(y-6)(y-1)=y^2-y-6y+6$
$=y^2-7y+6$

49) $(4p+5)(p-3)=4p^2-12p+5p-15$
$=4p^2-7p-15$

51) $(8n+3)(3n+4)=24n^2+32n+9n+12$
$=24n^2+41n+12$

53) $(6-5y)(2-y)=12-6y-10y+5y^2$
$=5y^2-16y+12$

55) $(m+7)(3-m)=3m-m^2+21-7m$
$=-m^2-4m+21$

57) $(4a-5b)(3a+4b)$
$=12a^2+16ab-15ab-20b^2$
$=12a^2+ab-20b^2$

59) $(6p+5q)(10p+3q)$
$=60p^2+18pq+50pq+15q^2$
$=60p^2+68pq+15q^2$

61) $\left(a+\frac{1}{4}\right)\left(a+\frac{4}{5}\right)$
$=a^2+\frac{4}{5}a+\frac{1}{4}a+\frac{4}{20}$
$=a^2+\frac{16}{20}a+\frac{5}{20}a+\frac{1}{5}$
$=a^2+\frac{21}{20}a+\frac{1}{5}$

63) a) $P=2(y+6)+2(y-2)$
$=2y+12+2y-4$
$=4y+8$ units

b) $A=(y+6)(y-2)$
$=y^2-2y+6y-12$
$=y^2+4y-12$ sq. units

65) a) $P=2(3a)+2(a^2-a+8)$
$=6a+2a^2-2a+16$
$=2a^2+4a+16$ units

b) $A=(3a)(a^2-a+8)$
$=3a^3-3a^2+24a$ sq. units

67) Both are correct.

69) $3(y+4)(5y-2)$
$=(3y+12)(5y-2)$
$=15y^2-6y+60y-24$
$=15y^2+54y-24$

71) $-18(3a-1)(a+2)$
$=(-54a+18)(a+2)$
$=-54a^2-108a+18a+36$
$=-54a^2-90a+36$

73) $-7r^2(r-9)(r-2)$
$=-7r^2(r^2-2r-9r+18)$
$=-7r^2(r^2-11r+18)$
$=-7r^4+77r^3-126r^2$

75) $(c+3)(c+4)(c-1)$
$=(c^2+4c+3c+12)(c-1)$
$=(c^2+7c+12)(c-1)$
$=c^3+7c^2+12c-c^2-7c-12$
$=c^3+6c^2+5c-12$

77) $(2p-1)(p-5)(p-2)$
$=(2p-1)(p^2-2p-5p+10)$
$=(2p-1)(p^2-7p+10)$
$=2p^3-14p^2+20p-p^2+7p-10$
$=2p^3-15p^2+27p-10$

79) $10n\left(\frac{1}{2}n^2+3\right)(n^2+5)$
$=(5n^3+30n)(n^2+5)$
$=5n^5+25n^3+30n^3+150n$
$=5n^5+55n^3+150n$

81) $(x+6)(x-6)=x^2-6^2=x^2-36$

83) $(t-3)(t+3)=(t+3)(t-3)$
$=t^2-3^2$
$=t^2-9$

85) $(2-r)(2+r)=(2+r)(2-r)$
$=2^2-r^2$
$=4-r^2$

87) $\left(n+\frac{1}{2}\right)\left(n-\frac{1}{2}\right)=n^2-\left(\frac{1}{2}\right)^2=n^2-\frac{1}{4}$

89) $\left(\frac{2}{3}-k\right)\left(\frac{2}{3}+k\right)=\left(\frac{2}{3}+k\right)\left(\frac{2}{3}-k\right)$
$=\left(\frac{2}{3}\right)^2-k^2$
$=\frac{4}{9}-k^2$

91) $(3m+2)(3m-2)=(3m)^2-2^2$
$=9m^2-4$

93) $-(6a-b)(6a+b)$
$=-(6a+b)(6a-b)$
$=-\left[(6a)^2-b^2\right]$
$=-(36a^2-b^2)$
$=b^2-36a^2$

95) $(y+8)^2=y^2+2(y)(8)+8^2$
$=y^2+16y+64$

97) $(t-11)^2=t^2-2(t)(11)+(-11)^2$
$=t^2-22t+121$

99) $(k-2)^2=k^2-2(k)(2)+(-2)^2$
$=k^2-4k+4$

101) $(4w+1)^2=(4w)^2+2(4w)(1)+1^2$
$=16w^2+8w+1$

103) $(2d-5)^2=(2d)^2-2(2d)(5)+(-5)^2$
$=4d^2-20d+25$

105) No. The order of operations tell us to perform exponents, $(t+3)^2$, before multiplying by 4.

107) $6(x+1)^2=6\left[x^2+2(x)(1)+1^2\right]$
$=6\left[x^2+2x+1\right]$
$=6x^2+12x+6$

109) $2a(a+3)^2=2a\left[a^2+2(a)(3)+3^2\right]$
$=2a\left[a^2+6a+9\right]$
$=2a^3+12a^2+18a$

111) $-3(m-1)^2$
$=-3\left[m^2-2(m)(1)+(-1)^2\right]$
$=-3\left[m^2-2m+1\right]$
$=-3m^2+6m-3$

113) $(r+5)^3=(r+5)^2(r+5)$
$=(r^2+10r+25)(r+5)$
$=r^3+10r^2+25r+5r^2$
$+50r+125$
$=r^3+15r^2+75r+125$

115) $(s-2)^3=(s-2)^2(s-2)$
$=(s^2-4s+4)(s-2)$
$=s^3-4s^2+4s-2s^2$
$+8s-8$
$=s^3-6s^2+12s-8$

117) $(c^2-9)^2=(c^2)^2-2(c^2)(9)+(-9)^2$
$=c^4-18c^2+81$

119) $\left(\frac{2}{3}n+4\right)^2=\left(\frac{2}{3}n\right)^2+2\left(\frac{2}{3}n\right)(4)+(4)^2$
$=\frac{4}{9}n^2+\frac{16}{3}n+16$

121) $(y+2)^4=(y+2)^2(y+2)^2$
$=(y^2+4y+4)(y^2+4y+4)$
$=y^4+4y^3+4y^2+4y^3+16y^2$
$+16y+4y^2+16y+16$
$=y^4+8y^3+24y^2+32y+16$

123) No; $(x+5)^2=x^2+10x+25$

125) $V=(h+2)^3$
$=(h+2)^2(h+2)$
$=(h^2+4h+4)(h+2)$
$=h^3+2h^2+4h^2+8h+4h+8$
$=h^3+6h^2+12h+8$ cubic units

127) $A=(5x+3)(2x+5)-(x-1)^2$
$=10x^2+25x+6x+15$
$-(x^2-2x+1)$
$=10x^2+31x+15-x^2+2x-1$
$=9x^2+33x+14$ sq. units

Section 6.4: Exercises

1) dividend: $12c^3+20c^2-4c$; divisor: $4c$; quotient: $3c^2+5c-1$

3) Answers may vary. Divide each term in the polynomial by the monomial and simplify.

5) $\frac{4a^5-10a^4+6a^3}{2a^3}=\frac{4a^5}{2a^3}-\frac{10a^4}{2a^3}+\frac{6a^3}{2a^3}=2a^2-5a+3$

7) $\frac{18u^7+18u^5+45u^4-72u^2}{9u^2}=\frac{18u^7}{9u^2}+\frac{18u^5}{9u^2}+\frac{45u^4}{9u^2}-\frac{72u^2}{9u^2}=2u^5+2u^3+5u^2-8$

9) $\left(35d^5-7d^2\right)\div\left(-7d^2\right)=\frac{35d^5-7d^2}{-7d^2}=\frac{35d^5}{-7d^2}-\frac{7d^2}{-7d^2}=-5d^3+1$

11) $\frac{9w^5+42w^4-6w^3+3w^2}{6w^3}=\frac{9w^5}{6w^3}+\frac{42w^4}{6w^3}-\frac{6w^3}{6w^3}+\frac{3w^2}{6w^3}=\frac{3}{2}w^2+7w-1+\frac{1}{2w}$

13) $\left(10v^7-36v^5-22v^4-5v^2+1\right)\div\left(4v^4\right)=\frac{10v^7-36v^5-22v^4-5v^2+1}{4v^4}$

$$=\frac{10v^7}{4v^4}-\frac{36v^5}{4v^4}-\frac{22v^4}{4v^4}-\frac{5v^2}{4v^4}+\frac{1}{4v^4}$$

$$=\frac{5}{2}v^3-9v-\frac{11}{2}-\frac{5}{4v^2}+\frac{1}{4v^4}$$

15) $\frac{90a^4b^3+60a^3b^3-40a^3b^2+100a^2b^2}{10ab^2}=\frac{90a^4b^3}{10ab^2}+\frac{60a^3b^3}{10ab^2}-\frac{40a^3b^2}{10ab^2}+\frac{100a^2b^2}{10ab^2}$

$$=9a^3b+6a^2b-4a^2+10a$$

17) $\left(9t^5u^4-63t^4u^4-108t^3u^4+t^3u^2\right)\div\left(-9tu^2\right)=\frac{9t^5u^4-63t^4u^4-108t^3u^4+t^3u^2}{-9tu^2}$

$$=\frac{9t^5u^4}{-9tu^2}-\frac{63t^4u^4}{-9tu^2}-\frac{108t^3u^4}{-9tu^2}+\frac{t^3u^2}{-9tu^2}$$

$$=-t^4u^2+7t^3u^2+12t^2u^2-\frac{1}{9}t^2$$

19) The answer is incorrect. When you divide $4t$ by $4t$, you get 1.
The quotient should be $4t^2-9t+1$.

21)
$$\begin{array}{r} g+4 \\ g+5\overline{\big)\, g^2+9g+20} \\ \underline{-(g^2+5g)} \\ 4g+20 \\ \underline{-(4g+20)} \\ 0 \end{array}$$

23)
$$\begin{array}{r} p+6 \\ p+2\overline{\big)\, p^2+8p+12} \\ \underline{-(p^2+2p)} \\ 6p+12 \\ \underline{-(6p+12)} \\ 0 \end{array}$$

25)
$$\begin{array}{r} k-5 \\ k+9\overline{\big)\, k^2+4k-45} \\ \underline{-(k^2+9k)} \\ -5k-45 \\ \underline{-(-5k-45)} \\ 0 \end{array}$$

27)
$$\begin{array}{r} h+8 \\ h-3\overline{\big)\, h^2+5h-24} \\ \underline{-(h^2-3h)} \\ 8h-24 \\ \underline{-(8h-24)} \\ 0 \end{array}$$

29)
$$\begin{array}{r} 2a^2-7a-3 \\ 2a-5\overline{\big)\, 4a^3-24a^2+29a+15} \\ \underline{-(4a^3-10a^2)} \\ -14a^2+29a \\ \underline{-(-14a^2+35a)} \\ -6a+15 \\ \underline{-(-6a+15)} \\ 0 \end{array}$$

31)
$$\begin{array}{r} 3p^2+7p-1 \\ 6p+1\overline{\big)\, 18p^3+45p^2+p-1} \\ \underline{-(18p^3+3p^2)} \\ 42p^2+p \\ \underline{-(42p^2+7p)} \\ -6p-1 \\ \underline{-(-6p-1)} \\ 0 \end{array}$$

33)
$$\begin{array}{r} 6t+23 \\ t-5\overline{\big)\, 6t^2-7t+4} \\ \underline{-(6t^2-30t)} \\ 23t+4 \\ \underline{-(23t-115)} \\ 119 \end{array}$$

$$\left(6t^2-7t+4\right)\div\left(t-5\right)=6t+23+\frac{119}{t-5}$$

35)
$$\begin{array}{r} 4z^2+8z\ \ +7 \\ 3z+5\overline{\big)\ 12z^3+44z^2+61z-37} \\ \underline{-(12z^3+20z^2)\qquad\qquad} \\ 24z^2+61z\qquad \\ \underline{-\ (24z^2+40z)\qquad} \\ 21z-37 \\ \underline{-\ (21z+35)} \\ -72 \end{array}$$

$$\left(12z^3+44z^2+61z-37\right)\div\left(3z+5\right)$$
$$=4z^2+8z+7-\frac{72}{3z+5}$$

37)
$$\begin{array}{r} w^2-4w+16 \\ w+4\overline{\big)\ w^3+0w^2+0w+64} \\ \underline{-(w^3+4w^2)\qquad\qquad} \\ -4w^2+0w\qquad \\ \underline{-(-4w^2-16w)\qquad} \\ 16w+64 \\ \underline{-(16w+64)} \\ 0 \end{array}$$

39)
$$\begin{array}{r} 2r^2+8r+3 \\ 8r-3\overline{\big)\ 16r^3+58r^2+0r-9} \\ \underline{-(16r^3-6r^2)\qquad\qquad} \\ 64r^2+0r\qquad \\ \underline{-\ (64r^2-24r)\qquad} \\ 24r-9 \\ \underline{-(24r-9)} \\ 0 \end{array}$$

41) Use synthetic division when the divisor is in the form $x-c$.

43) No. The divisor, $2x+5$, is not in the form $x-c$.

45)
$$\begin{array}{r|rrr} \underline{4} & 1 & 5 & -36 \\ & & 4 & 36 \\ \hline & 1 & 9 & 0 \end{array}$$

$$\left(t^2+5t-36\right)\div\left(t-4\right)=t+9$$

47)
$$\begin{array}{r|rrr} \underline{-3} & 5 & 21 & 20 \\ & & -15 & -18 \\ \hline & 5 & 6 & 2 \end{array}$$

$$\frac{5n^2+21n+20}{n+3}=5n+6+\frac{2}{n+3}$$

49)
$$\begin{array}{r|rrrr} \underline{-5} & 2 & 7 & -10 & 21 \\ & & -10 & 15 & -25 \\ \hline & 2 & -3 & 5 & -4 \end{array}$$

$$\left(2y^3+7y^2-10y+21\right)\div\left(y+5\right)$$
$$=2y^2-3y+5-\frac{4}{y+5}$$

51)
$$\begin{array}{r|rrrr} \underline{3} & 3 & -10 & 4 & -3 \\ & & 9 & -3 & 3 \\ \hline & 3 & -1 & 1 & 0 \end{array}$$

$$\left(3p^3-10p^2+4p-3\right)\div\left(p-3\right)$$
$$=3p^2-p+1$$

53)
$$\begin{array}{r|rrrrr} \underline{-1} & 5 & 7 & -1 & -8 & 2 \\ & & -5 & -2 & 3 & 5 \\ \hline & 5 & 2 & -3 & -5 & 7 \end{array}$$

$$\left(5x^4+7x^3-x^2-8x+2\right)\div\left(x+1\right)$$
$$=5x^3+2x^2-3x-5+\frac{7}{x+1}$$

55)
$$\begin{array}{r|rrrr} 2 & 1 & -3 & 0 & 4 \\ & & 2 & -2 & -4 \\ \hline & 1 & -1 & -2 & 0 \end{array}$$

$$\frac{r^3-3r^2+4}{r-2}=r^2-r-2$$

57)
$$\begin{array}{r|rrrrr} 3 & 1 & 0 & 0 & 0 & -81 \\ & & 3 & 9 & 27 & 81 \\ \hline & 1 & 3 & 9 & 27 & 0 \end{array}$$

$$\frac{m^4-81}{m-3}=m^3+3m^2+9m+27$$

59)
$$\begin{array}{r|rrrrrr} 2 & 2 & -3 & 0 & 0 & -11 & 0 \\ & & 4 & 2 & 4 & 8 & -6 \\ \hline & 2 & 1 & 2 & 4 & -3 & -6 \end{array}$$

$$\left(2c^5-3c^4-11c\right)\div(c-2)$$
$$=2c^4+c^3+2c^2+4c-3-\frac{6}{c-2}$$

61)
$$\begin{array}{r|rrrr} \frac{1}{2} & 2 & 7 & -16 & 6 \\ & & 1 & 4 & -6 \\ \hline & 2 & 8 & -12 & 0 \end{array}$$

$$\left(2x^3+7x^2-16x+6\right)\div\left(x-\frac{1}{2}\right)$$
$$=2x^2+8x-12$$

63)
$$\frac{6x^4y^4+30x^4y^3-x^2y^2+3xy}{6x^2y^2}$$
$$=\frac{6x^4y^4}{6x^2y^2}+\frac{30x^4y^3}{6x^2y^2}-\frac{x^2y^2}{6x^2y^2}+\frac{3xy}{6x^2y^2}$$
$$=x^2y^2+5x^2y-\frac{1}{6}+\frac{1}{2xy}$$

65)
$$\begin{array}{r} -2g^3-5g^2+g-4 \\ 4g-9 \overline{\big)\ -8g^4-2g^3+49g^2-25g+36} \\ -(-8g^4+18g^3)\qquad\qquad\qquad \\ \hline -20g^3+49g^2\qquad\qquad \\ -\ (-20g^3+45g^2)\qquad\qquad \\ \hline 4g^2-25g\qquad \\ -\ (4g^2-9g)\qquad \\ \hline -16g+36 \\ -\ (-16g+36) \\ \hline 0 \end{array}$$

67)
$$\begin{array}{r|rrr} 8 & 6 & -43 & -20 \\ & & 48 & 40 \\ \hline & 6 & 5 & 20 \end{array}$$

$$\frac{6t^2-43t-20}{t-8}=6t+5+\frac{20}{t-8}$$

69)
$$\begin{array}{r} 4n^2+10n+25 \\ 2n-5 \overline{\big)\ 8n^3+0n^2+0n-125} \\ -(8n^3-20n^2)\qquad\qquad \\ \hline 20n^2+0n\qquad \\ -\ (20n^2-50n)\qquad \\ \hline 50n-125 \\ -(50n-125) \\ \hline 0 \end{array}$$

71)
$$\begin{array}{r} 5x^2-7x+3 \\ x^2+2 \overline{\big)\ 5x^4-7x^3+13x^2-14x+6} \\ -(5x^4 \qquad 10x^2)\qquad\qquad \\ \hline -7x^3+3x^2-14x\qquad \\ -\ (-7x^3 \qquad -14x)\qquad \\ \hline 3x^2 \qquad +6 \\ -(3x^2 \qquad +6) \\ \hline 0 \end{array}$$

73) $\dfrac{-12a^3+9a^2-21a}{-3a}=\dfrac{-12a^3}{-3a}+\dfrac{9a^2}{-3a}-\dfrac{21a}{-3a}=4a^2-3a+7$

75)
$$\begin{array}{r}
5h^2-3h-2 \\
2h^2-9\overline{)\;10h^4-6h^3-49h^2+27h+19} \\
\underline{-(10h^4 \qquad -45h^2)} \\
-6h^3-4h^2+27h \\
\underline{-(-6h^3 \qquad +27h)} \\
-4h^2 \qquad +19 \\
\underline{-(-4h^2 \qquad +18)} \\
1
\end{array}$$

$$\frac{10h^4-6h^3-49h^2+27h+19}{2h^2-9}=5h^2-3h-2+\frac{1}{2h^2-9}$$

77)
$$\begin{array}{r}
3d^2-d-8 \\
2d^2+7d+5\overline{)\;6d^4+19d^3-8d^2-61d-40} \\
\underline{-(6d^4+21d^3+15d^2)} \\
-2d^3-23d^2-61d \\
\underline{-\;(-2d^3-7d^2-5d)} \\
-16d^2-56d-40 \\
\underline{-\;(-16d^2-56d-40)} \\
0
\end{array}$$

79)
$$\begin{array}{r}
9c^2+8c+3 \\
c^2-10c+4\overline{)\;9c^4-82c^3-41c^2+9c+16} \\
\underline{-\;(9c^4-90c^3+36c^2)} \\
8c^3-77c^2+9c \\
\underline{-\;(8c^3-80c^2+32c)} \\
3c^2-23c+16 \\
\underline{-\;(3c^2-30c+12)} \\
7c+4
\end{array}$$

$$\frac{9c^4-82c^3-41c^2+9c+16}{c^2-10c+4}=9c^2+8c+3+\frac{7c+4}{c^2-10c+4}$$

81) Long division of $k^4+0k^3+0k^2+0k-81$ by k^2+9:

$$\begin{array}{r} k^2 \qquad -9 \\ k^2+9\overline{)\;k^4+0k^3+0k^2+0k-81} \\ -(k^4 \qquad +9k^2) \\ \hline -9k^2+0k-81 \\ -(-9k^2 \qquad -81) \\ \hline 0 \end{array}$$

83) $\dfrac{49a^4-15a^2-14a^3+5a^6}{-7a^3}$

$=\dfrac{5a^6}{-7a^3}+\dfrac{49a^4}{-7a^3}-\dfrac{14a^3}{-7a^3}-\dfrac{15a^2}{-7a^3}$

$=-\dfrac{5}{7}a^3-7a+2+\dfrac{15}{7a}$

85) $l=$ length of the rectangle.

$4y^2-23y-6=l(y-6)$

$\dfrac{4y^2-23y-6}{y-6}=l$

$$\begin{array}{r|rrr} 6 & 4 & -23 & -6 \\ & & 24 & 6 \\ \hline & 4 & 1 & 0 \end{array}$$

$l=4y+1$

87) $w=$ width of the rectangle.

$18a^5-45a^4+9a^3=(9a^3)w$

$\dfrac{18a^5-45a^4+9a^3}{9a^3}=w$

$\dfrac{18a^5}{9a^3}-\dfrac{45a^4}{9a^3}-\dfrac{9a^3}{9a^3}=w$

$2a^2-5a+1=w$

89) $h=$ height of the triangle.

$b=$ base of the triangle.

$6h^3+3h^2+h=\dfrac{1}{2}\text{(base)(height)}$

$2\left(\dfrac{6h^3+3h^2+h}{\text{height}}\right)=\text{base}$

$2\left(\dfrac{6h^3+3h^2+h}{h}\right)=b$

$2\left(\dfrac{6h^3}{h}+\dfrac{3h^2}{h}+\dfrac{h}{h}\right)=b$

$2\left(6h^2+3h+1\right)=b$

$12h^2+6h+2=b$

Chapter 6 Review

1) $\dfrac{3^{10}}{3^6}=3^{10-6}=3^4=81$

3) $\left(\dfrac{5}{4}\right)^{-3}=\left(\dfrac{4}{5}\right)^3=\dfrac{64}{125}$

5) $\left(z^6\right)^3=z^{6\cdot 3}=z^{18}$

7) $\dfrac{70r^9}{10r^4}=7r^{9-4}=7r^5$

9) $(-9t)\left(6t^6\right)=-54t^{1+6}=-54t^7$

11) $\dfrac{k^3}{k^{11}}=k^{3-11}=k^{-8}=\dfrac{1}{k^8}$

13) $\left(-2a^2b\right)^3\left(5a^{-12}b\right)$

$=\left(-8a^6b^3\right)\left(5a^{-12}b\right)$

$=-40a^{6+(-12)}b^{3+1}$

$=-\dfrac{40b^4}{a^6}$

15) $\left(\frac{3pq^{-10}}{2p^{-2}q^{5}}\right)^{-2} = \left(\frac{2p^{-2}q^{5}}{3pq^{-10}}\right)^{2}$
$= \left(\frac{2q^{15}}{3p^{3}}\right)^{2}$
$= \frac{4q^{30}}{9p^{6}}$

17) $\frac{s^{-1}t^{9}}{st^{11}} = s^{-1-1}t^{9-11} = s^{-2}t^{-2} = \frac{1}{s^{2}t^{2}}$

19) $A = (2x)(4x)$
$= 8x^2$ sq. units

$P = 2(2x) + 2(4x)$
$= 4x + 8x$
$= 12x$ units

21) $x^{5t} \cdot x^{3t} = x^{5t+3t} = x^{8t}$

23) $\frac{r^{9a}}{r^{3a}} = r^{9a-3a} = r^{6a}$

25) $\left(y^{2p}\right)^{3} = y^{2p\cdot 3} = y^{6p}$

27) Degree of poly = 3

Term	Coeff	Degree
$4r^3$	4	3
$-7r^2$	-7	2
r	1	1
5	5	0

29) $-x^2y^2 - 7xy + 2x + 5, \quad x = -3;\ y = 2$
$-(-3)^2(2)^2 - 7(-3)(2) + 2(-3) + 5$
$= -(9)(4) - 7(-3)(2) + 2(-3) + 5$
$= -36 + 42 - 6 + 5 = 5$

31) a) $h(x) = 4x^2 - x - 7$
$h(-3) = 4(-3)^2 - (-3) - 7$
$h(-3) = 4(9) + 3 - 7$
$h(-3) = 36 - 4 = 32$

b) $h(x) = 4x^2 - x - 7$
$h(0) = 4(0)^2 - (0) - 7$
$h(0) = -7$

33) $\left(5t^2 + 11t - 4\right) - \left(7t^2 + t - 9\right)$
$= \left(5t^2 + 11t - 4\right) + \left(-7t^2 - t + 9\right)$
$= -2t^2 + 10t + 5$

35)
$$\begin{array}{r} 5.8p^3 - 1.2p^2 + \quad p - 7.5 \\ +\ \ 2.1p^3 + 6.3p^2 + 3.8p + 3.9 \\ \hline 7.9p^3 + 5.1p^2 + 4.8p - 3.6 \end{array}$$

37) $\left(\frac{7}{4}k^2 + \frac{1}{6}k + 5\right) - \left(\frac{1}{2}k^2 + \frac{5}{6}k - 2\right)$
$= \left(\frac{7}{4}k^2 + \frac{1}{6}k + 5\right) + \left(-\frac{2}{4}k^2 - \frac{5}{6}k + 2\right)$
$= \frac{5}{4}k^2 - \frac{4}{6}k + 7 = \frac{5}{4}k^2 - \frac{2}{3}k + 7$

39) $\left(a^2b^2+7a^2b-3ab+11\right)-\left(3a^2b^2-10a^2b+ab+6\right)$
$=\left(a^2b^2+7a^2b-3ab+11\right)+\left(-3a^2b^2+10a^2b-ab-6\right)$
$=-2a^2b^2+17a^2b-4ab+5$

41) $(4m+9n-19)+(-5m+6n+14)=-m+15n-5$

43) $\left[(4s-11)+\left(9s^2-19s+2\right)\right]-\left(4s^2+3s+17\right)=\left[9s^2-15s-9\right]+\left(-4s^2-3s-17\right)$
$=5s^2-18s-26$

45) $P=2\left(d^2+3d+5\right)+2\left(d^2-5d+2\right)=2d^2+6d+10+2d^2-10d+4=4d^2-4d+14$ units

47) $4r(7r-15)=28r^2-60r$

49) $(2w+5)\left(-12w^3+6w^2-2w+3\right)=-24w^4+12w^3-4w^2+6w-60w^3+30w^2-10w+15$
$=-24w^4-48w^3+26w^2-4w+15$

51) $(y-7)(y+8)=y^2+8y-7y-56=y^2+y-56$

53) $(3n-7)(2n-9)=6n^2-27n-14n+63=6n^2-41n+63$

55) $-(a-11)(a+12)=-\left(a^2+12a-11a-132\right)=-\left(a^2+a-132\right)=-a^2-a+132$

57) $7u^4v^2\left(-8u^2v+7uv^2+12u-3\right)=-56u^6v^3+49u^5v^4+84u^5v^2-21u^4v^2$

59) $(3x-8y)(2x+y)=6x^2+3xy-16xy-8y^2=6x^2-13xy-8y^2$

61) $(ab+5)(ab+6)=a^2b^2+6ab+5ab+30=a^2b^2+11ab+30$

63) $\left(x^2+4x-11\right)\left(12x^4-7x^2+9\right)=12x^6-7x^4+9x^2+48x^5-28x^3+36x-132x^4+77x^2-99$
$=12x^6+48x^5-139x^4-28x^3+86x^2+36x-99$

65) $6c^3(4c-5)(c-2)=6c^3\left(4c^2-8c-5c+10\right)=6c^3\left(4c^2-13c+10\right)=24c^5-78c^4+60c^3$

67) $(z+4)(z+1)(z+5)$
$=(z+4)(z^2+5z+z+5)$
$=(z+4)(z^2+6z+5)$
$=z^3+6z^2+5z+4z^2+24z+20$
$=z^3+10z^2+29z+20$

69) $\left(\frac{3}{5}m+2\right)\left(\frac{1}{3}m-4\right)$
$=\frac{1}{5}m^2-\frac{12}{5}m+\frac{2}{3}m-8$
$=\frac{1}{5}m^2-\frac{36}{15}m+\frac{10}{15}m-8$
$=\frac{1}{5}m^2-\frac{26}{15}m-8$

71) $(b+7)^2=b^2+14b+49$

73) $(5q-2)^2=25q^2-20q+4$

75) $(x-2)^3=(x-2)^2(x-2)$
$=(x^2-4x+4)(x-2)$
$=x^3-4x^2+4x-2x^2$
$+8x-8$
$=x^3-6x^2+12x-8$

77) $(z+9)(z-9)=(z)^2-(9)^2$
$=z^2-81$

79) $\left(\frac{1}{5}n-2\right)\left(\frac{1}{5}n+2\right)$
$=\left(\frac{1}{5}n+2\right)\left(\frac{1}{5}n-2\right)$
$=\frac{1}{25}n^2-4$

81) $\left(\frac{7}{8}-r^2\right)\left(\frac{7}{8}+r^2\right)=\left(\frac{7}{8}+r^2\right)\left(\frac{7}{8}-r^2\right)$
$=\frac{49}{64}-r^4$

83) $-2(3c-4)^2=-2(9c^2-24c+16)$
$=-18c^2+48c-32$

85) a) $A=(4m+5)(m-3)$
$=4m^2-12m+5m-15$
$=4m^2-7m-15$ sq. units

b) $P=2(4m+5)+2(m-3)$
$=8m+10+2m-6$
$=10m+4$ units

87) $\frac{8t^5-14t^4-20t^3}{2t^3}=\frac{8t^5}{2t^3}-\frac{14t^4}{2t^3}-\frac{20t^3}{2t^3}$
$=4t^2-7t-10$

89)
$$\begin{array}{r|l} & c+10 \\ \hline c-2 & c^2+8c-20 \\ & -(c^2-2c) \\ \hline & 10c-20 \\ & -(10c-20) \\ \hline & 0 \end{array}$$

91)
$$\begin{array}{r|l} & 4r^2-7r+3 \\ \hline 3r+2 & 12r^3-13r^2-5r+6 \\ & -(12r^3+8r^2) \\ \hline & -21r^2-5r \\ & -(-21r^2-14r) \\ \hline & 9r+6 \\ & -(9r+6) \\ \hline & 0 \end{array}$$

93) $\dfrac{30a^3+80a^2-15a+20}{10a^2}=\dfrac{30a^3}{10a^2}+\dfrac{80a^2}{10a^2}-\dfrac{15a}{10a^2}+\dfrac{20}{10a^2}=3a+8-\dfrac{3}{2a}+\dfrac{2}{a^2}$

95) $\left(15x^4y^4-42x^3y^4-6x^2y+10y\right)\div\left(-6x^2y\right)=\dfrac{15x^4y^4-42x^3y^4-6x^2y+10y}{-6x^2y}$

$$=\frac{15x^4y^4}{-6x^2y}-\frac{42x^3y^4}{-6x^2y}-\frac{6x^2y}{-6x^2y}+\frac{10y}{-6x^2y}$$

$$=-\frac{5}{2}x^2y^3+7xy^3+1-\frac{5}{3x^2}$$

97)
$$\begin{array}{r} 2q-4 \\ 3q+7\overline{\big)\ 6q^2+2q-35} \\ \underline{-(6q^2+14q)} \\ -12q-35 \\ \underline{-(-12q-28)} \\ -7 \end{array}$$

$\left(6q^2+2q-35\right)\div\left(3q+7\right)$

$=2q-4-\dfrac{7}{3q+7}$

99)
$$\begin{array}{r} 3a+7 \\ 5a-4\overline{\big)\ 15a^2+23a-7} \\ \underline{-(15a^2-12a)} \\ 35a-7 \\ \underline{-(35a-28)} \\ 21 \end{array}$$

$\left(15a^2+23a-7\right)\div\left(5a-4\right)$

$=3a+7+\dfrac{21}{5a-4}$

101)
$$\begin{array}{r} 3m^2+m-4 \\ 2m^2+5\overline{\big)\ 6m^4+2m^3+7m^2+5m-20} \\ \underline{-(6m^4\qquad+15m^2)} \\ 2m^3-8m^2+5m \\ \underline{-(2m^3\qquad+5m)} \\ -8m^2\qquad-20 \\ \underline{-(-8m^2\qquad-20)} \\ 0 \end{array}$$

103)
$$\begin{array}{r} b^2+4b+16 \\ b-4\overline{\big)\, b^3+0b^2+0b-64} \\ \underline{-(b^3-4b^2)} \\ 4b^2+0b \\ \underline{-(4b^2-16b)} \\ 16b-64 \\ \underline{-(16b-64)} \\ 0 \end{array}$$

105)
$$\begin{array}{r} 8w^2-6w-7 \\ 4w+3\overline{\big)\, 32w^3+0w^2-46w-23} \\ \underline{-(32w^3+24w^2)} \\ -24w^2-46w \\ \underline{-(-24w^2-18w)} \\ -28w-23 \\ \underline{-(-28w-21)} \\ -2 \end{array}$$

$$\frac{32w^3-46w-23}{4w+3}$$
$$=8w^2-6w-7-\frac{2}{4w+3}$$

107)
$$\begin{array}{r} 7u^2+u+4 \\ u^2-10u+3\overline{\big)\, 7u^4-69u^3+15u^2-37u+12} \\ \underline{-(7u^4-70u^3+21u^2)} \\ u^3-6u^2-37u \\ \underline{-(u^3-10u^2+3u)} \\ 4u^2-40u+12 \\ \underline{-(4u^2-40u+12)} \\ 0 \end{array}$$

109) Let $b=$ the base.

$$15y^2+12y=\frac{1}{2}b(6y)$$
$$15y^2+12y=b(3y)$$
$$\frac{15y^2+12y}{3y}=b$$
$$\frac{15y^2}{3y}+\frac{12y}{3y}=b$$
$$5y+4=b$$

Chapter 6 Test

1) $\left(\frac{5}{3}\right)^{-3} = \left(\frac{3}{5}\right)^{3} = \frac{27}{125}$

3) $\left(9d^4\right)\left(-3d^4\right) = -27d^8$

5) $\frac{a^{12}b^{-5}}{a^8b^7} = a^{12-8}b^{-5-7} = \frac{a^4}{b^{12}}$

7) a) -1 b) 3

9) $-m^2 + 3n,\ m = -5;\ n = 8$

$-(-5)^2 + 3(8) = -25 + 24 = -1$

11) $\left(10r^3s^2 + 7r^2s^2 - 11rs + 5\right) + \left(4r^3s^2 - 9r^2s^2 + 6rs + 3\right) = 14r^3s^2 - 2r^2s^2 - 5rs + 8$

13) $6\left(-n^3 + 4n - 2\right) - 3\left(2n^3 + 5n^2 + 8n - 1\right) = -6n^3 + 24n - 12 - 6n^3 - 15n^2 - 24n + 3$

$= -12n^3 - 15n^2 - 9$

15) $(3y+5)(2y+1) = 6y^2 + 3y + 10y + 5 = 6y^2 + 13y + 5$

17) $(2a - 5b)(3a + b) = 6a^2 + 2ab - 15ab - 5b^2 = 6a^2 - 13ab - 5b^2$

19) $3x(x+4)^2 = 3x\left(x^2 + 8x + 16\right) = 3x^3 + 24x^2 + 48x$

21) $(s-4)^3 = (s-4)^2(s-4) = \left(s^2 - 8s + 16\right)(s-4) = s^3 - 4s^2 - 8s^2 + 32s + 16s - 64$

$= s^3 - 12s^2 + 48s - 64$

23) $\frac{24t^5 - 60t^4 + 12t^3 - 8t^2}{12t^3} = \frac{24t^5}{12t^3} - \frac{60t^4}{12t^3} + \frac{12t^3}{12t^3} - \frac{8t^2}{12t^3} = 2t^2 - 5t + 1 - \frac{2}{3t}$

25) $x-2\overline{\smash{)}\,x^3+0x^2+0x-8}$ with quotient x^2+2x+4

$$\begin{array}{r} x^3+0x^2+0x-8 \\ -(x^3-2x^2) \\ \hline 2x^2+0x \\ -(2x^2-4x) \\ \hline 4x-8 \\ -(4x-8) \\ \hline 0 \end{array}$$

27) Let $b =$ the base.

$$12k^2+28k=\frac{1}{2}b(8k)$$
$$12k^2+28k=b(4k)$$
$$\frac{12k^2+28k}{4k}=b$$
$$\frac{12k^2}{4k}+\frac{28k}{4k}=b$$
$$3k+7=b$$

Cumulative Review: Chapters 1-6

1) a) $43, 0$

b) $-14, 43, 0$

c) $\frac{6}{11}, -14, 2.7, 43, 0.\overline{65}, 0$

3) $2\frac{6}{7}\div 1\frac{4}{21}=\frac{20}{7}\div\frac{25}{21}$
$$=\frac{20}{7}\cdot\frac{21}{25}$$
$$=\frac{4}{1}\cdot\frac{3}{5}$$
$$=\frac{12}{5}\text{ or }2\frac{2}{5}$$

5) $\left(\frac{2n^{-10}}{n^{-4}}\right)^3=\left(2n^{-10-(-4)}\right)^3$
$$=\left(2n^{-6}\right)^3$$
$$=8n^{-18}$$
$$=\frac{8}{n^{18}}$$

7) $-\frac{12}{5}c-7=20$
$$-\frac{12}{5}c=27$$
$$c=-\frac{5}{12}\cdot 27$$
$$c=-\frac{5\cdot 9}{4}$$
$$c=-\frac{45}{4},\ \left\{-\frac{45}{4}\right\}$$

9) $3y+16<4$ or $8-y\ge 7$
$$3y<-12 \text{ or } -y\ge -1$$
$$y<-4 \text{ or } y\le 1$$
$$(-\infty, 1]$$

11) $5x-2y=10$

x-int: Let $y=0$, and solve for x.
$$5x-2(0)=10$$
$$5x=10$$
$$x=2 \qquad (2, 0)$$

y-int: Let $x=0$, and solve for y.
$$5(0)-2y=10$$
$$-2y=10$$
$$y=-5 \qquad (0, -5)$$

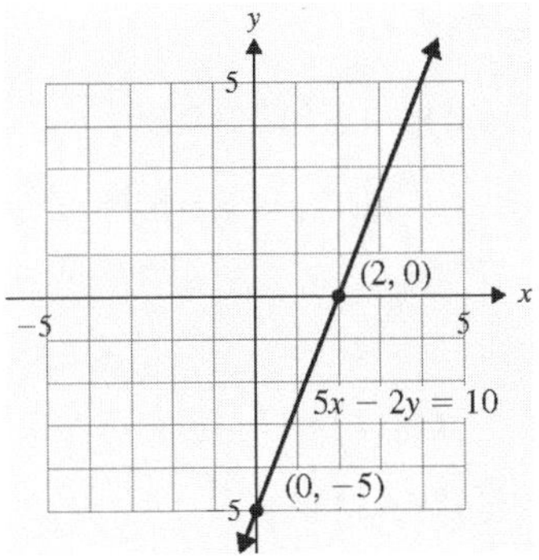

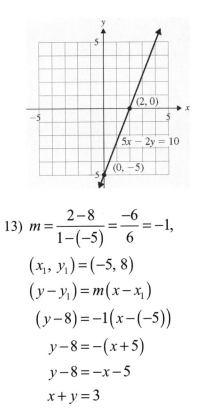

13) $m=\dfrac{2-8}{1-(-5)}=\dfrac{-6}{6}=-1,$

$(x_1, y_1)=(-5, 8)$

$(y-y_1)=m(x-x_1)$

$(y-8)=-1(x-(-5))$

$y-8=-(x+5)$

$y-8=-x-5$

$x+y=3$

15) Let l = the length of the rectangle.

Let w = the width of the rectangle.

$l=2w-7$

$2l+2w=76$

Use substitution.

$2(2w-7)+2w=76$

$4w-14+2w=76$

$6w-14=76$

$6w=90$

$w=15$

$l=2w-7; \quad w=15$

$l=2(15)-7$

$l=30-7$

$l=23$

width: 15 cm, length : 23 cm

17) $(k-9)(k+6)=k^2+6k-9k-54$

$=k^2-3k-54$

19) $\dfrac{8a^4b^4-20a^3b^2+56ab+8b}{8a^3b^3}$

$=\dfrac{8a^4b^4}{8a^3b^3}-\dfrac{20a^3b^2}{8a^3b^3}+\dfrac{56ab}{8a^3b^3}+\dfrac{8b}{8a^3b^3}$

$=ab-\dfrac{5}{2b}+\dfrac{7}{a^2b^2}+\dfrac{1}{a^3b^2}$

21) $$\begin{array}{r|l} & 2p^2-3p+1 \\ \hline p+4 & 2p^3+5p^2-11p+9 \\ & \underline{-(2p^3+8p^2)} \\ & \quad -3p^2-11p \\ & \quad \underline{-(-3p^2-12p)} \\ & \qquad p+9 \\ & \qquad \underline{-(p+4)} \\ & \qquad\quad 5 \end{array}$$

$(2p^3+5p^2-11p+9)\div(p+4)$

$=2p^2-3p+1+\dfrac{5}{p+4}$

23) $(2c-5)(c-3)^2$

$=(2c-5)(c^2-6c+9)$

$=2c^3-12c^2+18c-5c^2+30c-45$

$=2c^3-17c^2+48c-45$

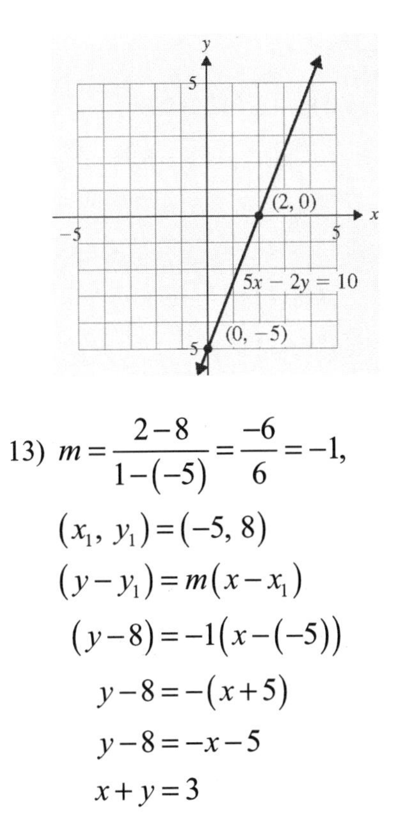

Section 7.1 Exercises

1) $45m^3 = 3\cdot3\cdot5\cdot m\cdot m\cdot m$,
$20m^2 = 2\cdot2\cdot5\cdot m\cdot m$
GCF of $45m^3$ and $20m^2$
$= 5\cdot m\cdot m = 5m^2$

3) $42k^5 = 2\cdot3\cdot7\cdot k\cdot k\cdot k\cdot k\cdot k$,
$54k^7 = 2\cdot3\cdot3\cdot3\cdot k\cdot k\cdot k\cdot k\cdot k\cdot k\cdot k$
$72k^9 = 2\cdot2\cdot2\cdot3\cdot3$
$\cdot k\cdot k\cdot k\cdot k\cdot k\cdot k\cdot k\cdot k\cdot k$
GCF of $42k^5$, $54k^7$ and $72k^9$
$= 2\cdot3\cdot k\cdot k\cdot k\cdot k\cdot k = 6k^5$

5) $9x^2y$

7) $4uv^3$

9) s^2t

11) $(n-7)$

13) Answers may vary.

15) GCF $= 6$
$30s+18 = (6)(5s)+(6)(3)$
$= 6(5s+3)$

17) GCF $= 4$
$24z-4 = (4)(6z)-(4)(1)$
$= 4(6z-1)$

19) GCF $= 3d$
$3d^2-6d = (3d)(d)-(3d)(2)$
$= 3d(d-2)$

21) GCF $= 7y^2$
$42y^2+35y^3 = (7y^2)(6)+(7y^2)(5y)$
$= 7y^2(6+5y)$

23) GCF $= t^4$
$t^5-t^4 = (t^4)(t)-(t^4)(1)$
$= t^4(t-1)$

25) GCF $= \frac{1}{2}c$
$\frac{1}{2}c^2+\frac{5}{2}c = \left(\frac{1}{2}c\right)(c)+\left(\frac{1}{2}c\right)(5)$
$= \frac{1}{2}c(c+5)$

27) GCF $= 5n^3$
$10n^5-5n^4+40n^3$
$= (5n^3)(2n^2)-(5n^3)(n)+(5n^3)(8)$
$= 5n^3(2n^2-n+8)$

29) GCF $= 2v^5$
$2v^8-18v^7-24v^6+2v^5$
$= (2v^5)(v^3)-(2v^5)(9v^2)$
$-(2v^5)(12v)+(2v^5)(1)$
$= 2v^5(v^3-9v^2-12v+1)$

31) GCF $=$ none
$8c^3+3d^2$ does not factor

33) GCF $= a^3b^2$
$a^4b^2+4a^3b^3$
$= (a^3b^2)(a)+(a^3b^2)(4b)$
$= a^3b^2(a+4b)$

35) $\text{GCF} = 10x^2y$
$$50x^3y^3 - 70x^3y^2 + 40x^2y$$
$$= (10x^2y)(5xy^2) - (10x^2y)(7xy) + (10x^2y)(4)$$
$$= 10x^2y(5xy^2 - 7xy + 4)$$

37) $\text{GCF} = n - 12$
$$m(n-12) + 8(n-12)$$
$$= (n-12)(m+8)$$

39) $\text{GCF} = 8r - 3$
$$p(8r-3) - q(8r-3)$$
$$= (8r-3)(p-q)$$

41) $\text{GCF} = z + 11$
$$y(z+11) + (z+11)$$
$$= y(z+11) + 1(z+11)$$
$$= (z+11)(y+1)$$

43) $\text{GCF} = 3r + 4$
$$2k^2(3r+4) - (3r+4)$$
$$= 2k^2(3r+4) - 1(3r+4)$$
$$= (3r+4)(2k^2-1)$$

45) $-64m - 40 = (-8)(8m) + (-8)(5)$
$$= -8(8m+5)$$

47) $-5t^3 + 10t^2 = (-5t^2)(t) + (-5t^2)(-2)$
$$= -5t^2(t-2)$$

49) $-3a^3 + 7a^2 - a$
$$= (-a)(3a^2) + (-a)(-7a) + (-a)(1)$$
$$= -a(3a^2 - 7a + 1)$$

51) $-b + 8 = (-1)(b) + (-1)(-8)$
$$= -1(b-8)$$

53) $kt + 3k + 8t + 24$
Factor out k from the first two terms.
Factor out 8 from the last two terms.
$$= k(t+3) + 8(t+3)$$
$$= (t+3)(k+8)$$

55) $fg - 7f + 4g - 28$
Factor out f from the first two terms.
Factor out 4 from the last two terms.
$$= f(g-7) + 4(g-7)$$
$$= (g-7)(f+4)$$

57) $2rs - 6r + 5s - 15$
Factor out $2r$ from the first two terms.
Factor out 5 from the last two terms.
$$= 2r(s-3) + 5(s-3)$$
$$= (s-3)(2r+5)$$

59) $3xy - 2y + 27x - 18$
Factor out y from the first two terms.
Factor out 9 from the last two terms.
$$= y(3x-2) + 9(3x-2)$$
$$= (3x-2)(y+9)$$

61) $8b^2 + 20bc + 2bc^2 + 5c^3$
$$= 4b(2b+5c) + c^2(2b+5c)$$
$$= (2b+5c)(4b+c^2)$$

63) $4a^3 - 12ab + a^2b - 3b^2$
$$= 4a(a^2-3b) + b(a^2-3b)$$
$$= (a^2-3b)(4a+b)$$

65) $kt + 7t - 5k - 35 = t(k+7) - 5(k+7)$
$$= (k+7)(t-5)$$

67) $mn-8m-10n+80$
$=m(n-8)-10(n-8)$
$=(n-8)(m-10)$

69) $dg-d+g-1=d(g-1)+1(g-1)$
$=(g-1)(d+1)$

71) $5tu+6t-5u-6$
$=t(5u+6)-1(5u+6)$
$=(5u+6)(t-1)$

73) $36g^4+3gh-96g^3h-8h^2$
$=3g(12g^3+h)-8h(12g^3+h)$
$=(12g^3+h)(3g-8h)$

75) Answers may vary.

77) $2ab+8a+6b+24$
$=2(ab+4a+3b+12)$
$=2[a(b+4)+3(b+4)]$
$=2(b+4)(a+3)$

79) $8s^2t-40st+16s^2-80s$
$=8s(st-5t+2s-10)$
$=8s[t(s-5)+2(s-5)]$
$=8s(s-5)(t+2)$

81) $7cd+12+28c+3d$
$=7cd+28c+3d+12$
$=7c(d+4)+3(d+4)$
$=(d+4)(7c+3)$

83) $42k^3+15d^2-18k^2d-35kd$
$=42k^3-35kd-18k^2d+15d^2$
$=7k(6k^2-5d)-3d(6k^2-5d)$
$=(7k-3d)(6k^2-5d)$

85) $9f^2j^2+45fj+9fj^2+45f^2j$
$=9fj(fj+5+j+5f)$
$=9fj(fj+j+5f+5)$
$=9fj[j(f+1)+5(f+1)]$
$=9fj(f+1)(j+5)$

87) $4x^4y-14x^3+28x^4-2x^3y$
$=2x^3(2xy-7+14x-y)$
$=2x^3(2xy-y+14x-7)$
$=2x^3[y(2x-1)+7(2x-1)]$
$=2x^3(2x-1)(y+7)$

Section 7.2 Exercises

1) a) 10, 3 b) 7, −4

c) −8, 1 d) −8, −7

3) They are negative.

5) Can I factor out a GCF?

7) Can I factor again?
If so, factor again.

9) $n^2+12n+27=(n+9)(n+3)$

11) $c^2-14c+45=(c-5)(c-9)$

13) $x^2+x-56=(x-7)(x+8)$

15) $g^2+8g+12=(g+6)(g+2)$

17) $w^2+13w+42=(w+7)(w+6)$

19) $c^2-13c+36=(c-4)(c-9)$

21) $b^2-2b-8=(b-4)(b+2)$

23) $u^2+u-132=(u+12)(u-11)$

25) $q^2-8q+15=(q-5)(q-3)$

27) $y^2+9y+10$ is prime

29) $w^2+4w-5=(w+5)(w-1)$

31) $p^2-20p+100$
$=(p-10)(p-10)$ or $(p-10)^2$

33) $24+14d+d^2=d^2+14d+24$
$=(d+12)(d+2)$

35) $-a^2-10a-16=-(a^2+10a+16)$
$=-(a+8)(a+2)$

37) $-h^2+2h+15=-(h^2-2h-15)$
$=-(h-5)(h+3)$

39) $-k^2+11k-28=-(k^2-11k+28)$
$=-(k-7)(k-4)$

41) $-x^2-x+90=-(x^2+x-90)$
$=-(x+10)(x-9)$

43) $-n^2-14n-49=-(n^2+14n+49)$
$=-(n+7)(n+7)$
or $-(n+7)^2$

45) $a^2+6ab+5b^2=(a+5b)(a+b)$

47) $m^2+4mn-21n^2=(m-3n)(m+7n)$

49) $x^2-15xy+36y^2=(x-12y)(x-3y)$

51) $f^2-10fg-11g^2=(f+g)(f-11g)$

53) $c^2+6cd-55d^2=(c-5d)(c+11d)$

55) $2r^2+8r+6=2(r^2+4r+3)$
$=2(r+3)(r+1)$

57) $4q^3-28q^2+48q=4q(q^2-7q+12)$
$=4q(q-3)(q-4)$

59) $m^4n+7m^3n^2-44m^2n^3$
$=m^2n(m^2+7mn-44n^2)$
$=m^2n(m-4n)(m+11n)$

61) $p^3q-17p^2q^2+70pq^3$
$=pq(p^2-17pq+70q^2)$
$=pq(p-7q)(p-10q)$

63) $18-11r+r^2=r^2-11r+18$
$=(r-9)(r-2)$

65) $7c^3d^2-7c^2d^2-14cd^2$
$=7cd^2(c^2-c-2)$
$=7cd^2(c-2)(c+1)$

67) $s^2+4st+5t^2$ is prime

69) $2r^4+26r^3+84r^2$
$=2r^2(r^2+13r+42)$
$=2r^2(r+6)(r+7)$

71) $8x^4y^5-16x^3y^4-64x^2y^3$
$=8x^2y^3(x^2y^2-2xy-8)$
$=8x^2y^3(xy-4)(xy+2)$

73) $(a+b)k^2+7(a+b)k-18(a+b)$
$=(a+b)(k^2+7k-18)$
$=(a+b)(k+9)(k-2)$

75) $(x+y)t^2-4(x+y)t-21(x+y)$
$=(x+y)(t^2-4t-21)$
$=(x+y)(t+3)(t-7)$

77) No; from $(2x+8)$ you can factor out a 2. The correct answer is $2(x+4)(x+5.)$

Section 7.3 Exercises

1) a) 10, −5 b) −27, −1

c) 6, 2 d) −12, 6

3) $2k^2+10k+9k+45$
$=2k(k+5)+9(k+5)$
$=(2k+9)(k+5)$

5) $7y^2-7y-6y+6$
$=7y(y-1)-6(y-1)$
$=(7y-6)(y-1)$

7) $8a^2-14ab+12ab-21b^2$
$=2a(4a-7b)+3b(4a-7b)$
$=(4a-7b)(2a+3b)$

9) Can I factor out a GCF?

11) The 2 can be factored out of $2x-4$, but cannot be factored out of $2x^2+13x-24$.

13) $4a^2+17a+18=(4a+9)(a+2)$

15) $6k^2-5k-21=(3k-7)(2k+3)$

17) $18x^2-17xy+4y^2=(2x-y)(9x-4y)$

19) $2r^2+11r+15=2r^2+6r+5r+15$
$=2r(r+3)+5(r+3)$
$=(2r+5)(r+3)$

21) $5p^2-21p+4=5p^2-20p-p+4$
$=5p(p-4)-1(p-4)$
$=(5p-1)(p-4)$

23) $11m^2-18m-8$
$=11m^2-22m+4m-8$
$=11m(m-2)+4(m-2)$
$=(11m+4)(m-2)$

25) $6v^2+11v-7=6v^2-3v+14v-7$
$=3v(2v-1)+7(2v-1)$
$=(3v+7)(2v-1)$

27) $10c^2+19c+6$
$=10c^2+15c+4c+6$
$=5c(2c+3)+2(2c+3)$
$=(5c+2)(2c+3)$

29) $9x^2-13xy+4y^2$
$=9x^2-9xy-4xy+4y^2$
$=9x(x-y)-4y(x-y)$
$=(9x-4y)(x-y)$

31) $5w^2+11w+6=(5w+6)(w+1)$

33) $3u^2-23u+30=(3u-5)(u-6)$

35) $7k^2+15k-18=(7k-6)(k+3)$

37) $8r^2+26r+15=(4r+3)(2r+5)$

39) $6v^2-19v+14=(6v-7)(v-2)$

41) $21d^2-22d-8=(7d+2)(3d-4)$

43) $48v^2+64v+5=(12v+1)(4v+5)$

45) $10a^2-13ab+4b^2$
$=(5a-4b)(2a-b)$

47) $(3t+4)(2t-1)$-the answer is the same.

49) $2y^2-19y+24=(2y-3)(y-8)$

51) $12c^3+15c^2-18c=3c(4c^2+5c-6)$
$=3c(4c-3)(c+2)$

53) $12t^2-28t-5=(2t-5)(6t+1)$

55) $45h^2+57h+18=3(15h^2+19h+6)$
$=3(5h+3)(3h+2)$

57) $3b^2-7b+5$ is prime.

59) $13t^2+17t-18=(13t-9)(t+2)$

61) $5c^2+23cd+12d^2$
$=(5c+3d)(c+4d)$

63) $2d^2+2d-40=2(d^2+d-20)$
$=2(d+5)(d-4)$

65) $8c^2d^3+4c^2d^2-60c^2d$
$=4c^2d(2d^2+d-15)$
$=4c^2d(2d-5)(d+3)$

67) $36a^2-12a+1=(6a-1)^2$

69) $3x^2(y+6)^2-11x(y+6)^2-20(y+6)^2$
$=(y+6)^2(3x^2-11x-20)$
$=(y+6)^2(3x+4)(x-5)$

71) $9y^2(z-10)^3+76y(z-10)^3+32(z-10)^3$
$=(z-10)^3(9y^2+76y+32)$
$=(z-10)^3(9y+4)(y+8)$

73) $8u^2(v+8)-38u(v+8)-33(v+8)$
$=(v+8)(8u^2-38u-33)$
$=(v+8)(2u-11)(4u+3)$

75) $-h^2-3h+54=-(h^2+3h-54)$
$=-(h+9)(h-6)$

77) $-10z^2+19z-6=-(10z^2-19z+6)$
$=-(5z-2)(2z-3)$

79) $-21v^2+54v+27$
$=-3(7v^2-18v-9)$
$=-3(7v+3)(v-3)$

81) $-2j^3-32j^2-120j$
$=-2j(j^2+16j+60)$
$=-2j(j+10)(j+6)$

83) $-16y^2-34y+15$
$=-(16y^2+34y-15)$
$=-(8y-3)(2y+5)$

85) $-6c^3d+27c^2d^2-12cd^3$
$=-3cd(2c^2-9cd+4d^2)$
$=-3cd(2c-d)(c-4d)$

Section 7.4 Exercises

1) a) $6^2=36$ b) $10^2=100$

c) $4^2=16$ d) $11^2=121$

e) $3^2=9$ f) $8^2=64$

g) $12^2=144$ h) $\left(\frac{1}{2}\right)^2=\frac{1}{4}$

i) $\left(\frac{3}{5}\right)^2=\frac{9}{25}$

3) a) $(n^2)^2=n^4$ b) $(5t)^2=25t^2$

c) $(7k)^2=49k^2$

d) $(4p^2)^2=16p^4$

e) $\left(\frac{1}{3}\right)^2=\frac{1}{9}$ f) $\left(\frac{5}{2}\right)^2=\frac{25}{4}$

5) $z^2+18z+81$

7) The middle term does not equal $2(3c)(-4)$. It would have to equal $-24c$ to be a perfect trinomial.

9) $t^2+16t+64=(t)^2+(2\cdot t\cdot 8)+(8)^2$
$=(t+8)^2$

11) $g^2-18g+81=(g)^2-(2\cdot g\cdot 9)+(9)^2$
$=(g-9)^2$

13) $4y^2+12y+9$
$=(2y)^2+(2\cdot 2y\cdot 3)+(3)^2$
$=(2y+3)^2$

15) $9k^2-24k+16$
$=(3k)^2-(2\cdot 3k\cdot 4)+(4)^2$
$=(3k-4)^2$

17) $a^2+\frac{2}{3}a+\frac{1}{9}$
$=(a)^2+\left(2\cdot a\cdot\frac{1}{3}\right)+\left(\frac{1}{3}\right)^2$
$=\left(a+\frac{1}{3}\right)^2$

19) $v^2 - 3v + \dfrac{9}{4}$

$= (v)^2 - \left(2 \cdot v \cdot \dfrac{3}{2}\right) + \left(\dfrac{3}{2}\right)^2$

$= \left(v - \dfrac{3}{2}\right)^2$

21) $x^2 + 6xy + 9y^2$

$= (x)^2 + (2 \cdot x \cdot 3y) + (3y)^2$

$= (x + 3y)^2$

23) $36t^2 - 60tu + 25u^2$

$= (6t)^2 - (2 \cdot 6t \cdot 5u) + (5u)^2$

$= (6t - 5u)^2$

25) $4f^2 + 24f + 36$

$= 4(f^2 + 6f + 9)$

$= 4\left[(f)^2 + (2 \cdot f \cdot 3) + (3)^2\right]$

$= 4(f + 3)^2$

27) $2p^4 - 24p^3 + 72p^2$

$= 2p^2(p^2 - 12p + 36)$

$= 2p^2\left[(p)^2 - (2 \cdot p \cdot 6) + (6)^2\right]$

$= 2p^2(p - 6)^2$

29) $-18d^2 - 60d - 50$

$= -2(9d^2 + 30d + 25)$

$= -2\left[(3d)^2 + (2 \cdot 3d \cdot 5) + (5)^2\right]$

$= -2(3d + 5)^2$

31) $12c^3 + 3c^2 + 27c = 3c(4c^2 + c + 9)$

33) a) $x^2 - 16$ b) $16 - x^2$

35) $x^2 - 9 = (x)^2 - (3)^2 = (x + 3)(x - 3)$

37) $n^2 - 121 = (n)^2 - (11)^2$

$= (n + 11)(n - 11)$

39) $m^2 + 64$ is prime.

41) $y^2 - \dfrac{1}{25} = (y)^2 - \left(\dfrac{1}{5}\right)^2$

$= \left(y + \dfrac{1}{5}\right)\left(y - \dfrac{1}{5}\right)$

43) $c^2 - \dfrac{9}{16} = (c)^2 - \left(\dfrac{3}{4}\right)^2$

$= \left(c + \dfrac{3}{4}\right)\left(c - \dfrac{3}{4}\right)$

45) $36 - h^2 = (6)^2 - (h)^2 = (6 + h)(6 - h)$

47) $169 - a^2 = (13)^2 - (a)^2$

$= (13 + a)(13 - a)$

49) $\dfrac{49}{64} - j^2 = \left(\dfrac{7}{8}\right)^2 - (j)^2$

$= \left(\dfrac{7}{8} + j\right)\left(\dfrac{7}{8} - j\right)$

51) $100m^2 - 49 = (10m)^2 - (7)^2$

$= (10m + 7)(10m - 7)$

53) $16p^2 - 81 = (4p)^2 - (9)^2$

$= (4p + 9)(4p - 9)$

55) $4t^2 + 25$ is prime.

57) $\frac{1}{4}k^2-\frac{4}{9}=\left(\frac{1}{2}k\right)^2-\left(\frac{2}{3}\right)^2$
$=\left(\frac{1}{2}k+\frac{2}{3}\right)\left(\frac{1}{2}k-\frac{2}{3}\right)$

59) $b^4-64=\left(b^2\right)^2-(8)^2$
$=\left(b^2+8\right)\left(b^2-8\right)$

61) $144m^2-n^4=(12m)^2-\left(n^2\right)^2$
$=\left(12m+n^2\right)\left(12m-n^2\right)$

63) $r^4-1=\left(r^2\right)^2-(1)^2$
$=\left(r^2+1\right)\left(r^2-1\right)$
$=\left(r^2+1\right)\left[(r)^2-(1)^2\right]$
$=\left(r^2+1\right)(r+1)(r-1)$

65) $16h^4-g^4$
$=\left(4h^2\right)^2-\left(g^2\right)^2$
$=\left(4h^2+g^2\right)\left(4h^2-g^2\right)$
$=\left(4h^2+g^2\right)\left[(2h)^2-(g)^2\right]$
$=\left(4h^2+g^2\right)(2h+g)(2h-g)$

67) $4a^2-100=4\left(a^2-25\right)$
$=4\left[(a)^2-(5)^2\right]$
$=4(a+5)(a-5)$

69) $2m^2-128=2\left(m^2-64\right)$
$=2\left[(m)^2-(8)^2\right]$
$=2(m+8)(m-8)$

71) $45r^4-5r^2=5r^2\left(9r^2-1\right)$
$=5r^2\left[(3r)^2-(1)^2\right]$
$=5r^2(3r+1)(3r-1)$

73) a) $4^3=64$ b) $1^3=1$

c) $10^3=1000$ d) $3^3=27$

e) $5^3=125$ f) $2^3=8$

75) a) $(y)^3=y^3$ b) $(2c)^3=8c^3$

c) $(5r)^3=125r^3$

d) $\left(x^2\right)^3=x^6$

77) $d^3+1=(d)^3+(1)^3$
$=(d+1)\left(d^2-d+1\right)$

79) $p^3-27=(p)^3-(3)^3$
$=(p-3)\left(p^2+3p+9\right)$

81) $k^3+64=(k)^3+(4)^3$
$=(k+4)\left(k^2-4k+16\right)$

83) $27m^3-125$
$=(3m)^3-(5)^3$
$=(3m-5)\left(9m^2+15m+25\right)$

85) $125y^3-8=(5y)^3-(2)^3$
$=(5y-2)\left(25y^2+10y+4\right)$

87) $1000c^3 - d^3 = (10c)^3 - (d)^3 = (10c - d)(100c^2 + 10cd + d^2)$

89) $8j^3 + 27k^3 = (2j)^3 + (3k)^3 = (2j + 3k)(4j^2 - 6jk + 9k^2)$

91) $64x^3 + 125y^3 = (4x)^3 + (5y)^3 = (4x + 5y)(16x^2 - 20xy + 25y^2)$

93) $6c^3 + 48 = 6(c^3 + 8) = 6\left[(c)^3 + (2)^3\right] = 6(c + 2)(c^2 - 2c + 4)$

95) $7v^3 - 7000w^3 = 7(v^3 - 1000w^3) = 7\left[(v)^3 - (10w)^3\right] = 7(v - 10w)(v^2 + 10vw + 100w^2)$

97) $h^6 - 64 = (h^3)^2 - (8)^2 = (h^3 + 8)(h^3 - 8) = \left[(h)^3 + (2)^3\right]\left[(h)^3 - (2)^3\right]$
$= (h + 2)(h^2 - 2h + 4)(h - 2)(h^2 + 2h + 4)$

99) $(x + 5)^2 - (x - 2)^2 = \left[(x + 5) + (x - 2)\right]\left[(x + 5) - (x - 2)\right]$
$= (2x + 3)(x + 5 - x + 2) = 7(2x + 3)$

101) $(2p + 3)^2 - (p + 4)^2 = \left[(2p + 3) + (p + 4)\right]\left[(2p + 3) - (p + 4)\right]$
$= (3p + 7)(2p + 3 - p - 4) = (3p + 7)(p - 1)$

103) $(t + 5)^3 + 8 = (t + 5)^3 + (2)^3 = \left[(t + 5) + 2\right]\left[(t + 5)^2 - 2(t + 5) + 4\right]$
$= (t + 7)(t^2 + 10t + 25 - 2t - 10 + 4) = (t + 7)(t^2 + 8t + 19)$

105) $(k - 9)^3 - 1 = (k - 9)^3 - (1)^3 = \left[(k - 9) - 1\right]\left[(k - 9)^2 + 1(k - 9) + 1\right]$
$= (k - 10)(k^2 - 18k + 81 + k - 9 + 1) = (k - 10)(k^2 - 17k + 73)$

Ch. 7 Mid-Chapter Summary Exercises

1) $m^2 + 16m + 60 = (m + 10)(m + 6)$

3) $uv + 6u + 9v + 54 = u(v + 6) + 9(v + 6) = (u + 9)(v + 6)$

5) $3k^2 - 14k + 8 = (3k - 2)(k - 4)$

7) $16d^6 + 8d^5 + 72d^4 = 8d^4(2d^2 + d + 9)$

9) $60w^3+70w^2-50w$
$=10w(6w^2+7w-5)$
$=10w(3w+5)(2w-1)$

11) $t^3+1000=(t)^3+(10)^3$
$=(t+10)(t^2-10t+100)$

13) $49-p^2=(7)^2-(p)^2$
$=(7+p)(7-p)$

15) $4x^2+4xy+y^2$
$=(2x)^2+(2\cdot 2x\cdot y)+(y)^2$
$=(2x+y)^2$

17) $3z^4-21z^3-24z^2=3z^2(z^2-7z-8)$
$=3z^2(z-8)(z+1)$

19) $4b^2+1$ is prime

21) $40x^3-135=5(8x^3-27)$
$=5\left[(2x)^3-(3)^3\right]$
$=5(2x-3)(4x^2+6x+9)$

23) $c^2-\frac{1}{4}=(c)^2-\left(\frac{1}{2}\right)^2$
$=\left(c+\frac{1}{2}\right)\left(c-\frac{1}{2}\right)$

25) $45s^2t+4-36s^2-5t$
$=45s^2t-36s^2-5t+4$
$=9s^2(5t-4)-1(5t-4)$
$=(9s^2-1)(5t-4)$
$=(3s+1)(3s-1)(5t-4)$

27) $k^2+9km+18m^2=(k+3m)(k+6m)$

29) $z^2-3z-88=(z-11)(z+8)$

31) $80y^2-40y+5$
$=5(16y^2-8y+1)$
$=5\left[(4y)^2-(2\cdot 4y\cdot 1)+(1)^2\right]$
$=5(4y-1)^2$

33) $20c^2+26cd+6d^2$
$=2(10c^2+13cd+3d^2)$
$=2(10c+3d)(c+d)$

35) $n^4-16m^4=(n^2)^2-(4m^2)^2$
$=(n^2+4m^2)(n^2-4m^2)$
$=(n^2+4m^2)\left[(n)^2-(2m)^2\right]$
$=(n^2+4m^2)(n+2m)(n-2m)$

37) $2a^2-10a-72=2(a^2-5a-36)$
$=2(a-9)(a+4)$

39) $r^2-r+\frac{1}{4}=(r)^2-\left(2\cdot r\cdot\frac{1}{2}\right)+\left(\frac{1}{2}\right)^2$
$=\left(r-\frac{1}{2}\right)^2$

41) $28gh+16g-63h-36$
$=4g(7h+4)-9(7h+4)$
$=(4g-9)(7h+4)$

43) $8b^2-14b-15=(4b+3)(2b-5)$

45) $55a^6b^3+35a^5b^3-10a^4b-20a^2b$
$=5a^2b\left(11a^4b^2+7a^3b^2-2a^2-4\right)$

47) $2d^2-9d+3$ is prime

49) $9p^2-24pq+16q^2$
$=(3p)^2-(2\cdot 3p\cdot 4q)+(4q)^2$
$=(3p-4q)^2$

51) $30y^2+37y-7=(6y-1)(5y+7)$

53) $80a^3-270b^3$
$=10\left(8a^3-27b^3\right)$
$=10\left[(2a)^3-(3b)^3\right]$
$=10(2a-3b)\left(4a^2+6ab+9b^2\right)$

55) $rt-r-t+1=r(t-1)-1(t-1)$
$=(r-1)(t-1)$

57) $4g^2-4=4\left(g^2-1\right)$
$=4\left[(g)^2-(1)^2\right]$
$=4(g+1)(g-1)$

59) $3c^2-24c+48$
$=3\left(c^2-8c+16\right)$
$=3\left[(c)^2-(2\cdot c\cdot 4)+(4)^2\right]$
$=3(c-4)^2$

61) $144k^2-121=(12k)^2-(11)^2$
$=(12k+11)(12k-11)$

63) $-48g^2-80g-12$
$=-4\left(12g^2+20g+3\right)$
$=-4(6g+1)(2g+3)$

65) $q^3+1=(q)^3+(1)^3$
$=(q+1)\left(q^2-q+1\right)$

67) $81u^4-v^4$
$=\left(9u^2\right)^2-\left(v^2\right)^2$
$=\left(9u^2+v^2\right)\left(9u^2-v^2\right)$
$=\left(9u^2+v^2\right)\left[(3u)^2+(v)^2\right]$
$=\left(9u^2+v^2\right)(3u+v)(3u-v)$

69) $11f^2+36f+9=(11f+3)(f+3)$

71) $2j^{11}-j^3=j^3\left(2j^8-1\right)$

73) $w^2-2w-48=(w-8)(w+6)$

75) k^2+100 is prime

77) $m^2+4m+4=(m)^2+(2\cdot m\cdot 2)+(2)^2$
$=(m+2)^2$

79) $9t^2-64=(3t)^2-(8)^2$
$=(3t+8)(3t-8)$

81) $(2z+1)y^2+6(2z+1)y-55(2z+1)$
$=(2z+1)\left(y^2+6y-55\right)$
$=(2z+1)(y+11)(y-5)$

83) $(r-4)^2+11(r-4)+28$
$=[(r-4)+4][(r-4)+7]$
$=r(r+3)$

85) $(x+y)^2-(2x-y)^2$
$=[(x+y)+(2x-y)]$
$\cdot[(x+y)-(2x-y)]$
$=(3x)(x+y-2x+y)$
$=3x(2y-x)$ or $-3x(x-2y)$

87) $n^2+12n+36-p^2$
$=(n^2+12n+36)-p^2$
$=(n+6)^2-p^2$
$=(n+p+6)(n-p+6)$

Section 7.5 Exercises

1) It says that if the product of two quantities equals 0, then one or both of the quantities must be zero.

3) $(m+9)(m-8)=0$
$m+9=0$ or $m-8=0$
$m=-9$ or $m=8$ $\{-9,8\}$

5) $(q-4)(q-7)=0$
$q-4=0$ or $q-7=0$
$q=4$ or $q=7$ $\{4,7\}$

7) $(4z+3)(z-9)=0$
$4z+3=0$ or $z-9=0$
$4z=-3$
$z=-\frac{3}{4}$ or $z=9$ $\left\{-\frac{3}{4},9\right\}$

9) $11s(s+15)=0$
$11s=0$ or $s+15=0$
$s=0$ or $s=-15$ $\{-15,0\}$

11) $(6x-5)^2=0$
$6x-5=0$
$6x=5$
$x=\frac{5}{6}$ $\left\{\frac{5}{6}\right\}$

13) $(4h+7)(h+3)=0$
$4h+7=0$ or $h+3=0$
$4h=-7$
$h=-\frac{7}{4}$ or $h=-3$ $\left\{-3,-\frac{7}{4}\right\}$

15) $\left(y+\frac{3}{2}\right)\left(y-\frac{1}{4}\right)=0$
$y+\frac{3}{2}=0$ or $y-\frac{1}{4}=0$
$y=-\frac{3}{2}$ or $y=\frac{1}{4}$ $\left\{-\frac{3}{2},\frac{1}{4}\right\}$

17) $q(q-2.5)=0$
$q=0$ or $q-2.5=0$
$q=2.5$ $\{0,2.5\}$

19) $v^2+15v+56=0$
$(v+8)(v+7)=0$
$v+8=0$ or $v+7=0$
$v=-8$ or $v=-7$ $\{-8,-7\}$

21) $k^2+12k-45=0$
$(k+15)(k-3)=0$
$k+15=0$ or $k-3=0$
$k=-15$ or $k=3$ $\{-15,3\}$

23) $3y^2 - y - 10 = 0$

$(3y+5)(y-2) = 0$

$3y+5=0$ or $y-2=0$

$3y = -5$

$y = -\frac{5}{3}$ or $y = 2$ $\left\{-\frac{5}{3}, 2\right\}$

25) $14w^2 + 8w = 0$

$2w(7w+4) = 0$

$2w = 0$ or $7w+4=0$

$w = 0$ $7w = -4$

$w = -\frac{4}{7}$ $\left\{-\frac{4}{7}, 0\right\}$

27) $d^2 - 15d = -54$

$d^2 - 15d + 54 = 0$

$(d-6)(d-9) = 0$

$d-6=0$ or $d-9=0$

$d = 6$ or $d = 9$ $\{6, 9\}$

29) $t^2 - 49 = 0$

$(t+7)(t-7) = 0$

$t+7=0$ or $t-7=0$

$t = -7$ or $t = 7$ $\{-7, 7\}$

31) $36 = 25n^2$

$0 = 25n^2 - 36$

$0 = (5n+6)(5n-6)$

$5n+6=0$ or $5n-6=0$

$5n = -6$ $5n = 6$

$n = -\frac{6}{5}$ or $n = \frac{6}{5}$ $\left\{-\frac{6}{5}, \frac{6}{5}\right\}$

33) $m^2 = 60 - 7m$

$m^2 + 7m - 60 = 0$

$(m+12)(m-5) = 0$

$m+12=0$ or $m-5=0$

$m = -12$ or $m = 5$ $\{-12, 5\}$

35) $55w = -20w^2 - 30$

$20w^2 + 55w + 30 = 0$ divide by 5

$4w^2 + 11w + 6 = 0$

$(4w+3)(w+2) = 0$

$4w+3=0$ or $w+2=0$

$4w = -3$

$w = -\frac{3}{4}$ or $w = -2$ $\left\{-2, -\frac{3}{4}\right\}$

37) $p^2 = 11p$

$p^2 - 11p = 0$

$p(p-11) = 0$

$p = 0$ or $p-11=0$

$p = 11$ $\{0, 11\}$

39) $45k + 27 = 18k^2$

$-18k^2 + 45k + 27 = 0$ divide by -9

$2k^2 - 5k - 3 = 0$

$(2k+1)(k-3) = 0$

$2k+1=0$ or $k-3=0$

$2k = -1$

$k = -\frac{1}{2}$ or $k = 3$ $\left\{-\frac{1}{2}, 3\right\}$

41)
$$b(b-4)=96$$
$$b^2-4b=96$$
$$b^2-4b-96=0$$
$$(b+8)(b-12)=0$$
$$b+8=0 \text{ or } b-12=0$$
$$b=-8 \text{ or } b=12 \quad \{-8,12\}$$

43)
$$-63=4j(j-8)$$
$$-63=4j^2-32j$$
$$0=4j^2-32j+63$$
$$0=(2j-7)(2j-9)$$
$$2j-7=0 \text{ or } 2j-9=0$$
$$2j=7 \qquad 2j=9$$
$$j=\frac{7}{2} \text{ or } j=\frac{9}{2} \quad \left\{\frac{7}{2},\frac{9}{2}\right\}$$

45)
$$10x(x+1)-6x=9(x^2+5)$$
$$10x^2+10x-6x=9x^2+45$$
$$10x^2+4x=9x^2+45$$
$$x^2+4x-45=0$$
$$(x+9)(x-5)=0$$
$$x+9=0 \text{ or } x-5=0$$
$$x=-9 \text{ or } x=5 \quad \{-9,5\}$$

47)
$$3(h^2-4)=5h(h-1)-9h$$
$$3h^2-12=5h^2-5h-9h$$
$$3h^2-12=5h^2-14h$$
$$-2h^2+14h-12=0 \qquad \text{divide by } -2$$
$$h^2-7h+6=0$$
$$(h-1)(h-6)=0$$
$$h-1=0 \text{ or } h-6=0$$
$$h=1 \text{ or } h=6 \quad \{1,6\}$$

49)
$$\frac{1}{2}(m+1)^2=-\frac{3}{4}m(m+5)-\frac{5}{2}$$
$$4\left[\frac{1}{2}(m+1)^2\right]=4\left[-\frac{3}{4}m(m+5)-\frac{5}{2}\right]$$
$$2(m+1)^2=-3m(m+5)-10$$
$$2(m^2+2m+1)=-3m^2-15m-10$$
$$2m^2+4m+2=-3m^2-15m-10$$
$$5m^2+19m+12=0$$
$$(5m+4)(m+3)=0$$
$$5m+4=0 \text{ or } m+3=0$$
$$5m=-4$$
$$m=-\frac{4}{5} \text{ or } m=-3 \quad \left\{-3,-\frac{4}{5}\right\}$$

51)
$$3t(t-5)+14=5-t(t+3)$$
$$3t^2-15t+14=5-t^2-3t$$
$$4t^2-12t+9=0$$
$$(2t-3)^2=0$$
$$2t-3=0$$
$$2t=3$$
$$t=\frac{3}{2} \quad \left\{\frac{3}{2}\right\}$$

53)
$$33=-m(14+m)$$
$$33=-14m-m^2$$
$$m^2+14m+33=0$$
$$(m+11)(m+3)=0$$
$$m+11=0 \text{ or } m+3=0$$
$$m=-11 \text{ or } m=-3 \quad \{-11,-3\}$$

55) $(3w+2)^2-(w-5)^2=0$

$(3w+2+w-5)(3w+2-(w-5))=0$

$(4w-3)(2w+7)=0$

$4w-3=0$ or $2w+7=0$

$4w=3 \qquad 2w=7$

$w=\frac{3}{4}$ o $\quad w=\frac{7}{2} \qquad \left\{-\frac{7}{2},\frac{3}{4}\right\}$

57) $(q+3)^2-(2q-5)^2=0$

$(q+3+2q-5)(q+3-(2q-5))=0$

$(3q-2)(-q+8)=0$

$-(3q-2)(q-8)=0$

$(3q-2)(q-8)=0$

$3q-2=0$ or $q-8=0$

$3q=2$

$q=\frac{2}{3}$ or $q=8 \qquad \left\{\frac{2}{3},8\right\}$

59) $8y(y+4)(2y-1)=0$

$8y=0$ or $y+4=0$ or $2y-1=0$

$y=0 \qquad y=-4 \qquad 2y=1$

$y=\frac{1}{2}$

$\left\{-4,0,\frac{1}{2}\right\}$

61) $(9p-2)(p^2-10p-11)=0$

$(9p-2)(p+1)(p-11)=0$

$9p-2=0$ or $p+1=0$ or $p-11=0$

$9p=2$

$p=\frac{2}{9}$ or $p=-1$ or $p=11$

$\left\{-1,\frac{2}{9},11\right\}$

63) $(2r-5)(r^2-6r+9)=0$

$(2r-5)(r-3)^2=0$

$2r-5=0$ or $r-3=0$

$2r=5$

$r=\frac{5}{2}$ or $r=3 \qquad \left\{\frac{5}{2},3\right\}$

65) $m^3=64m$

$m^3-64m=0$

$m(m^2-64)=0$

$m(m+8)(m-8)=0$

$m=0$ or $m+8=0$ or $m-8=0$

$m=-8 \qquad m=8$

$\{-8,0,8\}$

67) $5w^2+36w=w^3$

$-w^3+5w^2+36w=0$

$-w(w^2-5w-36)=0$

$-w(w+4)(w-9)=0$

$-w=0$ or $w+4=0$ or $w-9=0$

$w=0$ or $w=-4$ or $w=9$

$\{-4,0,9\}$

69) $2g^3=120g-14g^2$

$2g^3+14g^2-120g=0$

$2g(g^2+7g-60)=0$

$2g(g+12)(g-5)=0$

$2g=0$ or $g+12=0$ or $g-7=0$

$g=0$ or $g=-12$ or $g=7$

$\{-12,0,5\}$

71) $45h = 20h^3$

$$0 = 20h^3 - 45h$$
$$0 = 5h\left(4h^2 - 9\right)$$
$$0 = 5h(2h+3)(2h+3)$$
$$5h = 0 \text{ or } 2h+3=0 \text{ or } 2h-3=0$$
$$h = 0 \qquad 2h=-3 \qquad 2h=3$$
$$h = -\frac{3}{2} \qquad h=\frac{3}{2} \qquad \left\{-\frac{3}{2}, 0, \frac{3}{2}\right\}$$

73) $2s^2(3s+2)+3s(3s+2)-35(3s+2)=0$

$$\left(2s^2+3s-35\right)(3s+2)=0$$
$$(2s-7)(s+5)(3s+2)=0$$
$$2s-7=0 \text{ or } s+5=0 \text{ or } 3s+2=0$$
$$2s=7 \qquad 3s=-2$$
$$s=\frac{7}{2} \text{ or } s=-5 \text{ or } s=-\frac{2}{3} \qquad \left\{-5, -\frac{2}{3}, \frac{7}{2}\right\}$$

75) $10a^2(4a+3)+2(4a+3)=9a(4a+3)$

$$10a^2(4a+3)-9a(4a+3)+2(4a+3)=0$$
$$\left(10a^2-9a+2\right)(4a+3)=0$$
$$(2a-1)(5a+2)(4a+3)=0$$
$$2a-1=0 \text{ or } 5a-2=0 \text{ or } 4a+3=0$$
$$2a=1 \qquad 5a=2 \qquad 4a=-3$$
$$a=\frac{1}{2} \text{ or } a=\frac{2}{5} \text{ or } a=-\frac{3}{4} \qquad \left\{-\frac{3}{4}, \frac{2}{5}, \frac{1}{2}\right\}$$

77) $f(x)=0$

$$f(x)=x^2+10x+21$$
$$0=x^2+10x+21$$
$$0=(x+7)(x+3)$$
$$x+7=0 \text{ or } x+3=0$$
$$x=-7 \text{ or } x=-3$$

79) $g(a)=4$

$$g(a)=2a^2-13a+24$$
$$4=2a^2-13a+24$$
$$0=2a^2-13a+20$$
$$0=(2a-5)(a-4)$$
$$2a-5=0 \text{ or } a-4=0$$
$$2a=5$$
$$a=\frac{5}{2} \quad \text{or} \quad a=4$$

81) $H(b)=19$

$$H(b)=b^2+3$$
$$19=b^2+3$$
$$0=b^2-16$$
$$0=(b+4)(b-4)$$
$$b+4=0 \text{ or } b-4=0$$
$$b=-4 \quad \text{or} \quad b=4$$

83) $h(k)=0$

$$h(k)=5k^3-25k^2+20k$$
$$0=5k^3-25k^2+20k$$
$$0=5k(k^2-5k+4)$$
$$0=5k(k-1)(k-4)$$
$$5k=0 \text{ or } k-1=0 \text{ or } k-4=0$$
$$k=0 \quad \text{or} \quad k=1 \quad \text{or} \quad k=4$$

Section 7.6 Exercises

1) $x=$ length of rectangle

$x-9=$ width of rectangle

Area $=(\text{length})(\text{width})$

$$36=x(x-9)$$
$$36=x^2-9x$$
$$0=x^2-9x-36$$
$$0=(x-12)(x+3)$$
$$x-12=0 \text{ or } x+3=0$$
$$x=12 \qquad x=-3$$

length $=12$ in; width $=12-9=3$ in

3) $2x-1=$ base of triangle

$x+6=$ height of triangle

$$\text{Area}=\frac{1}{2}(\text{base})(\text{height})$$
$$12=\frac{1}{2}(2x-1)(x+6)$$
$$24=2x^2+11x-6$$
$$0=2x^2+11x-30$$
$$0=(2x+15)(x-2)$$
$$2x+15=0 \text{ or } x-2=0$$
$$2x=-15 \qquad x=2$$
$$x=-\frac{15}{2}$$

base $=2(2)-1=3$ cm;

height $=2+6=8$ cm

5) $x+1=$ base of parallelogram

$x-2=$ height of parallelogram

Area $=(\text{base})(\text{height})$

$$18=(x+1)(x-2)$$
$$18=x^2-x-2$$
$$0=x^2-x-20$$
$$0=(x-5)(x+4)$$
$$x-5=0 \text{ or } x+4=0$$
$$x=5 \qquad x=-4$$

base $=5+1=6$ in;

width $=5-2=3$ in

7) $3x+1=\text{length of box}$

$2x=\text{width of box}$

$\text{Volume}=(\text{length})(\text{width})(\text{height})$

$240=(3x+1)(2x)(4)$

$240=(6x^2+2x)4$

$60=6x^2+2x$

$0=6x^2+2x-60$

$0=3x^2+x-30$

$0=(3x+10)(x-3)$

$3x+10=0 \quad \text{or} \quad x-3=0$

$3x=-10 \qquad x=3$

$x=-\dfrac{10}{3}$

$\text{length}=3(3)+1=10 \text{ in};$

$\text{width}=2(3)=6 \text{ in}$

9) $w=\text{the width of the rug}$

$w+4=\text{the length of the rug}$

$\text{Area}=(\text{Length})(\text{Width})$

$45=(w+4)\cdot w$

$45=w^2+4w$

$0=w^2+4w-45$

$0=(w+9)(w-5)$

$w+9=0 \quad \text{or} \quad w-5=0$

$w=-9 \qquad w=5$

$\text{width}=5 \text{ ft; length}=5+4=9 \text{ ft}$

11) $l=\text{the length of the glass}$

$l-3=\text{the width of the glass}$

$\text{Area}=(\text{Length})(\text{Width})$

$54=l\cdot(l-3)$

$54=l^2-3l$

$0=l^2-3l-54$

$0=(l-9)(l+6)$

$l-9=0 \quad \text{or} \quad l+6=0$

$l=9 \qquad l=-6$

$\text{length}=9 \text{ in; width}=9-3=6 \text{ in}$

13) $w=\text{width of box}$

$\dfrac{w}{2}=\text{height of box}$

$\text{Volume}=(\text{length})(\text{width})(\text{height})$

$1440=20\cdot w\cdot\dfrac{w}{2}$

$1440=10w^2$

$144=w^2$

$0=w^2-144$

$0=(w+12)(w-12)$

$w+12=0 \quad \text{or} \quad w-12=0$

$w=-12 \qquad w=12$

$\text{width}=12 \text{ in; height}=\dfrac{12}{2}=6 \text{ in}$

15) $b=\text{base of triangle}$

$b+3=\text{height of triangle}$

$\text{Area}=\dfrac{1}{2}(\text{base})(\text{height})$

$35=\dfrac{1}{2}(b)(b+3)$

$70=b^2+3b$

$0=b^2+3b-70$

$0=(b+10)(b-7)$

$b+10=0 \quad \text{or} \quad b-7=0$

$b=-10 \qquad b=7$

$\text{base}=7 \text{ cm};$

$\text{height}=7+3=10 \text{ cm}$

17) $x =$ the first odd integer

$x+2 =$ the second odd integer

$$x(x+2)=3(x+x+2)-1$$
$$x^2+2x=3(2x+2)-1$$
$$x^2+2x=6x+6-1$$
$$x^2-4x-5=0$$
$$(x-5)(x+1)=0$$
$$x-5=0 \quad \text{or} \quad x+1=0$$
$$x=5 \qquad\qquad x=-1$$

$x=5$, then $5+2=7$

$x=-1$, then $-1+2=1$

5 and 7 or -1 and 1

19) $x =$ the first even integer

$x+2 =$ the second even integer

$x+4 =$ the third even integer

$$x+(x+2)=\frac{1}{4}(x+2)(x+4)$$
$$2x+2=\frac{1}{4}(x^2+6x+8)$$
$$8x+8=x^2+6x+8$$
$$0=x^2-2x$$
$$0=x(x-2)$$
$$x=0 \quad \text{or} \quad x-2=0$$
$$x=2$$

$x=0$, then $0+2=2$ and $0+4=4$

$x=2$, then $2+2=4$ and $2+4=6$

$0,2,4$ or $2,4,6$

21) $x =$ the first integer

$x+1 =$ the second integer

$x+2 =$ the third integer

$$(x+2)^2=x(x+1)+22$$
$$x^2+4x+4=x^2+x+22$$
$$3x-18=0$$
$$3x=18$$
$$x=6$$

$x=6$, then $6+1=7$ and $6+2=8$

$6,7,8$

23) Answers may vary.

25)
$$a^2+b^2=c^2$$
$$a^2+(12)^2=(15)^2$$
$$a^2+144=225$$
$$a^2-81=0$$
$$(a+9)(a-9)=0$$
$$a+9=0 \quad \text{or} \quad a-9=0$$
$$a=-9 \qquad\qquad a=9$$

The length of the missing side is 9.

27)
$$a^2+b^2=c^2$$
$$a^2+(8)^2=(17)^2$$
$$a^2+64=289$$
$$a^2-225=0$$
$$(a+15)(a-15)=0$$
$$a+15=0 \quad \text{or} \quad a-15=0$$
$$a=-15 \qquad\qquad a=15$$

The length of the missing side is 15.

29)
$$a^2+b^2=c^2$$
$$(8)^2+(6)^2=c^2$$
$$64+36=c^2$$
$$100=c^2$$
$$0=c^2-100$$
$$0=(c+10)(c-10)$$
$$c+10=0 \quad \text{or} \quad c-10=0$$
$$c=-10 \qquad\qquad c=10$$

The length of the missing side is 10.

31) $x =$ length of the longer leg

$x-2 =$ length of the shorter leg

$x+2 =$ length of the hypotenuse

$$(x-2)^2 + x^2 = (x+2)^2$$
$$x^2 - 4x + 4 + x^2 = x^2 + 4x + 4$$
$$2x^2 - 4x + 4 = x^2 + 4x + 4$$
$$x^2 - 8x = 0$$
$$x(x-8) = 0$$
$$x = 0 \quad \text{or} \quad x - 8 = 0$$
$$x = 8$$

The length of the longer leg is 8 in.

33) $x =$ distance from bottom of the ladder to the wall.

$x+7 =$ distance from the floor to the top of the ladder.

The length of the ladder is 13, so $c = 13$. Let $a = x$ and $b = x+7$.

$$a^2 + b^2 = c^2$$
$$(x)^2 + (x+7)^2 = (13)^2$$
$$x^2 + x^2 + 14x + 49 = 169$$
$$2x^2 + 14x + 49 = 169$$
$$2x^2 + 14x - 120 = 0$$
$$x^2 + 7x - 60 = 0$$
$$(x+12)(x-5) = 0$$
$$x + 12 = 0 \quad \text{or} \quad x - 5 = 0$$
$$x = -12 \qquad x = 5$$

The distance from the bottom of the ladder to the wall is 5 ft.

35) $a = x =$ Rana's distance from the bike shop

$b = 4 =$ Yasmeen's distance from the bike shop

$c = x + 2 =$ distance between Rana and Yasmeen

$$a^2 + b^2 = c^2$$
$$4^2 + x^2 = (x+2)^2$$
$$16 + x^2 = x^2 + 4x + 4$$
$$12 = 4x$$
$$3 = x$$
$$c = 3 + 2 = 5 \text{ miles}$$

They are 5 miles apart.

37) a) Let $t=0$ and solve for h.

$h=-16(0)^2+144$

$h=-16(0)+144=144$

The initial height is 144 ft.

b) Let $h=80$ and solve for t.

$80=-16t^2+144$

$-64=-16t^2$

$4=t^2$

$0=t^2-4$

$0=(t+2)(t-2)$

$t+2=0$ or $t-2=0$

$t=-2$ $\quad$ $t=2$

80 ft over water after 2 sec.

c) Let $h=0$ and solve for t.

$0=-16t^2+144$

$-144=-16t^2$

$9=t^2$

$0=t^2-9$

$0=(t+3)(t-3)$

$t+3=0$ or $t-3=0$

$t=-3$ $\quad$ $t=3$

It will hit the water after 3 sec.

39) a) Let $t=3$.

$y=-16(3)^2+144(3)$

$y=-16(9)+432=288$ ft

b) Let $t=3$.

$x=39(3)=117$ ft

c) Let $t=4.5$

$y=-16(4.5)^2+144(4.5)$

$y=-16(20.25)+648=324$ ft

d) Let $t=4.5$

$x=39(4.5)=175.5\approx 176$ ft

e) Let $t=3$.

$y=-16(3)^2+264(3)$

$y=-16(9)+792=648$ ft

f) Let $t=3$.

$x=71(3)=213$ ft

g) Let $t=8.25$

$y=-16(8.25)^2+264(8.25)$

$y=-16(68.0625)+188$

$y=-1089+2178=1089$ ft

h) Let $t=8.25$

$x=71(8.25)=585.75\approx 586$ ft

i) $648-288=360$ ft

j) $1089-324=765$ ft

k) The 10" shell would need to be 410 ft farther horizontally from he point of explosion than the 3" shell.

41) a) Let $p=40$

$R(40)=-5(40)^2+300(40)$

$R(40)=-5(1600)+12,000$

$R(40)=-8000+12,000=\$4000$

b) Let $p=25$

$R(25)=-5(25)^2+300(25)$

$R(25)=-5(625)+7500$

$R(40)=-3125+7500=\$4375$

c) Let $R(p)=4500$

$4500=-5p^2+300p$

$0=-5p^2+300p-4500$

$0=p^2-60p-900$

$0=(p-30)^2$

$0=p-30$

$30=p$ $30

Chapter 7 Review

1) GCF of 18 and 27 is 9.

3) GCF of $33p^5q^3$, $121p^4q^3$ and $44p^7q^4$
$=11\cdot p^4\cdot q^3=11p^4q^3$

5) $\text{GCF}=12$
$48y+84=(12)(4y)+(12)(7)$
$=12(4y+7)$

7) $\text{GCF}=7n^3$
$7n^5-21n^4+7n^3$
$=(7n^3)(n^2)-(7n^3)(3n)+(7n^3)(1)$
$=7n^3(n^2-3n+1)$

9) $\text{GCF}=(b+6)$
$a(b+6)-2(b+6)=(b+6)(a-2)$

11) $mn+2m+5n+10=m(n+2)+5(n+2)$
$=(n+2)(m+5)$

13) $5qr-10q-6r+12=5q(r-2)-6(r-2)$
$=(r-2)(5q-6)$

15) $-8x^3-12x^2+4x=-4x(2x^2+3x-1)$

17) $p^2+13p+40=(p+8)(p+5)$

19) $x^2-xy-20y^2=(x+5y)(x-4y)$

21) $3c^2-24c+36=3(c^2-8c+12)$
$=3(c-6)(c-2)$

23) $5y^2+11y+6=(5y+6)(y+1)$

25) $4m^2-16m+15=(2m-5)(2m-3)$

27) $56a^3+4a^2-16a=4a(14a^2+a-4)$
$=4a(7a+4)(2a-1)$

29) $3s^2+11st-4t^2=(3s-t)(s+4t)$

31) $n^2-25=(n+5)(n-5)$

33) $9t^2+16u^2$ is prime

35) $10q^2-810=10(q^2-81)$
$=10(q+9)(q-9)$

37) $a^2+16a+64=(a+8)^2$

39) $h^3+8=(h+2)(h^2-2h+4)$

41) $27p^3-64q^3$
$=(3p-4q)(9p^2+12pq+16q^2)$

43) $7r^2+8r-12=(7r-6)(r+2)$

45) $\frac{9}{25}-x^2=\left(\frac{3}{5}+x\right)\left(\frac{3}{5}-x\right)$

47) $st-5s-8t+40=s(t-5)-8(t-5)$
$=(s-8)(t-5)$

49) $w^5 - w^2 = w^2(w^3 - 1)$
$= w^2(w-1)(w^2 + w + 1)$

51) $a^2 + 3a - 14$ is prime

53) $(a-b)^2 - (a+b)^2$
$= [(a-b)+(a+b)][(a-b)-(a+b)]$
$= (2a)(-2b) = -4ab$

55) $c(2c-1) = 0$
$c = 0$ or $2c - 1 = 0$
$2c = 1$
$c = \frac{1}{2}$ $\quad \left\{0, \frac{1}{2}\right\}$

57) $3x^2 + x = 2$
$3x^2 + x - 2 = 0$
$(x+1)(3x-2) = 0$
$x + 1 = 0$ or $3x - 2 = 0$
$3x = 2$
$x = -1$ or $x = \frac{2}{3}$ $\quad \left\{-1, \frac{2}{3}\right\}$

59) $n^2 = 12n + 45$
$n^2 - 12n - 45 = 0$
$(n+3)(n-15) = 0$
$n + 3 = 0$ or $n - 15 = 0$
$n = -3$ or $n = 15$ $\quad \{-3, 15\}$

61) $36 = 49d^2$
$0 = 49d^2 - 36$
$0 = (7d+6)(7d-6)$
$7d + 6 = 0$ or $7d - 6 = 0$
$7d = -6$ $\quad 7d = 6$
$d = -\frac{6}{7}$ or $d = \frac{6}{7}$ $\quad \left\{-\frac{6}{7}, \frac{6}{7}\right\}$

63) $8b + 64 = 2b^2$
$-2b^2 + 8b + 64 = 0$
$b^2 - 4b - 32 = 0$
$(b+4)(b-8) = 0$
$b + 4 = 0$ or $b - 8 = 0$
$b = -4$ or $b = 8$ $\quad \{-4, 8\}$

65) $y(5y-9) = -4$
$5y^2 - 9y = -4$
$5y^2 - 9y + 4 = 0$
$(5y-4)(y-1) = 0$
$5y - 4 = 0$ or $y - 1 = 0$
$5y = 4$
$y = \frac{4}{5}$ or $y = 1$ $\quad \left\{\frac{4}{5}, 1\right\}$

67) $6a^3 - 3a^2 - 18a = 0$
$3a(2a^2 - a - 6) = 0$
$3a(2a+3)(a-2) = 0$
$3a = 0$ or $2a + 3 = 0$ or $a - 2 = 0$
$a = 0$ $\quad 2a = -3$ $\quad a = 2$
$a = -\frac{3}{2}$
$\left\{-\frac{3}{2}, 0, 2\right\}$

69) $c(5c-1)+8=4(20+c^2)$

$5c^2-c+8=80+4c^2$

$c^2-c-72=0$

$(c+8)(c-9)=0$

$c+8=0$ or $c-9=0$

$c=-8$ or $c=9$ $\{-8, 9\}$

71) $p^2(6p-1)-10p(6p-1)+21(6p-1)=0$

$(6p-1)(p^2-10p+21)=0$

$(6p-1)(p-3)(p-7)=0$

$6p-1=0$ or $p-3=0$ or $p-7=0$

$6p=1$

$p=\frac{1}{6}$ or $p=3$ or $p=7$ $\left\{\frac{1}{6},3,7\right\}$

73) $2x-3=$ base of triangle

$x+2=$ height of triangle

$\text{Area}=\frac{1}{2}(\text{base})(\text{height})$

$15=\frac{1}{2}(2x-3)(x+2)$

$30=2x^2+x-6$

$0=2x^2+x-36$

$0=(2x+9)(x-4)$

$2x+9=0$ or $x-4=0$

$2x=-9$ $x=4$

$x=-\frac{9}{2}$

base $=2(4)-3=5$ in;

height $=4+2=6$ in

75) $3x-1=$ length of box

$x=$ height of box

$\text{Volume}=(\text{length})(\text{width})(\text{height})$

$96=(3x-1)(4)(x)$

$96=(3x^2-x)4$

$24=3x^2-x$

$0=3x^2-x-24$

$0=(3x+8)(x-3)$

$3x+8=0$ or $x-3=0$

$3x=-8$ $x=3$

$x=-\frac{8}{3}$

height $=3$ in; length $=3(3)-1=8$ in

77) $a^2+b^2=c^2$

$a^2+(5)^2=(13)^2$

$a^2+25=169$

$a^2-144=0$

$(a+12)(a-12)=0$

$a+12=0$ or $a-12=0$

$a=-12$ $\quad a=12$

The length of the missing side is 12.

79) $l=$ length of the countertop

$l-3.5=$ width of the countertop

Area $=(\text{Length})(\text{Width})$

$15=l\cdot(l-3.5)$

$15=l^2-3.5l$

$0=l^2-3.5l-15$

$0=(l-6)(l+2.5)$

$l-6=0$ or $l+2.5=0$

$l=6$ $\quad l=-2.5$

width $=6-3.5=2.5$ ft; length $=6$ ft

81) $x=$ the first integer

$x+1=$ the second integer

$x+2=$ the third integer

$x+x+1+x+2=\frac{1}{3}(x+1)^2$

$3x+3=\frac{1}{3}(x^2+2x+1)$

$9x+9=x^2+2x+1$

$0=x^2-7x-8$

$0=(x-8)(x+1)$

$x-8=0$ or $x+1=0$

$x=8$ $\quad x=-1$

$x=8$, then $8+1=9$; $8+2=10$

$x=-1$, then $-1+1=0$; $-1+2=1$

$-1,0,1$ or $8,9,10$

83) $c=x=$ length of the ramp

$a=x-1=$ base of the ramp

$b=x-8=$ height of the ramp

$(x-1)^2+(x-8)^2=x^2$

$x^2-2x+1+x^2-16x+64=x^2$

$2x^2-18x+65=x^2$

$x^2-18x+65=0$

$(x-13)(x-5)=0$

$x-13=0$ or $x-5=0$

$\boxed{x=13}$ $\quad x=5$

height $=13-8=5$ in

85) a) Let $t=0$

$h=-16(0)^2+96(0)=0$ ft

b) Let $h=128$

$128=-16t^2+96t$

$0=-16t^2+96t-128$

$0=t^2-6t+8$

$0=(t-2)(t-4)$

$t-2=0$ or $t-4=0$

$t=2$ $\quad t=4$

when $t=2$ seconds; $t=4$ seconds

c) Let $t=3$

$h=-16(3)^2+96(3)$

$h=-16(9)+288$

$h=-144+288=144$ ft

d) Let $h=0$ and solve for t.

$0=-16t^2+96t$

$0=t^2-6t$

$0=t(t-6)$

$t=0$ or $t-6=0$

$t=6$

when $t=6$ seconds

Chapter 7 Test

1) See if you can factor out a GCF.

3) $16-b^2=(4+b)(4-b)$

5) $56p^6q^6-77p^4q^4+7p^2q^3$
$=7p^2q^3\left(8p^4q^3-11p^2q+1\right)$

7) $2d^3+14d^2-36d=2d\left(d^2+7d-18\right)$
$=2d(d+9)(d-2)$

9) $\left(9h^2+24h+16\right)=(3h+4)^2$

11) $s^2-3st-28t^2=(s-7t)(s+4t)$

13) $y^2(x+3)^2+15y(x+3)^2+56(x+3)^2$
$=(x+3)^2\left(y^2+15y+56\right)$
$=(x+3)^2(y+7)(y+8)$

15) $m^{12}+m^9=m^9\left(m^3+1\right)$
$=m^9(m+1)\left(m^2-m+1\right)$

17) $25k=k^3$
$-k^3+25k=0$
$-k\left(k^2-25\right)=0$
$-k(k+5)(k-5)=0$
$-k=0$ or $k+5=0$ or $k-5=0$
$k=-4$ $\quad k=-5$ $\quad k=5$
$\{-5,0,5\}$

19) $(c-5)(c+2)=18$
$c^2-3c-10=18$
$c^2-3c-28=0$
$(c+4)(c-7)=0$
$c+4=0$ or $c-7=0$
$b=-4$ o $\quad c=7$ $\quad \{-4,7\}$

21) $24y^2+80=88y$
$24y^2-88y+80=0$
$3y^2-11y+10=0$
$(3y-5)(y-2)=0$
$3y-5=0$ or $y-2=0$
$3y=5$
$y=\frac{5}{3}$ or $y=2$ $\quad \left\{\frac{5}{3},2\right\}$

23) $x=$ the first odd integer
$x+2=$ the second odd integer
$x+4=$ the third odd integer
$x+x+2+x+4=(x+4)^2-60$
$3x+6=\left(x^2+8x+16\right)-60$
$3x+6=x^2+8x-44$
$0=x^2+5x-50$
$0=(x+10)(x-5)$
$x+10=0$ or $x-5=0$
$x=-10$ $\quad \boxed{x=5}$
$x=5$, then $5+2=7$ and $5+4=9$
5, 7, 9

25) $w =$ the width of the run

$2w+4 =$ the length of the run

$$\text{Area} = (\text{Length})(\text{Width})$$
$$96 = (2w+4)w$$
$$96 = 2w^2+4w$$
$$0 = 2w^2+4w-96$$
$$0 = w^2+2w-48$$
$$0 = (w+8)(w-6)$$
$$w+8=0 \quad \text{or} \quad w-6=0$$
$$w=-8 \qquad \boxed{w=6}$$

length $= 2(6)+4 = 16$ ft; width $= 6$ ft

Cumulative Review: Chapters 1-7

1) $\dfrac{3}{8}-\dfrac{5}{6}+\dfrac{7}{12} = \dfrac{9}{24}-\dfrac{20}{24}+\dfrac{14}{24}$
$$= \frac{9-20+14}{24}$$
$$= \frac{3}{24} = \frac{1}{8}$$

3) $\dfrac{54t^5u^2}{36tu^8} = \dfrac{54}{36}t^{5-1}u^{2-8} = \dfrac{3}{2}t^4u^{-6} = \dfrac{3t^4}{2u^6}$

5) $4.813\times10^5:\ 481300 = 481,300$

7)
$$A = P+P\boxed{R}T$$
$$A-P = P\boxed{R}T$$
$$\frac{A-P}{PT} = \boxed{R}$$

9)

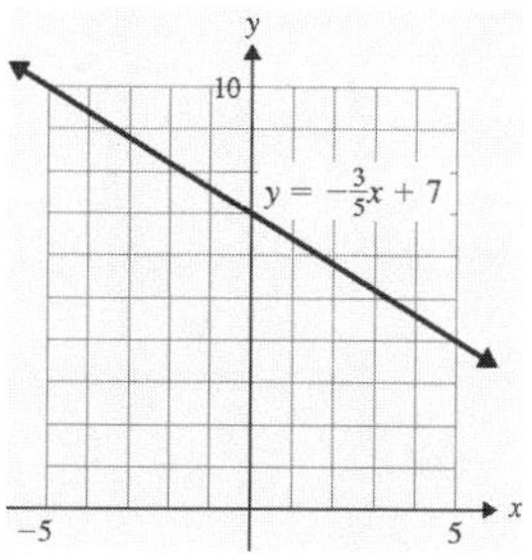

11) $6(x+2)+y = x-y-2$
$$6x+12+y = x-y-2$$
$$5x+2y=-14 \qquad (1)$$
$$5(2x-y+1) = 2(x-y)-5$$
$$10x-5y+5 = 2x-2y-5$$
$$8x-3y=-10 \qquad (2)$$

Use the elimination method.
$$\begin{array}{rl} 15x+6y=-42 & 3\cdot(1) \\ +\ 16x-6y=-20 & 2\cdot(2) \\ \hline 31x=-62 & \\ x=-2 & \end{array}$$

Substitute $x=-2$ into (1)

$x=-2$:
$$5(-2)+2y=-14$$
$$-10+2y=-14$$
$$2y=-4$$
$$y=-2 \qquad (-2,-2)$$

13) $(4p-7)(2p^2-9p+8)$
$$=8p^3-36p^2+32p-14p^2+63p-56$$
$$=8p^3-50p^2+95p-56$$

15)
$$\begin{array}{rl} & 4a^2b^2-17a^2b \qquad\qquad +12ab-11 \\ + & -a^2b^2+10a^2b-5ab^2+7ab\ \ +3 \\ \hline & 3a^2b^2-7a^2b-5ab^2+19ab-8 \end{array}$$

17) $\dfrac{12r^3+4r^2-10r+3}{4r^2}$
$$= \frac{12r^3}{4r^2}+\frac{4r^2}{4r^2}-\frac{10r}{4r^2}+\frac{3}{4r^2}$$
$$= 3r+1-\frac{5}{2r}+\frac{3}{4r^2}$$

19) $54q^2-114q+42 = 6(9q^2-19q+7)$
$$= 6(3q-7)(3q-1)$$

21) $t^4 - 81 = (t^2 + 9)(t^2 - 9)$

$= (t^2 + 9)(t + 3)(t - 3)$

23) $z^2 + 3z = 40$

$z^2 + 3z - 40 = 0$

$(z + 8)(z - 5) = 0$

$z + 8 = 0$ or $z - 5 = 0$

$z = -8$ or $z = 5$ $\{-8, 5\}$

Section 8.1: Exercises

1) when its numerator equals zero

3) Set the denominator equal to zero and solve for the variable. That value cannot be substituted into the expression because it will make the denominator equal to zero.

5) a) $\frac{7(2)+1}{3(2)-1}=\frac{14+1}{6-1}=\frac{15}{5}=3$

b) $\frac{7(-1)+1}{3(-1)-1}=\frac{-7+1}{-3-1}=\frac{-6}{-4}=\frac{3}{2}$

7) a) $\frac{[2(-3)]^2}{(-3)^2+5(-3)+6}=\frac{(-6)^2}{9-15+6}$

$=\frac{36}{-6+6}=\frac{36}{0}$

undefined

b) $\frac{[2(4)]^2}{(4)^2+5(4)+6}=\frac{(8)^2}{16+20+6}$

$=\frac{64}{36+6}=\frac{64}{42}$

$=\frac{32}{21}$

9) a) $\frac{12+4(-3)}{20-(-3)^2}=\frac{12-12}{20-9}=\frac{0}{11}=0$

b) $\frac{12+4(4)}{20-(4)^2}=\frac{12+16}{20-16}=\frac{28}{4}=7$

11) a) $x+2=0$
$x=-2$

b) $7x=0$
$x=0$

13) a) $3r+1=0$
$3r=-1$
$r=-\frac{1}{3}$

b) $2r-9=0$
$2r=9$
$r=\frac{9}{2}$

15) a) $10c-c^2=0$
$c(10-c)=0$
$c=0$ or $10-c=0$
$c=10$

b) $3c-4=0$
$3c=4$
$c=\frac{4}{3}$

17) a) It never equals zero.

b) $y=0$

19) a) $7a=0$
$a=0$

b) $a^2+9a-36=0$
$(a+12)(a-3)=0$
$a+12=0$ or $a-3=0$
$a=-12 \qquad a=3$

21) a) $r+10=0$
$r=-10$

b) $2r^2-5r-12=0$
$(2r+3)(r-4)=0$
$2r+3=0$ or $r-4=0$
$2r=-3 \qquad r=4$
$r=-\frac{3}{2}$

23) a) $t^2+8t+15=0$

$(t+3)(t+5)=0$

$t+3=0$ or $t+5=0$

$t=-3 \qquad t=-5$

b) $4t=0$

$t=0$

25) a) $4y=0$

$y=0$

b) y^2+9 never equals zero.

Never undefined

27) $\dfrac{2y(y-11)}{3(y-11)}=\dfrac{2y}{3}$

29) $\dfrac{12d^5}{30d^8}=\dfrac{2}{5d^3}$

31) $\dfrac{3c-12}{5c-20}=\dfrac{3(c-4)}{5(c-4)}=\dfrac{3}{5}$

33) $\dfrac{-18v-42}{15v+35}=\dfrac{-6(3v+7)}{5(3v+7)}=-\dfrac{6}{5}$

35) $\dfrac{39q^2+26}{30q^2+20}=\dfrac{13(3q^2+2)}{10(3q^2+2)}=\dfrac{13}{10}$

37) $\dfrac{b^2+b-56}{b+8}=\dfrac{(b+8)(b-7)}{b+8}=b-7$

39) $\dfrac{r-4}{r^2-16}=\dfrac{r-4}{(r-4)(r+4)}=\dfrac{1}{r+4}$

41) $\dfrac{3k^2+28k+32}{k^2+10k+16}=\dfrac{(3k+4)(k+8)}{(k+2)(k+8)}$

$=\dfrac{3k+4}{k+2}$

43) $\dfrac{p^2-16}{2p^2+7p-4}=\dfrac{(p+4)(p-4)}{(2p-1)(p+4)}$

$=\dfrac{p-4}{2p-1}$

45) $\dfrac{w^3+125}{5w^2-25w+125}$

$=\dfrac{(w+5)(w^2-5w+25)}{5(w^2-5w+25)}=\dfrac{w+5}{5}$

47) $\dfrac{8c+24}{c^3+27}=\dfrac{8(c+3)}{(c+3)(c^2-3c+9)}$

$=\dfrac{8}{c^2-3c+9}$

49) $\dfrac{4m^2-20m+4mn-20n}{11m+11n}$

$=\dfrac{4(m^2-5m+mn-5n)}{11(m+n)}$

$=\dfrac{4[m(m-5)+n(m-5)]}{11(m+n)}$

$=\dfrac{4(m-5)(m+n)}{11(m+n)}=\dfrac{4(m-5)}{11}$

51) $\dfrac{x^2-y^2}{x^3-y^3}=\dfrac{(x+y)(x-y)}{(x-y)(x^2+xy+y^2)}$

$=\dfrac{x+y}{x^2+xy+y^2}$

53) -1

55) $\dfrac{12-v}{v-12}=-1$

57) $$\frac{k^2-49}{7-k}=\frac{(k+7)(k-7)}{7-k}$$
$$=-(k+7)$$
$$=-k-7$$

59) $$\frac{30-35x}{7x^2+8x-12}=\frac{5(6-7x)}{(7x-6)(x+2)}$$
$$=-\frac{5}{x+2}$$

61) $$\frac{16-4b^2}{b-2}=\frac{4(4-b^2)}{b-2}$$
$$=\frac{4(2+b)(2-b)}{b-2}$$
$$=\frac{4(2+b)(2-b)}{b-1}$$
$$=-4(2+b)$$

63) $$\frac{8t^3-27}{9-4t^2}=\frac{(2t-3)(4t^2+6t+9)}{(3+2t)(3-2t)}$$
$$=-\frac{4t^2+6t+9}{2t+3}$$

65) $$\frac{-b-7}{b-2},\ \frac{-(b+7)}{b-2},\ \frac{b+7}{2-b},$$
$$\frac{b+7}{-(b-2)},\ \frac{b+7}{-b+2}$$

67) $$\frac{-9+5t}{2t-3},\ \frac{5t-9}{2t-3},\ \frac{-(9-5t)}{2t-3},$$
$$\frac{9-5t}{-2t+3},\ \frac{9-5t}{3-2t},\ \frac{9-5t}{-(2t-3)}$$

69) $$-\frac{w-6}{4w-7},\ \frac{6-w}{4w-7},$$
$$\frac{w-6}{7-4w},\ -\frac{6-w}{-4w+7}$$

71) a)
$$\begin{array}{r} 2x-7 \\ x+4\overline{)2x^2+x-28} \\ -\ \underline{(2x^2+8x)} \\ -7x-28 \\ -\ \underline{(-7x-28)} \\ 0 \end{array}$$

b) $$\frac{2x^2+x-28}{x+4}=\frac{(2x-7)(x+4)}{x+4}$$
$$=2x-7$$

73) a)
$$\begin{array}{r} 9t^2+6t+4 \\ 3t-2\overline{)27t^3+0t^2+0t-8} \\ -\ \underline{(27t^3-18t^2)} \\ 18t^2+0t \\ -\ \underline{(18t^2-12t)} \\ 12t-8 \\ -\ \underline{(12t-8)} \\ 0 \end{array}$$

b) $$\frac{27t^3-8}{3t-2}=\frac{(3t-2)(9t^2+6t+4)}{3t-2}$$
$$=9t^2+6t+4$$

75) $$l=\frac{5x^2+13x+6}{x+2}$$
$$=\frac{(5x+3)(x+2)}{x+2}=5x+3$$

77) $$w=\frac{c^3-2c^2+4c-8}{c^2+4}$$
$$=\frac{c^2(c-2)+4(c-2)}{c^2+4}$$
$$=\frac{(c-2)(c^2+4)}{c^2+4}=c-2$$

79) $p-7=0$

$p=7$

The domain contains all real numbers except 7. Domain: $(-\infty,7)\cup(7,\infty)$

81) $5r+2=0$

$5r=-2$

$r=-\frac{2}{5}$

The domain contains all real numbers except $-\frac{2}{5}$.

Domain: $\left(-\infty,-\frac{2}{5}\right)\cup\left(-\frac{2}{5},\infty\right)$

83) $t^2-9t+8=0$

$(t-8)(t-1)=0$

$t-8=0$ or $t-1=0$

$t=8$ $\quad t=1$

The domain contains all real numbers except 1 and 8.

Domain: $(-\infty,1)\cup(1,8)\cup(8,\infty)$

85) $w^2-81=0$

$(w+9)(w-9)=0$

$w+9=0$ or $w-9=0$

$w=-9$ $\quad w=9$

The domain contains all real numbers except -9 and 9.

Domain: $(-\infty,-9)\cup(-9,9)\cup(9,\infty)$

87) $\frac{8}{c^2+6}$ is defined for all real numbers since c^2+6 never equals zero. Domain: $(-\infty,\infty)$

Section 8.2: Exercises

1) $\frac{7}{8}\cdot\frac{1}{3}=\frac{7}{24}$

3) $\frac{9}{14}\cdot\frac{7}{6}=\frac{\overset{3}{\cancel{9}}}{\underset{2}{\cancel{14}}}\cdot\frac{\cancel{7}}{\underset{2}{\cancel{6}}}=\frac{3}{4}$

5) $\frac{22}{15r^2}\cdot\frac{5r^4}{2}=\frac{\overset{11}{\cancel{22}}}{\underset{3}{\cancel{15r^2}}}\cdot\frac{\overset{r^2}{\cancel{5r^4}}}{\cancel{2}}=\frac{11r^2}{3}$

7) $\frac{14u^5}{15v^2}\cdot\frac{20v^6}{7u^8}=\frac{\overset{2}{\cancel{14u^5}}}{\underset{3}{\cancel{15v^2}}}\cdot\frac{\overset{4v^4}{\cancel{20v^6}}}{\underset{u^3}{\cancel{7u^8}}}=\frac{8v^4}{3u^3}$

9) $-\frac{12x}{24x^4}\cdot\frac{8x^9}{9}=-\frac{\overset{4x}{\cancel{12x}}}{\underset{3}{\cancel{24x^4}}}\cdot\frac{\overset{x^5}{\cancel{8x^9}}}{\underset{3}{\cancel{9}}}=-\frac{4x^6}{9}$

11) $\frac{a-6}{10}\cdot\frac{2(a+1)}{(a-6)^2}=\frac{\cancel{a-6}}{\underset{5}{\cancel{10}}}\cdot\frac{\cancel{2}(a+1)}{(a-6)^{\cancel{2}}}$

$=\frac{a+1}{5(a-6)}$

13) $\frac{5t^2}{(3t-2)^2}\cdot\frac{3t-2}{10t^3}=\frac{\cancel{5t^2}}{(3t-2)^{\cancel{2}}}\cdot\frac{\cancel{3t-2}}{\underset{2t}{\cancel{10t^3}}}$

$=\frac{1}{2t(3t-2)}$

15) $\frac{8}{6p+3}\cdot\frac{4p^2-1}{12}$

$=\frac{\overset{2}{\cancel{8}}}{3\cancel{(2p+1)}}\cdot\frac{\cancel{(2p+1)}(2p-1)}{\underset{3}{\cancel{12}}}$

$=\frac{2(2p-1)}{9}$

17) $\dfrac{2v^2+15v+18}{3v+18}\cdot\dfrac{12v-3}{8v+12}$

$=\dfrac{\cancel{(2v+3)}\cancel{(v+6)}}{\cancel{3}\cancel{(v+6)}}\cdot\dfrac{\cancel{3}(4v-1)}{4\cancel{(2v+3)}}$

$=\dfrac{4v-1}{4}$

19) $\dfrac{4}{5}\div\dfrac{8}{3}=\dfrac{\cancel{4}}{5}\cdot\dfrac{3}{\cancelto{2}{8}}=\dfrac{3}{10}$

21) $\dfrac{12}{7}\div 6=\dfrac{\cancelto{2}{12}}{7}\cdot\dfrac{1}{\cancel{6}}=\dfrac{2}{7}$

23) $\dfrac{42k^6}{25}\div\dfrac{12k^2}{35}=\dfrac{\cancelto{7k^4}{42k^6}}{\cancelto{5}{25}}\cdot\dfrac{\cancelto{7}{35}}{\cancelto{2}{12k^2}}=\dfrac{49k^4}{10}$

25) $\dfrac{c^2}{6b}\div\dfrac{c^8}{b}=\dfrac{\cancel{c^2}}{6\cancel{b}}\cdot\dfrac{\cancel{b}}{\cancelto{c^6}{c^8}}=\dfrac{1}{6c^6}$

27) $\dfrac{30(x-7)}{x+8}\div\dfrac{18}{(x+8)^2}$

$=\dfrac{\cancelto{5}{30}(x-7)}{\cancel{x+8}}\cdot\dfrac{(x+8)^{\cancel{2}}}{\cancelto{3}{18}}$

$=\dfrac{5(x-7)(x+8)}{3}$

29) $\dfrac{2a-1}{8a^3}\div\dfrac{(2a-1)^2}{24a^5}$

$=\dfrac{\cancel{2a-1}}{\cancel{8a^3}}\cdot\dfrac{\cancelto{3a^2}{24a^5}}{(2a-1)^{\cancel{2}}}=\dfrac{3a^2}{2a-1}$

31) $\dfrac{18y-45}{18}\div\dfrac{4y^2-25}{10}$

$=\dfrac{18y-45}{18}\cdot\dfrac{10}{4y^2-25}$

$=\dfrac{\cancel{9}\cancel{(2y-5)}}{\cancelto{2}{18}}\cdot\dfrac{\cancelto{5}{10}}{(2y+5)\cancel{(2y-5)}}$

$=\dfrac{5}{2y+5}$

33) $\dfrac{j^2-25}{5j+25}\div\dfrac{7j-35}{5}$

$=\dfrac{j^2-25}{5j+25}\cdot\dfrac{5}{7j-35}$

$=\dfrac{\cancel{(j+5)}\cancel{(j-5)}}{\cancel{5}\cancel{(j+5)}}\cdot\dfrac{\cancel{5}}{7\cancel{(j-5)}}=\dfrac{1}{7}$

35) $\dfrac{4c-9}{2c^2-8c}\div\dfrac{12c-27}{c^2-3c-4}$

$=\dfrac{4c-9}{2c^2-8c}\cdot\dfrac{c^2-3c-4}{12c-27}$

$=\dfrac{\cancel{4c-9}}{2c\cancel{(c-4)}}\cdot\dfrac{\cancel{(c-4)}(c+1)}{3\cancel{(4c-9)}}=\dfrac{c+1}{6c}$

37) $\dfrac{\frac{4}{15}}{\frac{8}{35}}=\dfrac{4}{15}\div\dfrac{8}{35}=\dfrac{\cancel{4}}{\cancelto{3}{15}}\cdot\dfrac{\cancelto{7}{35}}{\cancelto{2}{8}}=\dfrac{7}{6}$

39) $\dfrac{\frac{2}{9}}{\frac{8}{3}}=\dfrac{2}{9}\div\dfrac{8}{3}=\dfrac{\cancel{2}}{\cancelto{3}{9}}\cdot\dfrac{\cancel{3}}{\cancelto{4}{8}}=\dfrac{1}{12}$

41) $\dfrac{\frac{6s-7}{4}}{\frac{6s-7}{12}}=\dfrac{6s-7}{4}\div\dfrac{6s-7}{12}$

$=\dfrac{\cancel{6s-7}}{\cancel{4}}\cdot\dfrac{\overset{3}{\cancel{12}}}{\cancel{6s-7}}=3$

43) $\dfrac{\frac{16r+24}{r^3}}{\frac{12r+18}{r}}=\dfrac{16r+24}{r^3}\div\dfrac{12r+18}{r}$

$=\dfrac{16r+24}{r^3}\cdot\dfrac{r}{12r+18}$

$=\dfrac{\overset{4}{\cancel{8}}\cancel{(2r+3)}}{\underset{r^2}{\cancel{r^3}}}\cdot\dfrac{\cancel{r}}{\underset{3}{\cancel{6}}\cancel{(2r+3)}}$

$=\dfrac{4}{3r^2}$

45) $\dfrac{\frac{4z-20}{z^2}}{\frac{z^2-25}{z^6}}$

$=\dfrac{4z-20}{z^2}\div\dfrac{z^2-25}{z^6}$

$=\dfrac{4z-20}{z^2}\cdot\dfrac{z^6}{z^2-25}$

$=\dfrac{4\cancel{(z-5)}}{\cancel{z^2}}\cdot\dfrac{\overset{z^4}{\cancel{z^6}}}{(z+5)\cancel{(z-5)}}=\dfrac{4z^4}{z+5}$

47) $\dfrac{\frac{12}{9a^2-4}}{\frac{16a^2}{3a^2+2a}}$

$=\dfrac{12}{9a^2-4}\div\dfrac{16a^2}{3a^2+2a}$

$=\dfrac{12}{9a^2-4}\cdot\dfrac{3a^2+2a}{16a^2}$

$=\dfrac{\overset{3}{\cancel{12}}}{\cancel{(3a+2)}(3a-2)}\cdot\dfrac{\cancel{a}\cancel{(3a+2)}}{\underset{4}{\cancel{16}}\,a^{\cancel{2}}}$

$=\dfrac{3}{4a(3a-2)}$

49) $\dfrac{d^2+d-56}{d-11}\cdot\dfrac{d^2-10d-11}{4d+32}$

$=\dfrac{\cancel{(d+8)}(d-7)}{\cancel{d-11}}\cdot\dfrac{\cancel{(d-11)}(d+1)}{4\cancel{(d+8)}}$

$=\dfrac{(d-7)(d+1)}{4}$

51) $\dfrac{b^2-4b+4}{4b-8}\div\dfrac{3b^2-4b-4}{3b+2}$

$=\dfrac{b^2-4b+4}{4b-8}\cdot\dfrac{3b+2}{3b^2-4b-4}$

$=\dfrac{\cancel{(b-2)^2}}{4\cancel{(b-2)}}\cdot\dfrac{\cancel{3b+2}}{\cancel{(3b+2)}\cancel{(b-2)}}=\dfrac{1}{4}$

53) $\dfrac{4n^2-1}{6n^5}\div\dfrac{2n^2-7n-4}{10n^3}$

$=\dfrac{4n^2-1}{6n^5}\cdot\dfrac{10n^3}{2n^2-7n-4}$

$=\dfrac{\cancel{(2n+1)}(2n-1)}{\underset{3n^2}{\cancel{6n^5}}}\cdot\dfrac{\overset{5}{\cancel{10n^3}}}{\cancel{(2n+1)}(n-4)}$

$=\dfrac{5(2n-1)}{3n^2(n-4)}$

55) $\dfrac{a^2-4a}{6a+54}\cdot\dfrac{a^2+13a+36}{16-a^2}$

$=\dfrac{a\cancel{(a-4)}^{-1}}{6\cancel{(a+9)}}\cdot\dfrac{\cancel{(a+9)}\cancel{(a+4)}}{\cancel{(4+a)}\cancel{(4-a)}}=-\dfrac{a}{6}$

57) $\dfrac{4t^2+12t+36}{4t-2}\cdot\dfrac{t^2-9}{t^3-27}$

$=\dfrac{\overset{2}{\cancel{4}}\cancel{(t^2+3t+9)}}{\cancel{2}(2t-1)}\cdot\dfrac{(t+3)\cancel{(t-3)}}{\cancel{(t-3)}\cancel{(t^2+3t+9)}}$

$=\dfrac{2(t+3)}{2t-1}$

59) $\dfrac{64-u^2}{40-5u}\div\dfrac{u^2+10u+16}{2u+3}$

$=\dfrac{64-u^2}{40-5u}\cdot\dfrac{2u+3}{u^2+10u+16}$

$=\dfrac{\cancel{(8+u)}\cancel{(8-u)}}{5\cancel{(8-u)}}\cdot\dfrac{2u+3}{\cancel{(u+8)}(u+2)}$

$=\dfrac{2u+3}{5(u+2)}$

61) $\dfrac{24x^3}{x^3+6x^2+5x-30}\cdot\dfrac{3x^2-20x+12}{18x^2-12x}$

$=\dfrac{\overset{4x^2}{\cancel{24x^3}}}{\cancel{(x-6)}(x^2+5)}\cdot\dfrac{\cancel{(3x-2)}\cancel{(x-6)}}{\cancel{6x}\cancel{(3x-2)}}$

$=\dfrac{4x^2}{x^2+5}$

63) $\dfrac{a^2-b^2}{a^3+b^3}\div\dfrac{9b-9a}{8}$

$=\dfrac{a^2-b^2}{a^3+b^3}\cdot\dfrac{8}{9b-9a}$

$=\dfrac{\cancel{(a+b)}\cancel{(a-b)}^{-1}}{\cancel{(a+b)}(a^2-ab+b^2)}\cdot\dfrac{8}{9\cancel{(b-a)}}$

$=-\dfrac{8}{9(a^2-ab+b^2)}$

65) $\dfrac{w^2-17w+72}{6w}\div(w-9)$

$=\dfrac{w^2-17w+72}{6w}\cdot\dfrac{1}{w-9}$

$=\dfrac{\cancel{(w-9)}(w-8)}{6w}\cdot\dfrac{1}{\cancel{w-9}}$

$=\dfrac{w-8}{6w}$

67) $\dfrac{a}{a^2+20a+100} \div \left(\dfrac{a^2-7a-18}{a^2-5a-14} \cdot \dfrac{a^2-7a}{a^2+a-90}\right)$

$= \dfrac{a}{(a+10)^2} \div \left(\dfrac{\cancel{(a-9)}\cancel{(a+2)}}{\cancel{(a+2)}\cancel{(a-7)}} \cdot \dfrac{a\cancel{(a-7)}}{(a+10)\cancel{(a-9)}}\right)$

$= \dfrac{a}{(a+10)^2} \div \dfrac{a}{a+10} = \dfrac{\cancel{a}}{(a+10)^{\cancel{2}}} \cdot \dfrac{\cancel{a+10}}{\cancel{a}} = \dfrac{1}{a+10}$

69) $\dfrac{t^3-1}{t-1} \div \left(\dfrac{5t+1}{12t-8} \cdot \dfrac{t^2+t+1}{t}\right) = \dfrac{\cancel{(t-1)}(t^2+t+1)}{\cancel{t-1}} \div \left(\dfrac{5t+1}{4(3t-2)} \cdot \dfrac{t^2+t+1}{t}\right)$

$= (t^2+t+1) \div \dfrac{(5t+1)(t^2+t+1)}{4t(3t-2)}$

$= \cancel{(t^2+t+1)} \cdot \dfrac{4t(3t-2)}{(5t+1)\cancel{(t^2+t+1)}} = \dfrac{4t(3t-2)}{(5t+1)}$

Section 8.3: Exercises

1) $8 = 2 \cdot 2 \cdot 2$
$20 = 2 \cdot 2 \cdot 5$
$\text{LCD} = 2^3 \cdot 5 = 8 \cdot 5 = 40$

3) $28 = 2 \cdot 2 \cdot 7$
$12 = 2 \cdot 2 \cdot 3$
$21 = 3 \cdot 7$
$\text{LCD} = 2^2 \cdot 7 \cdot 3 = 4 \cdot 21 = 84$

5) $\text{LCD} = c^4$

7) $\text{LCD} = 36p^8$

9) $\text{LCD} = 21w^7$

11) $\text{LCD} = 12k^5$

13) $\text{LCD} = 24a^3b^4$

15) $\text{LCD} = 2(n+4)$

17) $\text{LCD} = w(2w+1)$

19) $6k+30 = 6(k+5)$
$9k+45 = 9(k+5)$
$\text{LCD} = 18(k+5)$

21) $12a^2-4a = 4a(3a-1)$
$6a^4-2a^3 = 2a^3(3a-1)$
$\text{LCD} = 4a^3(3a-1)$

23) $r+7$ and $r-2$ are different factors. The LCD will be the product of these factors. $\text{LCD} = (r+7)(r-2)$

25) $x^2-10x+16 = (x-8)(x-2)$
$x^2-5x-24 = (x-8)(x+3)$
$\text{LCD} = (x-8)(x-2)(x+3)$

27) $w^2-3w-10=(w-5)(w+2)$
$w^2-2w-15=(w-5)(w+3)$
$w^2+5w+6=(w+3)(w+2)$
$\text{LCD}=(w-5)(w+2)(w+3)$

29) $\text{LCD}=b-4$ or $4-b$

31) $\text{LCD}=u-v$ or $v-u$

33) Multiply the numerator and denominator of the fraction by $x-2$.

35) $\frac{4}{3p^4}\cdot\frac{2p}{2p}=\frac{8p}{6p^5}$

37) $\frac{11}{4cd^2}\cdot\frac{6c^2d}{6c^2d}=\frac{66c^cd}{24c^3d^3}$

39) $\frac{5}{m-9}\cdot\frac{m}{m}=\frac{5m}{m(m-9)}$

41) $\frac{a}{8(3a+1)}\cdot\frac{2a}{2a}=\frac{2a^2}{16a(3a+1)}$

43) $\frac{3b}{b+2}\cdot\frac{b+5}{b+5}=\frac{3b(b+5)}{(b+2)(b+5)}$
$=\frac{3b^2+15b}{(b+2)(b+5)}$

45) $\frac{w+1}{4w-3}\cdot\frac{w-6}{w-6}=\frac{(w+1)(w-6)}{(4w-3)(w-6)}$
$=\frac{w^2-5w-6}{(4w-3)(w-6)}$

47) $\frac{8}{5-n}\cdot\frac{-1}{-1}=\frac{-8}{-(5-n)}=-\frac{8}{n-5}$

49) $-\frac{a}{3a-2}\cdot\frac{-1}{-1}=-\frac{-a}{-(3a-2)}=\frac{a}{2-3a}$

51) $\text{LCD}=t^3$
$\frac{3}{t}\cdot\frac{t^2}{t^2}=\frac{3t^2}{t^3}$
$\frac{8}{t^3}$ is already written with the LCD.

53) $\text{LCD}=24n^6$
$\frac{9}{8n^6}\cdot\frac{3}{3}=\frac{27}{24n^6}$
$\frac{2}{3n^2}\cdot\frac{8n^4}{8n^4}=\frac{16n^4}{24n^6}$

55) $\text{LCD}=5x^3y^5$
$\frac{1}{x^3y}\cdot\frac{5y^4}{5y^4}=\frac{5y^4}{5x^3y^5}$
$\frac{6}{5xy^5}\cdot\frac{x^2}{x^2}=\frac{6x^2}{5x^3y^5}$

57) $\text{LCD}=7(5t-6)$
$\frac{t}{5t-6}\cdot\frac{7}{7}=\frac{7t}{7(5t-6)}$
$\frac{10}{7}\cdot\frac{5t-6}{5t-6}=\frac{50t-60}{7(5t-6)}$

59) $\text{LCD}=c(c+1)$
$\frac{3}{c}\cdot\frac{c+1}{c+1}=\frac{3c+3}{c(c+1)}$
$\frac{2}{c+1}\cdot\frac{c}{c}=\frac{2c}{c(c+1)}$

61) $\text{LCD}=z(z-9)$
$\frac{z}{z-9}\cdot\frac{z}{z}=\frac{z^2}{z(z-9)}$
$\frac{4}{z}\cdot\frac{z-9}{z-9}=\frac{4z-36}{z(z-9)}$

63) $\dfrac{a}{24a+36}=\dfrac{a}{12(2a+3)}$

$\dfrac{1}{18a+27}=\dfrac{1}{9(2a+3)}$

$\text{LCD}=36(2a+3)$

$\dfrac{a}{12(2a+3)}\cdot\dfrac{3}{3}=\dfrac{3a}{36(2a+3)}$

$\dfrac{1}{9(2a+3)}\cdot\dfrac{4}{4}=\dfrac{4}{36(2a+3)}$

65) $\text{LCD}=(h+5)(h-3)$

$\dfrac{4}{h+5}\cdot\dfrac{h-3}{h-3}=\dfrac{4h-12}{(h+5)(h-3)}$

$\dfrac{7h}{h-3}\cdot\dfrac{h+5}{h+5}=\dfrac{7h^2+35h}{(h+5)(h-3)}$

67) $\text{LCD}=(3b-2)(b-9)$

$\dfrac{b}{3b-2}\cdot\dfrac{b-9}{b-9}=\dfrac{b^2-9b}{(3b-2)(b-9)}$

$\dfrac{1}{b-9}\cdot\dfrac{3b-2}{3b-2}=\dfrac{3b-2}{(3b-2)(b-9)}$

69) $\dfrac{9y}{y^2-y-42}=\dfrac{9y}{(y-7)(y+6)}$

$\dfrac{3}{2y^2+12y}=\dfrac{3}{2y(y+6)}$

$\text{LCD}=2y(y+6)(y-7)$

$\dfrac{9y}{(y-7)(y+6)}\cdot\dfrac{2y}{2y}=\dfrac{18y^2}{2y(y-7)(y+6)}$

$\dfrac{3}{2y(y+6)}\cdot\dfrac{(y-7)}{(y-7)}=\dfrac{3y-21}{2y(y-7)(y+6)}$

71) $\dfrac{z}{z^2-10z+25}=\dfrac{z}{(z-5)^2}$

$\dfrac{15z}{z^2-2z-15}=\dfrac{15z}{(z-5)(z+3)}$

$\text{LCD}=(z-5)^2(z+3)$

$\dfrac{z}{(z-5)^2}\cdot\dfrac{(z+3)}{(z+3)}=\dfrac{z^2+3z}{(z-5)^2(z+3)}$

$\dfrac{15z}{(z-5)(z+3)}\cdot\dfrac{(z-5)}{(z-5)}=\dfrac{15z^2-75z}{(z-5)^2(z+3)}$

73) $\dfrac{11}{g-3}$

$\dfrac{4}{9-g^2}=\dfrac{4}{(3+g)(3-g)}$

$=-\dfrac{4}{(g+3)(g-3)}$

$\text{LCD}=(g+3)(g-3)$

$\dfrac{11}{(g-3)}\cdot\dfrac{g+3}{g+3}=\dfrac{11g+33}{(g-3)(g+3)}$

$-\dfrac{4}{(g+3)(g-3)}$ written with LCD

75) $\dfrac{10}{4k-1}$

$\dfrac{k}{1-16k^2}=\dfrac{k}{(1+4k)(1-4k)}$

$=-\dfrac{k}{(4k+1)(4k-1)}$

$\text{LCD}=(4k+1)(4k-1)$

$\dfrac{10}{4k-1}\cdot\dfrac{4k+1}{4k+1}=\dfrac{40k+10}{(4k+1)(4k-1)}$

$-\dfrac{k}{(4k+1)(4k-1)}$ written with LCD

77) $\dfrac{4}{w^2-4w}=\dfrac{4}{w(w-4)}$

$\dfrac{6}{7w^2-28w}=\dfrac{6}{7w(w-4)}$

$\dfrac{11}{w^2-8w+16}=\dfrac{11}{(w-4)^2}$

$\text{LCD}=7w(w-4)^2$

$\dfrac{4}{w(w-4)}\cdot\dfrac{7(w-4)}{7(w-4)}=\dfrac{28(w-4)}{7w(w-4)^2}$

$=\dfrac{28w-112}{7w(w-4)^2}$

$\dfrac{6}{7w(w-4)}\cdot\dfrac{w-4}{w-4}=\dfrac{6w-24}{7w(w-4)^2}$

$\dfrac{11}{(w-4)^2}\cdot\dfrac{7w}{7w}=\dfrac{77w}{7w(w-4)^2}$

77) $-\dfrac{1}{a+4}$

$\dfrac{a}{a^2-16}=\dfrac{a}{(a+4)(a-4)}$

$\dfrac{3}{a^2+5a+4}=\dfrac{3}{(a+4)(a+1)}$

$\text{LCD}=(a+4)(a-4)(a+1)$

$-\dfrac{1}{a+4}\cdot\dfrac{(a-4)(a+1)}{(a-4)(a+1)}$

$=-\dfrac{a^2-3a-4}{(a+4)(a-4)(a+1)}$

$\dfrac{a}{(a+4)(a-4)}\cdot\dfrac{a+1}{a+1}$

$=-\dfrac{a^2+a}{(a+4)(a-4)(a+1)}$

$\dfrac{3}{(a+4)(a+1)}\cdot\dfrac{a-4}{a-4}$

$=\dfrac{3a-12}{(a+4)(a-4)(a+1)}$

Section 8.4: Exercises

1) $\dfrac{8}{11}-\dfrac{3}{11}=\dfrac{8-3}{11}=\dfrac{5}{11}$

3) $\dfrac{7}{20}+\dfrac{9}{20}=\dfrac{7+9}{20}=\dfrac{16}{20}=\dfrac{4}{5}$

5) $\dfrac{8}{a}+\dfrac{2}{a}=\dfrac{8+2}{a}=\dfrac{10}{a}$

7) $\dfrac{10}{3k^2}-\dfrac{2}{3k^2}=\dfrac{10-2}{3k^2}=\dfrac{8}{3k^2}$

9) $\dfrac{n}{n+6}-\dfrac{9}{n+6}=\dfrac{n-9}{n+6}$

11) $\dfrac{8}{x+4}+\dfrac{2x}{x+4}=\dfrac{8+2x}{x+4}=\dfrac{2\cancel{(4+x)}}{\cancel{x+4}}=2$

13) $\dfrac{7w-4}{w(3w-4)}-\dfrac{20-11w}{w(3w-4)}$

$=\dfrac{7w-4-20+11w}{w(3w-4)}$

$=\dfrac{18w-24}{w(3w-4)}$

$=\dfrac{6\cancel{(3w-4)}}{w\cancel{(3w-4)}}=\dfrac{6}{w}$

15) $\dfrac{2r+15}{(r-5)(r+2)}+\dfrac{r^2-10r}{(r-5)(r+2)}$

$=\dfrac{2r+15+r^2-10r}{(r-5)(r+2)}$

$=\dfrac{r^2-8r+15}{(r-5)(r+2)}$

$=\dfrac{\cancel{(r-5)}(r-3)}{\cancel{(r-5)}(r+2)}=\dfrac{(r-3)}{(r+2)}$

17) a) $x(x-3)$

b) Multiply $\dfrac{8}{x-3}$ by $\dfrac{x}{x}$, and multiply $\dfrac{2}{x}$ by $\dfrac{x-3}{x-3}$.

c) $\dfrac{8}{x-3}\cdot\dfrac{x}{x}=\dfrac{8x}{x(x-3)}$

$\dfrac{2}{x}\cdot\dfrac{x-3}{x-3}=\dfrac{2x-6}{x(x-3)}$

19) $\dfrac{5}{8}+\dfrac{1}{6}=\dfrac{15}{24}+\dfrac{4}{24}=\dfrac{19}{24}$

21) $\dfrac{5x}{12}-\dfrac{4x}{15}=\dfrac{25x}{60}-\dfrac{16x}{60}=\dfrac{9x}{60}=\dfrac{3x}{20}$

23) $\dfrac{5}{8u}-\dfrac{2}{3u^2}=\dfrac{15u}{24u^2}-\dfrac{16}{24u^2}=\dfrac{15u-16}{24u^2}$

25) $\dfrac{3}{2a}+\dfrac{6}{7a^2}=\dfrac{21a}{14a^2}+\dfrac{12}{14a^2}$

$=\dfrac{21a+12}{14a^2}=\dfrac{3(7a+4)}{14a^2}$

27) $\dfrac{2}{k}+\dfrac{9}{k+10}=\dfrac{2(k+10)}{k(k+10)}+\dfrac{9k}{k(k+10)}$

$=\dfrac{2k+20+9k}{k(k+10)}$

$=\dfrac{11k+20}{k(k+10)}$

29) $\dfrac{15}{d-8}-\dfrac{4}{d}=\dfrac{15d}{d(d-8)}-\dfrac{4(d-8)}{d(d-8)}$

$=\dfrac{15d-4d+32}{d(d-8)}$

$=\dfrac{11d+32}{d(d-8)}$

31) $\dfrac{1}{z+6}+\dfrac{4}{z+2}$

$=\dfrac{1(z+2)}{(z+6)(z+2)}+\dfrac{4(z+6)}{(z+6)(z+2)}$

$=\dfrac{z+2+4z+24}{(z+6)(z+2)}=\dfrac{5z+26}{(z+6)(z+2)}$

33) $\dfrac{x}{2x+1}-\dfrac{3}{x+5}$

$=\dfrac{x(x+5)}{(2x+1)(x+5)}-\dfrac{3(2x+1)}{(2x+1)(x+5)}$

$=\dfrac{x^2+5x-6x-3}{(2x+1)(x+5)}=\dfrac{x^2-x-3}{(2x+1)(x+5)}$

35) $\dfrac{t}{t+7}+\dfrac{11t-21}{t^2-49}$

$=\dfrac{t}{t+7}+\dfrac{11t-21}{(t+7)(t-7)}$

$=\dfrac{t(t-7)}{(t+7)(t-7)}+\dfrac{11t-21}{(t+7)(t-7)}$

$=\dfrac{t^2-7t+11t-21}{(t+7)(t-7)}=\dfrac{t^2+4t-21}{(t+7)(t-7)}$

$=\dfrac{\cancel{(t+7)}(t-3)}{\cancel{(t+7)}(t-7)}=\dfrac{(t-3)}{(t-7)}$

37) $\dfrac{b}{b^2-16}+\dfrac{10}{b^2-5b-36}=\dfrac{b}{(b+4)(b-4)}+\dfrac{10}{(b-9)(b+4)}$

$=\dfrac{b(b-9)}{(b+4)(b-4)(b-9)}+\dfrac{10(b-4)}{(b+4)(b-4)(b-9)}$

$=\dfrac{b^2-9b+10b-40}{(b+4)(b-4)(b-9)}=\dfrac{b^2+b-40}{(b+4)(b-4)(b-9)}$

39) $\dfrac{3c}{c^2+4c-12}-\dfrac{2c-5}{c^2+2c-24}=\dfrac{3c}{(c+6)(c-2)}-\dfrac{2c-5}{(c+6)(c-4)}$

$=\dfrac{3c(c-4)}{(c+6)(c-2)(c-4)}-\dfrac{(2c-5)(c-2)}{(c+6)(c-2)(c-4)}$

$=\dfrac{3c^2-12c-2c^2+9c-10}{(c+6)(c-2)(c-4)}$

$=\dfrac{c^2-3c-10}{(c+6)(c-2)(c-4)}=\dfrac{(c-5)(c+2)}{(c+6)(c-2)(c-4)}$

41) $\dfrac{4m}{m^2+m-6}-\dfrac{7}{m^2+4m-12}=\dfrac{4m}{(m+3)(m-2)}-\dfrac{7}{(m+6)(m-2)}$

$=\dfrac{4m(m+6)}{(m+3)(m-2)(m+6)}-\dfrac{7(m+3)}{(m+3)(m+6)(m-2)}$

$=\dfrac{4m^2+24m-7m-21}{(m+3)(m-2)(m+6)}$

$=\dfrac{4m^2+17m-21}{(m+3)(m-2)(m+6)}=\dfrac{(4m+21)(m-1)}{(m+3)(m-2)(m+6)}$

43) $\dfrac{4b+1}{3b-12}+\dfrac{5b}{b^2-b-12}=\dfrac{4b+1}{3(b-4)}+\dfrac{5b}{(b-4)(b+3)}=\dfrac{(4b+1)(b+3)}{3(b-4)(b+3)}+\dfrac{15b}{3(b-4)(b+3)}$

$=\dfrac{4b^2+13b+3+15b}{3(b-4)(b+3)}=\dfrac{4b^2+28b+3}{3(b-4)(b+3)}$

45) No. If the sum is rewritten as $\dfrac{5}{x-7}-\dfrac{2}{x-7}$. then the LCD $= x-7$.

If the sum is rewritten as $\dfrac{-5}{7-x}+\dfrac{2}{7-x}$, then the LCD $= 7-x$.

47) $\dfrac{9}{z-6}+\dfrac{2}{6-z}=\dfrac{9}{z-6}-\dfrac{2}{z-6}=\dfrac{7}{z-6}$

49) $\dfrac{5}{c-d}-\dfrac{3}{d-c}=\dfrac{5}{c-d}+\dfrac{3}{c-d}=\dfrac{8}{c-d}$

51) $\dfrac{10}{m-3}+\dfrac{m+11}{3-m}=\dfrac{10}{m-3}-\dfrac{m+11}{m-3}=\dfrac{10-m-11}{m-3}=\dfrac{-m-1}{m-3}=\dfrac{-1(m+1)}{m-3}=\dfrac{m+1}{3-m}$

53) $\dfrac{5}{2-n}+\dfrac{n+3}{n-2}=\dfrac{5}{2-n}-\dfrac{n+3}{2-n}=\dfrac{5-n-3}{2-n}=\dfrac{2-n}{2-n}=1$

55) $\dfrac{2c}{12b-7c}-\dfrac{13}{7c-12b}=\dfrac{2c}{12b-7c}+\dfrac{13}{12b-7c}=\dfrac{2c+13}{12b-7c}$

57) $\dfrac{5}{8-t}+\dfrac{10}{t^2-64}=\dfrac{5}{8-t}+\dfrac{10}{(t+8)(t-8)}=\dfrac{-5}{t-8}+\dfrac{10}{(t+8)(t-8)}$

$=\dfrac{-5(t+8)}{(t+8)(t-8)}+\dfrac{10}{(t+8)(t-8)}=\dfrac{-5t-40+10}{(t+8)(t-8)}$

$=\dfrac{-5t-30}{(t+8)(t-8)}=-\dfrac{5(t+6)}{(t+8)(t-8)}$

59) $$\frac{a}{4a^2-9}-\frac{4}{3-2a}=\frac{a}{(2a+3)(2a-3)}+\frac{4}{(2a-3)}=\frac{a}{(2a+3)(2a-3)}+\frac{4(2a+3)}{(2a+3)(2a-3)}$$
$$=\frac{a+8a+12}{(2a+3)(2a-3)}=\frac{9a+12}{(2a+3)(2a-3)}=\frac{3(3a+4)}{(2a+3)(2a-3)}$$

61) $$\frac{2}{j^2+8j}+\frac{2j}{j+8}-\frac{1}{3j}=\frac{2}{j(j+8)}+\frac{2j}{j+8}-\frac{1}{3j}=\frac{3\cdot 2}{3j(j+8)}+\frac{3j\cdot 2j}{3j(j+8)}-\frac{1(j+8)}{3j(j+8)}$$
$$=\frac{6+6j^2-j-8}{3j(j+8)}=\frac{6j^2-j-2}{3j(j+8)}$$

63) $$\frac{2k+7}{k^2-4k}+\frac{9k}{2k^2-15k+28}+\frac{15}{2k^2-7k}=\frac{2k+7}{k(k-4)}+\frac{9k}{(2k-7)(k-4)}+\frac{15}{k(2k-7)}$$
$$=\frac{(2k+7)(2k-7)}{k(k-4)(2k-7)}+\frac{k\cdot 9k}{k(2k-7)(k-4)}+\frac{15(k-4)}{k(k-4)(2k-7)}$$
$$=\frac{(4k^2-49)+9k^2+(15k-60)}{k(k-4)(2k-7)}=\frac{13k^2+15k-109}{k(k-4)(2k-7)}$$

65) $$\frac{c}{c^2-8c+16}-\frac{5}{c^2-c-12}=\frac{c}{(c-4)^2}-\frac{5}{(c-4)(c+3)}=\frac{c(c+3)}{(c-4)^2(c+3)}-\frac{5(c-4)}{(c-4)^2(c+3)}$$
$$=\frac{c^2+3c-5c+20}{(c-4)^2(c+3)}=\frac{c^2-2c+20}{(c-4)^2(c+3)}$$

67) $$\frac{1}{x+y}+\frac{x}{x^2-y^2}-\frac{4}{2x-2y}=\frac{1}{x+y}+\frac{x}{(x+y)(x-y)}-\frac{4}{2(x-y)}$$
$$=\frac{1}{x+y}+\frac{x}{(x+y)(x-y)}-\frac{2}{(x-y)}$$
$$=\frac{x-y}{(x+y)(x-y)}+\frac{x}{(x+y)(x-y)}-\frac{2(x+y)}{(x+y)(x-y)}$$
$$=\frac{x-y+x-2x-2y}{(x+y)(x-y)}=-\frac{3y}{(x+y)(x-y)}$$

69) $\dfrac{n+5}{4n^2+7n-2}-\dfrac{n-4}{3n^2+7n+2}=\dfrac{n+5}{(4n-1)(n+2)}-\dfrac{n-4}{(3n+1)(n+2)}$

$$=\frac{(n+5)(3n+1)}{(4n-1)(3n+1)(n+2)}-\frac{(4n-1)(n-4)}{(4n-1)(3n+1)(n+2)}$$

$$=\frac{3n^2+16n+5-\left(4n^2-17n+4\right)}{(4n-1)(3n+1)(n+2)}=\frac{-n^2+33n+1}{(4n-1)(3n+1)(n+2)}$$

71) $\dfrac{y+6}{y^2-4y}+\dfrac{y}{2y^2-13y+20}-\dfrac{1}{2y^2-5y}$

$$=\frac{y+6}{y(y-4)}+\frac{y}{(2y-5)(y-4)}-\frac{1}{y(2y-5)}$$

$$=\frac{(y+6)(2y-5)}{y(y-4)(2y-5)}+\frac{y\cdot y}{y(y-4)(2y-5)}-\frac{1\cdot(y-4)}{y(y-4)(2y-5)}$$

$$=\frac{2y^2+7y-30+y^2-y+4}{y(y-4)(2y-5)}=\frac{3y^2+6y-26}{y(y-4)(2y-5)}$$

73) a) $A=\left(\dfrac{x+1}{2}\right)\left(\dfrac{4}{x-3}\right)=\dfrac{2(x+1)}{x-3}$ sq. units b) $P=2\left(\dfrac{x+1}{2}\right)+2\left(\dfrac{4}{x-3}\right)$

$$=x+1+\frac{8}{x-3}=\frac{(x+1)(x-3)}{(x-3)}+\frac{8}{x-3}$$

$$=\frac{(x+1)(x-3)+8}{x-3}=\frac{x^2-2x+5}{x-3}\text{ units}$$

75) a) $A=\left(\dfrac{1}{w^2-4}\right)\left(\dfrac{w}{w+2}\right)=\left(\dfrac{1}{(w+2)(w-2)}\right)\left(\dfrac{w}{w+2}\right)=\dfrac{w}{(w+2)^2(w-2)}$ sq. units

b) $P=2\left(\dfrac{1}{w^2-4}\right)+2\left(\dfrac{w}{w+2}\right)=\dfrac{2}{(w+2)(w-2)}+\dfrac{2w}{w+2}$

$$=\frac{2}{(w+2)(w-2)}+\frac{2w(w-2)}{(w+2)(w-2)}=\frac{2+2w^2-4w}{(w+2)(w-2)}$$

$$=\frac{2w^2-4w+2}{(w+2)(w-2)}=\frac{2\left(w^2-2w+1\right)}{(w+2)(w-2)}=\frac{2(w-1)^2}{(w+2)(w-2)}\text{ units}$$

Mid-Chapter Summary

1) a) $\dfrac{-4}{2(-4)-6}=\dfrac{-4}{-8-6}=\dfrac{2}{7}$

b) $\dfrac{3}{2(3)-6}=\dfrac{3}{6-6}=\dfrac{3}{0}$ undefined

3) a) $-\dfrac{(-4)^2}{(-4)^2-10}=-\dfrac{16}{16-10}$

$=-\dfrac{16}{6}=-\dfrac{8}{3}$

b) $-\dfrac{(3)^2}{(3)^2-10}=-\dfrac{9}{9-10}=-\dfrac{9}{-1}=9$

5) a) $w^2-9=0$

$(w+3)(w-3)=0$

$w+3=0$ or $w-3=0$

$w=-3 \qquad w=3$

b) $4w=0$

$w=0$

7) a) $b^2+2b-8=0$

$(b-4)(b-2)=0$

$b+4=0$ or $b-2=0$

$b=-4 \qquad b=2$

b) $3-5b=0$

$-5b=-3$

$b=\dfrac{3}{5}$

9) a) Never undefined b) $t-6=0$

$t=6$

11) $\dfrac{\cancel{36}^{\,4}\,\cancel{n^9}}{\cancel{27n^{12}}_{\,3n^3}}=\dfrac{4}{3n^3}$

13) $\dfrac{2j+5}{2j^2-3j-20}=\dfrac{\cancel{2j+5}}{\cancel{(2j+5)}(j-4)}$

$=\dfrac{1}{j-4}$

15) $\dfrac{12-15n}{5n^2+6n-8}=\dfrac{3(4-5n)}{(5n-4)(n+2)}$

$=\dfrac{3\,\overset{-1}{\cancel{(4-5n)}}}{\cancel{(5n-4)}(n+2)}$

$=-\dfrac{3}{n+2}$

17) $\dfrac{5}{f+8}-\dfrac{2}{f}=\dfrac{5f}{f(f+8)}-\dfrac{2(f+8)}{f(f+8)}$

$=\dfrac{5f-2f-16}{f(f+8)}$

$=\dfrac{3f-16}{f(f+8)}$

19) $\dfrac{\cancel{9a^3}^{\,a^2}}{\cancel{10b}}\cdot\dfrac{\cancel{40b^2}^{\,4b}}{\cancel{81a}_{\,9}}=\dfrac{4a^2b}{9}$

21) $\dfrac{3}{q^2-q-20}+\dfrac{8q}{q^2+11q+28}$

$=\dfrac{3}{(q-5)(q+4)}+\dfrac{8q}{(q+7)(q+4)}$

$=\dfrac{3(q+7)}{(q-5)(q+4)(q+7)}+\dfrac{8q(q-5)}{(q-5)(q+4)(q+7)}$

$=\dfrac{3q+21+8q^2-40q}{(q-5)(q+4)(q+7)}$

$=\dfrac{8q^2-37q+21}{(q-5)(q+4)(q+7)}$

23) $\dfrac{16-m^2}{m+4}\div\dfrac{8m-32}{m+7}$

$=\dfrac{16-m^2}{m+4}\cdot\dfrac{m+7}{8m-32}$

$=\dfrac{(4+m)(4-m)}{m+4}\cdot\dfrac{m+7}{8(m-4)}$

$=\dfrac{\cancel{(4+m)}\,\overset{-1}{\cancel{(4-m)}}}{\cancel{m+4}}\cdot\dfrac{m+7}{8\cancel{(m-4)}}$

$=-\dfrac{m+7}{8}$

25) $\dfrac{13}{r-8}+\dfrac{4}{8-r}=\dfrac{13}{r-8}-\dfrac{4}{r-8}=\dfrac{9}{r-8}$

27) $\dfrac{\frac{10d}{d+11}}{\frac{5d^4}{2d+22}}=\dfrac{10d}{d+11}\div\dfrac{5d^4}{2d+22}$

$=\dfrac{\overset{2}{\cancel{10d}}}{\cancel{d+11}}\cdot\dfrac{2\cancel{(d+11)}}{\underset{d^3}{\cancel{5d^4}}}=\dfrac{4}{d^3}$

29) $\dfrac{a^2-4}{a^3+8}\cdot\dfrac{5a^2-10a+20}{3a-6}$

$=\dfrac{\cancel{(a+2)}\cancel{(a-2)}}{\cancel{(a+2)}\cancel{(a^2-2a+4)}}\cdot\dfrac{5\cancel{(a^2-2a+4)}}{3\cancel{(a-2)}}$

$=\dfrac{5}{3}$

31) $\dfrac{13}{5z}-\dfrac{1}{3z}=\dfrac{13\cdot3}{15z}-\dfrac{1\cdot5}{15z}=\dfrac{39-5}{15z}=\dfrac{34}{15z}$

33) $\dfrac{12a^4}{10a-20}\div\dfrac{3a}{a^3-2a^2+5a-10}$

$=\dfrac{12a^4}{10a-20}\cdot\dfrac{a^3-2a^2+5a-10}{3a}$

$=\dfrac{\overset{4a^3}{\cancel{12a^4}}}{10\cancel{(a-2)}}\cdot\dfrac{\cancel{(a-2)}(a^2+5)}{\cancel{3a}}$

$=\dfrac{4a^3(a^2+5)}{10}=\dfrac{2a^3(a^2+5)}{5}$

35) $\dfrac{10}{x-8}+\dfrac{4}{x+3}$

$=\dfrac{10(x+3)}{(x-8)(x+3)}+\dfrac{4(x-8)}{(x-8)(x+3)}$

$=\dfrac{10x+30+4x-32}{(x-8)(x+3)}$

$=\dfrac{14x-2}{(x-8)(x+3)}=\dfrac{2(7x-1)}{(x-8)(x+3)}$

37) $\dfrac{10q}{8p-10q}+\dfrac{8p}{10q-8p}$

$=\dfrac{10q}{8p-10q}-\dfrac{8p}{8p-10q}$

$=\dfrac{10q-8p}{8p-10q}=-1$

39) $\dfrac{6u+1}{3u^2-2u}-\dfrac{u}{3u^2+u-2}+\dfrac{10}{u^2+u}=\dfrac{6u+1}{u(3u-2)}-\dfrac{u}{(3u-2)(u+1)}+\dfrac{10}{u(u+1)}$

$$=\frac{(6u+1)(u+1)}{u(3u-2)(u+1)}-\frac{u^2}{u(3u-2)(u+1)}+\frac{10(3u-2)}{u(3u-2)(u+1)}$$

$$=\frac{6u^2+7u+1-u^2+30u-20}{u(3u-2)(u+1)}=\frac{5u^2+37u-19}{u(3u-2)(u+1)}$$

41) $\dfrac{\frac{6v-30}{4}}{\frac{v-5}{2}}=\dfrac{6v-30}{4}\div\dfrac{v-5}{2}=\dfrac{6v-30}{4}\cdot\dfrac{2}{v-5}=\dfrac{6\cancel{(v-5)}}{\cancel{4}_{2}}\cdot\dfrac{\cancel{2}}{\cancel{v-5}}=\dfrac{6}{2}=3$

43) $\dfrac{x}{2x^2-7x-4}-\dfrac{x+3}{4x^2+4x+1}=\dfrac{x}{(2x+1)(x-4)}-\dfrac{x+3}{(2x+1)^2}$

$$=\frac{x(2x+1)}{(2x+1)^2(x-4)}-\frac{(x+3)(x-4)}{(2x+1)^2(x-4)}$$

$$=\frac{2x^2+x-(x^2-x-12)}{(2x+1)^2(x-4)}=\frac{x^2+2x+12}{(2x+1)^2(x-4)}$$

45) $\left(\dfrac{2c}{c+8}+\dfrac{4}{c-2}\right)\div\dfrac{6}{5c+40}=\left(\dfrac{2c(c-2)}{(c+8)(c-2)}+\dfrac{4(c+8)}{(c+8)(c-2)}\right)\cdot\dfrac{5c+40}{6}$

$$=\left(\frac{2c^2-4c+4c+32}{(c+8)(c-2)}\right)\cdot\frac{5c+40}{6}=\frac{2c^2+32}{\cancel{(c+8)}(c-2)}\cdot\frac{5\cancel{(c+8)}}{6}$$

$$=\frac{\cancel{2}(c^2+16)}{(c-2)}\cdot\frac{5}{\cancel{6}_{3}}=\frac{5(c^2+16)}{3(c-2)}$$

47) $\dfrac{3}{w^2-w}+\dfrac{4}{5w}-\dfrac{3}{w-1}=\dfrac{3}{w(w-1)}+\dfrac{4}{5w}-\dfrac{3}{w-1}=\dfrac{15}{5w(w-1)}+\dfrac{4(w-1)}{5w(w-1)}-\dfrac{15w}{5w(w-1)}$

$$=\frac{15+4w-4-15w}{5w(w-1)}=\frac{-11w+11}{5w(w-1)}=\frac{-11(w-1)}{5w(w-1)}=-\frac{11}{5w}$$

49) a) $A=\left(\dfrac{x-3}{4}\right)\left(\dfrac{x}{2}\right)=\dfrac{x(x-3)}{8}$ sq. units

b) $P=2\left(\dfrac{x-3}{4}\right)+2\left(\dfrac{x}{2}\right)=\dfrac{x-3}{2}+\dfrac{2x}{2}=\dfrac{x-3+2x}{2}=\dfrac{3x-3}{2}=\dfrac{3(x-1)}{2}$ units

Section 8.5: Exercises

1) i) Rewrite it as a division problem, then simplify.

$$\frac{2}{9}\div\frac{5}{18}=\frac{2}{\cancel{9}}\cdot\frac{\overset{2}{\cancel{18}}}{5}=\frac{4}{5}$$

ii) Multiply the numerator and denominator by 18, the LCD of $\frac{2}{9}$ and $\frac{5}{18}$. Then, simplify

$$\frac{\overset{2}{\cancel{18}}\left(\frac{2}{\cancel{9}}\right)}{\cancel{18}\left(\frac{5}{\cancel{18}}\right)}=\frac{4}{5}$$

3) $$\frac{\frac{7}{10}}{\frac{5}{4}}=\frac{7}{\underset{5}{\cancel{10}}}\cdot\frac{\overset{2}{\cancel{4}}}{5}=\frac{14}{25}$$

5) $$\frac{\frac{a^2}{b}}{\frac{a}{b^3}}=\frac{\overset{a}{\cancel{a^2}}}{\cancel{b}}\cdot\frac{\overset{b^2}{\cancel{b^3}}}{\cancel{a}}=ab^2$$

7) $$\frac{\frac{s^3}{t^3}}{\frac{s^4}{t}}=\frac{s^3}{t^3}\div\frac{s^4}{t}=\frac{\cancel{s^3}}{\underset{t^2}{\cancel{t^3}}}\cdot\frac{\cancel{t}}{\underset{s}{\cancel{s^4}}}=\frac{1}{st^2}$$

9) $$\frac{\frac{14m^5n^4}{9}}{\frac{35mn^6}{3}}=\frac{14m^5n^4}{9}\div\frac{35mn^6}{3}$$

$$=\frac{\overset{2m^4}{\cancel{14m^5n^4}}}{\underset{3}{\cancel{9}}}\cdot\frac{\cancel{3}}{\underset{5n^2}{\cancel{35mn^6}}}=\frac{2m^4}{15n^2}$$

11) $$\frac{\frac{t-6}{5}}{\frac{t-6}{t}}=\frac{t-6}{5}\div\frac{t-6}{t}=\frac{t-6}{5}\cdot\frac{t}{t-6}=\frac{t}{5}$$

13) $$\frac{\frac{8}{y^2-64}}{\frac{6}{y+8}}=\frac{8}{y^2-64}\div\frac{6}{y+8}$$

$$=\frac{\overset{4}{\cancel{8}}}{\cancel{(y+8)}(y-8)}\cdot\frac{\cancel{y+8}}{\underset{3}{\cancel{6}}}$$

$$=\frac{4}{3(y-8)}$$

15) $$\frac{\frac{25w-35}{w^5}}{\frac{30w-42}{w}}=\frac{25w-35}{w^5}\div\frac{30w-42}{w}$$

$$=\frac{5\cancel{(5w-7)}}{\underset{w^4}{\cancel{w^5}}}\cdot\frac{\cancel{w}}{6\cancel{(5w-7)}}$$

$$=\frac{5}{6w^4}$$

17) $$\frac{\frac{2x}{x+7}}{\frac{2}{x^2+4x-21}}=\frac{2x}{x+7}\div\frac{2}{x^2+4x-21}$$

$$=\frac{\cancel{2}x}{\cancel{x+7}}\cdot\frac{\cancel{(x+7)}(x-3)}{\cancel{2}}$$

$$=x(x-3)$$

19) i) $$\frac{\frac{1}{4}+\frac{3}{2}}{\frac{2}{3}+\frac{1}{2}}=\frac{\frac{1}{4}+\frac{6}{4}}{\frac{4}{6}+\frac{3}{6}}=\frac{\frac{7}{4}}{\frac{7}{6}}$$

$$=\frac{7}{4}\div\frac{7}{6}=\frac{7}{4}\cdot\frac{6}{7}=\frac{6}{4}=\frac{3}{2}$$

ii) $$\frac{\frac{1}{4}+\frac{3}{2}}{\frac{2}{3}+\frac{1}{2}}=\frac{12\left(\frac{1}{4}+\frac{3}{2}\right)}{12\left(\frac{2}{3}+\frac{1}{2}\right)}$$
$$=\frac{3+18}{8+6}=\frac{21}{14}=\frac{3}{2}$$

21) i) $$\frac{\frac{7}{c}+\frac{2}{d}}{1-\frac{5}{c}}=\frac{\frac{7d}{cd}+\frac{2c}{cd}}{\frac{c}{c}-\frac{5}{c}}=\frac{\frac{7d+2c}{cd}}{\frac{c-5}{c}}$$
$$=\frac{7d+2c}{cd}\div\frac{c-5}{c}$$
$$=\frac{7d+2c}{\cancel{c}d}\cdot\frac{\cancel{c}}{c-5}$$
$$=\frac{7d+2c}{d(c-5)}$$

ii) $$\frac{\frac{7}{c}+\frac{2}{d}}{1-\frac{5}{c}}=\frac{cd\left(\frac{7}{c}+\frac{2}{d}\right)}{cd\left(1-\frac{5}{c}\right)}=\frac{7d+2c}{cd-5d}$$
$$=\frac{7d+2c}{d(c-5)}$$

23) i) $$\frac{\frac{5}{z-2}-\frac{1}{z+1}}{\frac{1}{z-2}+\frac{4}{z+1}}$$
$$=\frac{\frac{5(z+1)}{(z-2)(z+1)}-\frac{1(z-2)}{(z-2)(z+1)}}{\frac{(z+1)}{(z-2)(z+1)}+\frac{4(z-2)}{(z-2)(z+1)}}$$
$$=\frac{\frac{5z+5-z+2}{(z-2)(z+1)}}{\frac{z+1+4z-8}{(z-2)(z+1)}}=\frac{\frac{4z+7}{(z-2)(z+1)}}{\frac{5z-7}{(z-2)(z+1)}}$$
$$=\frac{4z+7}{(z-2)(z+1)}\div\frac{5z-7}{(z-2)(z+1)}$$
$$=\frac{4z+7}{(z-2)(z+1)}\cdot\frac{(z-2)(z+1)}{5z-7}$$
$$=\frac{4z+7}{5z-7}$$

ii) $$\frac{\frac{5}{z-2}-\frac{1}{z+1}}{\frac{1}{z-2}+\frac{4}{z+1}}$$
$$=\frac{(z-2)(z+1)\left(\frac{5}{z-2}-\frac{1}{z+1}\right)}{(z-2)(z+1)\left(\frac{1}{z-2}+\frac{4}{z+1}\right)}$$
$$=\frac{5(z+1)-(z-2)}{(z+1)+4(z-2)}$$
$$=\frac{5z+5-z+2}{z+1+4z-8}=\frac{4z+7}{5z-7}$$

25) $$\frac{9+\frac{5}{y}}{\frac{9y+5}{8}}=\frac{\frac{9y+5}{y}}{\frac{9y+5}{8}}=\frac{9y+5}{y}\div\frac{9y+5}{8}$$
$$=\frac{9y+5}{y}\cdot\frac{8}{9y+5}=\frac{8}{y}$$

27) $$\frac{x-\frac{7}{x}}{x-\frac{11}{x}}=\frac{x\left(x-\frac{7}{x}\right)}{x\left(x-\frac{11}{x}\right)}=\frac{x^2-7}{x^2-11}$$

29) $$\frac{30\left(\frac{4}{3}+\frac{2}{5}\right)}{30\left(\frac{1}{6}-\frac{2}{3}\right)}=\frac{40+12}{5-20}=-\frac{52}{15}$$

31) $$\frac{\frac{2}{a}-\frac{2}{b}}{\frac{1}{a^2}-\frac{1}{b^2}}=\frac{a^2b^2\left(\frac{2}{a}-\frac{2}{b}\right)}{a^2b^2\left(\frac{1}{a^2}-\frac{1}{b^2}\right)}$$

$$=\frac{2ab^2-2a^2b}{b^2-a^2}$$

$$=\frac{2ab(b-a)}{(b+a)(b-a)}=\frac{2ab}{(b+a)}$$

33) $$\frac{\frac{r}{s^2}+\frac{1}{rs}}{\frac{s}{r}+\frac{1}{r^2}}=\frac{r^2s^2\left(\frac{r}{s^2}+\frac{1}{rs}\right)}{r^2s^2\left(\frac{s}{r}+\frac{1}{r^2}\right)}$$

$$=\frac{r^3+rs}{rs^3+s^2}=\frac{r\left(r^2+s\right)}{s^2(rs+1)}$$

35) $$\frac{1-\frac{4}{t+5}}{\frac{4}{t^2-25}+\frac{t}{t-5}}$$

$$=\frac{(t+5)(t-5)\left(1-\frac{4}{t+5}\right)}{(t+5)(t-5)\left(\frac{4}{(t+5)(t-5)}+\frac{t}{t-5}\right)}$$

$$=\frac{(t+5)(t-5)-4(t-5)}{4+t(t+5)}$$

$$=\frac{t^2-25-4t+20}{4+t^2+5t}=\frac{t^2-4t-5}{t^2+5t+4}$$

$$=\frac{(t-5)(t+1)}{(t+4)(t+1)}=\frac{t-5}{t+4}$$

37) $$\frac{b+\frac{1}{b}}{b-\frac{3}{b}}=\frac{b\left(b+\frac{1}{b}\right)}{b\left(b-\frac{3}{b}\right)}=\frac{b^2+1}{b^2-3}$$

39) $$\frac{\frac{m}{n^2}}{\frac{m^4}{n}}=\frac{m}{n^2}\div\frac{m^4}{n}=\frac{\cancel{m}}{n^{\cancel{2}}}\cdot\frac{\cancel{n}}{\cancel{m^4}_{\,m^3}}=\frac{1}{m^3n}$$

41) $$\frac{\frac{6}{x+3}-\frac{4}{x-1}}{\frac{2}{x-1}+\frac{1}{x+2}}$$

$$=\frac{(x+3)(x+2)(x-1)\left(\frac{6}{x+3}-\frac{4}{x-1}\right)}{(x+3)(x+2)(x-1)\left(\frac{2}{x-1}+\frac{1}{x+2}\right)}$$

$$=\frac{6(x+2)(x-1)-4(x+3)(x+2)}{2(x+3)(x+2)+1(x+3)(x-1)}$$

$$=\frac{(x+2)(6x-6-4x-12)}{(x+3)(2x+4+x-1)}$$

$$=\frac{(x+2)(2x-18)}{(x+3)(3x+3)}=\frac{2(x+2)(x-9)}{3(x+3)(x+1)}$$

43) $$\frac{\frac{r^2-6}{20}}{r-\frac{6}{r}}=\frac{\frac{r^2-6}{20}}{\frac{r^2-6}{r}}=\frac{r^2-6}{20}\div\frac{r^2-6}{r}$$

$$=\frac{r^2-6}{20}\cdot\frac{r}{r^2-6}=\frac{r}{20}$$

45) $$\frac{\frac{a-4}{12}}{\frac{a-4}{a}}=\frac{a-4}{12}\div\frac{a-4}{a}$$

$$=\frac{a-4}{12}\cdot\frac{a}{a-4}=\frac{a}{12}$$

47) $$\frac{\frac{5}{6}}{\frac{9}{15}}=\frac{5}{6}\div\frac{9}{15}=\frac{5}{\cancel{6}_{2}}\cdot\frac{\cancel{15}^{5}}{9}=\frac{25}{18}$$

49) $$\frac{\frac{5}{2n+1}+1}{\frac{1}{n+3}+\frac{2}{2n+1}}$$

$$=\frac{(n+3)(2n+1)\left(\frac{5}{2n+1}+1\right)}{(n+3)(2n+1)\left(\frac{1}{n+3}+\frac{2}{2n+1}\right)}$$

$$=\frac{5(n+3)+(n+3)(2n+1)}{(2n+1)+2(n+3)}$$

$$=\frac{(n+3)(5+2n+1)}{2n+1+2n+6}=\frac{(n+3)(2n+6)}{4n+7}$$

$$=\frac{2(n+3)(n+3)}{4n+7}=\frac{2(n+3)^2}{4n+7}$$

Section 8.6: Exercises

1) Eliminate the denominators.

3) sum; $\frac{m}{8}+\frac{m-7}{4}=\frac{m}{8}+\frac{2(m-7)}{8}$

$$=\frac{2+2m-14}{8}$$

$$=\frac{3m-14}{8}$$

5) equation; $\frac{2f-19}{20}=\frac{f}{4}+\frac{2}{5}$

$$20\left(\frac{2f-19}{20}\right)=20\left(\frac{f}{4}+\frac{2}{5}\right)$$

$$2f-19=5f+8$$

$$-27=3f$$

$$-9=f \qquad \{-9\}$$

7) difference; $\dfrac{z}{z-6}-\dfrac{4}{z}$

$$=\frac{z\cdot z}{z(z-6)}-\frac{4(z-6)}{z(z-6)}$$

$$=\frac{z^2-4z+24}{z(z-6)}$$

9) equation; $1+\dfrac{4}{c+2}=\dfrac{9}{c+2}$

$$(c+2)\left(1+\frac{4}{c+2}\right)=(c+2)\left(\frac{9}{c+2}\right)$$

$$c+2+4=9$$

$$c+6=9$$

$$c=3 \qquad \{3\}$$

11) $t+10=0 \qquad t=0$

$t=-10$

13) $d^2-81=0$

$(d+9)(d-9)=0$

$d+9=0$ or $d-9=0$

$d=-9 \qquad d=9$

$d=0$

15) $v^2-13v+36=0$

$(v-9)(v-4)=0$

$v-9=0$ or $v-4=0$

$v=9 \qquad v=4$

$3v-12=0 \qquad v-9=0$

$3v=12 \qquad v=9$

$v=4$

17) $\dfrac{y}{3}-\dfrac{1}{2}=\dfrac{1}{6}$

$$6\left(\frac{y}{3}-\frac{1}{2}\right)=6\cdot\frac{1}{6}$$

$$2y-3=1$$

$$2y=4$$

$$y=2 \qquad \{2\}$$

19) $\dfrac{1}{2}h+h=-3$

$$2\left(\frac{1}{2}h+h\right)=2(-3)$$

$$h+2h=-6$$

$$3h=-6$$

$$h=-2 \qquad \{-2\}$$

21) $\dfrac{7u+12}{15}=\dfrac{2u}{5}-\dfrac{3}{5}$

$$15\left(\frac{7u+12}{15}\right)=15\left(\frac{2u}{5}-\frac{3}{5}\right)$$

$$7u+12=6u-9$$

$$u+12=-9$$

$$u=-21 \qquad \{-21\}$$

23) $\dfrac{4}{3t+2}=\dfrac{2}{2t-1}$

$$4(2t-1)=2(3t+2)$$

$$8t-4=6t+4$$

$$2t=8$$

$$t=4 \qquad \{4\}$$

25) $\frac{w}{3}=\frac{2w-5}{12}$

$$12w=3(2w-5)$$
$$12w=6w-15$$
$$6w=-15$$
$$w=-\frac{15}{6}=-\frac{5}{2} \qquad \left\{-\frac{5}{2}\right\}$$

27) $\frac{12}{a}-2=\frac{6}{a}$

$$a\left(\frac{12}{a}-2\right)=a\cdot\frac{6}{a}$$
$$12-2a=6$$
$$-2a=-6$$
$$a=3 \qquad \{3\}$$

29) $\frac{n}{n+2}+3=\frac{8}{n+2}$

$$(n+2)\left(\frac{n}{n+2}+3\right)=(n+2)\left(\frac{8}{n+2}\right)$$
$$n+3(n+2)=8$$
$$n+3n+6=8$$
$$4n=2$$
$$n=\frac{2}{4}=\frac{1}{2} \qquad \left\{\frac{1}{2}\right\}$$

31) $\frac{2}{s+6}+4=\frac{2}{s+6}$

$$(s+6)\left(\frac{2}{s+6}+4\right)=(s+6)\left(\frac{2}{s+6}\right)$$
$$2+(s+6)4=2$$
$$2+4s+24=2$$
$$4s+26=2$$
$$4s=-24$$
$$s=-6$$

If $s=-6$, the denominators = 0. $\varnothing$

33) $\frac{c}{c-7}-4=\frac{10}{c-7}$

$$(c-7)\left(\frac{c}{c-7}-4\right)=(c-7)\left(\frac{10}{c-7}\right)$$
$$c-4(c-7)=10$$
$$c-4c+28=10$$
$$-3c=-18$$
$$c=6 \qquad \{6\}$$

35) $\frac{32}{g}+10=-\frac{8}{g}$

$$g\left(\frac{32}{g}+10\right)=g\left(-\frac{8}{g}\right)$$
$$32+10g=-8$$
$$10g=-40$$
$$g=-4 \qquad \{-4\}$$

37)
$$\frac{1}{m-1}+\frac{2}{m+3}=\frac{4}{m+3}$$
$$(m-1)(m+3)\left(\frac{1}{m-1}+\frac{2}{m+3}\right)=(m-1)(m+3)\left(\frac{4}{m+3}\right)$$
$$(m+3)+2(m-1)=4(m-1)$$
$$m+3+2m-2=4m-4$$
$$3m+1=4m-4$$
$$5=m \qquad \{5\}$$

39)
$$\frac{4}{w-8}-\frac{10}{w+8}=\frac{40}{w^2-64}$$
$$\frac{4}{w-8}-\frac{10}{w+8}=\frac{40}{(w+8)(w-8)}$$
$$(w+8)(w-8)\left(\frac{4}{w-8}-\frac{10}{w+8}\right)=(w+8)(w-8)\left(\frac{40}{(w+8)(w-8)}\right)$$
$$4(w+8)-10(w-8)=40$$
$$4w+32-10w+80=40$$
$$-6w+110=40$$
$$-6w=-72$$
$$w=12 \qquad \{12\}$$

41)
$$\frac{3}{a+3}+\frac{14}{a^2-4a-21}=\frac{5}{a-7}$$
$$\frac{3}{a+3}+\frac{14}{(a+3)(a-7)}=\frac{5}{a-7}$$
$$(a+3)(a-7)\left(\frac{3}{a+3}+\frac{14}{(a+3)(a-7)}\right)=(a+3)(a-7)\left(\frac{5}{a-7}\right)$$
$$3(a-7)+14=5(a+3)$$
$$3a-21+14=5a+15$$
$$3a-7=5a+15$$
$$-22=2a$$
$$-11=a \qquad \{-11\}$$

43)
$$\frac{9}{t+4}+\frac{8}{t^2-16}=\frac{1}{t-4}$$
$$\frac{9}{t+4}+\frac{8}{(t+4)(t-4)}=\frac{8}{t-4}$$
$$(t+4)(t-4)\left(\frac{9}{t+4}+\frac{8}{(t+4)(t-4)}\right)=(t+4)(t-4)\left(\frac{1}{t-4}\right)$$
$$9(t-4)+8=t+4$$
$$9t-36+8=t+4$$
$$9t-28=t+4$$
$$8t=32$$
$$t=4$$
If $t=4$, two denominators = 0. $\varnothing$

45)
$$\frac{4}{x^2+2x-15}=\frac{8}{x-3}+\frac{2}{x+5}$$
$$\frac{4}{(x+5)(x-3)}=\frac{8}{x-3}+\frac{2}{x+5}$$
$$(x+5)(x-3)\left(\frac{4}{(x+5)(x-3)}\right)=(x+5)(x-3)\left(\frac{8}{x-3}+\frac{2}{x+5}\right)$$
$$4=8(x+5)+2(x-3)$$
$$4=8x+40+2x-6$$
$$4=10x+34$$
$$-30=10x$$
$$-3=x \qquad \{-3\}$$

47)
$$\frac{k^2}{3}=\frac{k^2+2k}{4}$$
$$4k^2=3(k^2+2k)$$
$$4k^2=3k^2+6k$$
$$k^2-6k=0$$
$$k(k-6)=0$$
$$k=0 \text{ or } k-6=0$$
$$k=6 \qquad \{0,6\}$$

49)
$$\frac{5}{m^2-25}=\frac{4}{m^2+5m}$$
$$\frac{5}{(m+5)(m-5)}=\frac{4}{m(m+5)}$$
$$m(m+5)(m-5)\frac{5}{(m+5)(m-5)}=m(m+5)(m-5)\frac{4}{m(m+5)}$$
$$5m=4(m-5)$$
$$5m=4m-20$$
$$m=-20 \qquad \{-20\}$$

51)
$$\frac{10v}{3v-12}-\frac{v+6}{v-4}=\frac{v}{3}$$
$$3(v-4)\left(\frac{10v}{3(v-4)}-\frac{v+6}{v-4}\right)=3(v-4)\left(\frac{v}{3}\right)$$
$$10v-3(v+6)=v(v-4)$$
$$10v-3v-18=v^2-4v$$
$$7v-18=v^2-4v$$
$$0=v^2-11v+18$$
$$0=(v-9)(v-2)$$
$$v-9=0 \text{ or } v-2=0$$
$$v=9 \qquad v=2 \qquad \{2,9\}$$

53)
$$\frac{w}{5}=\frac{w-3}{w+1}+\frac{12}{5w+5}$$
$$5(w+1)\left(\frac{w}{5}\right)=5(w+1)\left(\frac{w-3}{w+1}+\frac{12}{5(w+1)}\right)$$
$$w(w+1)=5(w-3)+12$$
$$w^2+w=5w-15+12$$
$$w^2+w=5w-3$$
$$w^2-4w+3=0$$
$$(w-1)(w-3)=0$$
$$w-1=0 \text{ or } w-3=0$$
$$w=1 \qquad w=3 \qquad \{1,3\}$$

55) $$\frac{8}{p+2}+\frac{p}{p+1}=\frac{5p+2}{p^2+3p+2}$$

$$(p+2)(p+1)\left(\frac{8}{p+2}+\frac{p}{p+1}\right)=(p+2)(p+1)\left(\frac{5p+2}{(p+2)(p+1)}\right)$$

$$(p+1)8+(p+2)p=5p+2$$

$$8p+8+p^2+2p=5p+2$$

$$p^2+5p+6=0$$

$$(p+3)(p+2)=0$$

$p+3=0$ or $p+2=0$

$p=-3$ $\qquad p=-2$

If $p=-2$, two denominators = 0. The solution set is $\{-3\}$.

57) $$\frac{11}{c+9}=\frac{c}{c-4}-\frac{36-8c}{c^2+5c-36}$$

$$(c+9)(c-4)\left(\frac{11}{c+9}\right)=(c+9)(c-4)\left(\frac{c}{c-4}-\frac{36-8c}{(c+9)(c-4)}\right)$$

$$11(c-4)=c(c+9)-36+8c$$

$$11c-44=c^2+9c-36+8c$$

$$11c-44=c^2+17c-36$$

$$0=c^2+6c+8$$

$$0=(c+4)(c+2)$$

$c+4=0$ or $c+2=0$

$x=-4$ $\qquad c=-2$ $\qquad \{-4,-2\}$

59) $$\frac{8}{3g^2-7g-6}+\frac{4}{g-3}=\frac{8}{3g+2}$$

$$(3g+2)(g-3)\left(\frac{8}{(3g+2)(g-3)}+\frac{4}{g-3}\right)=(3g+2)(g-3)\left(\frac{8}{3g+2}\right)$$

$$8+4(3g+2)=8(g-3)$$

$$8+12g+8=8g-24$$

$$12g+16=8g-24$$

$$4g=-40$$

$$g=-10 \qquad \{-10\}$$

61) $\dfrac{h}{h^2+2h-8}+\dfrac{4}{h^2+8h-20}=\dfrac{4}{h^2+14h+40}$

$$\frac{h}{(h+4)(h-2)}+\frac{4}{(h+10)(h-2)}=\frac{4}{(h+10)(h+4)}$$

$$(h+10)(h+4)(h-2)\left(\frac{h}{(h+4)(h-2)}+\frac{4}{(h+10)(h-2)}\right)$$

$$=(h+10)(h+4)(h-2)\left(\frac{4}{(h+10)(h+4)}\right)$$

$$h(h+10)+4(h+4)=4(h-2)$$

$$h^2+10h+4h+16=4h-8$$

$$h^2+14h+16=4h-8$$

$$h^2+10h+24=0$$

$$(h+6)(h+4)=0$$

$h+6=0$ or $h+4=0$

$h=-6$ $\quad$ $h=-4$

If $h=-4$, two denominators = 0. The solution set is $\{-6\}$.

63) $\dfrac{u}{8}=\dfrac{2}{10-u}$

$$u(10-u)=2(8)$$

$$10u-u^2=16$$

$$0=u^2-10u+16$$

$$0=(u-8)(u-2)$$

$u-8=0$ or $u-2=0$

$u=8$ $\quad$ $u=2$ $\quad$ $\{2,8\}$

65) $\dfrac{5}{r+4}-\dfrac{2}{r}=-1$

$$r(r+4)\left(\frac{5}{r+4}-\frac{2}{r}\right)=r(r+4)(-1)$$

$$5r-2(r+4)=-r(r+4)$$

$$5r-2r-8=-r^2-4r$$

$$3r-8=-r^2-4r$$

$$r^2+7r-8=0$$

$$(r+8)(r-1)=0$$

$r+8=0$ or $r-1=0$

$r=-8$ $\quad$ $r=1$ $\quad$ $\{-8,1\}$

67) $$\frac{q}{q^2+4q-32}+\frac{2}{q^2-14q+40}=\frac{6}{q^2-2q-80}$$

$$\frac{q}{(q+8)(q-4)}+\frac{2}{(q-10)(q-4)}=\frac{6}{(q-10)(q+8)}$$

$$(q+8)(q-4)(q-10)\left(\frac{q}{(q+8)(q-4)}+\frac{2}{(q-10)(q-4)}\right)$$

$$=(q+8)(q-4)(q-10)\left(\frac{6}{(q-10)(q+8)}\right)$$

$$q(q-10)+2(q+8)=6(q-4)$$

$$q^2-10q+2q+16=6q-24$$

$$q^2-8q+16=6q-24$$

$$q^2-14q+40=0$$

$$(q-10)(q-4)=0$$

$q-10=0$ or $q-4=0$

$q=10$ $\quad q=4$

If $q=10$ or 4, the denominators are undefined. $\varnothing$

69) $$V=\frac{nRT}{\boxed{P}}$$

$$\boxed{P}V=nRT$$

$$\boxed{P}=\frac{nRT}{V}$$

71) $$y=\frac{kx}{\boxed{z}}$$

$$\boxed{z}y=kx$$

$$\boxed{z}=\frac{kx}{y}$$

73) $$B=\frac{t+u}{3\boxed{x}}$$

$$3\boxed{x}B=t+u$$

$$x=\frac{t+u}{3B}$$

75) $$z=\frac{a}{\boxed{b}+c}$$

$$\left(\boxed{b}+c\right)z=a$$

$$\boxed{b}z+cz=a$$

$$\boxed{b}z=a-cz$$

$$\boxed{b}=\frac{a-cz}{z}$$

77) $$A=\frac{4r}{q-\boxed{t}}$$

$$\left(q-\boxed{t}\right)A=4r$$

$$Aq-A\boxed{t}=4r$$

$$Aq-4r=A\boxed{t}$$

$$\frac{Aq-4r}{A}=\boxed{t}$$

79) $w = \frac{na}{k\boxed{c}+b}$

$\left(k\boxed{c}+b\right)w = na$

$wk\boxed{c} + wb = na$

$wk\boxed{c} = na - wb$

$\boxed{c} = \frac{na - wb}{wk}$

81) $\frac{1}{t} = \frac{1}{\boxed{r}} - \frac{1}{s}$

$\boxed{r}st\left(\frac{1}{t}\right) = \boxed{r}st\left(\frac{1}{\boxed{r}} - \frac{1}{s}\right)$

$\boxed{r}s = st - \boxed{r}t$

$\boxed{r}s + \boxed{r}t = st$

$\boxed{r}(s+t) = st$

$\boxed{r} = \frac{st}{s+t}$

83) $\frac{2}{A} + \frac{1}{\boxed{C}} = \frac{3}{B}$

$\left(A\boxed{C}B\right)\left(\frac{2}{A} + \frac{1}{\boxed{C}}\right) = \left(A\boxed{C}B\right)\left(\frac{3}{B}\right)$

$2\boxed{C}B + AB = 3A\boxed{C}$

$AB = 3A\boxed{C} - 2\boxed{C}B$

$AB = \boxed{C}(3A - 2B)$

$\frac{AB}{3A-2B} = \boxed{C}$

Section 8.7: Exercises

1) $\frac{12}{7} = \frac{60}{x}$

$12x = 7 \cdot 60$

$12x = 420$

$x = \frac{420}{12} = 35 \qquad \{35\}$

3) $\frac{6}{13} = \frac{x}{x+56}$

$6(x+56) = 13x$

$6x + 336 = 13x$

$336 = 7x$

$\frac{336}{7} = x$

$48 = x \qquad \{48\}$

5) f = number of female spectators

$\frac{10}{3} = \frac{370}{f}$

$10f = 3(370)$

$10f = 1110$

$f = 111$

There were 111 female spectators.

7) t = cups of tapioca flour

$t + 3$ = cups of potato-starch flour

$\frac{2}{1} = \frac{t+3}{t}$

$2t = t + 3$

$t = 3$

$t + 3 = 3 + 3 = 6$

There are 3 cups of tapioca flour and 6 cups of potato-starch flour.

9) l = length of the floor

$l - 18$ = width of the floor

$\frac{8}{5} = \frac{l}{l-18}$

$8(l-18) = 5l$

$8l - 144 = 5l$

$3l = 144$

$l = 48$

$l - 18 = 48 - 18 = 30$

The length is 48 feet, and the width is 30 feet.

11) $s =$ dollars invested in stocks

$s + 4000 =$ dollars invested in bonds

$$\frac{3}{2} = \frac{s + 4000}{x}$$

$$3s = 2(s + 4000)$$

$$3s = 2s + 8000$$

$$s = 8000$$

$$s + 4000 = 8000 + 4000 = 12{,}000$$

She invested \$8000 in stocks and \$12,000 in bonds.

13) $n =$ households without pets

$n + 271 =$ households with pets

$$\frac{5}{4} = \frac{n + 271}{n}$$

$$5n = 4(n + 271)$$

$$5n = 4n + 1084$$

$$n = 1084$$

$$n + 271 = 1084 + 271 = 1355$$

1355 households have pets.

15) a) Speed against current $= 10 - 3$ $= 7$ mph

b) Speed with current $= 10 + 3$ $= 13$ mph

17) a) Speed with wind $= x + 30$ mph

b) Speed into wind $= x - 30$ mph

19) $s =$ speed of boat in still water

	$d =$	r	$\cdot\ t$
Downstream	20	$s+5$	$\frac{20}{s+5}$
Upstream	12	$s-5$	$\frac{12}{s-5}$

Solve $d = rt$ for t to get $t = \frac{d}{r}$.

Time with the current = Time against the current

$$\frac{20}{s+5} = \frac{15}{s-5}$$

$$20(s-5) = 12(s+5)$$

$$20s - 100 = 12s + 60$$

$$8s = 160$$

$$s = 20$$

The speed of the boat in still water is 20 mph.

21) $s =$ speed of the current

	$d =$	r	$\cdot\ t$
downstream	15	$16+s$	$\frac{15}{16+s}$
upstream	9	$16-s$	$\frac{9}{16-s}$

Solve $d = rt$ for t to get $t = \frac{d}{r}$.

Time with the current = Time against the current

$$\frac{15}{16+s} = \frac{9}{16-s}$$

$$15(16-s) = 9(16+s)$$

$$240 - 15s = 144 + 9s$$

$$96 = 24s$$

$$4 = s$$

The speed of the current is 4 mph.

23) s = speed of the plane

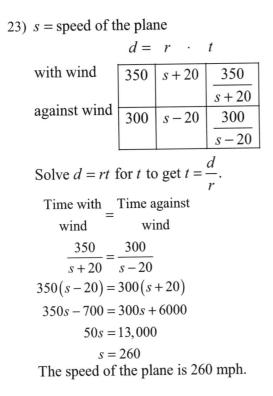

	$d=$	r	$\cdot\ t$
with wind	350	$s+20$	$\frac{350}{s+20}$
against wind	300	$s-20$	$\frac{300}{s-20}$

Solve $d = rt$ for t to get $t = \frac{d}{r}$.

$$\text{Time with wind} = \text{Time against wind}$$

$$\frac{350}{s+20} = \frac{300}{s-20}$$

$$350(s-20) = 300(s+20)$$

$$350s - 700 = 300s + 6000$$

$$50s = 13{,}000$$

$$s = 260$$

The speed of the plane is 260 mph.

25) s = speed of the current

	$d=$	r	$\cdot\ t$
against current	4	$10-s$	$\frac{4}{10-s}$
with current	6	$10+s$	$\frac{6}{10+s}$

Solve $d = rt$ for t to get $t = \frac{d}{r}$.

$$\text{Time against the current} = \text{Time with the current}$$

$$\frac{4}{10-s} = \frac{6}{10+s}$$

$$4(10+s) = 6(10-s)$$

$$40 + 4s = 60 - 6s$$

$$10s = 20$$

$$s = 2$$

The speed of the current is 2 mph.

27) rate $= \frac{1 \text{ job}}{4 \text{ hours}} = \frac{1}{4}$ job/hour

29) rate $= \frac{1 \text{ job}}{t \text{ hours}} = \frac{1}{t}$ job/hour

31) t = the number of hours together

$$\text{fractional part Arlene} + \text{fractional part Andre} = 1 \text{ job}$$

$$\frac{1}{2}t \quad + \quad \frac{1}{3}t \quad = 1$$

$$\frac{1}{2}t + \frac{1}{3}t = 1$$

$$6\left(\frac{1}{2}t + \frac{1}{3}t\right) = 6(1)$$

$$3t + 2t = 6$$

$$5t = 6$$

$$t = \frac{6}{5}$$

They could trim the bushes together in $1\frac{1}{5}$ hours.

33) t = the number of hours to assemble the notebook together

$$\text{fractional part Jermaine} + \text{fractional part Sue} = 1 \text{ job}$$

$$\frac{1}{5}t \quad + \quad \frac{1}{8}t \quad = 1$$

$$\frac{1}{5}t + \frac{1}{8}t = 1$$

$$40\left(\frac{1}{5}t + \frac{1}{8}t\right) = 40(1)$$

$$8t + 5t = 40$$

$$13t = 40$$

$$t = \frac{40}{13} = 3\frac{1}{13}$$

It would take $3\frac{1}{13}$ hours to assemble the notebooks together.

35) t = number of minutes it would take to fill a tub that has a leaky drain.

$$\begin{matrix}\text{fractional} \\ \text{part faucet}\end{matrix} + \begin{matrix}\text{fractional} \\ \text{part drain}\end{matrix} = 1 \text{ job}$$

$$\frac{1}{12}t \quad + \left(-\frac{1}{30}t\right) \quad = 1$$

$$\frac{1}{12}t - \frac{1}{30}t = 1$$

$$120\left(\frac{1}{12}t - \frac{1}{30}t\right) = 120(1)$$

$$10t - 4t = 120$$

$$6t = 120$$

$$t = 20$$

It takes 20 minutes to fill the tub.

37) t = number of hours for Fatima
$2t$ = number of hours for Antonio

$$\begin{matrix}\text{fractional} \\ \text{part Fatima}\end{matrix} + \begin{matrix}\text{fractional} \\ \text{part Antonio}\end{matrix} = 1 \text{ job}$$

$$\frac{1}{t}(2) \quad + \quad \frac{1}{2t}(2) \quad = 1$$

$$\frac{2}{t} + \frac{1}{t} = 1$$

$$\frac{3}{t} = 1$$

$$3 = t$$

It would take Fatima 3 hours to cut out the shapes.

39) t = number of hours for experienced painters

$$\text{new worker} + \begin{matrix}\text{experienced} \\ \text{worker}\end{matrix} = 1 \text{ job}$$

$$\frac{1}{6}(2) \quad + \quad \frac{1}{t}(2) \quad = 1$$

$$\frac{1}{3} + \frac{2}{t} = 1$$

$$3t\left(\frac{1}{3} + \frac{2}{t}\right) = 3t \cdot 1$$

$$t + 6 = 3t$$

$$6 = 2t$$

$$3 = t$$

It would take the experienced worker 3 hours to paint.

Chapter 8 Review

1) a) $\dfrac{5(4)-2}{(4)^2-9} = \dfrac{20-2}{16-9} = \dfrac{18}{7}$

b) $\dfrac{5(-3)-2}{(-3)^2-9} = \dfrac{-15}{9-9} = \dfrac{-15}{0}$

undefined

3) a) $k + 5 = 0$
$k = -5$

b) $k - 1 = 0$
$k = 1$

5) a)

$$2c^2 - 11c - 6 = 0$$

$$(2c+1)(c-6) = 0$$

$$2c + 1 = 0 \text{ or } c - 6 = 0$$

$$2c = -1 \qquad c = 6$$

$$c = -\frac{1}{2}$$

b)

$$c^2 - 3c = 0$$

$$c(c-3) = 0$$

$$c = 0 \text{ or } c - 3 = 0$$

$$c = 3$$

7) a) $14 - 7d = 0$
$14 = 7d$
$2 = d$

b) never undefined

9) $\frac{63a^2}{9a^{11}} = \frac{7}{a^9}$

11) $\frac{15c-55}{33c-121} = \frac{5\cancel{(3c-11)}}{11\cancel{(3c-11)}} = \frac{5}{11}$

13) $\frac{2z-7}{6z^2-19z-7} = \frac{\cancel{2z-7}}{(3z+1)\cancel{(2z-7)}}$

$= \frac{1}{3z+1}$

15) $\frac{10-x}{x^2-100} = \frac{\overset{-1}{\cancel{10-x}}}{(x+10)\cancel{(x-10)}}$

$= -\frac{1}{x+10}$

17) $\frac{-u+6}{u+2}, \frac{6-u}{u+2}, \frac{u-6}{-u-2},$

$\frac{-(u-6)}{u+2}, \frac{u-6}{-(u+2)}$

19) $l = \frac{2l^2-5l-3}{l-3} = \frac{(2l+1)(l-3)}{l-3}$

$= 2l+1$

21) $x-5=0$

$x=5$

The domain is all real numbers except 5. Domain: $(-\infty,5)\cup(5,\infty)$

23) $a^2-2a-24=0$

$(a+4)(a-6)=0$

$a+4=0$ or $a-6=0$

$a=-4 \qquad a=6$

The domain contains all real numbers except -4 and 6.

D: $(-\infty,-4)\cup(-4,6)\cup(6,\infty)$

25) $\frac{10}{9}\cdot\frac{6}{25} = \frac{\overset{2}{\cancel{10}}}{\underset{3}{\cancel{9}}}\cdot\frac{\overset{2}{\cancel{6}}}{\underset{5}{\cancel{25}}} = \frac{4}{15}$

27) $\frac{16k^4}{3m^2} \div \frac{4k^2}{27m} = \frac{\overset{4k^2}{\cancel{16k^4}}}{\underset{m}{\cancel{3m^2}}}\cdot\frac{\overset{9}{\cancel{27m}}}{\cancel{4k^2}} = \frac{36k^2}{m}$

29) $\frac{6w-1}{(6w^2+5w-1)}\cdot\frac{3w+3}{12w}$

$= \frac{\cancel{6w-1}}{\cancel{(6w-1)}\cancel{(w+1)}}\cdot\frac{\cancel{3}\cancel{(w+1)}}{\underset{4}{\cancel{12}}w} = \frac{1}{4w}$

31) $\frac{25-a^2}{4a^2+12a} \div \frac{a^3-125}{a^2+3a}$

$= \frac{25-a^2}{4a^2+12a}\cdot\frac{a^2+3a}{a^3-125}$

$= \frac{(5+a)\overset{-1}{\cancel{(5-a)}}}{4a\cancel{(a+3)}}\cdot\frac{\cancel{a(a+3)}}{\cancel{(a-5)}(a^2+5a+25)}$

$= -\frac{a+5}{4(a^2+5a+25)}$

33) $\frac{\frac{8}{9}}{\frac{12}{5}} = \frac{8}{9}\div\frac{12}{5} = \frac{\overset{2}{\cancel{8}}}{9}\cdot\frac{5}{\underset{3}{\cancel{12}}} = \frac{10}{27}$

35) $\dfrac{\frac{16m-8}{m^2}}{\frac{12m-6}{m^3}} = \dfrac{16m-8}{m^2} \div \dfrac{12m-6}{m^3}$

$= \dfrac{8(2m-1)}{m^2} \cdot \dfrac{m^3}{6(2m-1)}$

$= \dfrac{4m}{3}$

37) $\text{LCD} = 18$ 39) $\text{LCD} = k^2$

41) m and $m+5$ are different factors. The LCD will be the product of these factors. $\text{LCD} = m(m+5)$

43) $2d^2 - d = d(2d-1)$

$6d - 3 = 3(2d-1)$

$\text{LCD} = 3d(2d-1)$

45) $\text{LCD} = b-2$ or $2-b$

47) $c^2 + 10c + 24 = (c+4)(c+6)$

$c^2 - 3c - 28 = (c+4)(c-7)$

$\text{LCD} = (c+4)(c+6)(c-7)$

49) $x^2 + 8x = x(x+8)$

$3x^2 + 24x = 3x(x+8)$

$x^2 + 16x + 64 = (x+8)^2$

$\text{LCD} = 3x(x+8)^2$

51) $c^2 - d^2 = (c+d)(c-d)$

$d - c$

$\text{LCD} = (c+d)(c-d)$

or $(c+d)(d-c)$

53) $\dfrac{6}{5r} \cdot \dfrac{4r^2}{4r^2} = \dfrac{24r^2}{20r^3}$

55) $\dfrac{8}{3z+4} \cdot \dfrac{z}{z} = \dfrac{8z}{z(3z+4)}$

57) $\dfrac{t-3}{2t+1} \cdot \dfrac{t+5}{t+5} = \dfrac{t^2+2t-15}{(2t+1)(t+5)}$

59) $\text{LCD} = 36z^5$

$\dfrac{4}{9z^3} \cdot \dfrac{4z^2}{4z^2} = \dfrac{16z^2}{36z^5}$

$\dfrac{7}{12z^5} \cdot \dfrac{3}{3} = \dfrac{21}{36z^5}$

61) $\text{LCD} = p(p+7)$

$\dfrac{8}{p+7} \cdot \dfrac{p}{p} = \dfrac{8p}{p(p+7)}$

$\dfrac{2}{p} \cdot \dfrac{p+7}{p+7} = \dfrac{2p+14}{p(p+7)}$

63) $\text{LCD} = g-10$

$\dfrac{1}{g-10}$ is written with the LCD.

$\dfrac{3}{10-g} \cdot \dfrac{-1}{-1} = -\dfrac{3}{g-10}$

65) $\dfrac{5}{18} - \dfrac{7}{12} = \dfrac{10}{36} - \dfrac{21}{32} = -\dfrac{11}{36}$

67) $\dfrac{4m}{m-3} - \dfrac{5}{m-3} = \dfrac{4m-5}{m-3}$

69) $\dfrac{8}{t+4} + \dfrac{3}{t} = \dfrac{8t}{t(t+4)} + \dfrac{3(t+4)}{t(t+4)}$

$= \dfrac{8t+3t+12}{t(t+4)} = \dfrac{11t+12}{t(t+4)}$

71) $\dfrac{5}{y-2}-\dfrac{6}{y+3}$

$=\dfrac{5(y+3)}{(y-2)(y+3)}-\dfrac{6(y-2)}{(y-2)(y+3)}$

$=\dfrac{5y+15-6y+12}{(y-2)(y+3)}$

$=\dfrac{27-y}{(y-2)(y+3)}$

73) $\dfrac{10p+3}{4p+4}-\dfrac{8}{p^2-6p-7}$

$=\dfrac{10p+3}{4(p+1)}-\dfrac{8}{(p-7)(p+1)}$

$=\dfrac{(10p+3)(p-7)}{4(p+1)(p-7)}-\dfrac{4\cdot 8}{4(p+1)(p-7)}$

$=\dfrac{10p^2-67p-21-32}{4(p+1)(p-7)}$

$=\dfrac{10p^2-67p-53}{4(p+1)(p-7)}$

75) $\dfrac{2}{m-11}+\dfrac{19}{11-m}=\dfrac{2}{m-11}-\dfrac{19}{m-11}$

$=-\dfrac{17}{m-11}$ or $\dfrac{17}{11-m}$

77) $\dfrac{1}{8-r}+\dfrac{16}{r^2-64}$

$=\dfrac{1}{8-r}+\dfrac{16}{(r+8)(r-8)}$

$=\dfrac{-(r+8)}{-(r+8)(8-r)}+\dfrac{16}{(r+8)(r-8)}$

$=\dfrac{-r-8+16}{(r+8)(r-8)}=\dfrac{-r+8}{(r+8)(r-8)}$

$=-\dfrac{r-8}{(r+8)(r-8)}=-\dfrac{1}{r+8}$

79) $\dfrac{8}{w^2+7w}+\dfrac{3w}{w+7}+\dfrac{2}{5w}$

$=\dfrac{8}{w(w+7)}+\dfrac{3w}{w+7}+\dfrac{2}{5w}$

$=\dfrac{5\cdot 8}{5w(w+7)}+\dfrac{5w\cdot 3w}{5w(w+7)}+\dfrac{2(w+7)}{5w(w+7)}$

$=\dfrac{40+15w^2+2w+14}{5w(w+7)}=\dfrac{15w^2+2w+54}{5w(w+7)}$

81) $\dfrac{d+4}{d^2+2d}+\dfrac{d}{5d^2+7d-6}-\dfrac{10}{5d^2-3d}=\dfrac{d+4}{d(d+2)}+\dfrac{d}{(5d-3)(d+2)}-\dfrac{10}{d(5d-3)}$

$=\dfrac{(d+4)(5d-3)}{d(d+2)(5d-3)}+\dfrac{d\cdot d}{d(d+2)(5d-3)}-\dfrac{10(d+2)}{d(d+2)(5d-3)}$

$=\dfrac{5d^2+17d-12+d^2-10d-20}{d(d+2)(5d-3)}=\dfrac{6d^2+7d-32}{d(d+2)(5d-3)}$

83) a) $A=\left(\dfrac{12}{x-4}\right)\left(\dfrac{x}{8}\right)=\dfrac{\overset{3}{\cancel{12}}\,x}{\underset{2}{\cancel{8}}(x-4)}=\dfrac{3x}{2(x-4)}$ sq. units

b) $P = 2\left(\frac{12}{x-4}\right) + 2\left(\frac{x}{8}\right) = \frac{24}{x-4} + \frac{x}{4} = \frac{4\cdot 24}{4(x-4)} + \frac{x(x-4)}{4(x-4)}$

$= \frac{96 + x^2 - 4x}{4(x-4)} = \frac{x^2 - 4x + 96}{4(x-4)}$ units

85) $\dfrac{\frac{2}{5}}{\frac{7}{15}} = \dfrac{\overset{3}{\cancel{15}}\left(\frac{2}{\cancel{5}}\right)}{\cancel{15}\left(\frac{7}{\cancel{15}}\right)} = \frac{6}{7}$

87) $\dfrac{p + \frac{6}{p}}{\frac{8}{p} + p} = \dfrac{p\left(p + \frac{6}{p}\right)}{p\left(\frac{8}{p} + p\right)} = \frac{p^2 + 6}{8 + p^2}$

$= \frac{p^2 + 6}{p^2 + 8}$

89) $\dfrac{\frac{n}{6n+48}}{\frac{n^2}{4n+32}} = \dfrac{\frac{n}{6(n+8)}}{\frac{n^2}{4(n+8)}}$

$= \dfrac{12(n+8)\left(\frac{n}{6(n+8)}\right)}{12(n+8)\left(\frac{n^2}{4(n+8)}\right)}$

$= \frac{2n}{3n^2} = \frac{2}{3n}$

91) $\dfrac{1 - \frac{1}{y-9}}{\frac{2}{y+3} + 1} = \dfrac{(y+3)(y-9)\left(1 - \frac{1}{y-9}\right)}{(y+3)(y-9)\left(\frac{2}{y+3} + 1\right)}$

$= \frac{(y+3)(y-9) - (y+3)}{2(y-9) + (y+3)(y-9)}$

$= \frac{y^2 - 6y - 27 - y - 3}{2y - 18 + y^2 - 6y - 27}$

$= \frac{y^2 - 7y - 30}{y^2 - 4y - 45}$

$= \frac{(y+3)(y-10)}{(y-9)(y+5)}$

93) $\dfrac{\frac{c}{c+2} + \frac{1}{c^2-4}}{1 - \frac{3}{c+2}}$

$= \dfrac{(c+2)(c-2)\left(\frac{c}{c+2} + \frac{1}{(c+2)(c-2)}\right)}{(c+2)(c-2)\left(1 - \frac{3}{c+2}\right)}$

$= \frac{c(c-2) + 1}{(c+2)(c-2) - 3(c-2)}$

$= \frac{c^2 - 2c + 1}{c^2 - 4 - 3c + 6} = \frac{(c-1)^2}{c^2 - 3c + 2}$

$= \frac{(c-1)^2}{(c-1)(c-2)} = \frac{c-1}{c-2}$

95) $$\frac{5w}{6}-\frac{1}{2}=-\frac{1}{6}$$
$$6\left(\frac{5w}{6}-\frac{1}{2}\right)=6\left(-\frac{1}{6}\right)$$
$$5w-3=2$$
$$5w=2$$
$$w=\frac{2}{5} \qquad \left\{\frac{2}{5}\right\}$$

97) $$\frac{4}{y-6}=\frac{12}{y+2}$$
$$4(y+2)=12(y-6)$$
$$4y+8=12y-72$$
$$80=8y$$
$$10=y \qquad \{10\}$$

99) $$\frac{r}{r+6}+3=\frac{10}{r+6}$$
$$(r+6)\left(\frac{r}{r+6}+3\right)=(r+6)\left(\frac{10}{r+6}\right)$$
$$r+3(r+6)=10$$
$$r+3r+18=10$$
$$4r=-8$$
$$r=-2 \qquad \{-2\}$$

101) $$\frac{16}{9t-27}+\frac{2t-4}{t-3}=\frac{t}{9}$$
$$\frac{16}{9(t-3)}+\frac{2t-4}{t-3}=\frac{t}{9}$$
$$9(t-3)\left(\frac{16}{9(t-3)}+\frac{2t-4}{t-3}\right)=9(t-3)\left(\frac{t}{9}\right)$$
$$16+9(2t-4)=t(t-3)$$
$$16+18t-36=t^2-3t$$
$$18t-20=t^2-3t$$
$$0=t^2-21t+20$$
$$0=(t-20)(t-1)$$
$$t-20=0 \text{ or } t-1=0$$
$$t=20 \qquad t=1 \qquad \{1,20\}$$

103)
$$\frac{3}{b+2}=\frac{16}{b^2-4}-\frac{4}{b-2}$$
$$\frac{3}{b+2}=\frac{16}{(b+2)(b-2)}-\frac{4}{b-2}$$
$$(b+2)(b-2)\left(\frac{3}{b+2}\right)=(b+2)(b-2)\left(\frac{16}{(b+2)(b-2)}-\frac{4}{b-2}\right)$$
$$3(b-2)=16-4(b+2)$$
$$3b-6=16-4b-8$$
$$3b-6=8-4b$$
$$7b=14$$
$$b=2$$
If $b=2$, two denominators = 0. $\varnothing$

105)
$$\frac{c}{c^2+3c-28}-\frac{5}{c^2+15c+56}=\frac{5}{c^2+4c-32}$$
$$\frac{c}{(c+7)(c-4)}-\frac{5}{(c+8)(c+7)}=\frac{5}{(c+8)(c-4)}$$
$$(c+7)(c-4)(c+8)\left(\frac{c}{(c+7)(c-4)}-\frac{5}{(c+8)(c+7)}\right)=(c+7)(c-4)(c+8)\left(\frac{5}{(c+8)(c-4)}\right)$$
$$c(c+8)-5(c-4)=5(c+7)$$
$$c^2+8c-5c+20=5c+35$$
$$c^2+3c+20=5c+35$$
$$c^2-2c-15=0$$
$$(c+3)(c-5)=0$$
$$c=-3 \quad \text{or} \quad c=5 \qquad \{-3,5\}$$

107)
$$A=\frac{2p}{\boxed{c}}$$
$$\boxed{c}A=2p$$
$$\boxed{c}=\frac{2p}{A}$$

109)
$$n=\frac{t}{\boxed{a}+b}$$
$$n\left(\boxed{a}+b\right)=t$$
$$n\boxed{a}+nb=t$$
$$n\boxed{a}=t-nb$$
$$\boxed{a}=\frac{t-nb}{n}$$

111)
$$\frac{1}{r}=\frac{1}{\boxed{s}}+\frac{1}{t}$$
$$\left(r\boxed{s}t\right)\left(\frac{1}{r}\right)=\left(r\boxed{s}t\right)\left(\frac{1}{\boxed{s}}+\frac{1}{t}\right)$$
$$\boxed{s}t=rt+r\boxed{s}$$
$$\boxed{s}t-r\boxed{s}=rt$$
$$\boxed{s}(t-r)=rt$$
$$\boxed{s}=\frac{rt}{t-r}$$

113) $s=$ grams of saturated fat

$s+4=$ grams of total fat
$$\frac{2}{3}=\frac{s}{s+4}$$
$$2(s+8)=3s$$
$$2s+8=3s$$
$$12=s$$
$$s+4=8+4=12$$
There are 12 g of total fat.

115) $s=$ speed of the plane

	$d=$	r ·	t
with wind	800	$s+40$	$\frac{80}{s+40}$
against wind	600	$s-40$	$\frac{600}{s-40}$

Solve $d=rt$ for t to get $t=\frac{d}{r}$.

time with wind = time against wind
$$\frac{800}{s+40}=\frac{600}{s-40}$$
$$800(s-40)=600(s+40)$$
$$800s-32{,}000=600s+24{,}000$$
$$200s=56{,}000$$
$$s=280$$
The speed of the plane is 280 mph.

Chapter 8 Test

1) $$\frac{3(-5)+5}{(-5)^2+25}=\frac{-15+5}{25+25}=\frac{-10}{50}=-\frac{1}{5}$$

3) a) $$k^2+2k-48=0$$
$$(k+8)(k-6)=0$$
$$k+8=0 \text{ or } k-6=0$$
$$k=-8 \qquad k=6$$

b) It never equals zero.

5) $$\frac{7v^2+55v-8}{v^2-64}=\frac{(7v-1)\cancel{(v+8)}}{\cancel{(v+8)}(v-8)}$$
$$=\frac{7v-1}{v-8}$$

7) b and $b+8$ are different factors. The LCD will be the product of these factors. LCD $=b(b+8)$

9) $$-\frac{21m^4}{n}\div\frac{12m^8}{n^3}=-\frac{\overset{7}{\cancel{21m^4}}}{\cancel{n}}\cdot\frac{\overset{n^2}{\cancel{n^3}}}{\underset{4m^4}{\cancel{12m^8}}}$$
$$=-\frac{7n^2}{4m^4}$$

11) $$\frac{r}{2r+1}+\frac{3}{r+5}=\frac{r(r+5)+3(2r+1)}{(2r+1)(r+5)}$$
$$=\frac{r^2+5r+6r+3}{(2r+1)(r+5)}$$
$$=\frac{r^2+11r+3}{(2r+1)(r+5)}$$

13) $$\frac{c-3}{c-15}+\frac{c+8}{15-c}=\frac{c-3}{c-15}-\frac{c+8}{c-15}$$
$$=-\frac{11}{c-15}\text{ or }\frac{11}{15-c}$$

15) $$\frac{1+\frac{2}{d-3}}{\frac{-2d}{d-3}-d}=\frac{(d-3)\left(1+\frac{2}{d-3}\right)}{(d-3)\left(\frac{-2d}{d-3}-d\right)}$$
$$=\frac{d-3+2}{-2d-d(d-3)}$$
$$=\frac{d-1}{-2d-d^2+3d}=\frac{d-1}{-d^2+d}$$
$$=\frac{\cancel{d-1}}{-d\cancel{(d-1)}}=-\frac{1}{d}$$

17) $$\frac{7t}{12}+\frac{t-4}{6}=\frac{7}{3}$$
$$12\left(\frac{7t}{12}+\frac{t-4}{6}\right)=12\left(\frac{7}{3}\right)$$
$$7t+2(t-4)=4\cdot 7$$
$$7t+2t-8=28$$
$$9t-8=28$$
$$9t=36$$
$$t=4 \qquad \{4\}$$

19) $$\frac{5}{n^2+10n+24}+\frac{5}{n^2+3n-18}=\frac{n}{n^2+n-12}$$
$$\frac{5}{(n+6)(n+4)}+\frac{5}{(n+6)(n-3)}=\frac{n}{(n+4)(n-3)}$$
$$5(n-3)+5(n+4)=n(n+6)$$
$$5n-15+5n+20=n^2+6n$$
$$10n+5=n^2+6n$$
$$0=n^2-4n-5$$
$$0=(n-5)(n+1)$$
$n-5=0$ or $n+1=0$

$n=5 \qquad n=-1 \qquad \{-1,5\}$

21) $t=$ the number of hours to assemble swing set together

Leticia + Betty = 1 job

$$\frac{1}{3}t \quad + \quad \frac{1}{5}t \quad =1$$
$$15\left(\frac{1}{3}t+\frac{1}{5}t\right)=15\cdot 1$$
$$5t+3t=15$$
$$8t=15$$
$$t=\frac{15}{8}=1\frac{7}{8}$$

It would take $1\frac{7}{8}$ hours to assemble.

Cumulative Review: Chapters 1-8

1) $A=\frac{1}{2}bh$ $\quad P=a+b+c$

$=\frac{1}{2}(8)(4)$ $\quad =7+5+8=20\text{ cm}$

$=16\text{ cm}^2$

3) $(3y^2)^4=3^4y^8=81y^8$

5) $-9\le 4c+5\le 9$

$-14\le 4c\le 4$

$-\frac{14}{4}\le c\le \frac{4}{4}$

$-\frac{7}{2}\le c\le 1$

$\left[-\frac{7}{2},1\right]$

7) $3x+5y=10$

x-int: Let $y=0$, and solve for x.

$3x+5(0)=10$

$3x+0=10$

$3x=10$

$x=\frac{10}{3}$ $\quad \left(\frac{10}{3},0\right)$

y-int: Let $x=0$, and solve for y.

$3(x)+5y=10$

$0+5y=10$

$5y=10$

$y=2$ $\quad (0, 2)$

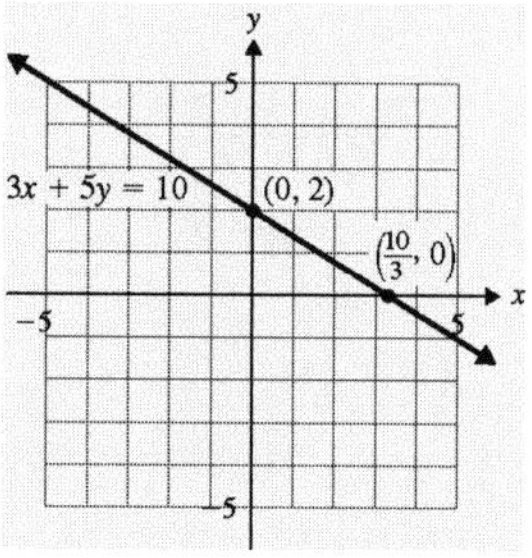

9) $3(10x+3y)=3(-3)$

$30x+9y=-9$

$-5(6x+4y)=-5\cdot 7$

$-30x-20y=-35$

Add the equations.

$30x+9y=-9$

$+\ -30x-20y=-35$

$-11y=-44$

$y=4$

Substitute $y=4$ into

$6x+4(4)=7$

$6x+16=7$

$6x=-9$

$x=\frac{-9}{6}$

$x=-\frac{3}{2}$ $\quad \left(-\frac{3}{2},4\right)$

11) $(m-7)^2=m^2-14m+49$

13) $\frac{36c^4+56c^3-8c^2+6c}{8c^2}$

$=\frac{\overset{9c^2}{\cancel{36c^4}}}{\underset{2}{\cancel{8c^2}}}+\frac{\overset{7c}{\cancel{56c^3}}}{\cancel{8c^2}}-\frac{\cancel{8c^2}}{\cancel{8c^2}}+\frac{\overset{3}{\cancel{6c}}}{\underset{4c}{\cancel{8c^2}}}$

$=\frac{9}{2}c^2+7c-1+\frac{3}{4c}$

15) $8h^3+125=(2h+5)(4h^2-10h+25)$

17) $ab+3b-2a-6=b(a+3)-2(a+3)$

$=(a+3)(b-2)$

19) a) $z^2-4z=0$

$z(z-4)=0$

$z=0$ or $z-4=0$

$z=4$

b) $9z+1=0$

$$9z=-1$$

$$z=-\frac{1}{9}$$

21) $$\frac{9-d^2}{d^3+3d^2}\cdot\frac{14d^5}{2d^2-5d-3}$$

$$=\frac{\cancel{(3+d)}\,\overset{-1}{\cancel{(3-d)}}}{\cancel{d^2}\,\cancel{(d+3)}}\cdot\frac{14\overset{d^3}{\cancel{d^5}}}{(2d+1)\cancel{(d-3)}}$$

$$=-\frac{14d^3}{2d+1}$$

23) $$\frac{\frac{2}{x-9}-1}{1+\frac{5}{x-9}}=\frac{(x-9)\left(\frac{2}{x-9}-1\right)}{(x-9)\left(1+\frac{5}{x-9}\right)}$$

$$=\frac{2-x+9}{x-9+5}$$

$$=\frac{-x+11}{x-4}=\frac{11-x}{x-4}$$

25) $$\frac{10}{m+7}+\frac{m}{m-10}=-\frac{40}{m^2-3m-70}$$

$$\frac{10}{m+7}+\frac{m}{m-10}=-\frac{40}{(m-10)(m+7)}$$

$$10(m-10)+m(m+7)=-40$$

$$10m-100+m^2+7m=-40$$

$$m^2+17m-100=-40$$

$$m^2+17m-60=0$$

$$(m+20)(m-3)=0$$

$m+20=0$ or $m-3=0$

$m=-20$ $\quad m=3$ $\quad \{-20,3\}$

Section 9.1: Exercises

1) The number's distance from zero.

3) $|q|=6$

$q=6$ or $q=-6$ $\{-6,6\}$

5) $|q-5|=3$

$q-5=3$ or $q-5=-3$

$q=8$ or $q=2$ $\{2,8\}$

7) $|4t-5|=7$

$4t-5=7$ or $4t-5=-7$

$4t=12$ $4t=-2$

$t=3$ or $t=-\frac{1}{2}$ $\left\{-\frac{1}{2},3\right\}$

9) $|12c+5|=1$

$12c+5=1$ or $12c+5=-1$

$12c=-4$ $12c=-6$

$c=\frac{-4}{12}$ $c=\frac{-6}{12}$

$c=-\frac{1}{3}$ or $c=-\frac{1}{2}$

$\left\{-\frac{1}{2},-\frac{1}{3}\right\}$

11) $|1-8m|=9$

$1-8m=9$ or $1-8m=-9$

$-8m=8$ $-8m=-10$

$m=\frac{8}{-8}$ $m=\frac{-10}{-8}$

$m=-1$ or $m=\frac{5}{4}$

$\left\{-1,\frac{5}{4}\right\}$

13) $\left|\frac{2}{3}b+3\right|=13$

$\frac{2}{3}b+3=13$ or $\frac{2}{3}b+3=-13$

$2b+9=39$ $2b+9=-39$

$2b=30$ $2b=-48$

$b=15$ or $b=-24$

$\{-24, 15\}$

15) $\left|4-\frac{3}{5}d\right|=6$

$4-\frac{3}{5}d=6$ or $4-\frac{3}{5}d=-6$

$20-3d=30$ $20-3d=-30$

$-3d=10$ $-3d=-50$

$d=-\frac{10}{3}$ or $d=\frac{50}{3}$

$\left\{-\frac{10}{3},\frac{50}{3}\right\}$

17) $\left|\frac{3}{4}y-2\right|=\frac{3}{5}$

$\frac{3}{4}y-2=\frac{3}{5}$ or $\frac{3}{4}y-2=-\frac{3}{5}$

$15y-40=12$ $15y-40=-12$

$15y=52$ $15y=28$

$y=\frac{52}{15}$ or $y=\frac{28}{15}$

$\left\{\frac{28}{15},\frac{52}{15}\right\}$

19) $\varnothing$, the absolute value of a quantity cannot be negative.

21) $|10p+2|=0$

$10p+2=0$

$10p=-2$

$p=-\frac{2}{10}$

$p=-\frac{1}{5}$ $\left\{-\frac{1}{5}\right\}$

23) $|z-6|+4=20$

$|z-6|=16$

$z-6=16$ or $z-6=-16$

$z=22$ or $z=-10$

$\{-10,22\}$

25) $|2a+5|+8=13$

$|2a+5|=5$

$2a+5=5$ or $2a+5=-5$

$2a=0$ $\quad$ $2a=-10$

$a=0$ or $a=-5$

$\{-5,0\}$

27) $|w+14|=0$

$w+14=0$

$w=-14$ $\{-14\}$

29) $\varnothing$, the absolute value of a quantity cannot be negative.

31) $|5b+3|+6=19$

$|5b+3|=13$

$5b+3=13$ or $5b+3=-13$

$5b=10$ $\quad$ $5b=-16$

$b=2$ or $b=-\frac{16}{5}$

$\left\{-\frac{16}{5},2\right\}$

33) $|3m-1|+5=2$

$|3m-1|=-3$

$\varnothing$, the absolute value of a quantity cannot be negative.

35) $|s+9|=|2s+5|$

$s+9=2s+5$ or $s+9=-2s-5$

$3s=-14$

$4=s$ or $s=-\frac{14}{3}$

$\left\{-\frac{14}{3},4\right\}$

37) $|3z+2|=|6-5z|$

$3z+2=6-5z$ or $3z+2=-6+5z$

$8z=4$ $\quad$ $8=2z$

$z=\frac{1}{2}$ or $4=z$

$\left\{\frac{1}{2},4\right\}$

39) $\left|\frac{3}{2}x-1\right|=|x|$

$\frac{3}{2}x-1=x$ or $\frac{3}{2}x-1=-x$

$3x-2=2x$ $\quad$ $3x-2=-2x$

$5x=2$

$x=2$ or $x=\frac{2}{5}$ $\left\{\frac{2}{5},2\right\}$

41) $|7c+10|=|5c+2|$

$7c+10=5c+2$ or $7c+10=-5c-2$

$2c=-8$ $\quad$ $12c=-12$

$c=-4$ or $c=-1$

$\{-4,-1\}$

43) $\left|\frac{1}{4}t-\frac{5}{2}\right|=\left|5-\frac{1}{2}t\right|$

$\frac{1}{4}t-\frac{5}{2}=5-\frac{1}{2}t$ or $\frac{1}{4}t-\frac{5}{2}=-5+\frac{1}{2}t$

$t-10=20-2t$ $\quad$ $t-10=-20+2t$

$3t=30$ $\quad$ $-t=-10$

$t=10$ or $t=10$

$\{10\}$

45) $|x|=9$

47) $|x|=\frac{1}{2}$

Section 9.2: Exercises

1) $[-1,5]$

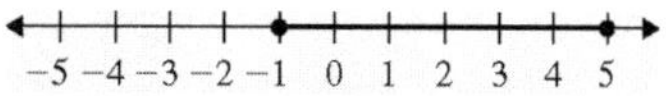

3) $(-\infty,2)\cup(9,\infty)$

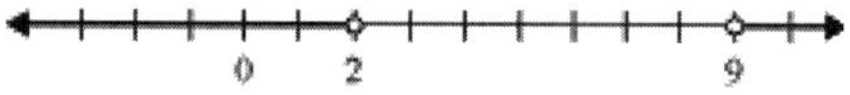

5) $\left(-\infty,-\frac{9}{2}\right]\cup\left[\frac{3}{5},\infty\right)$

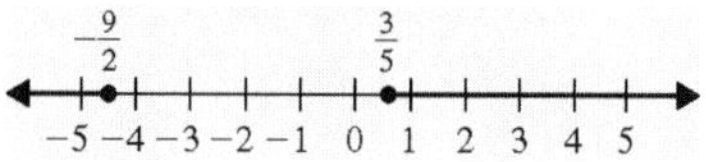

7) $\left(4,\frac{17}{3}\right)$

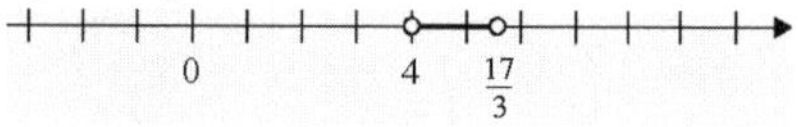

9) $|m|\le 7$

$-7\le m\le 7$ $\quad[7,7]$

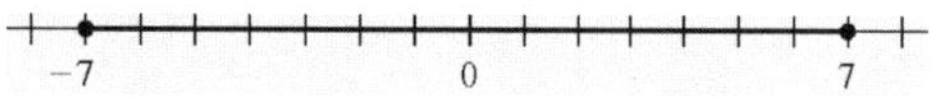

11) $|3k|<12$

$-12<3k<12$

$-4<k<4$ $\quad(-4,4)$

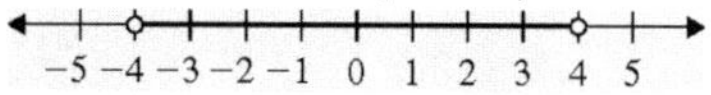

13) $|w-2|<4$

$-4<w-2<4$

$-2<w<6$ $\quad(-2,6)$

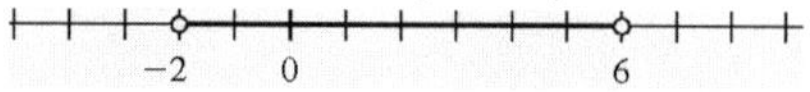

15) $|3r+10|\le 4$

$-4\le 3r+10\le 4$

$-14\le 3r\le -6$

$-\frac{14}{3}\le r\le -2$ $\quad\left[-\frac{14}{3},-2\right]$

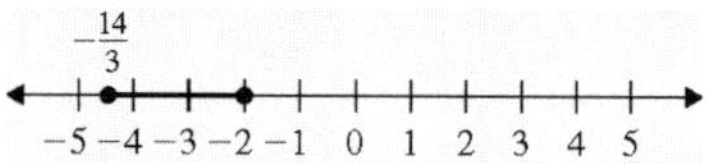

17) $|7-6p|\le 3$

$-3\le 7-6p\le 3$

$-10\le -6p\le -4$

$\frac{-10}{-6}\ge p\ge\frac{-4}{-6}$

$\frac{5}{3}\ge p\ge\frac{2}{3}$ $\quad\left[\frac{2}{3},\frac{5}{3}\right]$

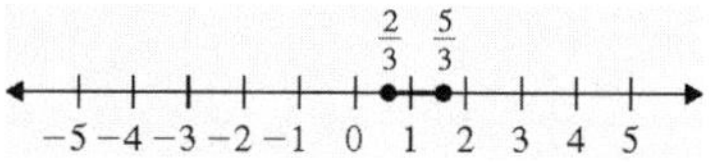

19) $\varnothing$, the absolute value of a quantity cannot be less than 0.

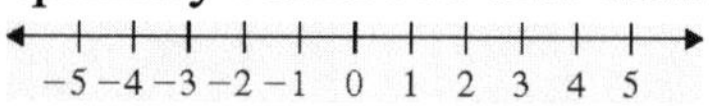

21) $\varnothing$, the absolute value of a quantity can't be less than a negative number.

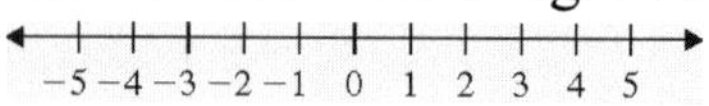

23) $|8c-3|+15<20$

$|8c-3|<5$

$-5<8c-3<5$

$-2<8c<8$

$-\frac{1}{4}<c<1 \qquad \left(-\frac{1}{4},1\right)$

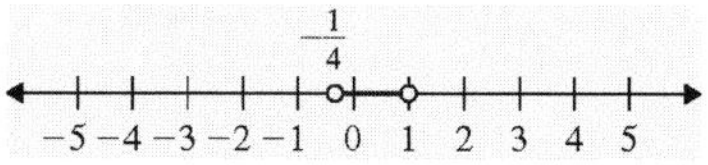

25) $\left|\frac{3}{2}h+6\right|-2\leq 10$

$\left|\frac{3}{2}h+6\right|\leq 12$

$-12\leq\frac{3}{2}h+6\leq 12$

$-18\leq\frac{3}{2}h\leq 6$

$-36\geq 3h\geq 12$

$-12\geq h\geq 4 \qquad [-12,4]$

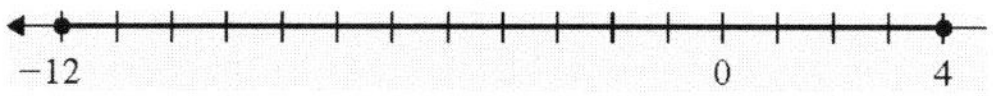

27) $|t|\geq 7$

$t\geq 7$ or $t\leq -7 \quad (-\infty,-7]\cup[7,\infty)$

29) $|5a|>2$

$5a>2$ or $5a<-2$

$a>\frac{2}{5} \qquad a<-\frac{2}{5}$

$\left(-\infty,-\frac{2}{5}\right)\cup\left(\frac{2}{5},\infty\right)$

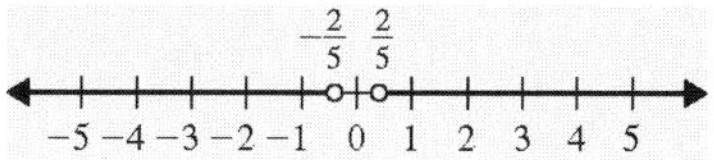

31) $|d+10|\geq 4$

$d+10\geq 4$ or $d+10\leq -4$

$d\geq -6 \qquad d\leq -14$

$(-\infty,-14]\cup[-6,\infty)$

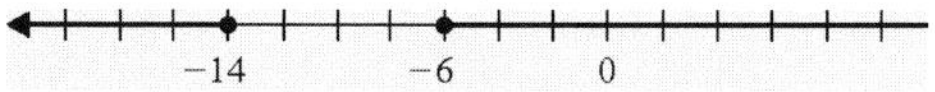

33) $|4v-3|\geq 9$

$4v-3\geq 9$ or $4v-3\leq -9$

$4v\geq 12 \qquad 4v\leq -6$

$v\geq 3 \qquad v\leq -\frac{3}{2}$

$\left(-\infty,-\frac{3}{2}\right]\cup[3,\infty)$

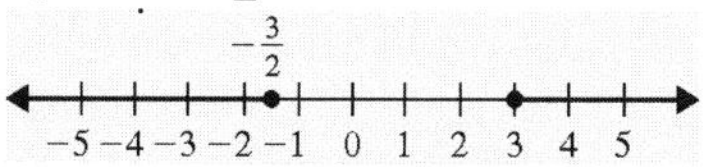

35) $|17-6x|>5$

$17-6x>5$ or $17-6x<-5$

$-6x>-12 \qquad -6x<-22$

$x<\frac{-12}{-6} \qquad x>\frac{-22}{-6}$

$x<2 \qquad x>\frac{11}{3}$

$(-\infty,2)\cup\left(\frac{11}{3},\infty\right)$

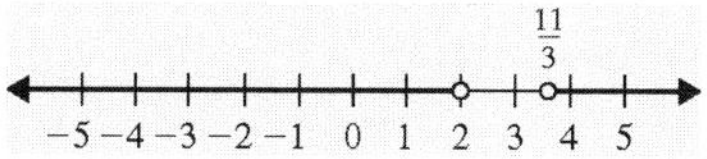

37) $|8k+5|\geq 0 \quad (-\infty,\infty)$

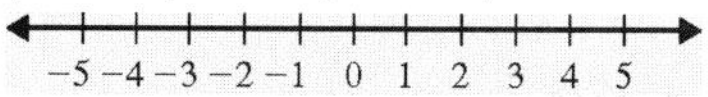

39) $|z-3|\geq -5 \quad (-\infty,\infty)$

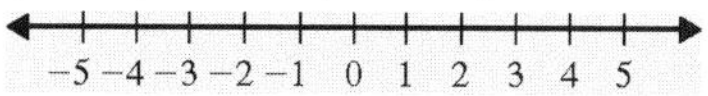

41) $|2m-1|+4>5$

$$|2m-1|>1$$

$$2m-1>1 \text{ or } 2m-1<-1$$

$$2m>2 \qquad 2m<0$$

$$m>1 \qquad m<0$$

$(-\infty,0)\cup(1,\infty)$

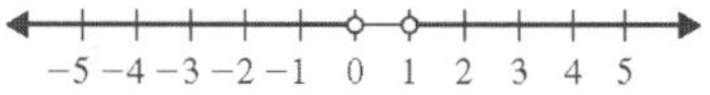

43)
$$-3+\left|\frac{5}{6}n+\frac{1}{2}\right|\geq 1$$

$$\left|\frac{5}{6}n+\frac{1}{2}\right|\geq 4$$

$$\frac{5}{6}n+\frac{1}{2}\geq 4 \text{ or } \frac{5}{6}n+\frac{1}{2}\leq -4$$

$$5n+3\geq 24 \qquad 5n+3\leq -24$$

$$5n\geq 21 \qquad 5n\leq -27$$

$$n\geq \frac{21}{5} \qquad n\leq -\frac{27}{5}$$

$$\left(-\infty,-\frac{27}{5}\right]\cup\left[\frac{21}{5},\infty\right)$$

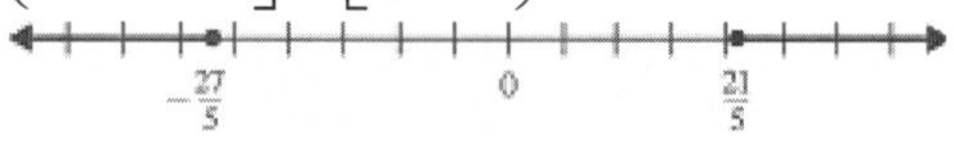

45) The absolute value of a quantity is always 0 or positive; it cannot be less than 0.

47) The absolute value of a quantity is always 0 or positive; so for any real number, x, the quantity $|2x+1|$ will be greater than -3.

49) $|2v+9|>3$

$$2v+9>3 \text{ or } 2v+9<-3$$

$$2v>-6 \qquad 2v<-12$$

$$v>-3 \qquad v<-6$$

$(-\infty,-6)\cup(-3,\infty)$

51) $3=|4t+5|$

$$3=4t+5 \text{ or } -3=4t+5$$

$$-2=4t \qquad -8=4t$$

$$-\frac{1}{2}=t \qquad -2=t$$

$$\left\{-2,-\frac{1}{2}\right\}$$

53) $9\leq|7-8q|$

$$9\leq 7-8q \text{ or } -9\geq 7-8q$$

$$2\leq -8q \qquad -16\geq -8q$$

$$\frac{2}{-8}\geq q \qquad \frac{-16}{-8}\leq q$$

$$-\frac{1}{4}\geq q \qquad 2\leq q$$

$$\left(-\infty,-\frac{1}{4}\right]\cup[2,\infty)$$

55) $2(x-8)+10<4x$

$$2x-16+10<4x$$

$$-6<2x$$

$$-3<x \quad (-3,\infty)$$

57) $|8-r|\leq 5$

$$-5\leq 8-r\leq 5$$

$$-13\leq -r\leq -3$$

$$13\geq r\geq 3 \quad [3,13]$$

59) $\varnothing$, the absolute value of a quantity cannot be less than or equal to a negative number.

61) $\left|\frac{4}{3}x+1\right|=\left|\frac{5}{3}x+8\right|$

$\frac{4}{3}x+1=\frac{5}{3}x+8$ or $\frac{4}{3}x+1=-\frac{5}{3}x-8$

$4x+3=5x+24 \qquad 4x+3=-5x-24$

$-21=x \qquad 9x=-27$

$x=-3$

$\{-21,-3\}$

63) $|3m-8|-11>-3$

$|3m-8|>8$

$3m-8>8$ or $3m-8<-8$

$3m>16 \qquad 3m<0$

$m>\frac{16}{3} \qquad m<0$

$(-\infty,0)\cup\left(\frac{16}{3},\infty\right)$

65) $|4-9t|+2=1$

$|4-9t|=-1$

$\varnothing$, the absolute value of a quantity cannot be negative.

67) $-\frac{3}{5}\geq\frac{5}{2}a-\frac{1}{2}$

$-6\geq 25a-5$

$-1\geq 25a$

$-\frac{1}{25}\geq a \qquad \left(-\infty,-\frac{1}{25}\right]$

69) $|6k+17|>-4 \quad (-\infty,\infty)$

71) $5\geq|c+8|-2$

$7\geq|c+8|$

$7\geq c+8\geq -7$

$-1\geq c\geq -15 \quad [-15,-1]$

73) $|5h-8|>7$

$5h-8>7$ or $5h-8<-7$

$5h>15 \qquad 5h<1$

$h>3 \qquad h<\frac{1}{5}$

$\left(-\infty,\frac{1}{5}\right)\cup(3,\infty)$

75) $\left|\frac{1}{2}d-4\right|+7=13$

$\left|\frac{1}{2}d-4\right|=6$

$\frac{1}{2}d-4=6$ or $\frac{1}{2}d-4=-6$

$\frac{1}{2}d=10 \qquad \frac{1}{2}d=-2$

$d=20 \qquad d=-4$

$\{-4,20\}$

77) $|5j+3|+1\leq 9$

$|5j+3|\leq 8$

$-8\leq 5j+3\leq 8$

$-11\leq 5j+3\leq 5$

$-\frac{11}{5}\leq j\leq 1 \qquad \left[-\frac{11}{5},1\right]$

79) $|a-128| \le 0.75$

$127.25 \le a \le 128.25$

There is between 127.25 oz and 128.75 oz of milk in the container.

81) $|b-38| \le 5$

$33 \le b \le 43$

He will spend between \$33 and \$43 on his daughter's gift.

Section 9.3: Exercises

1) -6) Answers may vary.

7)

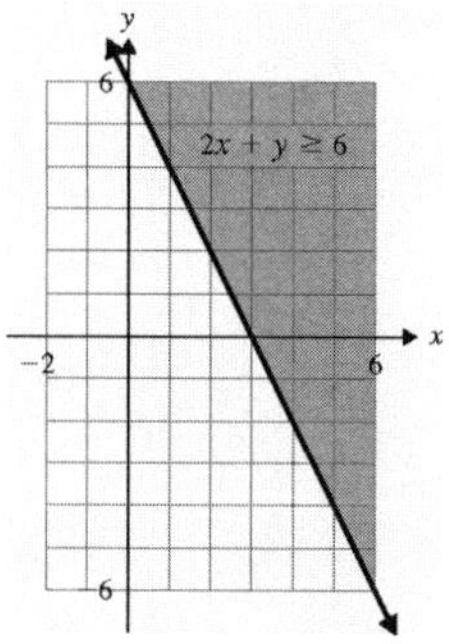

9)

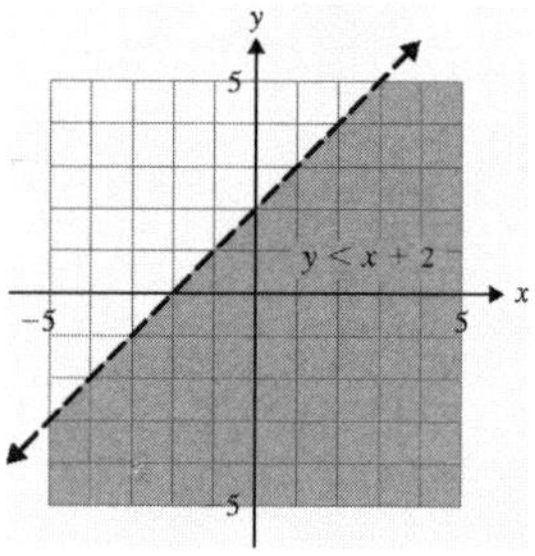

11)

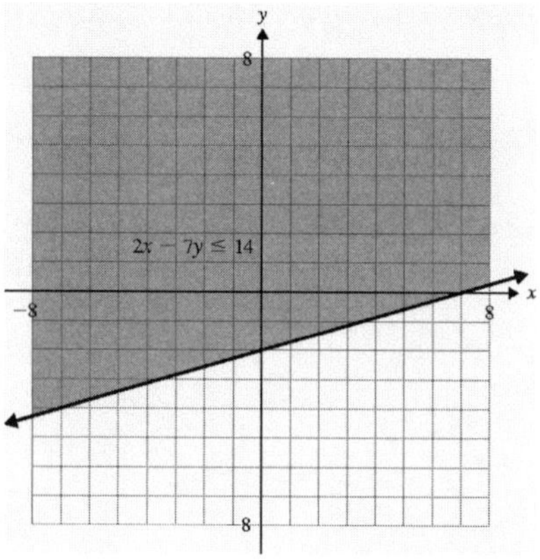

13)

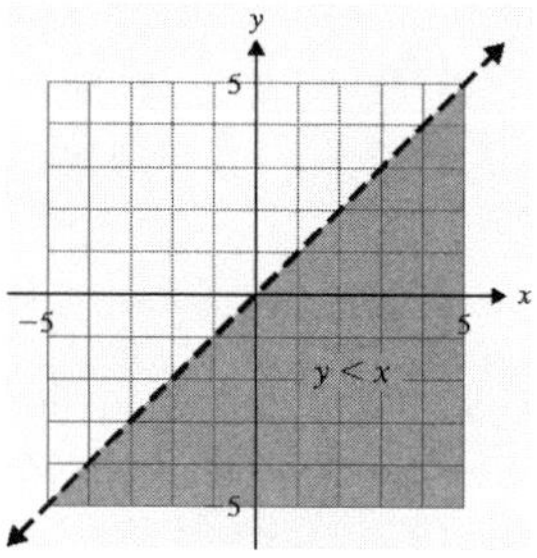

15)

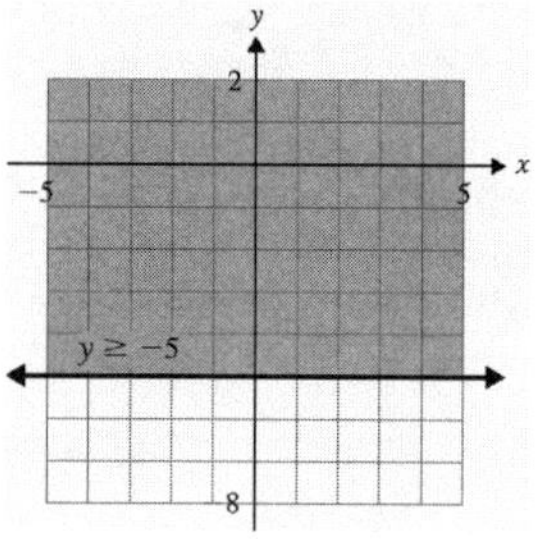

17)

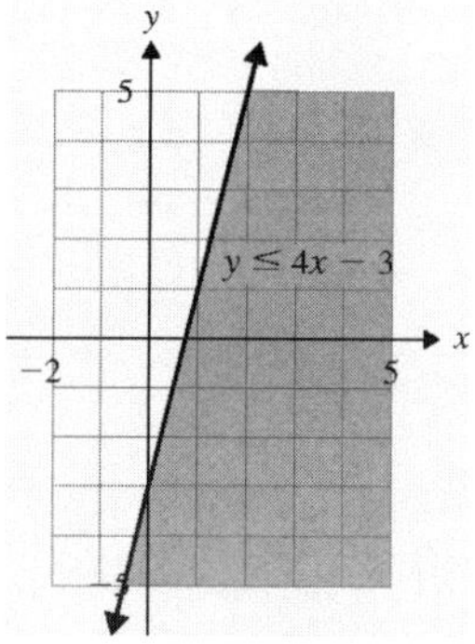

19)

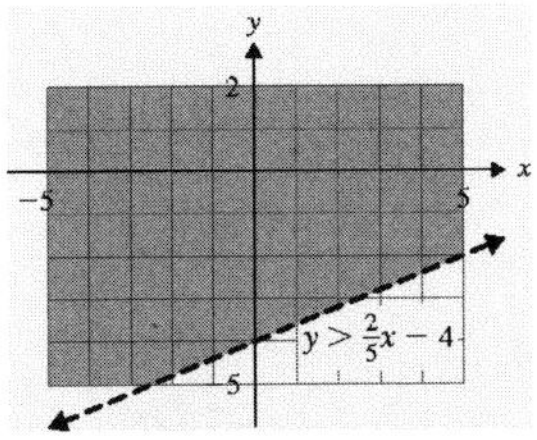

21)

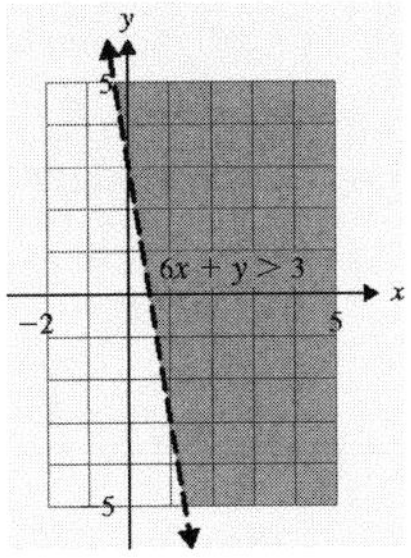

23)

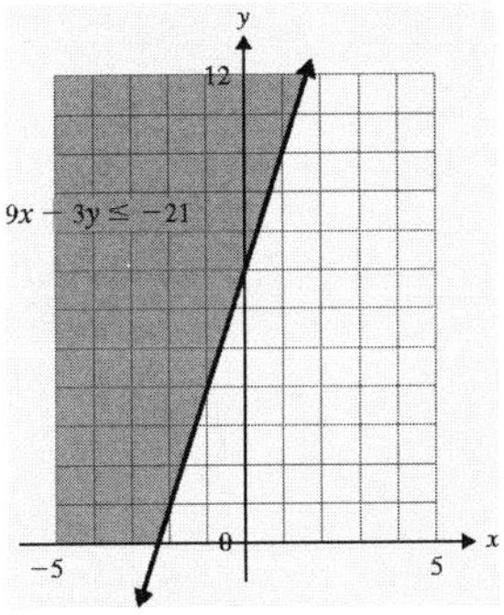

25)

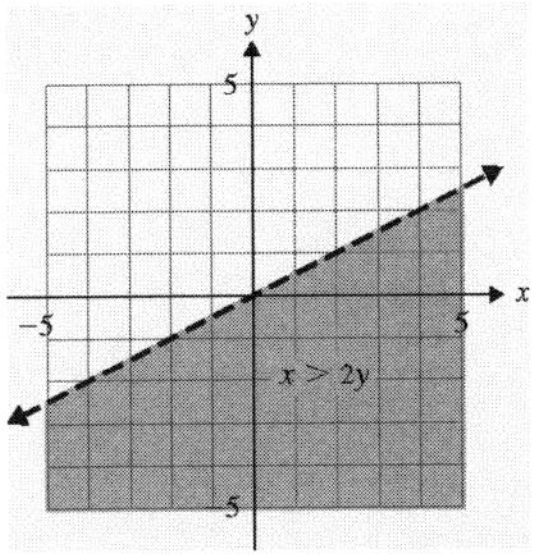

27) Answers may vary.

29)

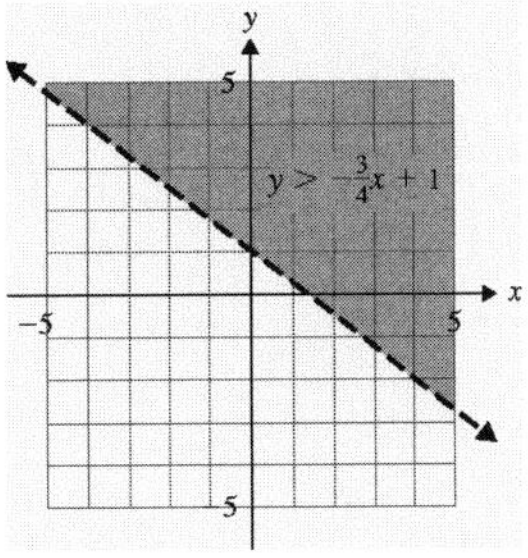

31)

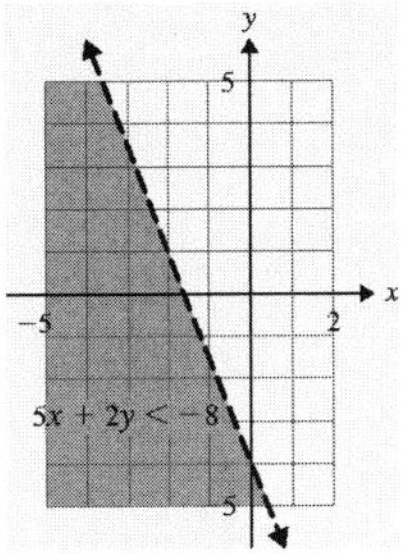

33)

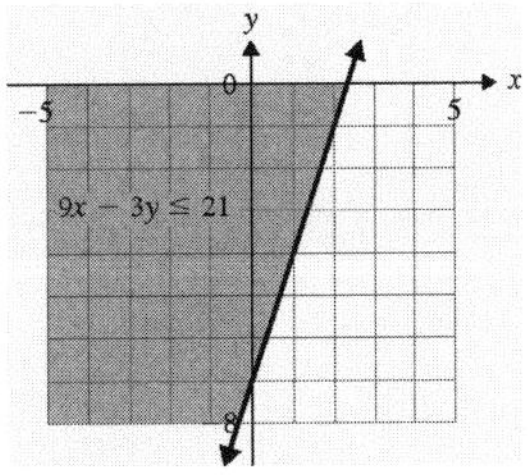

35)

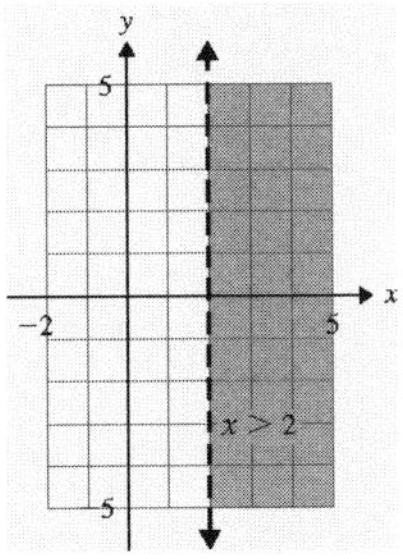

37)

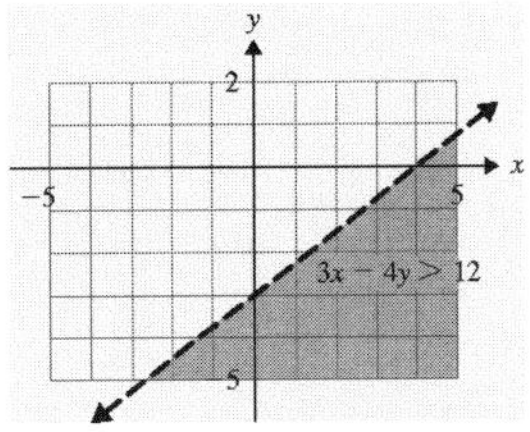

39) 42) Answers may vary.

43) No; $(3,5)$ satisfies $x-y\ge -6$ but not $2x+y<7$. Since the inequality contains *and*, it must satisfy *both* inequalities.

45)

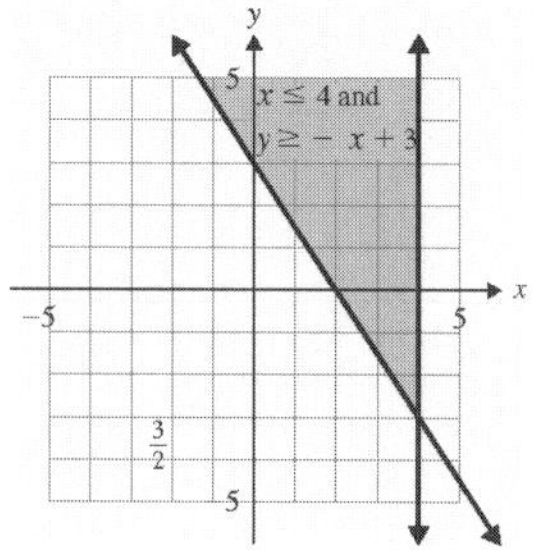

47)

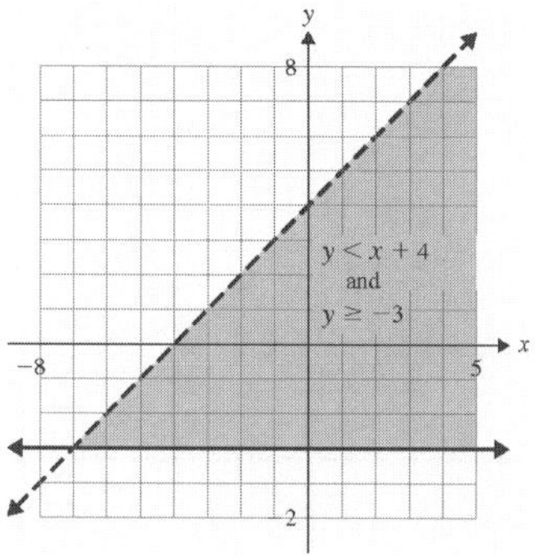

49)

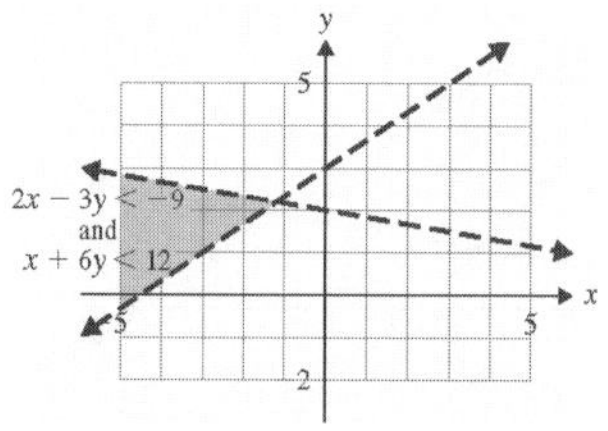

51)

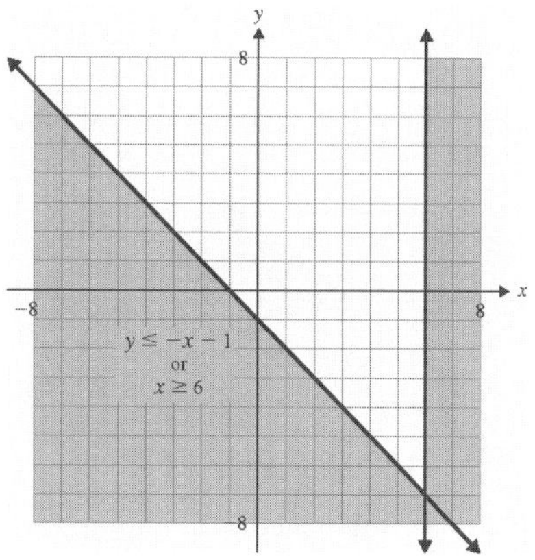

53)

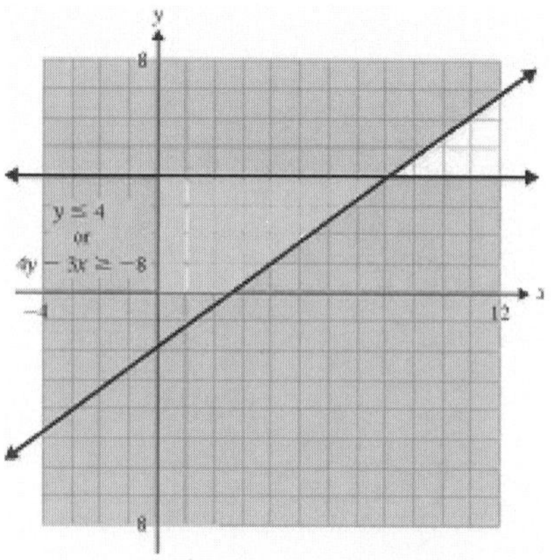

55)

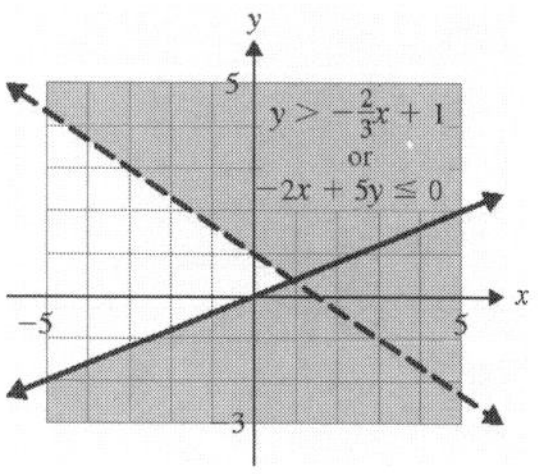

57)

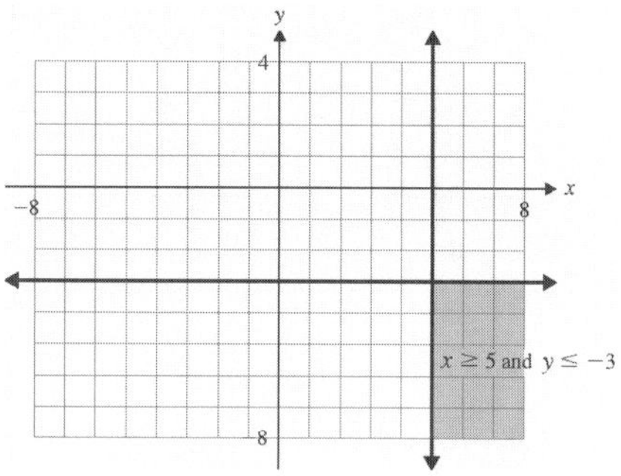

59)

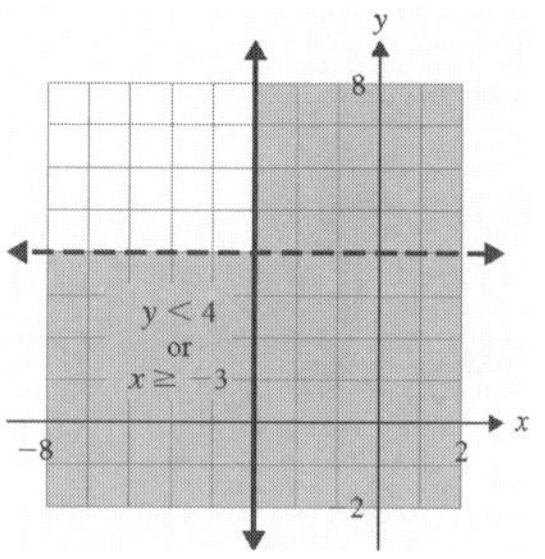

61)

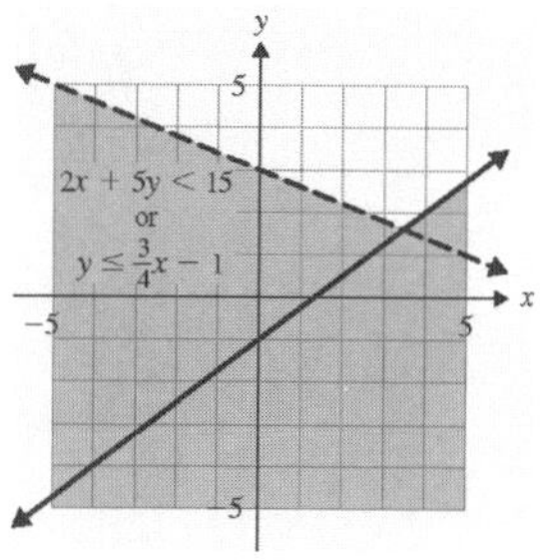

63)

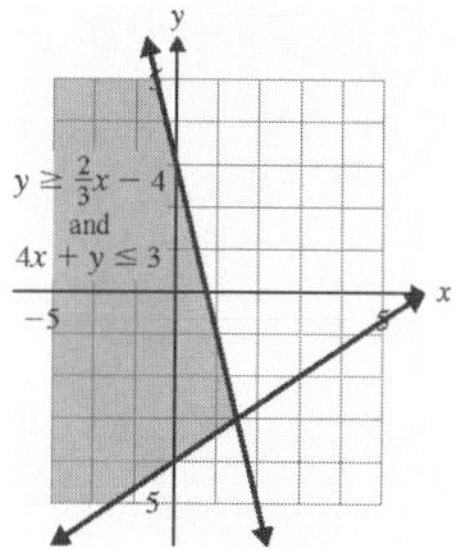

65) a)

$$x \ge 0$$
$$y \ge 0$$
$$x + y \le 15$$

b)

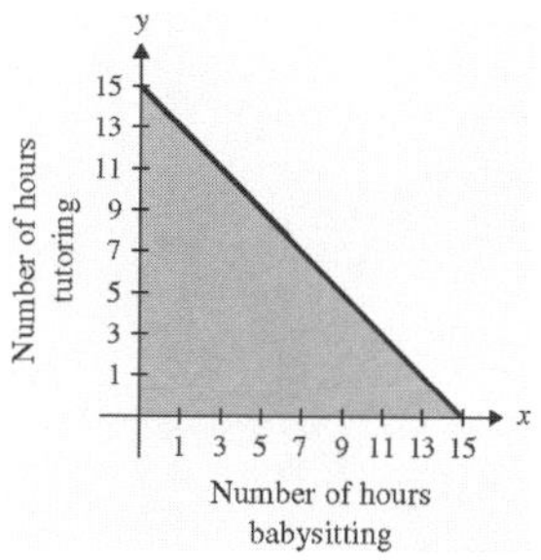

c-d) Answers may vary.

Section 9.4: Exercises

1) $\left[\begin{array}{cc|c} 1 & -7 & 15 \\ 4 & 3 & -1 \end{array}\right]$

3) $\left[\begin{array}{ccc|c} 1 & 6 & -1 & -2 \\ 3 & 1 & 4 & 7 \\ -1 & -2 & 3 & 8 \end{array}\right]$

5) $\begin{aligned} 3x + 10y &= -4 \\ x - 2y &= 5 \end{aligned}$

7) $\begin{aligned} x - 6y &= 8 \\ y &= -2 \end{aligned}$

9) $\begin{aligned} x - 3y + 2z &= 7 \\ 4x - y + 3z &= 0 \\ -2x + 2y - 3z &= -9 \end{aligned}$

11) $\begin{aligned} x + 5y + 2z &= 14 \\ y - 8z &= 2 \\ z &= -3 \end{aligned}$

13) $\left[\begin{array}{cc|c} 1 & 4 & -1 \\ 3 & 5 & 4 \end{array}\right] \xrightarrow{-3R_1 + R_2 \to R_2}$

$\left[\begin{array}{cc|c} 1 & 4 & -1 \\ 0 & -7 & 7 \end{array}\right] \xrightarrow{-\frac{1}{7}R_2 \to R_2} \left[\begin{array}{cc|c} 1 & 4 & -1 \\ 0 & 1 & -1 \end{array}\right]$

$y = -1 \qquad x + 4y = -1$

$x + 4(-1) = -1$

$x - 4 = -1$

$x = 3 \qquad (3, -1)$

15) $\left[\begin{array}{cc|c} 1 & -3 & 9 \\ -6 & 5 & 11 \end{array}\right] \xrightarrow{6R_1 + R_2 \to R_2}$

$\left[\begin{array}{cc|c} 1 & -3 & 9 \\ 0 & -13 & 65 \end{array}\right] \xrightarrow{-\frac{1}{13}R_2 \to R_2}$

$\left[\begin{array}{cc|c} 1 & -3 & 9 \\ 0 & 1 & -5 \end{array}\right]$

$y = -5 \qquad x - 3y = 9$

$x - 3(-5) = 9$

$x + 15 = 9$

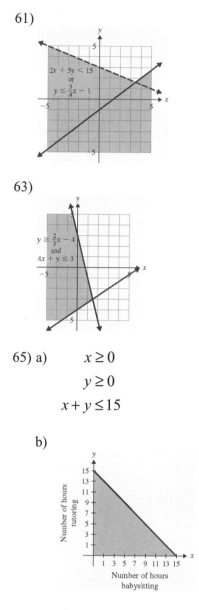

$x = -6 \qquad (-6, -5)$

17) $\left[\begin{array}{cc|c} 4 & -3 & 6 \\ 1 & 1 & -2 \end{array}\right] \xrightarrow{R_1 \leftrightarrow R_2} \left[\begin{array}{cc|c} 1 & 1 & -2 \\ 4 & -3 & 6 \end{array}\right] \xrightarrow{-4R_1+R_2 \to R_2} \left[\begin{array}{cc|c} 1 & 1 & -2 \\ 0 & -7 & 14 \end{array}\right] \xrightarrow{-\frac{1}{7}R_2 \to R_2} \left[\begin{array}{cc|c} 1 & 1 & -2 \\ 0 & 1 & -2 \end{array}\right]$

$y=-2 \qquad x+y=-2$

$x+(-2)=-2$

$x=0 \qquad (0,-2)$

19) $\left[\begin{array}{ccc|c} 1 & 1 & -1 & -5 \\ 4 & 5 & -2 & 0 \\ 8 & -3 & 2 & -4 \end{array}\right] \xrightarrow[-8R_1+R_3 \to R_3]{-4R_1+R_2 \to R_2} \left[\begin{array}{ccc|c} 1 & 1 & -1 & -5 \\ 0 & 1 & 2 & 20 \\ 0 & -11 & 10 & 36 \end{array}\right]$

$\xrightarrow{11R_2+R_3 \to R_3} \left[\begin{array}{ccc|c} 1 & 1 & -1 & -5 \\ 0 & 1 & 2 & 20 \\ 0 & 0 & 32 & 256 \end{array}\right] \xrightarrow{\frac{1}{32}R_3 \to R_3} \left[\begin{array}{ccc|c} 1 & 1 & -1 & -5 \\ 0 & 1 & 2 & 20 \\ 0 & 0 & 1 & 8 \end{array}\right]$

$x+y-z=-5 \qquad y+2z=20 \qquad x+y-z=-5$

$y+2z=20 \qquad y+2(8)=20 \qquad x+4-8=-5$

$z=8 \qquad y+16=20 \qquad x-4=-5$

$y=4 \qquad x=-1 \qquad (-1, 4, 8)$

21) $\left[\begin{array}{ccc|c} 1 & -3 & 2 & -1 \\ 3 & -8 & 4 & 6 \\ -2 & -3 & -6 & 1 \end{array}\right] \xrightarrow[2R_1+R_3 \to R_3]{-3R_1+R_2 \to R_2} \left[\begin{array}{ccc|c} 1 & -3 & 2 & -1 \\ 0 & 1 & -2 & 9 \\ 0 & -9 & -2 & -1 \end{array}\right] \xrightarrow{9R_2+R_3 \to R_3}$

$\left[\begin{array}{ccc|c} 1 & -3 & 2 & -1 \\ 0 & 1 & -2 & 9 \\ 0 & 0 & -20 & 80 \end{array}\right] \xrightarrow{-\frac{1}{20}R_3 \to R_3} \left[\begin{array}{ccc|c} 1 & -3 & 2 & -1 \\ 0 & 1 & -2 & 9 \\ 0 & 0 & 1 & -4 \end{array}\right]$

$x-3y+2z=-1 \qquad y-2z=9 \qquad x-3y+2z=-1$

$y-2z=9 \qquad y-2(-4)=9 \qquad x-3(1)+2(-4)=-1$

$z=-4 \qquad y+8=9 \qquad x-3-8=-1$

$y=1 \qquad x-11=-1$

$x=10 \qquad (10, 1, -4)$

23) $\left[\begin{array}{ccc|c} -4 & -3 & 1 & 5 \\ 1 & 1 & -1 & -7 \\ 6 & 4 & 1 & 12 \end{array}\right] \xrightarrow{R_1 \leftrightarrow R_2} \left[\begin{array}{ccc|c} 1 & 1 & -1 & -7 \\ -4 & -3 & 1 & 5 \\ 6 & 4 & 1 & 12 \end{array}\right]$

$\xrightarrow[-6R_1+R_3 \to R_3]{4R_1+R_2 \to R_2} \left[\begin{array}{ccc|c} 1 & 1 & -1 & -7 \\ 0 & 1 & -3 & -23 \\ 0 & -2 & 7 & 54 \end{array}\right] \xrightarrow{2R_2+R_3 \to R_3} \left[\begin{array}{ccc|c} 1 & 1 & -1 & -7 \\ 0 & 1 & -3 & -23 \\ 0 & 0 & 1 & 8 \end{array}\right]$

$x+y-z=-7$	$y-3z=-23$	$x+y-z=-7$
$y-3z=-23$	$y-3(8)=-23$	$x+1-8=-7$
$z=8$	$y-24=-23$	$x-7=-7$
	$y=1$	$x=0$ $\quad (0,1,8)$

25) $\left[\begin{array}{ccc|c} 1 & -3 & 1 & -4 \\ 4 & 5 & -1 & 0 \\ 2 & -6 & 2 & 1 \end{array}\right] \xrightarrow[-2R_1+R_3 \to R_3]{-4R_1+R_2 \to R_2} \left[\begin{array}{ccc|c} 1 & -3 & 1 & -4 \\ 0 & 17 & -5 & 16 \\ 0 & 0 & 0 & 9 \end{array}\right] \varnothing$

Chapter 9 Review

1) $|m|=9$

$m=9$ or $m=-9 \quad \{-9,9\}$

3) $|7t+3|=4$

$7t+3=4$ or $7t+3=-4$

$7t=1 \qquad 7t=-7$

$t=\frac{1}{7}$ or $t=-1 \quad \left\{-1,\frac{1}{7}\right\}$

5) $|8p+11|-7=-3$

$|8p+11|=4$

$8p+11=4$ or $8p+11=-4$

$8p=-7 \qquad 8p=-15$

$p=-\frac{7}{8}$ or $p=-\frac{15}{8}$

$\left\{-\frac{15}{8},-\frac{7}{8}\right\}$

7) $\left|4-\frac{5}{3}x\right|=\frac{1}{3}$

$4-\frac{5}{3}x=\frac{1}{3}$ or $4-\frac{5}{3}x=-\frac{1}{3}$

$12-5x=1 \qquad 12-5x=-1$

$-5x=-11 \qquad -5x=-13$

$x=\frac{11}{5}$ or $x=\frac{13}{5}$

$\left\{\frac{11}{5},\frac{13}{5}\right\}$

9) $|7r-6|=|8r+2|$

$7r-6=8r+2$ or $7r-6=-8r-2$

$15r=4$

$-8=r$ or $r=\frac{4}{15}$

$\left\{-8,\frac{4}{15}\right\}$

11) $\varnothing$, the absolute value of a quantity cannot be negative.

13) $|9d+4|=0$

$9d+4=0$

$9d=-4$

$d=-\frac{4}{9}$ $\quad \left\{-\frac{4}{9}\right\}$

15) $|a|=4$

17) $|c|\le 3$

$-3\le c\le 3 \quad [-3,3]$

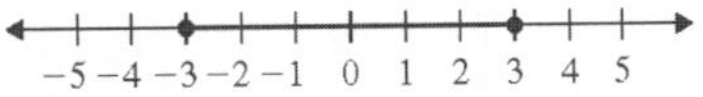

19) $|4t|>8$

$4t>8$ or $4t<-8$

$t>2 \qquad t<-2$

$(-\infty,-2)\cup(2,\infty)$

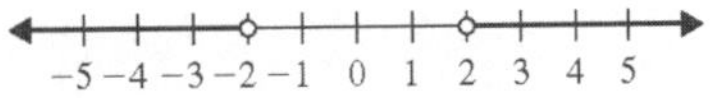

21) $|12r+5|\ge 7$

$12r+5\ge 7$ or $12r+5\le -7$

$12r\ge 2 \qquad 12r\le -12$

$r\ge \frac{1}{6} \qquad r\le -1$

$(-\infty,-1]\cup\left[\frac{1}{6},\infty\right)$

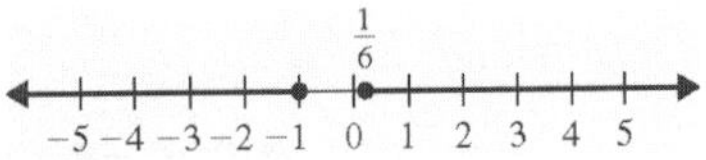

23) $|4-a|<9$

$-9<4-a<9$

$-13<-a<5$

$13>a>-5 \quad (-5,13)$

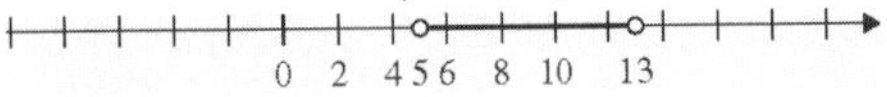

25) $|4c+9|-8\le -2$

$|4c+9|\le 6$

$-6\le 4c+9\le 6$

$-15\le 4c\le -3$

$-\frac{15}{4}\le c\le -\frac{3}{4} \quad \left[-\frac{15}{4},-\frac{3}{4}\right]$

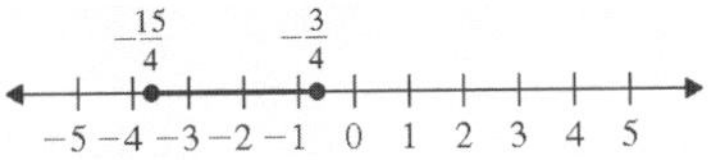

27) $|5y+12|-15\ge -8$

$|5y+12|\ge 7$

$5y+12\ge 7$ or $5y+12\le -7$

$5y\ge -5 \qquad 5y\le -19$

$y\ge -1 \qquad y\le -\frac{19}{5}$

$\left(-\infty,-\frac{19}{5}\right]\cup[-1,\infty)$

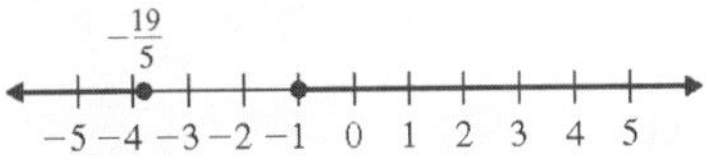

29) $(-\infty,\infty)$

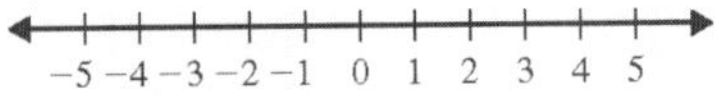

31) $|12s+1|\le 0$

$12s+1=0$

$12s=-1$

$s=-\frac{1}{12}$ $\quad \left\{-\frac{1}{12}\right\}$

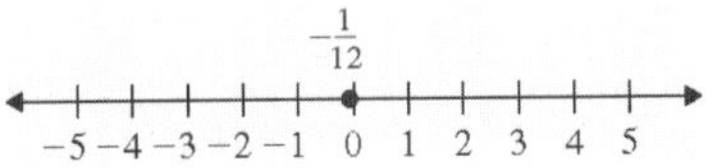

33)

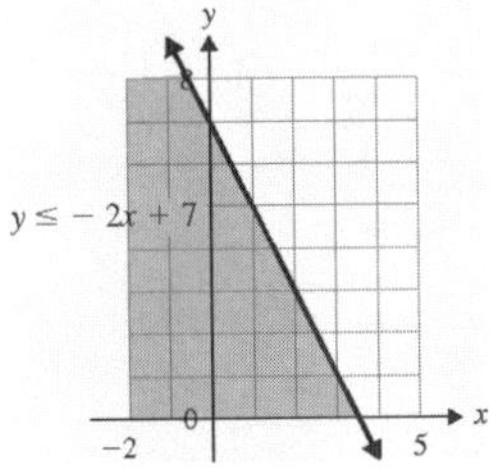

35)

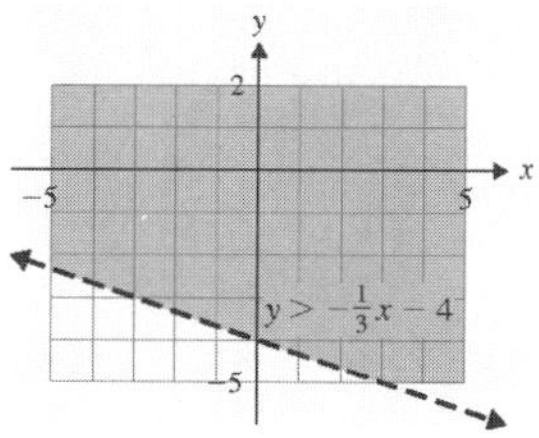

37)

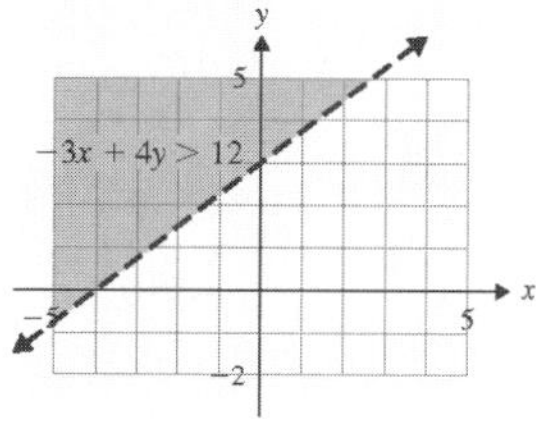

39)

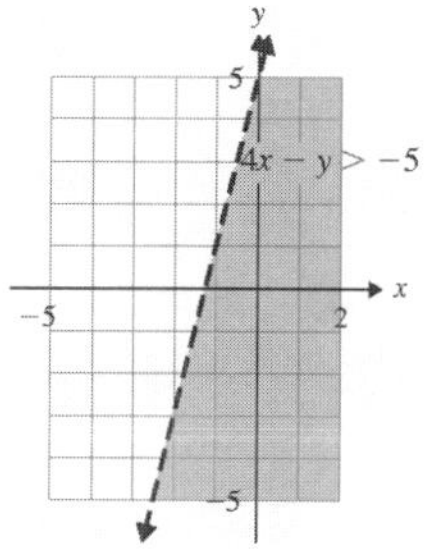

41)

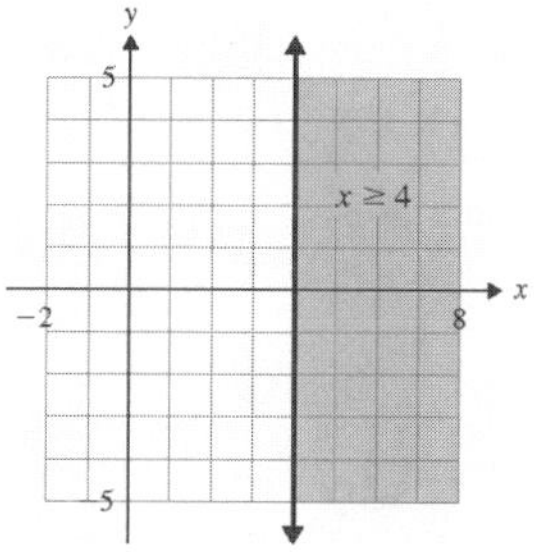

43)

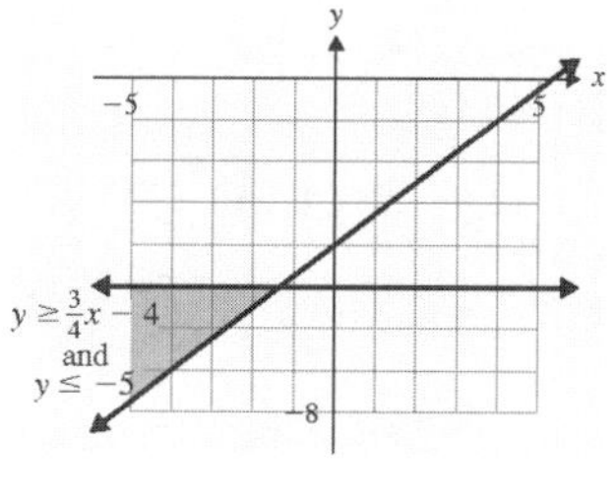

45)

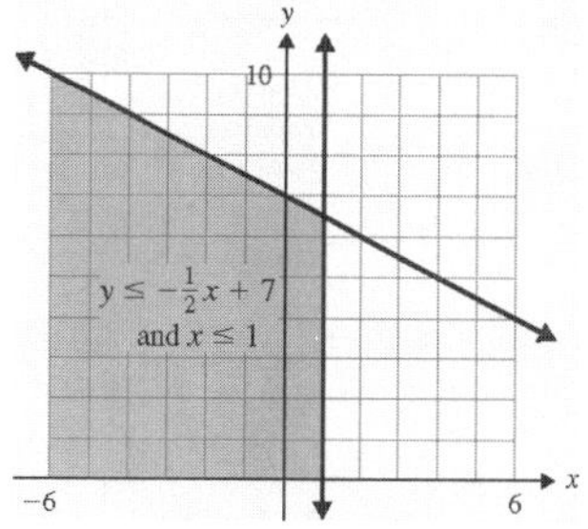

47)

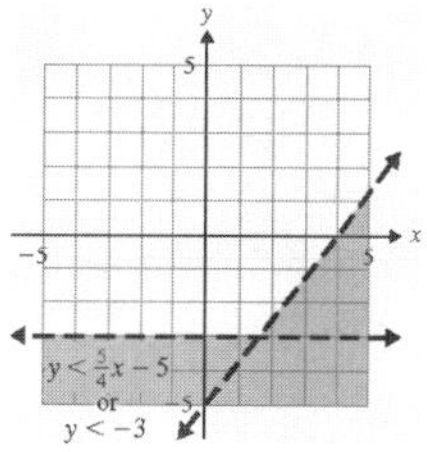

49)

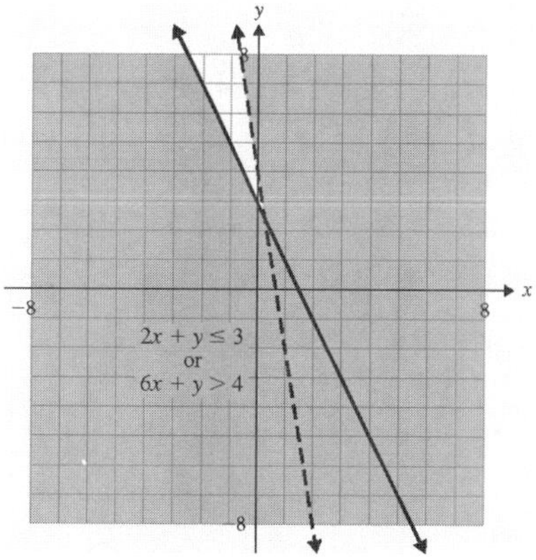

51)

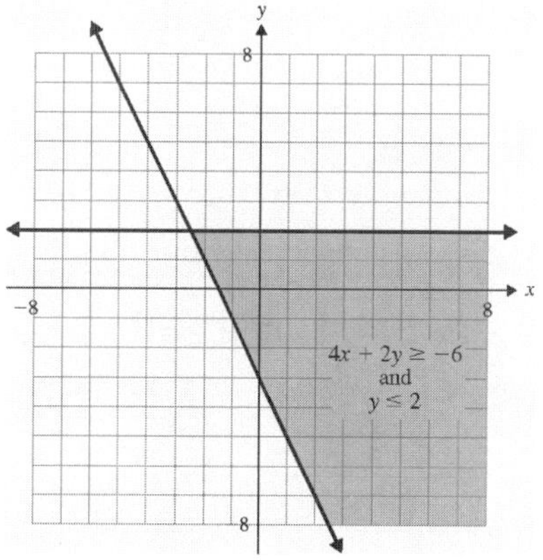

53) $\left[\begin{array}{cc|c} 1 & -1 & -11 \\ 2 & 9 & 0 \end{array}\right] \xrightarrow{-2R_1+R_2 \to R_2} \left[\begin{array}{cc|c} 1 & -1 & -11 \\ 0 & 11 & 22 \end{array}\right] \xrightarrow{\frac{1}{11}R_2 \to R_2} \left[\begin{array}{cc|c} 1 & -1 & -11 \\ 0 & 1 & 2 \end{array}\right]$

$y=2 \quad x-y=-11$

$x-2=-11$

$x=-9 \quad (-9,2)$

55) $\left[\begin{array}{cc|c} 5 & 3 & 5 \\ -1 & 8 & -1 \end{array}\right] \xrightarrow{R_1 \leftrightarrow R_1} \left[\begin{array}{cc|c} -1 & 8 & -1 \\ 5 & 3 & 5 \end{array}\right] \xrightarrow{-R_1 \to R_1} \left[\begin{array}{cc|c} 1 & -8 & 1 \\ 5 & 3 & 5 \end{array}\right]$

$\xrightarrow{-5R_1+R_2 \to R_2} \left[\begin{array}{cc|c} 1 & -8 & 1 \\ 0 & 43 & 0 \end{array}\right] \xrightarrow{\frac{1}{43}R_2 \to R_2} \left[\begin{array}{cc|c} 1 & -8 & 1 \\ 0 & 1 & 0 \end{array}\right]$

$y=0 \quad x-8y=1$

$x-8(0)=1$

$x=1 \quad (1,0)$

57) $\left[\begin{array}{ccc|c} 1 & -3 & -3 & -7 \\ 2 & -5 & -3 & 2 \\ -3 & 5 & 4 & -1 \end{array}\right] \xrightarrow[3R_1+R_3 \to R_3]{-2R_1+R_2 \to R_2} \left[\begin{array}{ccc|c} 1 & -3 & -3 & -7 \\ 0 & 1 & 3 & 16 \\ 0 & -4 & -5 & -22 \end{array}\right] \xrightarrow{4R_2+R_3 \to R_3}$

$\left[\begin{array}{ccc|c} 1 & -3 & -3 & -7 \\ 0 & 1 & 3 & 16 \\ 0 & 0 & 7 & 42 \end{array}\right] \xrightarrow{\frac{1}{7}R_3 \to R_3} \left[\begin{array}{ccc|c} 1 & -3 & -3 & -7 \\ 0 & 1 & 3 & 16 \\ 0 & 0 & 1 & 6 \end{array}\right]$

$x-3y-3z=-7 \quad y+3z=16 \quad x-3y-3z=-7$

$y+3z=16 \quad y+3(6)=16 \quad x-3(-2)-3(6)=-7$

$z=6 \quad y+18=16 \quad x+6-18=-7$

$y=-2 \quad x-12=-7$

$x=5 \quad (5,-2,6)$

Chapter 9 Test

1) $|4y-9|=11$

$4y-9=11$ or $4y-9=-11$

$4y=20 \qquad 4y=-2$

$y=5$ or $y=-\frac{1}{2}$

$\left\{-\frac{1}{2},5\right\}$

3) $|3k+5|=|k-11|$

$3k+5=k-11$ or $3k+5=-k+11$

$2k=-16 \qquad 4k=6$

$k=-8$ or $k=\frac{3}{2}$

$\left\{-8,\frac{3}{2}\right\}$

5) $|x| = 8$

7)
$$|2z-7| \le 9$$
$$-9 \le 2z-7 \le 9$$
$$-2 \le 2z \le 16$$
$$-1 \le z \le 8 \qquad [-1,8]$$

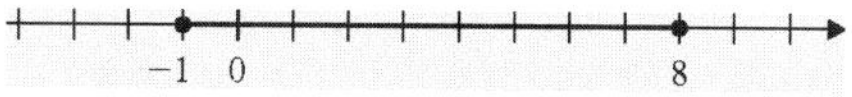

9)
$$|w-168| \le 0.75$$
$$167.25 \le w \le 168.75$$
Thanh's weight is between 167.25 lb and 168.75 lb.

11)

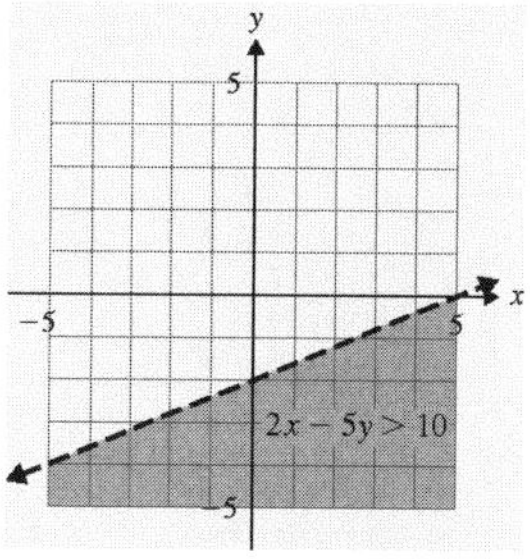

13)

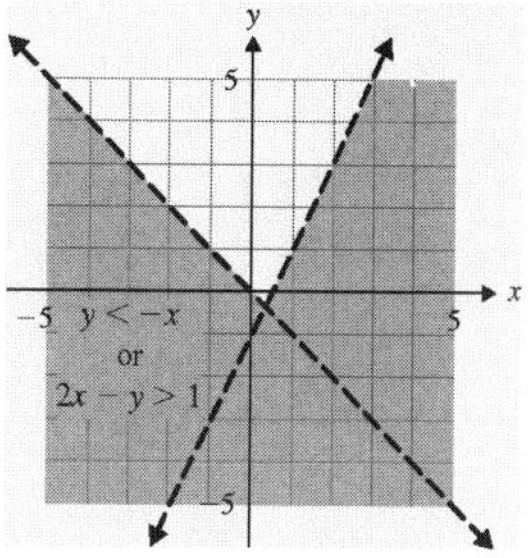

15)
$$\left[\begin{array}{ccc|c} -3 & 5 & 8 & 0 \\ 1 & -3 & 4 & 8 \\ 2 & -4 & -3 & 3 \end{array}\right] \xrightarrow{R_1 \leftrightarrow R_2} \left[\begin{array}{ccc|c} 1 & -3 & 4 & 8 \\ -3 & 5 & 8 & 0 \\ 2 & -4 & -3 & 3 \end{array}\right] \xrightarrow[-2R_1+R_3 \to R_3]{3R_1+R_2 \to R_2} \left[\begin{array}{ccc|c} 1 & -3 & 4 & 8 \\ 0 & -4 & 20 & 24 \\ 0 & 2 & -11 & -13 \end{array}\right]$$

$$\xrightarrow{-\frac{1}{4}R_2 \to R_2} \left[\begin{array}{ccc|c} 1 & -3 & 4 & 8 \\ 0 & 1 & -5 & -6 \\ 0 & 2 & -11 & -13 \end{array}\right] \xrightarrow{-2R_2+R_3 \to R_3} \left[\begin{array}{ccc|c} 1 & -3 & 4 & 8 \\ 0 & 1 & -5 & -6 \\ 0 & 0 & -1 & -1 \end{array}\right] \xrightarrow{-R_3 \to R_3} \left[\begin{array}{ccc|c} 1 & -3 & 4 & 8 \\ 0 & 1 & -5 & -6 \\ 0 & 0 & 1 & 1 \end{array}\right]$$

$$x-3y+4z=8 \qquad y-5z=-6 \qquad x-3y+4z=8$$
$$y-5z=-6 \qquad y-5(1)=-6 \qquad x-3(-1)+4(1)=8$$
$$z=1 \qquad y-5=-6 \qquad x+3+4=8$$
$$y=-1 \qquad x+7=8$$
$$x=1 \quad (1,-1,1)$$

Cumulative Review: Chapters 1-9

1) $5\times6-36\div3^2 = 5\times6-36\div9$
$$= 30-4 = 26$$

3) $3^4 = 81$

5) $\left(\dfrac{1}{8}\right)^2 = \dfrac{1}{64}$

7) $0.00000914 = 9.14\times10^{-6}$

9)
$$3-\frac{2}{7}n \ge 9$$
$$-\frac{2}{7}n \ge 6$$
$$n \le -\frac{7}{2}\cdot6$$
$$n \le -21 \qquad (-\infty,-21]$$

11) $m=\dfrac{y_2-y_1}{x_2-x_1}=\dfrac{2-4}{4-(-6)}=\dfrac{-2}{10}$

$=-\dfrac{1}{5}$

13) $y=mx+b$

$2=\left(\dfrac{1}{3}\right)7+b$

$2=\dfrac{7}{3}+b$

$-\dfrac{1}{3}=b \qquad y=\dfrac{1}{3}x-\dfrac{1}{3}$

15) $-4p^2\left(3p^2-7p-1\right)$

$=-12p^4+28p^3+4p^2$

17) $(t+8)^2=t^2+16t+64$

19) $9m^2-121=(3m+11)(3m-11)$

21) $a^2+6a+9=0$

$(a+3)^2=0$

$a=-3 \qquad \{-3\}$

23) $\dfrac{1}{r^2-25}-\dfrac{r+3}{2r+10}$

$=\dfrac{1}{(r+5)(r-5)}-\dfrac{r+3}{2(r+5)}$

$=\dfrac{2-(r-5)(r+3)}{2(r+5)(r-5)}$

$=\dfrac{2-\left(r^2-2r-15\right)}{2(r+5)(r-5)}$

$=\dfrac{-r^2+2r+17}{2(r+5)(r-5)}$

25) $\left|\dfrac{1}{4}q-7\right|-8=-5$

$\left|\dfrac{1}{4}q-7\right|=3$

$\dfrac{1}{4}q-7=3$ or $\dfrac{1}{4}q-7=-3$

$\dfrac{1}{4}q=10 \qquad \dfrac{1}{4}q=4$

$q=40$ or $q=16 \quad \{16,40\}$

27)

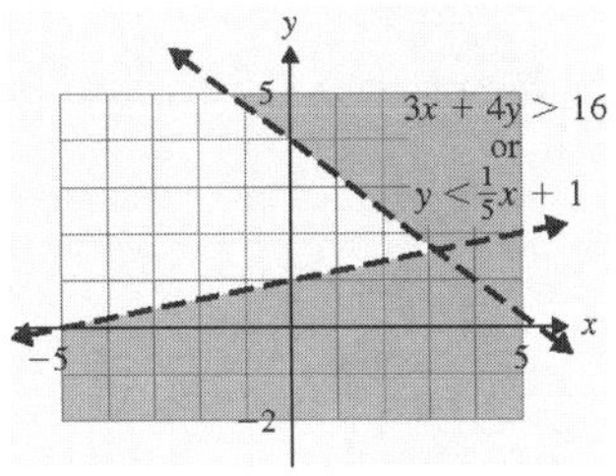

Chapter 10: Radicals and Rational Exponents

Section 10.1: Exercises

1) False; the $\sqrt{\ }$ symbol means to find only the positive square root of 121. $\sqrt{121}=11$

3) True

5) False; the even root of a negative number is not a real number.

7) 7 and -7

9) 1 and -1

11) 20 and -20

13) 50 and -50

15) $\frac{2}{3}$ and $-\frac{2}{3}$

17) $\frac{1}{9}$ and $-\frac{1}{9}$

19) $\sqrt{49}=7$

21) $\sqrt{1}=1$

23) $\sqrt{169}=13$

25) $\sqrt{-4}$ is not real

27) $\sqrt{\frac{81}{25}}=\frac{9}{5}$

29) $\sqrt{\frac{49}{64}}=\frac{7}{8}$

31) $-\sqrt{36}=-6$

33) $-\sqrt{\frac{1}{121}}=-\frac{1}{11}$

35) Since 11 is between 9 and 16,

$$\sqrt{9}<\sqrt{11}<\sqrt{16}$$
$$3<\sqrt{11}<4$$
$$\sqrt{11}\approx 3.3$$

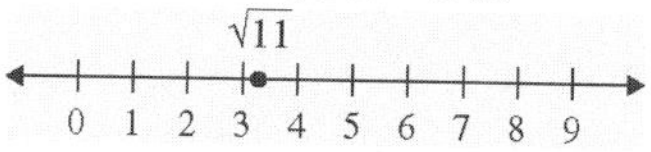

37) Since 46 is between 36 and 49,

$$\sqrt{36}<\sqrt{46}<\sqrt{49}$$
$$6<\sqrt{46}<7$$
$$\sqrt{46}\approx 6.8$$

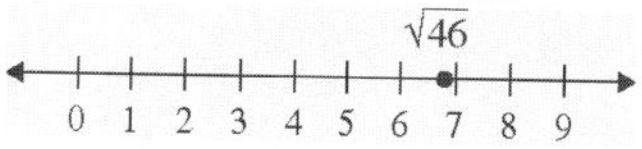

39) Since 17 is between 16 and 25,

$$\sqrt{16}<\sqrt{17}<\sqrt{25}$$
$$4<\sqrt{17}<5$$
$$\sqrt{17}\approx 4.1$$

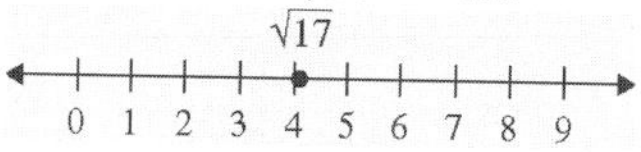

41) Since 5 is between 4 and 9,

$$\sqrt{4}<\sqrt{5}<\sqrt{9}$$
$$2<\sqrt{5}<3$$
$$\sqrt{5}\approx 2.2$$

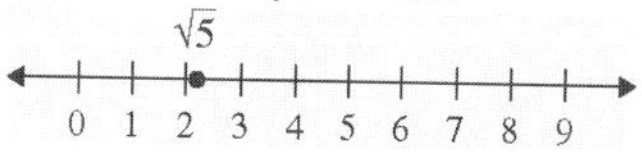

43) Since 61 is between 49 and 64,

$$\sqrt{49}<\sqrt{61}<\sqrt{64}$$
$$7<\sqrt{61}<8$$
$$\sqrt{61}\approx 7.8$$

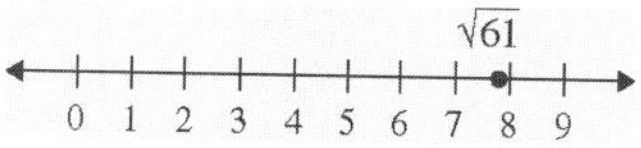

45) $\sqrt[3]{64}$ is the number you cube to get 64. $\sqrt[3]{64}=4$

47) No; the even root of a negative number is not a real number.

49) $\sqrt[3]{8}=2$

51) $\sqrt[3]{125}=5$

53) $\sqrt[3]{-1}=-1$

55) $\sqrt[4]{81}=3$

57) not real

59) $-\sqrt[4]{16}=-2$

61) $\sqrt[5]{-32}=-2$

63) $-\sqrt[3]{-27}=-(-3)=3$

65) not real

67) $\sqrt[3]{\dfrac{8}{125}}=\dfrac{2}{5}$

69) $\sqrt{60-11}=\sqrt{49}=7$

71) $\sqrt[3]{100+25}=\sqrt[3]{125}=5$

73) $\sqrt{1-9}=\sqrt{-8}$; not real

75) $\sqrt{5^2+12^2}=\sqrt{25+144}=\sqrt{169}=13$

Section 10.2: Exercises

1) The denominator of 2 becomes the index of the radical. $25^{1/2}=\sqrt{25}$

3) $9^{1/2}=\sqrt{9}=3$

5) $1000^{1/3}=\sqrt[3]{1000}=10$

7) $32^{1/5}=\sqrt[5]{32}=2$

9) $-125^{1/3}=-\sqrt[3]{125}=-5$

11) $\left(\dfrac{4}{121}\right)^{1/2}=\sqrt{\dfrac{4}{121}}=\dfrac{2}{11}$

13) $\left(\dfrac{125}{64}\right)^{1/3}=\sqrt[3]{\dfrac{125}{64}}=\dfrac{5}{4}$

15) $-\left(\dfrac{36}{169}\right)^{1/2}=-\sqrt{\dfrac{36}{169}}=-\dfrac{6}{13}$

17) The denominator of 4 becomes the index of the radical. The numerator of 3 is the power to which we raise the radical expression.
$16^{3/4}=\left(\sqrt[4]{16}\right)^3$

19) $8^{4/3}=\left(8^{1/3}\right)^4=\left(\sqrt[3]{8}\right)^4=2^4=16$

21) $125^{2/3}=\left(125^{1/3}\right)^2=\left(\sqrt[3]{125}\right)^2$
$=5^2=25$

23) $64^{5/6}=\left(64^{1/6}\right)^5=\left(\sqrt[6]{64}\right)^5=2^5=32$

25) $-27^{4/3}=-\left(\sqrt[3]{27}\right)^4=-(3)^4=-81$

27) $\left(\dfrac{16}{81}\right)^{3/4}=\left(\sqrt[4]{\dfrac{16}{81}}\right)^3=\left(\dfrac{2}{3}\right)^3=\dfrac{8}{27}$

29) $-\left(\dfrac{1000}{27}\right)^{2/3}=-\left(\sqrt[3]{\dfrac{1000}{27}}\right)^2$
$=-\left(\dfrac{10}{3}\right)^2=-\dfrac{100}{9}$

31) False; the negative exponent does not make the result negative.
$81^{-1/2}=\dfrac{1}{9}$

33) $49^{-1/2}=\left(\dfrac{1}{49}\right)^{1/2}=\sqrt{\dfrac{1}{49}}=\dfrac{1}{7}$

35) $1000^{-1/3}=\left(\dfrac{1}{1000}\right)^{1/3}=\sqrt[3]{\dfrac{1}{1000}}=\dfrac{1}{10}$

37) $\left(\frac{1}{81}\right)^{-1/4}=(81)^{1/4}=\sqrt[4]{81}=3$

39) $-\left(\frac{1}{64}\right)^{-1/3}=-(64)^{1/3}=-\sqrt[3]{64}=-4$

41) $64^{-5/6}=\left(\frac{1}{64}\right)^{5/6}=\left(\sqrt[6]{\frac{1}{64}}\right)^{5}$
$=\left(\frac{1}{2}\right)^{5}=\frac{1}{32}$

43) $125^{-2/3}=\left(\frac{1}{125}\right)^{2/3}=\left(\sqrt[3]{\frac{1}{125}}\right)^{2}$
$=\left(\frac{1}{5}\right)^{2}=\frac{1}{25}$

45) $\left(\frac{25}{4}\right)^{-3/2}=\left(\frac{4}{25}\right)^{3/2}=\left(\sqrt{\frac{4}{25}}\right)^{3}$
$=\left(\frac{2}{5}\right)^{3}=\frac{8}{125}$

47) $\left(\frac{64}{125}\right)^{-2/3}=\left(\frac{125}{64}\right)^{2/3}=\left(\sqrt[3]{\frac{125}{64}}\right)^{2}$
$=\left(\frac{5}{4}\right)^{2}=\frac{25}{16}$

49) $2^{2/3}\cdot 2^{7/3}=2^{2/3+7/3}=2^{9/3}=2^{3}=8$

51) $\left(9^{1/4}\right)^{2}=9^{\frac{1}{4}\cdot 2}=9^{1/2}=\sqrt{9}=3$

53) $8^{7/5}\cdot 8^{-3/5}=8^{7/5+(-3/5)}=8^{4/5}$

55) $\frac{4^{10/3}}{4^{4/3}}=4^{10/3-4/3}=4^{6/3}=4^{2}=16$

57) $\frac{5^{3/2}}{5^{9/2}}=5^{3/2-9/2}=5^{-6/2}=5^{-3}$
$=\left(\frac{1}{5}\right)^{3}=\frac{1}{125}$

59) $\frac{6^{-1}}{6^{1/2}\cdot 6^{-5/2}}=\frac{6^{-1}}{6^{1/2+(-5/2)}}=\frac{6^{-1}}{6^{-4/2}}=\frac{6^{-1}}{6^{-2}}$
$=6^{-1-(-2)}=6^{-1+2}=6^{1}=6$

61) $\frac{7^{4/9}\cdot 7^{1/9}}{7^{2/9}}=\frac{7^{4/9+1/9}}{7^{2/9}}=\frac{7^{5/9}}{7^{2/9}}$
$=7^{5/9-2/9}=7^{3/9}=7^{1/3}$

63) $k^{7/4}\cdot k^{3/4}=k^{7/4+3/4}=k^{10/4}=k^{5/2}$

65) $j^{-3/5}\cdot j^{3/10}=j^{-3/5+3/10}=j^{-6/10+3/10}$
$=j^{-3/10}=\frac{1}{j^{3/10}}$

67) $\left(-9v^{5/8}\right)\left(8v^{3/4}\right)=-72v^{5/8+3/4}$
$=-72v^{5/8+6/8}$
$=-72v^{11/8}$

69) $\frac{a^{5/9}}{a^{4/9}}=a^{5/9-4/9}=a^{1/9}$

71) $\frac{20c^{-2/3}}{72c^{5/6}}=\frac{5}{18}c^{-2/3-5/6}=\frac{5}{18}c^{-4/6-5/6}$
$=\frac{5}{18}c^{-9/6}=\frac{5}{18}c^{-3/2}=\frac{5}{18c^{3/2}}$

73) $\left(q^{4/5}\right)^{10}=q^{\frac{4}{5}\cdot 10}=q^{8}$

75) $\left(x^{-2/9}\right)^{3}=x^{-\frac{2}{9}\cdot 3}=x^{-2/3}=\frac{1}{x^{2/3}}$

77) $\left(z^{1/5}\right)^{2/3}=z^{\frac{1}{5}\cdot\frac{2}{3}}=z^{2/15}$

79) $\left(81u^{8/3}v^4\right)^{3/4}$

$=81^{3/4}\cdot\left(u^{8/3}\right)^{3/4}\cdot\left(v^4\right)^{3/4}$

$=\left(\sqrt[4]{81}\right)^3\cdot u^2\cdot v^3$

$=(3)^3\cdot u^2v^3=27u^2v^3$

81) $\left(32r^{1/3}s^{4/9}\right)^{3/5}$

$=32^{3/5}\cdot\left(r^{1/3}\right)^{3/5}\cdot\left(s^{4/9}\right)^{3/5}$

$=\left(\sqrt[5]{32}\right)^3\cdot r^{1/5}\cdot s^{4/15}$

$=(2)^3\cdot r^{1/5}s^{4/15}=8r^{1/5}s^{4/15}$

83) $\left(\dfrac{f^{6/7}}{27g^{-5/3}}\right)^{1/3}=\dfrac{\left(f^{6/7}\right)^{1/3}}{(27)^{1/3}\left(g^{-5/3}\right)^{1/3}}$

$=\dfrac{f^{2/7}}{3g^{-5/9}}=\dfrac{f^{2/7}g^{5/9}}{3}$

85) $\left(\dfrac{x^{-5/3}}{w^{3/2}}\right)^{-6}=\left(\dfrac{w^{3/2}}{x^{-5/3}}\right)^{6}=\dfrac{\left(w^{3/2}\right)^6}{\left(x^{-5/3}\right)^6}$

$=\dfrac{w^9}{x^{-10}}=x^{10}w^9$

87) $\dfrac{y^{1/2}\cdot y^{-1/3}}{y^{5/6}}=\dfrac{y^{3/6}\cdot y^{-2/6}}{y^{5/6}}=\dfrac{y^{1/6}}{y^{5/6}}$

$=y^{1/6-5/6}=y^{-4/6}$

$=y^{-2/3}=\dfrac{1}{y^{2/3}}$

89) $\left(\dfrac{a^4b^3}{32a^{-2}b^4}\right)^{2/5}$

$=\left(\dfrac{1}{32}a^{4-(-2)}b^{3-4}\right)^{2/5}$

$=\left(\dfrac{1}{32}a^6b^{-1}\right)^{2/5}$

$=\left(\dfrac{1}{32}\right)^{2/5}\cdot\left(a^6\right)^{2/5}\cdot\left(b^{-1}\right)^{2/5}$

$=\left(\sqrt[5]{\dfrac{1}{32}}\right)^2\cdot a^{12/5}\cdot b^{-2/5}$

$=\left(\dfrac{1}{2}\right)^2\cdot a^{12/5}b^{-2/5}=\dfrac{a^{12/5}}{4b^{2/5}}$

91) $\sqrt[6]{49^3}=\left(49^3\right)^{1/6}=49^{1/2}=\sqrt{49}=7$

93) $\sqrt[6]{1000^2}=\left(1000^2\right)^{1/6}=1000^{1/3}$

$=\sqrt[3]{1000}=10$

95) $\left(\sqrt{5}\right)^2=5$ 97) $\left(\sqrt{3}\right)^2=3$

99) $\left(\sqrt[3]{12}\right)^3=12$ 101) $\left(\sqrt[4]{15}\right)^4=15$

103) $\left(\sqrt[3]{x^{12}}\right)=x^{12/3}=x^4$

105) $\left(\sqrt[6]{k^2}\right)=k^{2/6}=k^{1/3}=\sqrt[3]{k}$

107) $\left(\sqrt[4]{z^2}\right)=z^{2/4}=z^{1/2}=\sqrt{z}$

109) $\sqrt{d^4}=d^{4/2}=d^2$

Section 10.3: Exercises

1) $\sqrt{3}\cdot\sqrt{7}=\sqrt{3\cdot 7}=\sqrt{21}$

3) $\sqrt{10}\cdot\sqrt{3}=\sqrt{10\cdot 3}=\sqrt{30}$

5) $\sqrt{6}\cdot\sqrt{y}=\sqrt{6\cdot y}=\sqrt{6y}$

7) False; 20 contains the factor 4 which is a perfect square.

9) True; 42 does not have any factors (other than 1) that are perfect squares.

11) $\sqrt{20}=\sqrt{4\cdot 5}=\sqrt{4}\cdot\sqrt{5}=2\sqrt{5}$

13) $\sqrt{54}=\sqrt{9\cdot 6}=\sqrt{9}\cdot\sqrt{6}=3\sqrt{6}$

15) $\sqrt{33}$; simplified

17) $\sqrt{8}=\sqrt{4\cdot 2}=\sqrt{4}\cdot\sqrt{2}=2\sqrt{2}$

19) $\sqrt{80}=\sqrt{16\cdot 5}=\sqrt{16}\cdot\sqrt{5}=4\sqrt{5}$

21) $\sqrt{98}=\sqrt{49\cdot 2}=\sqrt{49}\cdot\sqrt{2}=7\sqrt{2}$

23) $\sqrt{38}$; simplified

25) $\sqrt{400}=20$

27) $\sqrt{\frac{144}{25}}=\frac{\sqrt{144}}{\sqrt{25}}=\frac{12}{5}$

29) $\sqrt{\frac{4}{49}}=\frac{\sqrt{4}}{\sqrt{49}}=\frac{2}{7}$

31) $\sqrt{\frac{8}{2}}=\sqrt{4}=2$

33) $\frac{\sqrt{54}}{\sqrt{6}}=\sqrt{\frac{54}{6}}=\sqrt{9}=3$

35) $\sqrt{\frac{60}{5}}=\sqrt{12}=\sqrt{4\cdot 3}=\sqrt{4}\cdot\sqrt{3}=2\sqrt{3}$

37) $\frac{\sqrt{120}}{\sqrt{6}}=\sqrt{\frac{120}{6}}=\sqrt{20}=\sqrt{4\cdot 5}$
$=\sqrt{4}\cdot\sqrt{5}=2\sqrt{5}$

39) $\frac{\sqrt{30}}{\sqrt{2}}=\sqrt{\frac{30}{2}}=\sqrt{15}$

41) $\sqrt{\frac{6}{49}}=\frac{\sqrt{6}}{\sqrt{49}}=\frac{\sqrt{6}}{7}$

43) $\sqrt{\frac{45}{16}}=\frac{\sqrt{45}}{\sqrt{16}}=\frac{\sqrt{9\cdot 5}}{4}=\frac{\sqrt{9}\cdot\sqrt{5}}{4}$
$=\frac{3\sqrt{5}}{4}$

45) $\sqrt{x^8}=x^{8/2}=x^4$

47) $\sqrt{m^{10}}=m^{10/2}=m^5$

49) $\sqrt{w^{14}}=w^{14/2}=w^7$

51) $\sqrt{100c^2}=\sqrt{100}\cdot\sqrt{c^2}=10c^{2/2}=10c$

53) $\sqrt{64k^6}=\sqrt{64}\cdot\sqrt{k^6}=8k^{6/2}=8k^3$

55) $\sqrt{28r^4}=\sqrt{28}\cdot\sqrt{r^4}=\sqrt{4}\cdot\sqrt{7}\cdot r^{4/2}$
$=2\sqrt{7}\cdot r^2=2r^2\sqrt{7}$

57) $\sqrt{300q^{22}}=\sqrt{300}\cdot\sqrt{q^{22}}=\sqrt{100}\cdot\sqrt{3}\cdot q^{22/2}$
$=10\sqrt{3}\cdot q^{11}=10q^{11}\sqrt{3}$

59) $\sqrt{\frac{81}{c^6}}=\frac{\sqrt{81}}{\sqrt{c^6}}=\frac{9}{c^{6/2}}=\frac{9}{c^3}$

61) $\frac{\sqrt{40}}{\sqrt{t^8}}=\frac{\sqrt{4}\cdot\sqrt{10}}{t^{8/2}}=\frac{2\sqrt{10}}{t^4}$

63) $\sqrt{\frac{75}{y^{12}}}=\frac{\sqrt{75}}{\sqrt{y^{12}}}=\frac{\sqrt{25}\cdot\sqrt{3}}{y^{12/2}}=\frac{5\sqrt{3}}{y^{6}}$

65) $\sqrt{a^5}=\sqrt{a^4}\cdot\sqrt{a}=a^{4/2}\cdot\sqrt{a}=a^2\sqrt{a}$

67) $\sqrt{g^{13}}=\sqrt{g^{12}}\cdot\sqrt{g}=g^6\sqrt{g}$

69) $\sqrt{b^{25}}=\sqrt{b^{24}}\cdot\sqrt{b}=b^{12}\sqrt{b}$

71) $\sqrt{72x^3}=\sqrt{72}\cdot\sqrt{x^3}$
$=\sqrt{36}\cdot\sqrt{2}\cdot\sqrt{x^2}\cdot\sqrt{x}$
$=6\sqrt{2}\cdot x\sqrt{x}=6x\sqrt{2x}$

73) $\sqrt{13q^7}=\sqrt{13}\cdot\sqrt{q^7}$
$=\sqrt{13}\cdot\sqrt{q^6}\cdot\sqrt{q}$
$=\sqrt{13}\cdot q^3\sqrt{q}=q^3\sqrt{13q}$

75) $\sqrt{75t^{11}}=\sqrt{75}\cdot\sqrt{t^{11}}$
$=\sqrt{25}\cdot\sqrt{3}\cdot\sqrt{t^{10}}\cdot\sqrt{t}$
$=5\sqrt{3}\cdot t^5\sqrt{t}=5t^5\sqrt{3t}$

77) $\sqrt{c^8d^2}=\sqrt{c^8}\cdot\sqrt{d^2}=c^4d$

79) $\sqrt{a^4b^3}=\sqrt{a^4}\cdot\sqrt{b^3}=a^2\cdot\sqrt{b^2}\cdot\sqrt{b}$
$=a^2b\sqrt{b}$

81) $\sqrt{u^5v^7}=\sqrt{u^5}\cdot\sqrt{v^7}$
$=\sqrt{u^4}\cdot\sqrt{u}\cdot\sqrt{v^6}\cdot\sqrt{v}$
$=u^2\sqrt{u}\cdot v^3\sqrt{v}=u^2v^3\sqrt{uv}$

83) $\sqrt{36m^9n^4}=\sqrt{36}\cdot\sqrt{m^9}\cdot\sqrt{n^4}$
$=6\cdot\sqrt{m^8}\cdot\sqrt{m}\cdot n^2$
$=6m^4n^2\sqrt{m}$

85) $\sqrt{44x^{12}y^5}=\sqrt{44}\cdot\sqrt{x^{12}}\cdot\sqrt{y^5}$
$=\sqrt{4}\cdot\sqrt{11}\cdot x^6\cdot\sqrt{y^4}\cdot\sqrt{y}$
$=2\sqrt{11}\cdot x^6\cdot y^2\sqrt{y}$
$=2x^6y^2\sqrt{11y}$

87) $\sqrt{32t^5u^7}$
$=\sqrt{32}\cdot\sqrt{t^5}\cdot\sqrt{u^7}$
$=\sqrt{16}\cdot\sqrt{2}\cdot\sqrt{t^4}\cdot\sqrt{t}\cdot\sqrt{u^6}\cdot\sqrt{u}$
$=4\sqrt{2}\cdot t^2\sqrt{t}\cdot u^3\sqrt{u}=4t^2u^3\sqrt{2tu}$

89) $\sqrt{\frac{a^7}{81b^6}}=\frac{\sqrt{a^6}\cdot\sqrt{a}}{\sqrt{81b^6}}=\frac{a^3\sqrt{a}}{9b^3}$

91) $\sqrt{\frac{3r^9}{s^2}}=\frac{\sqrt{3r^9}}{\sqrt{s^2}}=\frac{\sqrt{3}\cdot\sqrt{r^9}}{s}$
$=\frac{\sqrt{3}\cdot\sqrt{r^8}\cdot\sqrt{r}}{s}=\frac{r^4\sqrt{3r}}{s}$

93) $\sqrt{5}\cdot\sqrt{10}=\sqrt{50}=\sqrt{25}\cdot\sqrt{2}=5\sqrt{2}$

95) $\sqrt{21}\cdot\sqrt{3}=\sqrt{63}=\sqrt{9}\cdot\sqrt{7}=3\sqrt{7}$

97) $\sqrt{w}\cdot\sqrt{w^5}=\sqrt{w^6}=w^3$

99) $\sqrt{n^3}\cdot\sqrt{n^4}=\sqrt{n^2}\cdot\sqrt{n}\cdot n^2$
$=n\cdot\sqrt{n}\cdot n^2=n^3\sqrt{n}$

101) $\sqrt{2k}\cdot\sqrt{8k^5}=\sqrt{16k^6}=\sqrt{16}\cdot\sqrt{k^6}$
$=4k^3$

103) $\sqrt{6x^4y^3}\cdot\sqrt{2x^5y^2}$
$=\sqrt{12x^9y^5}=\sqrt{12}\cdot\sqrt{x^9}\cdot\sqrt{y^5}$
$=\sqrt{4}\cdot\sqrt{3}\cdot\sqrt{x^8}\cdot\sqrt{x}\cdot\sqrt{y^4}\cdot\sqrt{y}$
$=2\sqrt{3}\cdot x^4\sqrt{x}\cdot y^2\sqrt{y}=2x^4y^2\sqrt{3xy}$

105) $\sqrt{8c^9d^2}\cdot\sqrt{5cd^7}$
$=\sqrt{40c^{10}d^9}=\sqrt{40}\cdot\sqrt{c^{10}}\cdot\sqrt{d^9}$
$=\sqrt{4}\cdot\sqrt{10}\cdot c^5\cdot\sqrt{d^8}\cdot\sqrt{d}$
$=2\sqrt{10}\cdot c^5\cdot d^4\sqrt{d}=2c^5d^4\sqrt{10d}$

107) $\dfrac{\sqrt{18k^{11}}}{\sqrt{2k^3}}=\sqrt{\dfrac{18k^{11}}{2k^3}}=\sqrt{9k^8}$
$=\sqrt{9}\cdot\sqrt{k^8}=3k^4$

109) $\dfrac{\sqrt{120h^8}}{\sqrt{3h^2}}=\sqrt{\dfrac{120h^8}{3h^2}}=\sqrt{40h^6}$
$=\sqrt{4}\cdot\sqrt{10}\cdot\sqrt{h^6}$
$=2\sqrt{10}\cdot h^3=2h^3\sqrt{10}$

111) $\dfrac{\sqrt{50a^{16}}}{\sqrt{5a^7}}=\sqrt{\dfrac{50a^{16}}{5a^7}}=\sqrt{10a^9}$
$=\sqrt{10}\cdot\sqrt{a^8}\cdot\sqrt{a}$
$=\sqrt{10}\cdot a^4\sqrt{a}=a^4\sqrt{10a}$

Section 10.4: Exercises

1) To multiply radicals with the same indices, multiply the radicands and put the product under a radical with the same index.

3) i) Its radicand will not contain any factors that are perfect cubes.
ii) There will be no radical in the denominator of a fraction.
iii) The radicand will not contain fractions.

5) $\sqrt[3]{5}\cdot\sqrt[3]{4}=\sqrt[3]{20}$

7) $\sqrt[5]{9}\cdot\sqrt[5]{m^2}=\sqrt[5]{9m^2}$

9) $\sqrt[3]{a^2}\cdot\sqrt[3]{b}=\sqrt[3]{a^2b}$

11) $\sqrt[3]{24}=\sqrt[3]{8}\cdot\sqrt[3]{3}=2\sqrt[3]{3}$

13) $\sqrt[4]{64}=\sqrt[4]{16}\cdot\sqrt[4]{4}=2\sqrt[4]{4}$

15) $\sqrt[3]{54}=\sqrt[3]{27}\cdot\sqrt[3]{2}=3\sqrt[3]{2}$

17) $\sqrt[3]{2000}=\sqrt[3]{1000}\cdot\sqrt[3]{2}=10\sqrt[3]{2}$

19) $\sqrt[5]{64}=\sqrt[5]{32}\cdot\sqrt[5]{2}=2\sqrt[5]{2}$

21) $\sqrt[4]{\dfrac{1}{16}}=\dfrac{\sqrt[4]{1}}{\sqrt[4]{16}}=\dfrac{1}{2}$

23) $\sqrt[3]{-\dfrac{54}{2}}=\sqrt[3]{-27}=-3$

25) $\dfrac{\sqrt[3]{48}}{\sqrt[3]{2}}=\sqrt[3]{\dfrac{48}{2}}=\sqrt[3]{24}=\sqrt[3]{8}\cdot\sqrt[3]{3}=2\sqrt[3]{3}$

27) $\dfrac{\sqrt[4]{240}}{\sqrt[4]{3}}=\sqrt[4]{\dfrac{240}{3}}=\sqrt[4]{80}$
$=\sqrt[4]{16}\cdot\sqrt[4]{5}=2\sqrt[4]{5}$

29) $\sqrt[3]{d^6}=d^{6/3}=d^2$

31) $\sqrt[4]{n^{20}}=n^{20/4}=n^5$

33) $\sqrt[5]{x^5y^{15}}=x^{5/5}y^{15/5}=xy^3$

35) $\sqrt[3]{w^{14}}=\sqrt[3]{w^{12}}\cdot\sqrt[3]{w^2}=w^{12/3}\cdot\sqrt[3]{w^2}$
$=w^4\sqrt[3]{w^2}$

37) $\sqrt[4]{y^9}=\sqrt[4]{y^8}\cdot\sqrt[4]{y}=y^{8/4}\cdot\sqrt[4]{y}=y^2\sqrt[4]{y}$

39) $\sqrt[3]{d^5}=\sqrt[3]{d^3}\cdot\sqrt[3]{d^2}=d\sqrt[3]{d^2}$

41) $\sqrt[3]{u^{10}v^{15}} = \sqrt[3]{u^{10}} \cdot \sqrt[3]{v^{15}}$
$= \sqrt[3]{u^9} \cdot \sqrt[3]{u} \cdot v^{15/3}$
$= u^{9/3}\sqrt[3]{u} \cdot v^5 = u^3v^5\sqrt[3]{u}$

43) $\sqrt[3]{b^{16}c^5} = \sqrt[3]{b^{16}} \cdot \sqrt[3]{c^5}$
$= \sqrt[3]{b^{15}} \cdot \sqrt[3]{b} \cdot \sqrt[3]{c^3} \cdot \sqrt[3]{c^2}$
$= b^{15/3} \cdot \sqrt[3]{b} \cdot c^{3/3} \cdot \sqrt[3]{c^2}$
$= b^5c\sqrt[3]{bc^2}$

45) $\sqrt[4]{m^3n^{18}} = \sqrt[4]{m^3} \cdot \sqrt[4]{n^{18}}$
$= \sqrt[4]{m^3} \cdot \sqrt[4]{n^{16}} \cdot \sqrt[4]{n^2}$
$= \sqrt[4]{m^3} \cdot n^{16/4} \cdot \sqrt[4]{n^2}$
$= n^4\sqrt[4]{m^3n^2}$

47) $\sqrt[3]{24x^{10}y^{12}} = \sqrt[3]{24} \cdot \sqrt[3]{x^{10}} \cdot \sqrt[3]{y^{12}}$
$= \sqrt[3]{8} \cdot \sqrt[3]{3} \cdot \sqrt[3]{x^9} \cdot \sqrt[3]{x} \cdot y^{12/3}$
$= 2\sqrt[3]{3} \cdot x^{9/3}\sqrt[3]{x} \cdot y^4$
$= 2x^3y^4\sqrt[3]{3x}$

49) $\sqrt[3]{250w^4x^{16}}$
$= \sqrt[3]{250} \cdot \sqrt[3]{w^4} \cdot \sqrt[3]{x^{16}}$
$= \sqrt[3]{125} \cdot \sqrt[3]{2} \cdot \sqrt[3]{w^3} \cdot \sqrt[3]{w} \cdot \sqrt[3]{x^{15}} \cdot \sqrt[3]{x}$
$= 5\sqrt[3]{2} \cdot w^{3/3}\sqrt[3]{w} \cdot x^{15/3}\sqrt[3]{x}$
$= 5wx^5\sqrt[3]{2wx}$

51) $\sqrt[4]{\dfrac{m^8}{81}} = \dfrac{\sqrt[4]{m^8}}{\sqrt[4]{81}} = \dfrac{m^{8/4}}{3} = \dfrac{m^2}{3}$

53) $\sqrt[5]{\dfrac{32a^{23}}{b^{15}}} = \dfrac{\sqrt[5]{32a^{23}}}{\sqrt[5]{b^{15}}}$
$= \dfrac{\sqrt[5]{32} \cdot \sqrt[5]{a^{20}} \cdot \sqrt[5]{a^3}}{b^3}$
$= \dfrac{2a^4\sqrt[5]{a^3}}{b^3}$

55) $\sqrt[4]{\dfrac{t^9}{81s^{24}}} = \dfrac{\sqrt[4]{t^9}}{\sqrt[4]{81s^{24}}} = \dfrac{\sqrt[4]{t^8} \cdot \sqrt[4]{t}}{\sqrt[4]{81} \cdot \sqrt[4]{s^{24}}}$
$= \dfrac{t^2\sqrt[4]{t}}{3s^6}$

57) $\sqrt[3]{\dfrac{u^{28}}{v^3}} = \dfrac{\sqrt[3]{u^{28}}}{\sqrt[3]{v^3}} = \dfrac{\sqrt[3]{u^{27}} \cdot \sqrt[3]{u}}{v} = \dfrac{u^9\sqrt[3]{u}}{v}$

59) $\sqrt[3]{6} \cdot \sqrt[3]{4} = \sqrt[3]{24} = \sqrt[3]{8} \cdot \sqrt[3]{3} = 2\sqrt[3]{3}$

61) $\sqrt[3]{9} \cdot \sqrt[3]{12} = \sqrt[3]{9} \cdot \sqrt[3]{3} \cdot \sqrt[3]{4} = \sqrt[3]{27} \cdot \sqrt[3]{4}$
$= 3\sqrt[3]{4}$

63) $\sqrt[3]{20} \cdot \sqrt[3]{4} = \sqrt[3]{80} = \sqrt[3]{8} \cdot \sqrt[3]{10} = 2\sqrt[3]{10}$

65) $\sqrt[3]{m^4} \cdot \sqrt[3]{m^5} = \sqrt[3]{m^9} = m^3$

67) $\sqrt[4]{k^7} \cdot \sqrt[4]{k^9} = \sqrt[4]{k^{16}} = k^4$

69) $\sqrt[3]{r^7} \cdot \sqrt[3]{r^4} = \sqrt[3]{r^{11}} = \sqrt[3]{r^9} \cdot \sqrt[3]{r^2}$
$= r^3\sqrt[3]{r^2}$

71) $\sqrt[5]{p^{14}} \cdot \sqrt[5]{p^9} = \sqrt[5]{p^{23}} = \sqrt[5]{p^{20}} \cdot \sqrt[5]{p^3}$
$= p^4\sqrt[5]{p^3}$

73) $\sqrt[3]{9z^{11}} \cdot \sqrt[3]{3z^8} = \sqrt[3]{27z^{19}} = \sqrt[3]{27} \cdot \sqrt[3]{z^{19}}$
$= 3 \cdot \sqrt[3]{z^{18}} \cdot \sqrt[3]{z} = 3z^6\sqrt[3]{z}$

75) $\sqrt[3]{\dfrac{h^{14}}{h^2}} = \sqrt[3]{h^{12}} = h^4$

77) $\sqrt[3]{\dfrac{c^{11}}{c^4}} = \sqrt[3]{c^7} = \sqrt[3]{c^6} \cdot \sqrt[3]{c} = c^2\sqrt[3]{c}$

79) $\sqrt[4]{\dfrac{162d^{21}}{2d^2}} = \sqrt[4]{81d^{19}} = \sqrt[4]{81} \cdot \sqrt[4]{d^{19}}$

$= 3 \cdot \sqrt[4]{d^{16}} \cdot \sqrt[4]{d^3} = 3d^4\sqrt[4]{d^3}$

81) $\sqrt{p} \cdot \sqrt[3]{p} = p^{1/2} \cdot p^{1/3} = p^{3/6} \cdot p^{2/6}$

$= p^{5/6} = \sqrt[6]{p^5}$

83) $\sqrt[4]{n^3} \cdot \sqrt{n} = n^{3/4} \cdot n^{1/2} = n^{3/4} \cdot n^{2/4}$

$= n^{5/4} = n^{4/4} \cdot n^{1/4} = n\sqrt[4]{n}$

85) $\sqrt[5]{c^3} \cdot \sqrt[3]{c^2} = c^{3/5} \cdot c^{2/3} = c^{9/15} \cdot c^{10/15}$

$= c^{19/15} = c^{15/15} \cdot c^{4/15}$

$= c\sqrt[15]{c^4}$

87) $\dfrac{\sqrt{w}}{\sqrt[4]{w}} = \dfrac{w^{1/2}}{w^{1/4}} = \dfrac{w^{2/4}}{w^{1/4}} = w^{1/4} = \sqrt[4]{w}$

89) $\dfrac{\sqrt[5]{t^4}}{\sqrt[3]{t^2}} = \dfrac{t^{4/5}}{t^{2/3}} = \dfrac{t^{12/15}}{t^{10/15}} = t^{2/15} = \sqrt[15]{t^2}$

Section 10.5: Exercises

1) They have the same index and the same radicand.

3) $5\sqrt{2} + 9\sqrt{2} = 14\sqrt{2}$

5) $4\sqrt{3} - 9\sqrt{3} = -5\sqrt{3}$

7) $7\sqrt[3]{4} + 8\sqrt[3]{4} = 15\sqrt[3]{4}$

9) $6 - \sqrt{13} + 5 - 2\sqrt{3} = 11 - 3\sqrt{13}$

11) $15\sqrt[3]{z^2} - 20\sqrt[3]{z^2} = -5\sqrt[3]{z^2}$

13) $2\sqrt[3]{n^2} + 9\sqrt[5]{n^2} - 11\sqrt[3]{n^2} + \sqrt[5]{n^2}$

$= -9\sqrt[3]{n^2} + 10\sqrt[5]{n^2}$

15) $\sqrt{5c} - 8\sqrt{6c} + \sqrt{5c} + 6\sqrt{6c}$

$= 2\sqrt{5c} - 2\sqrt{6c}$

17) $6\sqrt{3} - \sqrt{12} = 6\sqrt{3} - 2\sqrt{3} = 4\sqrt{3}$

19) $\sqrt{48} + \sqrt{3} = 4\sqrt{3} + \sqrt{3} = 5\sqrt{3}$

21) $\sqrt{20} + 4\sqrt{45} = 2\sqrt{5} + 4\left(3\sqrt{5}\right)$

$= 2\sqrt{5} + 12\sqrt{5} = 14\sqrt{5}$

23) $3\sqrt{98} + 4\sqrt{50} = 3\left(7\sqrt{2}\right) + 4\left(5\sqrt{2}\right)$

$= 21\sqrt{2} + 20\sqrt{2}$

$= 41\sqrt{2}$

25) $\sqrt{32} - 3\sqrt{8} = 4\sqrt{2} - 3\left(2\sqrt{2}\right)$

$= 4\sqrt{2} - 6\sqrt{2} = -2\sqrt{2}$

27) $\sqrt{12} + \sqrt{75} - \sqrt{3}$

$= 2\sqrt{3} + 5\sqrt{3} - \sqrt{3} = 6\sqrt{3}$

29) $\sqrt{20} - 2\sqrt{45} - \sqrt{80}$

$= 2\sqrt{5} - 2\left(3\sqrt{5}\right) - 4\sqrt{5}$

$= 2\sqrt{5} - 6\sqrt{5} - 4\sqrt{5} = -8\sqrt{5}$

31) $8\sqrt[3]{9} + \sqrt[3]{72} = 8\sqrt[3]{9} + \sqrt[3]{8} \cdot \sqrt[3]{9}$

$= 8\sqrt[3]{9} + 2\sqrt[3]{9} = 10\sqrt[3]{9}$

33) $2\sqrt[3]{81} - 14\sqrt[3]{3} = 2\left(3\sqrt[3]{3}\right) - 14\sqrt[3]{3}$

$= 6\sqrt[3]{3} - 14\sqrt[3]{3} = -8\sqrt[3]{3}$

35) $\sqrt[3]{6}-\sqrt[3]{48}=\sqrt[3]{6}-\left(2\sqrt[3]{6}\right)=-\sqrt[3]{6}$

37) $6q\sqrt{q}+7\sqrt{q^3}=6q\sqrt{q}+7q\sqrt{q}$
$=13q\sqrt{q}$

39) $4d^2\sqrt{d}-24\sqrt{d^5}=4d^2\sqrt{d}-24d^2\sqrt{d}$
$=-20d^2\sqrt{d}$

41) $9\sqrt{n^5}-4n\sqrt{n^3}$
$=9\left(n^2\sqrt{n}\right)-4n\left(n\sqrt{n}\right)$
$=9n^2\sqrt{n}-4n^2\sqrt{n}=5n^2\sqrt{n}$

43) $9t^3\sqrt[3]{t}-5\sqrt[3]{t^{10}}=9t^3\sqrt[3]{t}-5t^3\sqrt[3]{t}$
$=4t^3\sqrt[3]{t}$

45) $5a\sqrt[4]{a^7}+\sqrt[4]{a^{11}}$
$=5a\left(a\sqrt[4]{a^3}\right)+\left(a^2\sqrt[4]{a^3}\right)$
$=5a^2\sqrt[4]{a^3}+a^2\sqrt[4]{a^3}=6a^2\sqrt[4]{a^3}$

47) $11\sqrt{5z}+2\sqrt{20z}=11\sqrt{5z}+2\left(2\sqrt{5z}\right)$
$=11\sqrt{5z}+4\sqrt{5z}=15\sqrt{5z}$

49) $2\sqrt{8p}-6\sqrt{2p}=2\left(2\sqrt{2p}\right)-6\sqrt{2p}$
$=4\sqrt{2p}-6\sqrt{2p}=-2\sqrt{2p}$

51) $7\sqrt[3]{81a^5}+4a\sqrt[3]{3a^2}$
$=7\left(3a\sqrt[3]{3a^2}\right)+4a\sqrt[3]{3a^2}$
$=21a\sqrt[3]{3a^2}+4a\sqrt[3]{3a^2}=25a\sqrt[3]{3a^2}$

53) $4c^2\sqrt[3]{108c}-15\sqrt[3]{32c^7}$
$=4c^2\left(3\sqrt[3]{4c}\right)-15\left(2c^2\sqrt[3]{4c}\right)$
$=12c^2\sqrt[3]{4c}-30c^2\sqrt[3]{4c}=-18c^2\sqrt[3]{4c}$

55) $\sqrt{xy^3}+3y\sqrt{xy}=y\sqrt{xy}+3y\sqrt{xy}$
$=4y\sqrt{xy}$

57) $6c^2\sqrt{8d^3}-9d\sqrt{2c^4d}$
$=6c^2\left(2d\sqrt{2d}\right)-9d\left(c^2\sqrt{2d}\right)$
$=12c^2d\sqrt{2d}-9c^2d\sqrt{2d}=3c^2d\sqrt{2d}$

59) $3\sqrt{75m^3n}+m\sqrt{12mn}$
$=3\left(5m\sqrt{3mn}\right)+m\left(2\sqrt{3mn}\right)$
$=15m\sqrt{3mn}+2m\sqrt{3mn}=17m\sqrt{3mn}$

61) $18a^5\sqrt[3]{7a^2b}+2a^3\sqrt[3]{7a^8b}$
$=18a^5\sqrt[3]{7a^2b}+2a^3\left(a^2\sqrt[3]{7a^2b}\right)$
$=18a^5\sqrt[3]{7a^2b}+2a^5\sqrt[3]{7a^2b}$
$=20a^5\sqrt[3]{7a^2b}$

63) $15cd\sqrt[4]{9cd}-\sqrt[4]{9c^5d^5}$
$=15cd\sqrt[4]{9cd}-\left(\sqrt[4]{c^4d^4}\cdot\sqrt[4]{9cd}\right)$
$=15cd\sqrt[4]{9cd}-cd\sqrt[4]{9cd}$
$=14cd\sqrt[4]{9cd}$

65) $\sqrt{m^5}+\sqrt{mn^2}=m^2\sqrt{m}+n\sqrt{m}$
$=\sqrt{m}\left(m^2+n\right)$

67) $\sqrt[3]{a^9b}-\sqrt[3]{b^7}=a^3\sqrt[3]{b}-b^2\sqrt[3]{b}$
$=\sqrt[3]{b}\left(a^3-b^2\right)$

69) $\sqrt[3]{u^2v^6}+\sqrt[3]{u^2}=v^2\sqrt[3]{u^2}+\sqrt[3]{u^2}$
$=\sqrt[3]{u^2}\left(v^2+1\right)$

Section 10.6: Exercises

1) $3(x+5)=3x+15$

3) $7\left(\sqrt{6}+2\right)=7\sqrt{6}+14$

5) $\sqrt{10}\left(\sqrt{3}-1\right)=\sqrt{30}-\sqrt{10}$

7) $-6\left(\sqrt{32}+\sqrt{2}\right)=-6\left(4\sqrt{2}+\sqrt{2}\right)$
$=-6\left(5\sqrt{2}\right)=-30\sqrt{2}$

9) $4\left(\sqrt{45}-\sqrt{20}\right)=4\left(3\sqrt{5}-2\sqrt{5}\right)$
$=4\left(\sqrt{5}\right)=4\sqrt{5}$

11) $\sqrt{5}\left(\sqrt{24}-\sqrt{54}\right)=\sqrt{5}\left(2\sqrt{6}-3\sqrt{6}\right)$
$=\sqrt{5}\left(-\sqrt{6}\right)$
$=-\sqrt{30}$

13) $\sqrt{3}\left(4+\sqrt{6}\right)=4\sqrt{3}+\sqrt{18}$
$=4\sqrt{3}+3\sqrt{2}$

15) $\sqrt{7}\left(\sqrt{24}+\sqrt{5}\right)=\sqrt{7}\left(2\sqrt{6}+\sqrt{5}\right)$
$=2\sqrt{42}+\sqrt{35}$

17) $\sqrt{t}\left(\sqrt{t}-\sqrt{81u}\right)=\sqrt{t}\left(\sqrt{t}-9\sqrt{u}\right)$
$=t-9\sqrt{tu}$

19) $\sqrt{ab}\left(\sqrt{5a}+\sqrt{27b}\right)$
$=\sqrt{ab}\left(\sqrt{5a}+3\sqrt{3b}\right)$
$=\sqrt{5a^2b}+3\sqrt{3ab^2}=a\sqrt{5b}+3b\sqrt{3a}$

21) Both are examples of multiplication of two binomials. They can be multiplied using FOIL.

23) $(a+b)(a-b)=a^2-b^2$

25) $(p+7)(p+6)=p^2+6p+7p+42$
$=p^2+13p+42$

27) $\left(6+\sqrt{7}\right)\left(2+\sqrt{7}\right)$
$=12+6\sqrt{7}+2\sqrt{7}+7=19+8\sqrt{7}$

29) $\left(\sqrt{2}+8\right)\left(\sqrt{2}-3\right)$
$=2-3\sqrt{2}+8\sqrt{2}-24=-22+5\sqrt{2}$

31) $\left(\sqrt{5}-4\sqrt{3}\right)\left(2\sqrt{5}-\sqrt{3}\right)$
$=2(5)-\sqrt{15}-8\sqrt{15}+4(3)$
$=10-9\sqrt{15}+12=22-9\sqrt{15}$

33) $\left(3\sqrt{6}-2\sqrt{2}\right)\left(\sqrt{2}+5\sqrt{6}\right)$
$=3\sqrt{12}+15(6)-2(2)-10\sqrt{12}$
$=-7\sqrt{12}+90-4$
$=-7\left(\sqrt{4}\cdot\sqrt{3}\right)+86$
$=-7\left(2\sqrt{3}\right)+86=-14\sqrt{3}+86$

35) $\left(5+2\sqrt{3}\right)\left(\sqrt{7}+\sqrt{2}\right)$
$=5\sqrt{7}+5\sqrt{2}+2\sqrt{21}+2\sqrt{6}$

37) $\left(\sqrt{x}+\sqrt{2y}\right)\left(\sqrt{x}+5\sqrt{2y}\right)$
$=x+5\sqrt{2xy}+\sqrt{2xy}+5(2y)$
$=x+6\sqrt{2xy}+10y$

39) $\left(\sqrt{6p}-2\sqrt{q}\right)\left(8\sqrt{q}+5\sqrt{6p}\right)$
$=8\sqrt{6pq}+5(6p)-16q-10\sqrt{6pq}$
$=-2\sqrt{6pq}+30p-16q$

41) $(5y-4)^2=(5y)^2-2(5y)(4)+(4)^2$
$=25y^2-40y+16$

43) $\left(\sqrt{3}+1\right)^2=\left(\sqrt{3}\right)^2+2\left(\sqrt{3}\right)(1)+(1)^2$
$=3+2\sqrt{3}+1=4+2\sqrt{3}$

45) $\left(\sqrt{11}-\sqrt{5}\right)^2$
$=\left(\sqrt{11}\right)^2-2\left(\sqrt{11}\right)\left(\sqrt{5}\right)+\left(\sqrt{5}\right)^2$
$=11-2\sqrt{55}+5=16-2\sqrt{55}$

47) $\left(2\sqrt{3}+\sqrt{10}\right)^2$
$=\left(2\sqrt{3}\right)^2+2\left(2\sqrt{3}\right)\left(\sqrt{10}\right)+\left(\sqrt{10}\right)^2$
$=4(3)+4\sqrt{30}+10=12+4\sqrt{30}+10$
$=22+4\sqrt{30}$

49) $\left(\sqrt{2}-4\sqrt{6}\right)^2$
$=\left(\sqrt{2}\right)^2-2\left(\sqrt{2}\right)\left(4\sqrt{6}\right)+\left(4\sqrt{6}\right)^2$
$=2-8\sqrt{12}+16(6)$
$=2-8\left(2\sqrt{3}\right)+96=98-16\sqrt{3}$

51) $\left(\sqrt{h}+\sqrt{7}\right)^2$
$=\left(\sqrt{h}\right)^2+2\left(\sqrt{h}\right)\left(\sqrt{7}\right)+\left(\sqrt{7}\right)^2$
$=h+2\sqrt{7h}+7$

53) $\left(\sqrt{x}-\sqrt{y}\right)^2$
$=\left(\sqrt{x}\right)^2-2\left(\sqrt{x}\right)\left(\sqrt{y}\right)+\left(\sqrt{y}\right)^2$
$=x-2\sqrt{xy}+y$

55) $(c+9)(c-9)=c^2-(9)^2=c^2-81$

57) $\left(\sqrt{2}+3\right)\left(\sqrt{2}-3\right)=\left(\sqrt{2}\right)^2-3^2$
$=2-9=-7$

59) $\left(6-\sqrt{5}\right)\left(6+\sqrt{5}\right)=6^2-\left(\sqrt{5}\right)^2$
$=36-5=31$

61) $\left(4\sqrt{3}+\sqrt{2}\right)\left(4\sqrt{3}-\sqrt{2}\right)$
$=\left(4\sqrt{3}\right)^2-\left(\sqrt{2}\right)^2$
$=16(3)-2=48-2=46$

63) $\left(\sqrt{11}+5\sqrt{3}\right)\left(\sqrt{11}-5\sqrt{3}\right)$
$=\left(\sqrt{11}\right)^2-\left(5\sqrt{3}\right)^2$
$=11-25(3)=11-75=-64$

65) $\left(\sqrt{c}+\sqrt{d}\right)\left(\sqrt{c}-\sqrt{d}\right)$
$=\left(\sqrt{c}\right)^2-\left(\sqrt{d}\right)^2=c-d$

67) $\left(5-\sqrt{t}\right)\left(5+\sqrt{t}\right)=(5)^2-\left(\sqrt{t}\right)^2$
$=25-t$

69) $\left(8\sqrt{f}-\sqrt{g}\right)\left(8\sqrt{f}+\sqrt{g}\right)$
$=\left(8\sqrt{f}\right)^2-\left(\sqrt{g}\right)^2=64f-g$

71) $\left(\sqrt[3]{2}-3\right)\left(\sqrt[3]{2}+3\right)=\left(\sqrt[3]{2}\right)^2-3^2$
$=\sqrt[3]{4}-9$

73) $\left(1+2\sqrt[3]{5}\right)\left(1-2\sqrt[3]{5}+4\sqrt[3]{25}\right)$
$=1-2\sqrt[3]{5}+4\sqrt[3]{25}+2\sqrt[3]{5}$
$-4\sqrt[3]{25}+8\sqrt[3]{125}$
$=1+8\sqrt[3]{125}=1+8(5)=1+40=41$

75) $\left[\left(\sqrt{3}+\sqrt{6}\right)+\sqrt{2}\right]\left[\left(\sqrt{3}+\sqrt{6}\right)-\sqrt{2}\right]$
$=\left(\sqrt{3}+\sqrt{6}\right)^2-\left(\sqrt{2}\right)^2$
$=\left[\left(\sqrt{3}\right)^2+2\left(\sqrt{3}\right)\left(\sqrt{6}\right)+\left(\sqrt{6}\right)^2\right]-2$
$=3+2\sqrt{18}+6-2$
$=7+2\left(3\sqrt{2}\right)=7+6\sqrt{2}$

Section 10.7: Exercises

1) Eliminate the radical of the denominator.

3) $\dfrac{1}{\sqrt{5}}=\dfrac{1}{\sqrt{5}}\cdot\dfrac{\sqrt{5}}{\sqrt{5}}=\dfrac{\sqrt{5}}{\sqrt{25}}=\dfrac{\sqrt{5}}{5}$

5) $\dfrac{3}{\sqrt{2}}=\dfrac{3}{\sqrt{2}}\cdot\dfrac{\sqrt{2}}{\sqrt{2}}=\dfrac{3\sqrt{2}}{\sqrt{4}}=\dfrac{3\sqrt{2}}{2}$

7) $\dfrac{9}{\sqrt{6}}=\dfrac{9}{\sqrt{6}}\cdot\dfrac{\sqrt{6}}{\sqrt{6}}=\dfrac{9\sqrt{6}}{\sqrt{36}}=\dfrac{9\sqrt{6}}{6}=\dfrac{3\sqrt{6}}{2}$

9) $-\dfrac{20}{\sqrt{8}}=-\dfrac{20}{2\sqrt{2}}=-\dfrac{10}{\sqrt{2}}=-\dfrac{10}{\sqrt{2}}\cdot\dfrac{\sqrt{2}}{\sqrt{2}}$
$=-\dfrac{10\sqrt{2}}{\sqrt{4}}=-\dfrac{10\sqrt{2}}{2}=-5\sqrt{2}$

11) $\dfrac{\sqrt{3}}{\sqrt{28}}=\dfrac{\sqrt{3}}{2\sqrt{7}}=\dfrac{\sqrt{3}}{2\sqrt{7}}\cdot\dfrac{\sqrt{7}}{\sqrt{7}}=\dfrac{\sqrt{21}}{2\sqrt{49}}$
$=\dfrac{\sqrt{21}}{2(7)}=\dfrac{\sqrt{21}}{14}$

13) $\sqrt{\dfrac{20}{60}}=\sqrt{\dfrac{1}{3}}=\dfrac{1}{\sqrt{3}}=\dfrac{1}{\sqrt{3}}\cdot\dfrac{\sqrt{3}}{\sqrt{3}}=\dfrac{\sqrt{3}}{3}$

15) $\sqrt{\dfrac{18}{26}}=\sqrt{\dfrac{9}{13}}=\dfrac{\sqrt{9}}{\sqrt{13}}=\dfrac{3}{\sqrt{13}}$
$=\dfrac{3}{\sqrt{13}}\cdot\dfrac{\sqrt{13}}{\sqrt{13}}=\dfrac{3\sqrt{13}}{13}$

17) $\sqrt{\dfrac{56}{48}}=\dfrac{2\sqrt{14}}{4\sqrt{3}}=\dfrac{\sqrt{14}}{2\sqrt{3}}=\dfrac{\sqrt{14}}{2\sqrt{3}}\cdot\dfrac{\sqrt{3}}{\sqrt{3}}$
$=\dfrac{\sqrt{42}}{2(3)}=\dfrac{\sqrt{42}}{6}$

19) $\sqrt{\dfrac{10}{7}}\cdot\sqrt{\dfrac{7}{3}}=\sqrt{\dfrac{10}{7}\cdot\dfrac{7}{3}}=\sqrt{\dfrac{10}{3}}$
$=\dfrac{\sqrt{10}}{\sqrt{3}}\cdot\dfrac{\sqrt{3}}{\sqrt{3}}=\dfrac{\sqrt{30}}{3}$

21) $\sqrt{\dfrac{6}{5}}\cdot\sqrt{\dfrac{1}{8}}=\sqrt{\dfrac{6}{5}\cdot\dfrac{1}{8}}=\sqrt{\dfrac{3}{20}}=\dfrac{\sqrt{3}}{\sqrt{20}}$
$=\dfrac{\sqrt{3}}{2\sqrt{5}}\cdot\dfrac{\sqrt{5}}{\sqrt{5}}=\dfrac{\sqrt{15}}{2(5)}=\dfrac{\sqrt{15}}{10}$

23) $\sqrt{\dfrac{6}{7}}\cdot\sqrt{\dfrac{7}{3}}=\sqrt{\dfrac{6}{7}\cdot\dfrac{7}{3}}=\sqrt{2}$

25) $\dfrac{8}{\sqrt{y}}=\dfrac{8}{\sqrt{y}}\cdot\dfrac{\sqrt{y}}{\sqrt{y}}=\dfrac{8\sqrt{y}}{\sqrt{y^2}}=\dfrac{8\sqrt{y}}{y}$

27) $\dfrac{\sqrt{5}}{\sqrt{t}}=\dfrac{\sqrt{5}}{\sqrt{t}}\cdot\dfrac{\sqrt{t}}{\sqrt{t}}=\dfrac{\sqrt{5t}}{\sqrt{t^2}}=\dfrac{\sqrt{5t}}{t}$

29) $\sqrt{\dfrac{10f^3}{g}} = \dfrac{f\sqrt{10f}}{\sqrt{g}} = \dfrac{f\sqrt{10f}}{\sqrt{g}} \cdot \dfrac{\sqrt{g}}{\sqrt{g}}$

$= \dfrac{f\sqrt{10fg}}{\sqrt{g^2}} = \dfrac{f\sqrt{10fg}}{g}$

31) $\sqrt{\dfrac{64v^7}{5w}} = \dfrac{8v^3\sqrt{v}}{\sqrt{5w}} = \dfrac{8v^3\sqrt{v}}{\sqrt{5w}} \cdot \dfrac{\sqrt{5w}}{\sqrt{5w}}$

$= \dfrac{8v^3\sqrt{5vw}}{5w}$

33) $\sqrt{\dfrac{a^3b^3}{3ab^4}} = \sqrt{\dfrac{a^2}{3b}} = \dfrac{a}{\sqrt{3b}}$

$= \dfrac{a}{\sqrt{3b}} \cdot \dfrac{\sqrt{3b}}{\sqrt{3b}} = \dfrac{a\sqrt{3b}}{3b}$

35) $-\dfrac{\sqrt{75}}{\sqrt{b^3}} = -\dfrac{5\sqrt{3}}{b\sqrt{b}} = -\dfrac{5\sqrt{3}}{b\sqrt{b}} \cdot \dfrac{\sqrt{b}}{\sqrt{b}}$

$= -\dfrac{5\sqrt{3b}}{b(b)} = -\dfrac{5\sqrt{3b}}{b^2}$

37) $\dfrac{\sqrt{13}}{\sqrt{j^5}} = \dfrac{\sqrt{13}}{j^2\sqrt{j}} = \dfrac{\sqrt{13}}{j^2\sqrt{j}} \cdot \dfrac{\sqrt{j}}{\sqrt{j}}$

$= \dfrac{\sqrt{13j}}{j^2(j)} = \dfrac{\sqrt{13j}}{j^3}$

39) 2^2 or 4 41) 3

43) c^2 45) 2^3 or 8 47) m

49) $\dfrac{4}{\sqrt[3]{3}} = \dfrac{4}{\sqrt[3]{3}} \cdot \dfrac{\sqrt[3]{3^2}}{\sqrt[3]{3^2}} = \dfrac{4\sqrt[3]{9}}{\sqrt[3]{3^3}} = \dfrac{4\sqrt[3]{9}}{3}$

51) $\dfrac{12}{\sqrt[3]{2}} = \dfrac{12}{\sqrt[3]{2}} \cdot \dfrac{\sqrt[3]{2^2}}{\sqrt[3]{2^2}} = \dfrac{12\sqrt[3]{4}}{\sqrt[3]{2^3}}$

$= \dfrac{12\sqrt[3]{4}}{2} = 6\sqrt[3]{4}$

53) $\dfrac{9}{\sqrt[3]{25}} = \dfrac{9}{\sqrt[3]{5^2}} = \dfrac{9}{\sqrt[3]{5^2}} \cdot \dfrac{\sqrt[3]{5}}{\sqrt[3]{5}}$

$= \dfrac{9\sqrt[3]{5}}{\sqrt[3]{5^3}} = \dfrac{9\sqrt[3]{5}}{5}$

55) $\sqrt[4]{\dfrac{5}{9}} = \dfrac{\sqrt[4]{5}}{\sqrt[4]{3^2}} = \dfrac{\sqrt[4]{5}}{\sqrt[4]{3^2}} \cdot \dfrac{\sqrt[4]{3^2}}{\sqrt[4]{3^2}}$

$= \dfrac{\sqrt[4]{5} \cdot \sqrt[4]{9}}{\sqrt[4]{3^4}} = \dfrac{\sqrt[4]{45}}{3}$

57) $\sqrt[5]{\dfrac{3}{8}} = \dfrac{\sqrt[5]{3}}{\sqrt[5]{2^3}} = \dfrac{\sqrt[5]{3}}{\sqrt[5]{2^3}} \cdot \dfrac{\sqrt[5]{2^2}}{\sqrt[5]{2^2}}$

$= \dfrac{\sqrt[5]{3} \cdot \sqrt[5]{4}}{\sqrt[5]{2^5}} = \dfrac{\sqrt[5]{12}}{2}$

59) $\sqrt[4]{\dfrac{2}{9}} = \dfrac{\sqrt[4]{2}}{\sqrt[4]{3^2}} = \dfrac{\sqrt[4]{2}}{\sqrt[4]{3^2}} \cdot \dfrac{\sqrt[4]{3^2}}{\sqrt[4]{3^2}}$

$= \dfrac{\sqrt[4]{2} \cdot \sqrt[4]{9}}{\sqrt[4]{3^4}} = \dfrac{\sqrt[4]{18}}{3}$

61) $\dfrac{10}{\sqrt[3]{z}} = \dfrac{10}{\sqrt[3]{z}} \cdot \dfrac{\sqrt[3]{z^2}}{\sqrt[3]{z^2}} = \dfrac{10\sqrt[3]{z^2}}{\sqrt[3]{z^3}} = \dfrac{10\sqrt[3]{z^2}}{z}$

63) $\sqrt[3]{\dfrac{3}{n^2}} = \dfrac{\sqrt[3]{3}}{\sqrt[3]{n^2}} \cdot \dfrac{\sqrt[3]{n}}{\sqrt[3]{n}} = \dfrac{\sqrt[3]{3n}}{\sqrt[3]{n^3}} = \dfrac{\sqrt[3]{3n}}{n}$

65) $\dfrac{\sqrt[3]{7}}{\sqrt[3]{2k^2}} = \dfrac{\sqrt[3]{7}}{\sqrt[3]{2k^2}} \cdot \dfrac{\sqrt[3]{2^2k}}{\sqrt[3]{2^2k}} = \dfrac{\sqrt[3]{7} \cdot \sqrt[3]{4k}}{\sqrt[3]{2^3k^3}}$

$= \dfrac{\sqrt[3]{28k}}{2k}$

67) $\dfrac{9}{\sqrt[5]{a^3}} = \dfrac{9}{\sqrt[5]{a^3}} \cdot \dfrac{\sqrt[5]{a^2}}{\sqrt[5]{a^2}} = \dfrac{9\sqrt[5]{a^2}}{\sqrt[5]{a^5}} = \dfrac{9\sqrt[5]{a^2}}{a}$

69) $\sqrt[4]{\dfrac{c}{d^3}} = \dfrac{\sqrt[4]{c}}{\sqrt[4]{d^3}} \cdot \dfrac{\sqrt[4]{d}}{\sqrt[4]{d}} = \dfrac{\sqrt[4]{cd}}{\sqrt[4]{d^4}} = \dfrac{\sqrt[4]{cd}}{d}$

71) $\sqrt[4]{\dfrac{5}{2m}}=\dfrac{\sqrt[4]{5}}{\sqrt[4]{2m}}\cdot\dfrac{\sqrt[4]{2^3m^3}}{\sqrt[4]{2^3m^3}}=\dfrac{\sqrt[4]{5}\cdot\sqrt[4]{8m^3}}{\sqrt[4]{2^4m^4}}$

$=\dfrac{\sqrt[4]{40m^3}}{2m}$

73) Change the sign between the two terms.

75) $(5+\sqrt{2})(5-\sqrt{2})=(5)^2-(\sqrt{2})^2$

$=25-2=23$

77) $(\sqrt{2}+\sqrt{6})(\sqrt{2}-\sqrt{6})$

$=(\sqrt{2})^2-(\sqrt{6})^2=2-6=-4$

79) $(\sqrt{t}-8)(\sqrt{t}+8)=(\sqrt{t})^2-(8)^2$

$=t-64$

81) $\dfrac{3}{2+\sqrt{3}}=\dfrac{3}{2+\sqrt{3}}\cdot\dfrac{2-\sqrt{3}}{2-\sqrt{3}}$

$=\dfrac{3(2-\sqrt{3})}{(2)^2-(\sqrt{3})^2}=\dfrac{3(2-\sqrt{3})}{4-3}$

$=6-3\sqrt{3}$

83) $\dfrac{10}{9-\sqrt{2}}=\dfrac{10}{9-\sqrt{2}}\cdot\dfrac{9+\sqrt{2}}{9+\sqrt{2}}$

$=\dfrac{10(9+\sqrt{2})}{(9)^2-(\sqrt{2})^2}=\dfrac{10(9+\sqrt{2})}{81-2}$

$=\dfrac{90+10\sqrt{2}}{79}$

85) $\dfrac{\sqrt{8}}{\sqrt{3}+\sqrt{2}}=\dfrac{2\sqrt{2}}{\sqrt{3}+\sqrt{2}}\cdot\dfrac{\sqrt{3}-\sqrt{2}}{\sqrt{3}-\sqrt{2}}$

$=\dfrac{2\sqrt{2}(\sqrt{3}-\sqrt{2})}{(\sqrt{3})^2-(\sqrt{2})^2}$

$=\dfrac{2\sqrt{6}-2(2)}{3-2}=2\sqrt{6}-4$

87) $\dfrac{\sqrt{3}-\sqrt{5}}{\sqrt{10}-\sqrt{3}}=\dfrac{\sqrt{3}-\sqrt{5}}{\sqrt{10}-\sqrt{3}}\cdot\dfrac{\sqrt{10}+\sqrt{3}}{\sqrt{10}+\sqrt{3}}$

$=\dfrac{(\sqrt{3}-\sqrt{5})(\sqrt{10}+\sqrt{3})}{(\sqrt{10})^2-(\sqrt{3})^2}$

$=\dfrac{\sqrt{30}+3-\sqrt{50}-\sqrt{15}}{10-3}$

$=\dfrac{\sqrt{30}+3-5\sqrt{2}-\sqrt{15}}{7}$

89) $\dfrac{\sqrt{m}}{\sqrt{m}+\sqrt{n}}=\dfrac{\sqrt{m}}{\sqrt{m}+\sqrt{n}}\cdot\dfrac{\sqrt{m}-\sqrt{n}}{\sqrt{m}-\sqrt{n}}$

$=\dfrac{\sqrt{m}(\sqrt{m}-\sqrt{n})}{(\sqrt{m})^2-(\sqrt{m})^2}=\dfrac{m-\sqrt{mn}}{m-n}$

91) $\dfrac{b-25}{\sqrt{b}-5}=\dfrac{b-25}{\sqrt{b}-5}\cdot\dfrac{\sqrt{b}+5}{\sqrt{b}+5}$

$=\dfrac{(b-25)(\sqrt{b}+5)}{(\sqrt{b})^2-(5)^2}$

$=\dfrac{(b-25)(\sqrt{b}+5)}{b-25}=\sqrt{b}+5$

93) $\dfrac{\sqrt{x}+\sqrt{y}}{\sqrt{x}-\sqrt{y}}=\dfrac{\sqrt{x}+\sqrt{y}}{\sqrt{x}-\sqrt{y}}\cdot\dfrac{\sqrt{x}+\sqrt{y}}{\sqrt{x}+\sqrt{y}}$

$$=\frac{\left(\sqrt{x}+\sqrt{y}\right)^2}{\left(\sqrt{x}\right)^2-\left(\sqrt{y}\right)^2}$$

$$=\frac{x+2\sqrt{xy}+y}{x-y}$$

95) $\dfrac{5+10\sqrt{3}}{5}=\dfrac{5\left(1+2\sqrt{3}\right)}{5}=1+2\sqrt{3}$

97) $\dfrac{30-18\sqrt{5}}{4}=\dfrac{6\left(5-3\sqrt{5}\right)}{4}$

$$=\frac{3\left(5-3\sqrt{5}\right)}{2}=\frac{15-9\sqrt{5}}{2}$$

99) $\dfrac{\sqrt{45}+6}{9}=\dfrac{3\sqrt{5}+6}{9}=\dfrac{3\left(\sqrt{5}+2\right)}{9}$

$$=\frac{\sqrt{5}+2}{3}$$

101) $\dfrac{-10-\sqrt{50}}{5}=\dfrac{-10-5\sqrt{2}}{5}$

$$=\frac{5\left(-2-\sqrt{2}\right)}{5}=-2-\sqrt{2}$$

Section 10.8: Exercises

1) Sometimes these are extraneous solutions.

3) $\sqrt{q}=7$

$\left(\sqrt{q}\right)^2=7^2$

$q=49$

Check $\sqrt{49}=7$ $\{49\}$

5) $\sqrt{w}-\dfrac{2}{3}=0$

$\sqrt{w}=\dfrac{2}{3}$

$\left(\sqrt{w}\right)^2=\left(\dfrac{2}{3}\right)^2$

$w=\dfrac{4}{9}$ $\left\{\dfrac{4}{9}\right\}$

Check is left to the student.

7) $\sqrt{a}+5=3$

$\sqrt{a}=-2$

$\left(\sqrt{a}\right)^2=(-2)^2$

$a=4$

Check $\sqrt{4}+5=3$

$2+5\neq 3$ $\varnothing$

9) $\sqrt[3]{y}=5$

$\left(\sqrt[3]{y}\right)^3=5^3$

$y=125$

Check is left to the student. $\{125\}$

11) $\sqrt[3]{m}=-4$

$\left(\sqrt[3]{m}\right)^3=(-4)^3$

$m=-64$

Check is left to the student. $\{-64\}$

13) $\sqrt{b-11}-3=0$

$\sqrt{b-11}=3$

$\left(\sqrt{b-11}\right)^2=(3)^2$

$b-11=9$

$b=20$

Check is left to the student. $\{20\}$

15) $\sqrt{4g-1}+7=1$

$$\sqrt{4g-1}=-6$$
$$\left(\sqrt{4g-1}\right)^2=(-6)^2$$
$$4g-1=36$$
$$4g=37$$
$$g=\frac{37}{4}$$

$\frac{37}{4}$ is an extraneous solution. $\varnothing$

17) $\sqrt{3f+2}+9=11$

$$\sqrt{3f+2}=2$$
$$\left(\sqrt{3f+2}\right)^2=(2)^2$$
$$3f+2=4$$
$$3f=2$$
$$f=\frac{2}{3}$$

Check is left to the student. $\left\{\frac{2}{3}\right\}$

19) $\sqrt[3]{2x-5}+3=1$

$$\sqrt[3]{2x-5}=-2$$
$$\left(\sqrt[3]{2x-5}\right)^3=(-2)^3$$
$$2x-5=-8$$
$$2x=-3$$
$$x=-\frac{3}{2}$$

Check is left to the student. $\left\{-\frac{3}{2}\right\}$

21) $\sqrt{2c+3}=\sqrt{5c}$

$$\left(\sqrt{2c+3}\right)^2=\left(\sqrt{5c}\right)^2$$
$$2c+3=5c$$
$$3=3c$$
$$1=c$$

Check is left to the student. $\{1\}$

23) $\sqrt[3]{6j-2}=\sqrt[3]{j-7}$

$$\left(\sqrt[3]{6j-2}\right)^3=\left(\sqrt[3]{j-7}\right)^3$$
$$6j-2=j-7$$
$$5j=-5$$
$$j=-1$$

Check is left to the student. $\{-1\}$

25) $5\sqrt{1-5h}=4\sqrt{1-8h}$

$$\left(5\sqrt{1-5h}\right)^2=\left(4\sqrt{1-8h}\right)^2$$
$$25(1-5h)=16(1-8h)$$
$$25-125h=16-128h$$
$$3h=-9$$
$$h=-3$$

Check is left to the student. $\{-3\}$

27) $3\sqrt{3x+6}=2\sqrt{9x-9}$

$$\left(3\sqrt{3x+6}\right)^2=\left(2\sqrt{9x-9}\right)^2$$
$$9(3x+6)=4(9x-9)$$
$$27x+54=36x-36$$
$$90=9x$$
$$10=x$$

Check is left to the student. $\{10\}$

29) $(x+3)^2=x^2+2(x)(3)+3^2$

$$=x^2+6x+9$$

31) $m=\sqrt{m^2-3m+6}$

$$m^2=\left(\sqrt{m^2-3m+6}\right)^2$$
$$m^2=m^2-3m+6$$
$$0=-3m+6$$
$$3m=6$$
$$m=2$$

Check is left to the student. $\{2\}$

33) $p+6=\sqrt{12+p}$

$$(p+6)^2=\left(\sqrt{12+p}\right)^2$$
$$p^2+12p+36=12+p$$
$$p^2+11p+24=0$$
$$(p+8)(p+3)=0$$
$$p+8=0 \text{ or } p+3=0$$
$$p=-8 \qquad p=-3$$

-8 is an extraneous solution. $\{-3\}$

35) $\sqrt{r^2-8r-19}=r-9$

$$\left(\sqrt{r^2-8r-19}\right)^2=(r-9)^2$$
$$r^2-8r-19=r^2-18r+81$$
$$-8r-19=-18r+81$$
$$10r=100$$
$$r=10$$

Check is left to the student. $\{10\}$

37) $6+\sqrt{c^2+3c-9}=c$

$$\sqrt{c^2+3c-9}=c-6$$
$$\left(\sqrt{c^2+3c-9}\right)^2=(c-6)^2$$
$$c^2+3c-9=c^2-12c+36$$
$$3c-9=-12c+36$$
$$15c=45$$
$$c=3$$

3 is an extraneous solution. $\varnothing$

39) $w-\sqrt{10w+6}=-3$

$$w+3=\sqrt{10w+6}$$
$$(w+3)^2=\left(\sqrt{10w+6}\right)^2$$
$$w^2+6w+9=10w+6$$
$$w^2-4w+3=0$$
$$(w-3)(w-1)=0$$
$$w-3=0 \text{ or } w-1=0$$
$$w=3 \qquad w=1$$

Check is left to the student. $\{1,3\}$

41) $3v=8+\sqrt{3v+4}$

$$3v-8=\sqrt{3v+4}$$
$$(3v-8)^2=\left(\sqrt{3v+4}\right)^2$$
$$9v^2-48v+64=3v+4$$
$$9v^2-51v+60=0 \qquad \text{Divide by 3.}$$
$$3v^2-17v+20=0$$
$$(3v-5)(v-4)=0$$
$$3v-5=0 \text{ or } v-4=0$$
$$3v=5 \qquad v=4$$
$$v=\frac{5}{3}$$

$\frac{5}{3}$ is an extraneous solution. $\{4\}$

43) $\left(\sqrt{x}+5\right)^2=\left(\sqrt{x}\right)^2+2\left(\sqrt{x}\right)(5)+(5)^2$

$$=x+10\sqrt{x}+25$$

45) $\left(9-\sqrt{a+4}\right)^2$

$$=(9)^2-2(9)\left(\sqrt{a+4}\right)+\left(\sqrt{a+4}\right)^2$$
$$=81-18\sqrt{a+4}+a+4$$
$$=85-18\sqrt{a+4}+a$$

47) $\left(2\sqrt{3n-1}+7\right)^2$

$=\left(2\sqrt{3n-1}\right)^2+2\left(2\sqrt{3n-1}\right)(7)+7^2$

$=4(3n-1)+28\sqrt{3n-1}+49$

$=12n-4+28\sqrt{3n-1}+49$

$=12n+28\sqrt{3n-1}+45$

49) $\sqrt{2y-1}=2+\sqrt{y-4}$

$\left(\sqrt{2y-1}\right)^2=\left(2+\sqrt{y-4}\right)^2$

$2y-1=4+4\sqrt{y-4}+y-4$

$2y-1=y+4\sqrt{y-4}$

$y-1=4\sqrt{y-4}$

$(y-1)^2=\left(4\sqrt{y-4}\right)^2$

$y^2-2y+1=16(y-4)$

$y^2-2y+1=16y-64$

$y^2-18y+65=0$

$(y-13)(y-5)=0$

$y-13=0$ or $y-5=0$

$y=13 \qquad y=5$

Check is left to the student. $\{5,13\}$

51) $1+\sqrt{3s-2}=\sqrt{2s+5}$

$\left(1+\sqrt{3s-2}\right)^2=\left(\sqrt{2s+5}\right)^2$

$1+2\sqrt{3s-2}+3s-2=2s+5$

$3s-1+2\sqrt{3s-2}=2s+5$

$2\sqrt{3s-2}=6-s$

$\left(2\sqrt{3s-2}\right)^2=(6-s)^2$

$4(3s-2)=36-12s+s^2$

$12s-8=s^2-12s+36$

$0=s^2-24s+44$

$0=(s-2)(s-22)$

$s-2=0$ or $s-22=0$

$s=2 \qquad s=22$

22 is an extraneous solution. $\{2\}$

53) $\sqrt{3k+1}-\sqrt{k-1}=2$

$\sqrt{3k+1}=2+\sqrt{k-1}$

$\left(\sqrt{3k+1}\right)^2=\left(2+\sqrt{k-1}\right)^2$

$3k+1=4+4\sqrt{k-1}+k-1$

$3k+1=k+3+4\sqrt{k-1}$

$2k-2=4\sqrt{k-1}$

$k-1=2\sqrt{k-1}$

$(k-1)^2=\left(2\sqrt{k-1}\right)^2$

$k^2-2k+1=4(k-1)$

$k^2-2k+1=4k-4$

$k^2-6k+5=0$

$(k-1)(k-5)=0$

$k-1=0$ or $k-5=0$

$k=1 \qquad k=5$

Check is left to the student. $\{1,5\}$

55) $\sqrt{3x+4}-5=\sqrt{3x-11}$

$\left(\sqrt{3x+4}-5\right)^2=\left(\sqrt{3x-11}\right)^2$

$3x+4-10\sqrt{3x+4}+25=3x-11$

$3x+29-10\sqrt{3x+4}=3x-11$

$-10\sqrt{3x+4}=-40$

$\sqrt{3x+4}=4$

$\left(\sqrt{3x+4}\right)^2=(4)^2$

$3x+4=16$

$3x=12$

$x=4$

4 is an extraneous solution. $\varnothing$

57) $\sqrt{3v+3}-\sqrt{v-2}=3$

$$\sqrt{3v+3}=\sqrt{v-2}+3$$
$$\left(\sqrt{3v+3}\right)^2=\left(\sqrt{v-2}+3\right)^2$$
$$3v+3=9+6\sqrt{v-2}+v-2$$
$$3v+3=7+6\sqrt{v-2}+v$$
$$2v-4=6\sqrt{v-2}$$
$$v-2=3\sqrt{v-2}$$
$$(v-2)^2=\left(3\sqrt{v-2}\right)^2$$
$$v^2-4v+4=9(v-2)$$
$$v^2-4v+4=9v-18$$
$$v^2-13v+22=0$$
$$(v-2)(v-11)=0$$
$$v-2=0 \text{ or } v-11=0$$
$$v=2 \qquad v=11$$

Check is left to the student. $\{2, 11\}$

59) $\sqrt{5a+19}-\sqrt{a+12}=1$

$$\sqrt{5a+19}=1+\sqrt{a+12}$$
$$\left(\sqrt{5a+19}\right)^2=\left(1+\sqrt{a+12}\right)^2$$
$$5a+19=1+2\sqrt{a+12}+a+12$$
$$5a+19=a+13+2\sqrt{a+12}$$
$$4a+6=2\sqrt{a+12}$$
$$2a+3=\sqrt{a+12}$$
$$(2a+3)^2=\left(\sqrt{a+12}\right)^2$$
$$4a^2+12a+9=a+12$$
$$4a^2+11a-3=0$$
$$(4a-1)(a+3)=0$$
$$4a-1=0 \text{ or } a+3=0$$
$$4a=1$$
$$a=\frac{1}{4} \qquad a=-3$$

-3 is an extraneous solution. $\left\{\frac{1}{4}\right\}$

61) $\sqrt{13+\sqrt{r}}=\sqrt{r+7}$

$$\left(\sqrt{13+\sqrt{r}}\right)^2=\left(\sqrt{r+7}\right)^2$$
$$13+\sqrt{r}=r+7$$
$$\sqrt{r}=r-6$$
$$\left(\sqrt{r}\right)^2=(r-6)^2$$
$$r=r^2-12r+36$$
$$0=r^2-13r+36$$
$$0=(r-9)(r-4)$$
$$r-9=0 \text{ or } r-4=0$$
$$r=9 \qquad r=4$$

4 is an extraneous solution. $\{9\}$

63) $\sqrt{y+\sqrt{y+5}}=\sqrt{y+2}$

$$\left(\sqrt{y+\sqrt{y+5}}\right)^2=\left(\sqrt{y+2}\right)^2$$
$$y+\sqrt{y+5}=y+2$$
$$\sqrt{y+5}=2$$
$$\left(\sqrt{y+5}\right)^2=(2)^2$$
$$y+5=4$$
$$y=-1$$

Check is left to the student. $\{-1\}$

65) $V = \sqrt{\frac{2E}{m}}$

$V^2 = \left(\sqrt{\frac{2E}{m}}\right)^2$

$V^2 = \frac{2E}{m}$

$mV^2 = 2E$

$\frac{mV^2}{2} = E$

67) $c = \sqrt{a^2 + b^2}$

$c^2 = \left(\sqrt{a^2 + b^2}\right)^2$

$c^2 = a^2 + b^2$

$c^2 - a^2 = b^2$

69) $T = \sqrt[4]{\frac{E}{\sigma}}$

$T^4 = \left(\sqrt[4]{\frac{E}{\sigma}}\right)^4$

$T^4 = \frac{E}{\sigma}$

$\sigma T^4 = E$

$\sigma = \frac{E}{T^4}$

71) a) Let $T = -17$

$V_s = 20\sqrt{-17 + 273}$

$= 20\sqrt{256}$

$= 20(16)$

$= 320$ 320 m/s

b) Let $T = 16$

$V_s = 20\sqrt{16 + 273}$

$= 20\sqrt{289}$

$= 20(17)$

$= 340$ 340 m/s

c) The speed of sound increases.

d) $V_s = 20\sqrt{T + 273}$

$V_s^2 = \left(20\sqrt{T + 273}\right)^2$

$V_s^2 = 400(T + 273)$

$\frac{V_s^2}{400} = T + 273$

$\frac{V_s^2}{400} - 273 = T$

73) a) Let $V = 28\pi$ and $h = 7$, solve for r.

$r = \sqrt{\frac{28\pi}{\pi(7)}} = \sqrt{\frac{28}{7}} = \sqrt{4} = 2$

2 in.

b) $r = \sqrt{\frac{V}{\pi h}}$

$r^2 = \left(\sqrt{\frac{V}{\pi h}}\right)^2$

$r^2 = \frac{V}{\pi h}$

$\pi r^2 h = V$

Chapter 10 Review

1) $\sqrt{25} = 5$ 3) $-\sqrt{81} = -9$

5) $\sqrt[3]{64} = 4$ 7) $\sqrt[3]{-1} = -1$

9) $\sqrt[6]{-64}$ is not real

11) Since 34 is between 25 and 36,

$\sqrt{25} < \sqrt{34} < \sqrt{36}$

$5 < \sqrt{34} < 6$

$\sqrt{34} \approx 5.8$

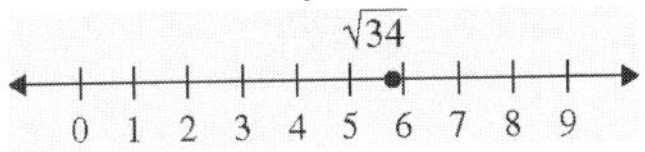

13) The denominator of the fractional exponent becomes the index on the radical. The numerator is the power to which we raise the radical expression. $8^{2/3}=\left(\sqrt[3]{8}\right)^2$

15) $36^{1/2}=\sqrt{36}=6$

17) $\left(\dfrac{27}{125}\right)^{1/3}=\sqrt[3]{\dfrac{27}{125}}=\dfrac{3}{5}$

19) $-16^{1/4}=-\sqrt[4]{16}=-2$

21) $125^{2/3}=\left(\sqrt[3]{125}\right)^2=(5)^2=25$

23) $\left(\dfrac{64}{27}\right)^{2/3}=\left(\sqrt[3]{\dfrac{64}{27}}\right)^2=\left(\dfrac{4}{3}\right)^2=\dfrac{16}{9}$

25) $81^{-1/2}=\left(\dfrac{1}{81}\right)^{1/2}=\sqrt{\dfrac{1}{81}}=\dfrac{1}{9}$

27) $81^{-3/4}=\left(\dfrac{1}{81}\right)^{3/4}=\left(\sqrt[4]{\dfrac{1}{81}}\right)^3$
$=\left(\dfrac{1}{3}\right)^3=\dfrac{1}{27}$

29) $\left(\dfrac{27}{1000}\right)^{-2/3}=\left(\dfrac{1000}{27}\right)^{2/3}=\left(\sqrt[3]{\dfrac{1000}{27}}\right)^2$
$=\left(\dfrac{10}{3}\right)^2=\dfrac{100}{9}$

31) $3^{6/7}\cdot 3^{8/7}=3^{6/7+8/7}=3^{14/7}=3^2=9$

33) $\left(8^{1/5}\right)^{10}=8^{\frac{1}{5}\cdot 10}=8^2=64$

35) $\dfrac{7^2}{7^{5/3}\cdot 7^{1/3}}=\dfrac{7^2}{7^{5/3+1/3}}=\dfrac{7^2}{7^{6/3}}=\dfrac{7^2}{7^2}=1$

37) $\left(64a^4b^{12}\right)^{5/6}=64^{5/6}\cdot\left(a^4\right)^{5/6}\cdot\left(b^{12}\right)^{5/6}$
$=\left(\sqrt[6]{64}\right)^5\cdot a^{4\cdot\frac{5}{6}}\cdot b^{12\cdot\frac{5}{6}}$
$=2^5\cdot a^{10/3}\cdot b^{10}$
$=32a^{10/3}b^{10}$

39) $\left(\dfrac{81c^{-5}d^9}{16c^{-1}d^2}\right)^{-1/4}=\left(\dfrac{81d^7}{16c^4}\right)^{-1/4}$
$=\left(\dfrac{16c^4}{81d^7}\right)^{1/4}=\dfrac{16^{1/4}\cdot c^{4\cdot\frac{1}{4}}}{81^{1/4}d^{7\cdot\frac{1}{4}}}=\dfrac{2c}{3d^{7/4}}$

41) $\sqrt[12]{27^4}=\left(27^4\right)^{1/12}=27^{1/3}=\sqrt[3]{27}=3$

43) $\sqrt[3]{7^3}=7^{3/3}=7$

45) $\sqrt[4]{k^{28}}=k^{28/4}=k^7$

47) $\sqrt{w^6}=w^{6/2}=w^3$

49) $\sqrt{1000}=\sqrt{100\cdot 10}=\sqrt{100}\cdot\sqrt{10}$
$=10\sqrt{10}$

51) $\sqrt{\dfrac{18}{49}}=\dfrac{\sqrt{18}}{\sqrt{49}}=\dfrac{\sqrt{9\cdot 2}}{7}=\dfrac{3\sqrt{2}}{7}$

53) $\sqrt{k^{12}}=k^{12/2}=k^6$

55) $\sqrt{x^9}=\sqrt{x^8}\cdot\sqrt{x}=x^{8/2}\cdot\sqrt{x}=x^4\sqrt{x}$

57) $\sqrt{45t^2}=\sqrt{45}\cdot\sqrt{t^2}=\sqrt{9}\cdot\sqrt{5}\cdot t^{2/2}$
$=3\sqrt{5}\cdot t=3t\sqrt{5}$

59) $\sqrt{72x^7y^{13}}$
$= \sqrt{72} \cdot \sqrt{x^7} \cdot \sqrt{y^{13}}$
$= 6\sqrt{2} \cdot x^3\sqrt{x} \cdot y^6\sqrt{y} = 6x^3y^6\sqrt{2xy}$

61) $\sqrt{5} \cdot \sqrt{3} = \sqrt{5 \cdot 3} = \sqrt{15}$

63) $\sqrt{2} \cdot \sqrt{12} = \sqrt{2 \cdot 12} = \sqrt{24}$
$= \sqrt{4} \cdot \sqrt{6} = 2\sqrt{6}$

65) $\sqrt{11x^5} \cdot \sqrt{11x^8} = \sqrt{121x^{13}}$
$= 11 \cdot \sqrt{x^{12}} \cdot \sqrt{x}$
$= 11x^6\sqrt{x}$

67) $\dfrac{\sqrt{200k^{21}}}{\sqrt{2k^5}} = \sqrt{\dfrac{200k^{21}}{2k^5}} = \sqrt{100k^{21-5}}$
$= \sqrt{100k^{16}} = 10k^8$

69) $\sqrt[3]{16} = \sqrt[3]{8} \cdot \sqrt[3]{2} = 2\sqrt[3]{2}$

71) $\sqrt[4]{48} = \sqrt[4]{16} \cdot \sqrt[4]{3} = 2\sqrt[4]{3}$

73) $\sqrt[4]{z^{24}} = z^{24/4} = z^6$

75) $\sqrt[3]{a^{20}} = \sqrt[3]{a^{18}} \cdot \sqrt[3]{a^2} = a^{18/3} \cdot \sqrt[3]{a^2}$
$= a^6\sqrt[3]{a^2}$

77) $\sqrt[3]{16z^{15}} = \sqrt[3]{16} \cdot \sqrt[3]{z^{15}} = 2\sqrt[3]{2} \cdot z^{15/3}$
$= 2\sqrt[3]{2} \cdot z^5 = 2z^5\sqrt[3]{2}$

79) $\sqrt[4]{\dfrac{h^{12}}{81}} = \dfrac{\sqrt[4]{h^{12}}}{\sqrt[4]{81}} = \dfrac{h^{12/4}}{3} = \dfrac{h^3}{3}$

81) $\sqrt[3]{3} \cdot \sqrt[3]{7} = \sqrt[3]{3 \cdot 7} = \sqrt[3]{21}$

83) $\sqrt[4]{4t^7} \cdot \sqrt[4]{8t^{10}} = \sqrt[4]{32t^{17}} = \sqrt[4]{32} \cdot \sqrt[4]{t^{17}}$
$= 2\sqrt[4]{2} \cdot t^4\sqrt[4]{t} = 2t^4\sqrt[4]{2t}$

85) $\sqrt[3]{n} \cdot \sqrt{n} = n^{1/3} \cdot n^{1/2} = n^{2/6+3/6}$
$= n^{5/6} = \sqrt[6]{n^5}$

87) $8\sqrt{5} + 3\sqrt{5} = 11\sqrt{5}$

89) $\sqrt{80} - \sqrt{48} + \sqrt{20}$
$= 4\sqrt{5} - 4\sqrt{3} + 2\sqrt{5} = 6\sqrt{5} - 4\sqrt{3}$

91) $3p\sqrt{p} - 7\sqrt{p^3}$
$= 3p\sqrt{p} - 7\left(p\sqrt{p}\right)$
$= 3p\sqrt{p} - 7p\sqrt{p} = -4p\sqrt{p}$

93) $10d^2\sqrt{8d} - 32d\sqrt{2d^3}$
$= 10d^2\left(2\sqrt{2d}\right) - 32d\left(d\sqrt{2d}\right)$
$= 20d^2\sqrt{22} - 32d^2\sqrt{2d}$
$= -12d^2\sqrt{2d}$

95) $3\sqrt{k}\left(\sqrt{20k} + \sqrt{2}\right)$
$= 3\sqrt{k}\left(2\sqrt{5k} + \sqrt{2}\right)$
$= 6\sqrt{5k^2} + 3\sqrt{2k} = 6k\sqrt{5} + 3\sqrt{2k}$

97) $\left(\sqrt{2r} + 5\sqrt{s}\right)\left(3\sqrt{s} + 4\sqrt{2r}\right)$
$= 3\sqrt{2rs} + 4\sqrt{4r^2} + 15\sqrt{s^2} + 20\sqrt{2rs}$
$= 23\sqrt{2rs} + 4(2r) + 15s$
$= 23\sqrt{2rs} + 8r + 15s$

99) $\left(1 + \sqrt{y+1}\right)^2$
$= 1^2 + 2(1)\left(\sqrt{y+1}\right) + \left(\sqrt{y+1}\right)^2$
$= 1 + 2\sqrt{y+1} + y + 1$
$= 2 + 2\sqrt{y+1} + y$

101) $\frac{14}{\sqrt{3}} = \frac{14}{\sqrt{3}} \cdot \frac{\sqrt{3}}{\sqrt{3}} = \frac{14\sqrt{3}}{3}$

103) $\frac{\sqrt{18k}}{\sqrt{n}} = \frac{3\sqrt{2k}}{\sqrt{n}} = \frac{3\sqrt{2k}}{\sqrt{n}} \cdot \frac{\sqrt{n}}{\sqrt{n}}$

$= \frac{3\sqrt{2kn}}{n}$

105) $\frac{7}{\sqrt[3]{2}} = \frac{7}{\sqrt[3]{2}} \cdot \frac{\sqrt[3]{2^2}}{\sqrt[3]{2^2}} = \frac{7\sqrt[3]{2^2}}{\sqrt[3]{2^3}} = \frac{7\sqrt[3]{4}}{2}$

107) $\frac{\sqrt[3]{x^2}}{\sqrt[3]{y}} = \frac{\sqrt[3]{x^2}}{\sqrt[3]{y}} \cdot \frac{\sqrt[3]{y^2}}{\sqrt[3]{y^2}} = \frac{\sqrt[3]{x^2y^2}}{\sqrt[3]{y^3}}$

$= \frac{\sqrt[3]{x^2y^2}}{y}$

109) $\frac{2}{3+\sqrt{3}} = \frac{2}{3+\sqrt{3}} \cdot \frac{3-\sqrt{3}}{3-\sqrt{3}}$

$= \frac{2(3-\sqrt{3})}{(3)^2-(\sqrt{3})^2}$

$= \frac{2(3-\sqrt{3})}{9-3}$

$= \frac{2(3-\sqrt{3})}{6} = \frac{3-\sqrt{3}}{3}$

111) $\frac{8-24\sqrt{2}}{8} = \frac{8(1-3\sqrt{2})}{8} = 1-3\sqrt{2}$

113) $\sqrt{x+8} = 3$

$(\sqrt{x+8})^2 = 3^2$

$x+8 = 9$

$x = 1$

Check $\sqrt{1+8} = 1$

$\sqrt{9} = 3 \quad \{1\}$

115) $\sqrt{3j+4} = -\sqrt{4j-1}$

$(\sqrt{3j+4})^2 = (-\sqrt{4j-1})^2$

$3j+4 = 4j-1$

$5 = j$

Check $\sqrt{3(5)+4} = -\sqrt{4(5)-1}$

$\sqrt{19} \neq -\sqrt{19} \qquad \varnothing$

117) $a = \sqrt{a+8} - 6$

$a+6 = \sqrt{a+8}$

$(a+6)^2 = (\sqrt{a+8})^2$

$a^2+12a+36 = a+8$

$a^2+11a+28 = 0$

$(a+7)(a+4) = 0$

$a+7=0$ or $a+4=0$

$a=-7 \qquad a=-4 \qquad \{-4\}$

-7 is an extraneous solution.

119) $\sqrt{4a+1} - \sqrt{a-2} = 3$

$\sqrt{4a+1} = 3+\sqrt{a-2}$

$(\sqrt{4a+1})^2 = (3+\sqrt{a-2})^2$

$4a+1 = 9+6\sqrt{a-2}+a-2$

$4a+1 = 7+a+6\sqrt{a-2}$

$3a-6 = 6\sqrt{a-2}$

$a-2 = 2\sqrt{a-2}$

$(a-2)^2 = (2\sqrt{a-2})^2$

$a^2-4a+4 = 4(a-2)$

$a^2-4a+4 = 4a-8$

$a^2-8a+12 = 0$

$(a-6)(a-2) = 0$

$a-6=0$ or $a-2=0$

$a=6 \qquad a=2 \qquad \{2,6\}$

Check is left to the student.

121) $r=\sqrt{\dfrac{3V}{\pi h}}$

$r^2=\left(\sqrt{\dfrac{3V}{\pi h}}\right)^2$

$r^2=\dfrac{3V}{\pi h}$

$\pi r^2 h=3V$

$\dfrac{1}{3}\pi r^2 h=V$

Chapter 10 Test

1) $\sqrt{144}=12$ 3) $\sqrt[3]{-27}=-3$

5) $16^{1/4}=\sqrt[4]{16}=2$

7) $49^{-1/2}=\left(\dfrac{1}{49}\right)^{1/2}=\sqrt{\dfrac{1}{49}}=\dfrac{1}{7}$

9) $m^{3/8}\cdot m^{1/4}=m^{3/8+2/8}=m^{5/8}$

11) $\left(2x^{3/10}y^{-2/5}\right)^{-5}=\left(\dfrac{2x^{3/10}}{y^{2/5}}\right)^{-5}$

$=\left(\dfrac{y^{2/5}}{2x^{3/10}}\right)^{5}=\dfrac{y^2}{32x^{3/2}}$

13) $\sqrt[3]{48}=\sqrt[3]{8}\cdot\sqrt[3]{6}=2\sqrt[3]{6}$

15) $\sqrt{y^6}=y^{6/2}=y^3$

17) $\sqrt{t^9}=\sqrt{t^8}\cdot\sqrt{t}=t^4\sqrt{t}$

19) $\sqrt[3]{c^{23}}=\sqrt[3]{c^{21}}\cdot\sqrt[3]{c^2}=c^7\sqrt[3]{c^2}$

21) $\sqrt{3}\cdot\sqrt{12}=\sqrt{36}=6$

23) $\dfrac{\sqrt{120w^{15}}}{\sqrt{2w^4}}=\sqrt{\dfrac{120w^{15}}{2w^4}}=\sqrt{60w^{11}}$

$=\sqrt{60}\cdot\sqrt{w^{11}}$

$=2\sqrt{15}\cdot w^5\sqrt{w}=2w^5\sqrt{15w}$

25) $\sqrt{12}-\sqrt{108}+\sqrt{18}$

$=2\sqrt{3}-6\sqrt{3}+3\sqrt{2}=3\sqrt{2}-4\sqrt{3}$

27) $\sqrt{6}\left(\sqrt{2}-5\right)=\sqrt{12}-5\sqrt{6}$

$=2\sqrt{3}-5\sqrt{6}$

29) $\left(\sqrt{7}+\sqrt{3}\right)\left(\sqrt{7}-\sqrt{3}\right)$

$=\left(\sqrt{7}\right)^2-\left(\sqrt{3}\right)^2=7-3=4$

31) $2\sqrt{t}\left(\sqrt{t}-\sqrt{3u}\right)=2\sqrt{t^2}-2\sqrt{3tu}$

$=2t-2\sqrt{3tu}$

33) $\dfrac{8}{\sqrt{7}+3}=\dfrac{8}{\sqrt{7}+3}\cdot\dfrac{\sqrt{7}-3}{\sqrt{7}-3}$

$=\dfrac{8\left(\sqrt{7}-3\right)}{\left(\sqrt{7}\right)^2-(3)^2}=\dfrac{8\left(\sqrt{7}-3\right)}{7-9}$

$=\dfrac{8\left(\sqrt{7}-3\right)}{-2}=-4\left(\sqrt{7}-3\right)$

$=12-4\sqrt{7}$

35) $\dfrac{5}{\sqrt[3]{9}}=\dfrac{5}{\sqrt[3]{3^2}}\cdot\dfrac{\sqrt[3]{3}}{\sqrt[3]{3}}=\dfrac{5\sqrt[3]{3}}{\sqrt[3]{3^3}}=\dfrac{5\sqrt[3]{3}}{3}$

37) $\sqrt{5h+4}=3$

$$\left(\sqrt{5h+4}\right)^2=3^2$$
$$5h+4=9$$
$$5h=5$$
$$h=1 \qquad \{1\}$$

Check is left to the student.

39) $\sqrt{3k+1}-\sqrt{2k-1}=1$

$$\sqrt{3k+1}=1+\sqrt{2k-1}$$
$$\left(\sqrt{3k+1}\right)^2=\left(1+\sqrt{2k-1}\right)^2$$
$$3k+1=1+2\sqrt{2k-1}+2k-1$$
$$3k+1=2k+2\sqrt{2k-1}$$
$$k+1=2\sqrt{2k-1}$$
$$(k+1)^2=\left(2\sqrt{2k-1}\right)^2$$
$$k^2+2k+1=4(2k-1)$$
$$k^2+2k+1=8k-4$$
$$k^2-6k+5=0$$
$$(k-5)(k-1)=0$$
$$k-5=0 \text{ or } k-1=0$$
$$k=5 \qquad k=1 \qquad \{1,5\}$$

Check is left to the student.

Cumulative Review: Chapters 1-10

1) $4x-3y+9-\frac{2}{3}x+y-1$

$$=\frac{12}{3}x-\frac{2}{3}x-3y+y+9-1$$
$$=\frac{10}{3}x-2y+8$$

3) $3(2c-1)+7=9c+5(c+2)$

$$6c-3+7=9c+5c+10$$
$$6c+4=14c+10$$
$$-6=8c$$
$$-\frac{6}{8}=c$$
$$-\frac{3}{4}=c$$
$$\left\{-\frac{3}{4}\right\}$$

5) $m=\frac{-2-3}{1-5}=\frac{-5}{-4}=\frac{5}{4}$

$$y-y_1=m(x-x_1)$$
$$y-3=\frac{5}{4}(x-5)$$
$$y-3=\frac{5}{4}x-\frac{25}{4}$$
$$y=\frac{5}{4}x-\frac{13}{4}$$

7) $(5p^2-2)(3p^2-4p-1)$

$$=15p^4-20p^3-5p^2-6p^2+8p+2$$
$$=15p^4-20p^3-11p^2+8p+2$$

9) $4w^2+5w-6=(4w-3)(w+2)$

11) $3(k^2+20)-4k=2k^2+11k+6$

$$3k^2+60-4k=2k^2+11k+6$$
$$k^2-15k+54=0$$
$$(k-9)(k-6)=0$$
$$k-9=0 \text{ or } k-6=0$$
$$k=9 \qquad k=6 \qquad \{6,9\}$$

13) $\dfrac{5a^2+3}{a^2+4a}-\dfrac{3a-2}{a+4}=\dfrac{5a^2+3}{a(a+4)}-\dfrac{3a-2}{a+4}$

$=\dfrac{5a^2+3-a(3a-2)}{a(a+4)}$

$=\dfrac{5a^2+3-3a^2+2a}{a(a+4)}$

$=\dfrac{2a^2+2a+3}{a(a+4)}$

15) $\dfrac{3}{r^2+8r+15}-\dfrac{4}{r+3}=1$

$\dfrac{3}{(r+3)(r+5)}-\dfrac{4}{r+3}=1$

$3-4(r+5)=r^2+8r+15$

$3-4r-20=r^2+8r+15$

$0=r^2+12r+32$

$0=(r+8)(r+4)$

$r+8=0$ or $r+4=0$

$r=-8 \qquad r=-4 \qquad \{-8,-4\}$

17) $\left[\begin{array}{ccc|c} 1 & 3 & 1 & 3 \\ 2 & -1 & -5 & -1 \\ -1 & 2 & 3 & 0 \end{array}\right] \xrightarrow[R_1+R_3\to R_3]{-2R_1+R_2\to R_2} \left[\begin{array}{ccc|c} 1 & 3 & 1 & 3 \\ 0 & -7 & -7 & -7 \\ 0 & 5 & 4 & 3 \end{array}\right] \xrightarrow{-\frac{1}{7}R_2\to R_2}$

$\left[\begin{array}{ccc|c} 1 & 3 & 1 & 3 \\ 0 & 1 & 1 & 1 \\ 0 & 5 & 4 & 3 \end{array}\right] \xrightarrow{-5R_2+R_3\to R_3} \left[\begin{array}{ccc|c} 1 & 3 & 1 & 3 \\ 0 & 1 & 1 & 1 \\ 0 & 0 & -1 & -2 \end{array}\right] \xrightarrow{-R_3\to R_3} \left[\begin{array}{ccc|c} 1 & 3 & 1 & 3 \\ 0 & 1 & 1 & 1 \\ 0 & 0 & 1 & 2 \end{array}\right]$ $\begin{aligned} x+3y+z&=3 \\ y+z&=1 \\ z&=2 \end{aligned}$

$y+z=1 \qquad x+3y+z=3$

$y+2=1 \qquad x+3(-1)+2=3$

$y=-1 \qquad x-3+2=3$

$x-1=3$

$x=4 \quad (4,-1,2)$

19) a) $81^{1/2}=\sqrt{81}=9$

b) $8^{4/3}=\left(\sqrt[3]{8}\right)^4=(2)^4=16$

c) $27^{-1/3}=\left(\dfrac{1}{27}\right)^{1/3}=\sqrt[3]{\dfrac{1}{27}}=\dfrac{1}{3}$

21) a) $\sqrt{\dfrac{20}{50}}=\dfrac{\sqrt{2}}{\sqrt{5}}=\dfrac{\sqrt{2}}{\sqrt{5}}\cdot\dfrac{\sqrt{5}}{\sqrt{5}}=\dfrac{\sqrt{10}}{5}$

b) $\dfrac{6}{\sqrt[3]{2}}=\dfrac{6}{\sqrt[3]{2}}\cdot\dfrac{\sqrt[3]{2^2}}{\sqrt[3]{2^2}}$

$=\dfrac{6\sqrt[3]{2^2}}{\sqrt[3]{2^3}}=\dfrac{6\sqrt[3]{4}}{2}=3\sqrt[3]{4}$

c) $\dfrac{x}{\sqrt[3]{y^2}}=\dfrac{x}{\sqrt[3]{y^2}}\cdot\dfrac{\sqrt[3]{y}}{\sqrt[3]{y}}=\dfrac{x\sqrt[3]{y}}{\sqrt[3]{y^3}}=\dfrac{x\sqrt[3]{y}}{y}$

d) $\dfrac{\sqrt{a}-2}{1-\sqrt{a}}$

$=\dfrac{\sqrt{a}-2}{1-\sqrt{a}}\cdot\dfrac{1+\sqrt{a}}{1+\sqrt{a}}$

$=\dfrac{\left(\sqrt{a}-2\right)\left(1+\sqrt{a}\right)}{(1)^2-\left(\sqrt{a}\right)^2}$

$=\dfrac{\sqrt{a}+a-2-2\sqrt{a}}{1-a}=\dfrac{a-2-\sqrt{a}}{1-a}$

Section 11.1: Exercises

1) $(t+7)(t-6)=0$

$t+7=0$ or $t-6=0$

$t=-7 \qquad t=6 \qquad \{-7, 6\}$

3) $u^2+15u+44=0$

$(u+11)(u+4)=0$

$u+11=0$ or $u+4=0$

$u=-11 \qquad u=-4 \qquad \{-11, -4\}$

5) $x^2=x+56$

$x^2-x-56=0$

$(x-8)(x+7)=0$

$x-8=0$ or $x+7=0$

$x=8 \qquad x=-7 \qquad \{-7, 8\}$

7) $1-100w^2=0$

$(1+10w)(1-10w)=0$

$1+10w=0$ or $1-10w=0$

$10w=-1 \qquad -10w=-1$

$w=-\frac{1}{10} \qquad w=\frac{1}{10}$

$\left\{-\frac{1}{10}, \frac{1}{10}\right\}$

9) $5m^2+8=22m$

$5m^2-22m+8=0$

$(5m-2)(m-4)=0$

$5m-2=0$ or $m-4=0$

$5m=2 \qquad m=4$

$m=\frac{2}{5} \qquad \left\{\frac{2}{5}, 4\right\}$

11) $23d=-10-6d^2$

$6d^2+23d+10=0$

$(3d+10)(2d+1)=0$

$3d+10=0$ or $2d+1=0$

$3d=-10 \qquad 2d=-1$

$a=-\frac{10}{3} \qquad 2d=-\frac{1}{2}$

$\left\{-\frac{10}{3}, -\frac{1}{2}\right\}$

13) $2r=7r^2$

$0=7r^2-2r$

$0=r(7r-2)$

$7r-2=0$ or $r=0$

$7r=2$

$r=\frac{2}{7} \qquad \left\{0, \frac{2}{7}\right\}$

15) quadratic

17) linear

19) quadratic

21) linear

23) $13c=2c^2+6$

$0=2c^2-13c+6$

$0=(2c-1)(c-6)$

$2c-1=0$ or $c-6=0$

$2c=1 \qquad c=6$

$c=\frac{1}{2} \qquad \left\{\frac{1}{2}, 6\right\}$

25) $2p(p+4)=p^2+5p+10$

$2p^2+8p=p^2+5p+10$

$p^2+3p-10=0$

$(p+5)(p-2)=0$

$p+5=0$ or $p-2=0$

$p=-5 \qquad p=2 \qquad \{-5, 2\}$

27) $5(3n-2)-11n=2n-1$
$$15n-10-11n=2n-1$$
$$4n-10=2n-1$$
$$2n=9$$
$$n=\frac{9}{2} \qquad \left\{\frac{9}{2}\right\}$$

29) $3t^3+5t=-8t^2$
$$3t^3+8t^2+5t=0$$
$$t(3t^2+8t+5)=0$$
$$t(3t+5)(t+1)=0$$
$$t=0 \text{ or } 3t+5=0 \text{ or } t+1=0$$
$$3t=-5 \qquad t=-1$$
$$t=-\frac{5}{3}$$
$$\left\{-\frac{5}{3}, -1, 0\right\}$$

31) $2(r+5)=10-4r^2$
$$2r+10=10-4r^2$$
$$2r=-4r^2$$
$$4r^2+2r=0$$
$$2r(2r+1)=0$$
$$2r+1=0 \text{ or } 2r=0$$
$$2r=-1 \qquad r=0$$
$$r=-\frac{1}{2} \qquad \left\{-\frac{1}{2}, 0\right\}$$

33) $9y-6(y+1)=12-5y$
$$9y-6y-6=12-5y$$
$$3y-6=12-5y$$
$$8y=18$$
$$y=\frac{18}{8}=\frac{9}{4} \qquad \left\{\frac{9}{4}\right\}$$

35) $\frac{1}{16}w^2+\frac{1}{8}w=\frac{1}{2}$
$$16\left(\frac{1}{16}w^2+\frac{1}{8}w\right)=16\cdot\frac{1}{2}$$
$$w^2+2w=8$$
$$w^2+2w-8=0$$
$$(w+4)(w-2)=0$$
$$w+4=0 \text{ or } w-2=0$$
$$w=-4 \qquad w=2 \qquad \{-4, 2\}$$

37) $12n+3=-12n^2$
$$12n^2+12n+3=0$$
$$4n^2+4n+1=0 \qquad \text{divide by 3}$$
$$(2n+1)^2=0$$
$$2n+1=0$$
$$2n=-1$$
$$n=-\frac{1}{2} \qquad \left\{-\frac{1}{2}\right\}$$

39) $3b^2-b-7=4b(2b+3)-1$
$$3b^2-b-7=8b^2+12b-1$$
$$0=5b^2+13b+6$$
$$0=(5b+3)(b+2)$$
$$5b+3=0 \text{ or } b+2=0$$
$$5b=-3 \qquad b=-2$$
$$b=-\frac{3}{5} \qquad \left\{-2, -\frac{3}{5}\right\}$$

41) $t^3+7t^2-4t-28=0$
$$t^2(t+7)-4(t+7)=0$$
$$(t+7)(t^2-4)=0$$
$$(t+7)(t+2)(t-2)=0$$
$$t+7=0 \text{ or } t+2=0 \text{ or } t-2=0$$
$$t=-7 \qquad t=-2 \qquad t=2$$
$$\{-7, -2, 2\}$$

43) $w = \text{width}$

$w+5 = \text{length}$

$\text{Area} = (\text{width})(\text{length})$

$14 = w(w+5)$

$14 = w^2 + 5w$

$0 = w^2 + 5w - 14$

$0 = (w+7)(w-2)$

$w+7=0$ or $w-2=0$

$w=-7$ $\quad \boxed{w=2}$

$\text{width} = 2$ in.

$\text{length} = 7$ in.

45) $w = \text{width}$

$2w-1 = \text{length}$

$\text{Area} = (\text{width})(\text{length})$

$45 = w(2w-1)$

$45 = 2w^2 - w$

$0 = 2w^2 - w - 45$

$0 = (2w+9)(w-5)$

$2w+9=0$ or $w-5=0$

$w=-9$ $\quad \boxed{w=5}$

$w = -\dfrac{9}{2}$

$\text{width} = 5$ cm

$\text{length} = 2(5)-1 = 9$ cm

47) $A = \dfrac{1}{2}(\text{base})(\text{height})$

$18 = \dfrac{1}{2}(x+6)(x+1)$

$36 = (x+6)(x+1)$

$36 = x^2 + 7x + 6$

$0 = x^2 + 7x - 30$

$0 = (x+10)(x-3)$

$x+10=0$ or $x-3=0$

$x=-10$ $\quad \boxed{x=3}$

$\text{base} = x+6 = 3+6 = 9$ in.

$\text{length} = x+1 = 3+1 = 4$ in.

49) $A = \dfrac{1}{2}(\text{base})(\text{height})$

$36 = \dfrac{1}{2}\left(\dfrac{1}{2}x\right)(x)$

$36 = \dfrac{1}{4}x^2$

$144 = x^2$

$0 = x^2 - 144$

$0 = (x+12)(x-12)$

$x+12=0$ or $x-12=0$

$x=-12$ $\quad \boxed{x=12}$

$\text{base} = \dfrac{1}{2}x = \dfrac{1}{2}(12) = 6$ cm

$\text{length} = x = 12$ cm

51) $x^2 + (x-7)^2 = (x+1)^2$

$x^2 + x^2 - 14x + 49 = x^2 + 2x + 1$

$2x^2 - 14x + 49 = x^2 + 2x + 1$

$x^2 - 16x + 48 = 0$

$(x-12)(x-4) = 0$

$x-12=0$ or $x-4=0$

$\boxed{x=12}$ $\quad x=4$

Reject $x=4$ because if $x=4$ then the length of the leg labelled $x-7$ would be -3 units.

$\text{one leg} = x = 12$ units

$\text{other leg} = x-7 = 12-7 = 5$ units

$\text{hypotenuse} = x+1 = 12+1 = 13$ units

53) $(2x)^2+(x+5)^2=(3x+1)^2$

$4x^2+(x^2+10x+25)=9x^2+6x+1$

$5x^2+10x+25=9x^2+6x+1$

$0=4x^2-4x-24$

$0=x^2-x-6$

$0=(x-3)(x+2)$

$x-3=0$ or $x+2=0$

$\boxed{x=3}$ $\quad x=-2$

one leg $=2x=2(3)=6$ units

other leg $=x+5=3+5=8$ units

hypotenuse $=3x+1=3(3)+1=10$ units

Section 11.2: Exercises

1) Factoring:

$y^2-16=0$

$(y+4)(y-4)=0$

$y+4=0$ or $y-4=0$

$y=-4 \quad y=4$

Square Root Property:

$y^2-16=0$

$y^2=16$

$y=\pm\sqrt{16}$

$y=\pm4 \quad \{-4, 4\}$

3) $b^2=36$

$b=\pm\sqrt{36}$

$b=\pm6 \quad \{-6, 6\}$

5) $t^2=-25$

$t=\pm\sqrt{-25}$

no real number solution

7) $r^2-27=0$

$r^2=27$

$r=\pm\sqrt{27}$

$r=\pm3\sqrt{3} \quad \{-3\sqrt{3}, 3\sqrt{3}\}$

9) $n^2=\frac{4}{9}$

$n=\pm\sqrt{\frac{4}{9}}=\pm\frac{2}{3} \quad \left\{-\frac{2}{3}, \frac{2}{3}\right\}$

11) $z^2+5=19$

$z^2=14$

$z=\pm\sqrt{14} \quad \{-\sqrt{14}, \sqrt{14}\}$

13) $2d^2+5=55$

$2d^2=50$

$d^2=25$

$d=\pm\sqrt{25}=\pm5 \quad \{-5, 5\}$

15) $4p^2+9=89$

$4p^2=80$

$p^2=20$

$p=\pm\sqrt{20}$

$p=\pm2\sqrt{5} \quad \{-2\sqrt{5}, 2\sqrt{5}\}$

17) $1=7-6h^2$

$-6=-6h^2$

$1=h^2$

$\pm\sqrt{1}=h$

$\pm1=h \quad \{-1, 1\}$

19) $2=11+9x^2$

$-9=9x^2$

$-1=x^2$

$\pm\sqrt{-1}=x$

no real number solution

21) $(r+10)^2=4$

$r+10=\pm\sqrt{4}$

$r+10=\pm 2$

$r+10=2$ or $r+10=-2$

$r=-8$ $\quad$ $r=-12$

$\{-12, -8\}$

23) $(q-7)^2=1$

$q-7=\pm\sqrt{1}$

$q-7=\pm 1$

$q-7=1$ or $q-7=-1$

$q=8$ $\quad$ $q=6$ $\quad$ $\{6, 8\}$

25) $(a+1)^2=22$

$a+1=\pm\sqrt{22}$

$a=-1\pm\sqrt{22}$

$\{-1-\sqrt{22}, -1+\sqrt{22}\}$

27) $(p+4)^2-18=0$

$(p+4)^2=18$

$p+4=\pm\sqrt{18}$

$p+4=\pm 3\sqrt{2}$

$p=-4\pm 3\sqrt{2}$

$\{-4-3\sqrt{2}, -4+3\sqrt{2}\}$

29) $(5y-2)^2+6=22$

$(5y-2)^2=16$

$5y-2=\pm\sqrt{16}$

$5y-2=\pm 4$

$5y-2=-4$ or $5y-2=4$

$5y=-2$ $\quad$ $5y=6$

$y=-\frac{2}{5}$ $\quad$ $y=\frac{6}{5}$

$\left\{-\frac{2}{5}, \frac{6}{5}\right\}$

31) $20=(2w+1)^2$

$\pm\sqrt{20}=2w+1$

$-1\pm 2\sqrt{5}=2w$

$\frac{-1\pm 2\sqrt{5}}{2}=w$

$\left\{\frac{-1-2\sqrt{5}}{2}, \frac{-1+2\sqrt{5}}{2}\right\}$

33) $8=(3q-10)^2-6$

$14=(3q-10)^2$

$\pm\sqrt{14}=3q-10$

$10\pm\sqrt{14}=3q$

$\frac{10\pm\sqrt{14}}{3}=q$

$\left\{\frac{10-\sqrt{14}}{3}, \frac{10+\sqrt{14}}{3}\right\}$

35) $(p+3)^2+4=2$

$(p+3)^2=-2$

$p+3=\pm\sqrt{-2}$

no real number solution

37) $(10v+7)^2-9=15$

$(10v+7)^2=24$

$10v+7=\pm\sqrt{24}$

$10v=-7\pm2\sqrt{6}$

$v=\dfrac{-7\pm2\sqrt{6}}{10}$

$\left\{\dfrac{-7\pm2\sqrt{6}}{10},\dfrac{-7\pm2\sqrt{6}}{10}\right\}$

39) $\left(\dfrac{3}{4}n-8\right)^2=4$

$\dfrac{3}{4}n-8=\pm\sqrt{4}$

$\dfrac{3}{4}n-8=\pm2$

$\dfrac{3}{4}n-8=2$ or $\dfrac{3}{4}n-8=-2$

$\dfrac{3}{4}n=10 \qquad \dfrac{3}{4}n=6$

$n=\dfrac{40}{3} \qquad n=8$

$\left\{8,\dfrac{40}{3}\right\}$

41) $14=10+(5c+4)^2$

$4=(5c+4)^2$

$\pm\sqrt{4}=5c+4$

$\pm2=5c+4$

$5c+4=2$ or $5c+4=-2$

$5c=-2 \qquad 5c=-6$

$c=-\dfrac{2}{5} \qquad c=-\dfrac{6}{5}$

$\left\{-\dfrac{6}{5},-\dfrac{2}{5}\right\}$

43) $a^2+8^2=10^2$

$a^2+64=100$

$a^2=36$

$a=\pm\sqrt{36}=\pm6$

Reject $a=-6$. $\quad a=6$

45) $2^2+5^2=c^2$

$4+25=c^2$

$29=c^2$

$\pm\sqrt{29}=c$

Reject $c=-\sqrt{29}$. $\quad c=\sqrt{29}$

47) Let a = length of rectangle

$a^2+4^2=\left(2\sqrt{13}\right)^2$

$a^2+16=4(13)$

$a^2+16=52$

$a^2=36$

$a=\pm\sqrt{36}=\pm6$

Reject -6 for the length.

The length is 6 inches.

49) Let c = length of the diagonal

$\left(4\sqrt{2}\right)^2+5^2=c^2$

$16(2)+25=c^2$

$32+25=c^2$

$57=c^2$

$\pm\sqrt{57}=c$

Reject $-\sqrt{57}$ for the length of the diagonal. The length of the diagonal is $\sqrt{57}$ cm.

51) Let $h =$ the height the ladder reaches

$$5^2 + h^2 = 13^2$$
$$25 + h^2 = 169$$
$$h^2 = 144$$
$$h = \pm\sqrt{144} = \pm 12$$

Reject -12 for the height.

The ladder reaches 12 feet.

53) $d = \sqrt{(x_2 - x_1)^2 + (y_2 - y_1)^2}$

$$d = \sqrt{(3-7)^2 + [2-(-1)]^2}$$
$$d = \sqrt{(-4)^2 + (3)^2}$$
$$d = \sqrt{16+9} = \sqrt{25} = 5$$

55) $d = \sqrt{(x_2 - x_1)^2 + (y_2 - y_1)^2}$

$$d = \sqrt{[-2-(-5)]^2 + [-8-(-6)]^2}$$
$$d = \sqrt{3^2 + (-2)^2}$$
$$d = \sqrt{9+4} = \sqrt{13}$$

57) $d = \sqrt{(x_2 - x_1)^2 + (y_2 - y_1)^2}$

$$d = \sqrt{(0-0)^2 + (7-13)^2}$$
$$d = \sqrt{0^2 + (-6)^2}$$
$$d = \sqrt{36} = 6$$

59) $d = \sqrt{(x_2 - x_1)^2 + (y_2 - y_1)^2}$

$$d = \sqrt{[2-(-4)]^2 + (6-11)^2}$$
$$d = \sqrt{6^2 + (-5)^2}$$
$$d = \sqrt{36-25} = \sqrt{61}$$

61) $d = \sqrt{(x_2 - x_1)^2 + (y_2 - y_1)^2}$

$$d = \sqrt{(5-3)^2 + [-7-(-3)]^2}$$
$$d = \sqrt{2^2 + (-4)^2}$$
$$d = \sqrt{4+16} = \sqrt{20} = 2\sqrt{5}$$

Section 11.3: Exercises

1) False 3) True

5) $\sqrt{-81} = \sqrt{-1} \cdot \sqrt{81} = i \cdot 9 = 9i$

7) $\sqrt{-25} = \sqrt{-1} \cdot \sqrt{25} = i \cdot 5 = 5i$

9) $\sqrt{-6} = \sqrt{-1} \cdot \sqrt{6} = i\sqrt{6}$

11) $\sqrt{-27} = \sqrt{-1} \cdot \sqrt{27} = i \cdot 3\sqrt{3} = 3i\sqrt{3}$

13) $\sqrt{-60} = \sqrt{-1} \cdot \sqrt{60} = i \cdot 2\sqrt{15} = 2i\sqrt{15}$

15) Write each radical in terms of i before multiplying.

$$\sqrt{-5} \cdot \sqrt{-10} = i\sqrt{5} \cdot i\sqrt{10} = i^2\sqrt{50}$$
$$= -1\sqrt{25} \cdot \sqrt{2} = -5\sqrt{2}$$

17) $\sqrt{-1} \cdot \sqrt{-5} = (i)(i\sqrt{5}) = i^2\sqrt{5} = -\sqrt{5}$

19) $\sqrt{-12} \cdot \sqrt{-3} = (i\sqrt{12})(i\sqrt{3})$

$$= i^2\sqrt{36} = -1(6) = -6$$

21) $\dfrac{\sqrt{-60}}{\sqrt{-15}} = \dfrac{i\sqrt{60}}{i\sqrt{15}} = \sqrt{\dfrac{60}{15}} = \sqrt{4} = 2$

23) $(\sqrt{-13})^2 = (i\sqrt{13})^2 = i^2(13)$

$$= -1(13) = -13$$

25) Add the real parts and add the imaginary parts.

27) -1

29) $(-4+9i)+(7+2i)=3+11i$

31) $(13-8i)-(9+i)=4-9i$

33) $\left(-\frac{3}{4}-\frac{1}{6}i\right)-\left(-\frac{1}{2}+\frac{2}{3}i\right)$

$=\left(-\frac{3}{4}+\frac{1}{2}\right)+\left(-\frac{1}{6}i-\frac{2}{3}i\right)$

$=\left(-\frac{3}{4}+\frac{2}{4}\right)+\left(-\frac{1}{6}i-\frac{4}{6}i\right)=-\frac{1}{4}-\frac{5}{6}i$

35) $16i-(3+10i)+(3+i)$

$=16i-3-10i+3+i=7i$

37) $3(8-5i)=24-15i$

39) $\frac{2}{3}(-9+2i)=-6+\frac{4}{3}i$

41) $6i(5+6i)=30i+36i^2$

$=30i+36(-1)=-36+30i$

43) $(2+5i)(1+6i)=2+12i+5i+30i^2$

$=2+17i+30(-1)$

$=2+17i-30$

$=-28+17i$

45) $(-1+3i)(4-6i)=-4+6i+12i-18i^2$

$=-4+18i-18(-1)$

$=-4+18i+18$

$=14+18i$

47) $(5-3i)(9-3i)=45-15i-27i+9i^2$

$=45-42i+9(-1)$

$=45-42i-9$

$=36-42i$

49) $\left(\frac{3}{4}+\frac{3}{4}i\right)\left(\frac{2}{5}+\frac{1}{5}i\right)$

$=\frac{3}{10}+\frac{3}{20}i+\frac{3}{10}i+\frac{3}{20}i^2$

$=\frac{3}{10}+\frac{9}{20}i+\frac{3}{20}(-1)$

$=\frac{3}{10}+\frac{9}{20}i-\frac{3}{20}=\frac{3}{20}+\frac{9}{20}i$

51) $(11+4i)(11-4i)$

$=121-44i+44i-16i^2$

$=121-16(-1)=121+16=137$

53) $(-3-7i)(-3+7i)$

$=9-21i+21i-49i^2=9-49(-1)$

$=9+49=58$

55) $(-6+4i)(-6-4i)$

$=36+24i-24i-16i^2$

$=36-16(-1)=36+16=52$

57) $\frac{4}{2-3i}=\frac{4}{2-3i}\cdot\frac{2+3i}{2+3i}=\frac{8+12i}{2^2+3^2}$

$=\frac{8+12i}{4+9}=\frac{8+12i}{13}=\frac{8}{13}+\frac{12}{13}i$

59) $\frac{8i}{4+i}=\frac{8i}{4+i}\cdot\frac{4-i}{4-i}=\frac{32i-8i^2}{4^2+1^2}$

$=\frac{32i-8(-1)}{16+1}=\frac{8}{17}+\frac{32}{17}i$

61) $\dfrac{2i}{-3+7i}=\dfrac{2i}{-3+7i}\cdot\dfrac{-3-7i}{-3-7i}$
$=\dfrac{-6i-14i^2}{(-3)^2+7^2}=\dfrac{-6i-14(-1)}{9+49}$
$=\dfrac{-6i+14}{58}=\dfrac{14}{58}-\dfrac{6}{58}i$
$=\dfrac{7}{29}-\dfrac{3}{29}i$

63) $\dfrac{3-8i}{-6+7i}=\dfrac{3-8i}{-6+7i}\cdot\dfrac{-6-7i}{-6-7i}$
$=\dfrac{-18-21i+48i+56i^2}{(-6)^2+7^2}$
$=\dfrac{-18+27i+56(-1)}{36+49}$
$=\dfrac{-74+27i}{85}=-\dfrac{74}{85}+\dfrac{27}{85}i$

65) $\dfrac{2+3i}{5-6i}=\dfrac{2+3i}{5-6i}\cdot\dfrac{5+6i}{5+6i}$
$=\dfrac{20+12i+15i+18i^2}{5^2+6^2}$
$=\dfrac{20+27i+18(-1)}{25+36}$
$=\dfrac{-8+27i}{61}=-\dfrac{8}{61}+\dfrac{27}{61}i$

67) $\dfrac{9}{i}=\dfrac{9}{i}\cdot\dfrac{-i}{-i}=\dfrac{-9i}{1^2}=-9i$

69) $q^2=-4$
$q=\pm\sqrt{-4}=\pm 2i \qquad \{-2i,\ 2i\}$

71) $z^2+3=0$
$z^2=-3$
$z=\pm\sqrt{-3}=\pm i\sqrt{3} \qquad \{-i\sqrt{3},\ i\sqrt{3}\}$

73) $5f^2+39=-21$
$5f^2=-60$
$f^2=-12$
$f=\pm\sqrt{-12}$
$f=\pm 2i\sqrt{3} \qquad \{-2i\sqrt{3},\ 2i\sqrt{3}\}$

75) $63=7x^2$
$9=x^2$
$\pm\sqrt{9}=x$
$\pm 3=x \qquad \{-3,3\}$

77) $(c+3)^2-4=-29$
$(c+3)^2=-25$
$c+3=\pm\sqrt{-25}$
$c+3=\pm 5i$
$c=-3\pm 5i \qquad \{-3-5i,-3+5i\}$

79) $1=15+(k-2)^2$
$-14=(k-2)^2$
$\pm\sqrt{-14}=k-2$
$\pm i\sqrt{14}=k-2$
$2\pm i\sqrt{14}=k$
$\{2-i\sqrt{14},2+i\sqrt{14}\}$

81) $-3=(m-9)^2+5$
$-8=(m-9)^2$
$\pm 2i\sqrt{2}=m-9$
$9\pm 2i\sqrt{2}=m$
$\{9-2i\sqrt{2},9+2i\sqrt{2}\}$

83) $36+(4p-5)^2=6$

$$(4p-5)^2=-30$$
$$4p-5=\pm\sqrt{-30}$$
$$4p-5=\pm i\sqrt{30}$$
$$4p=5\pm i\sqrt{30}$$
$$p=\frac{5}{4}\pm\frac{\sqrt{30}}{4}i$$
$$\left\{\frac{5}{4}-\frac{\sqrt{30}}{4}i, \frac{5}{4}+\frac{\sqrt{30}}{4}i\right\}$$

85) $(6g+11)^2+50=1$

$$(6g+11)^2=-49$$
$$6g+11=\pm\sqrt{-49}$$
$$6g+11=\pm 7i$$
$$6g=-11\pm 7i$$
$$g=\frac{-11\pm 7i}{6}$$
$$\left\{-\frac{11}{6}-\frac{7}{6}i, -\frac{11}{6}+\frac{7}{6}i\right\}$$

87) $n^3+49n=0$

$$n(n^2+49)=0$$
$$n=0 \text{ or } n^2+49=0$$
$$n^2=-49$$
$$n=\pm\sqrt{-49}$$
$$=\pm 7i$$
$$\{0, -7i, 7i\}$$

89) $q^3=-54q$

$$q^3+54q=0$$
$$q(q^2+54)=0$$
$$q=0 \text{ or } q^2+54=0$$
$$q^2=-54$$
$$q=\pm\sqrt{-54}=\pm 3i\sqrt{6}$$
$$\{0, -3i\sqrt{6}, 3i\sqrt{6}\}$$

Section 11.4: Exercises

1) It is a trinomial whose factored form is the square of a binomial.
x^2-6x+9

3) No, because the coefficient of y^2 is not 1.

5) 1) $\frac{1}{2}(12)=6$ 2) $6^2=36$
$a^2+12a+36;\quad (a+6)^2$

7) 1) $\frac{1}{2}(-18)=-9$ 2) $(-9)^2=81$
$c^2-18c+81;\quad (c-9)^2$

9) 1) $\frac{1}{2}(3)=\frac{3}{2}$ 2) $\left(\frac{3}{2}\right)^2=\frac{9}{4}$
$r^2+3r+\frac{9}{4};\quad \left(r+\frac{3}{2}\right)^2$

11) 1) $\frac{1}{2}(-9)=-\frac{9}{2}$ 2) $\left(-\frac{9}{2}\right)^2=\frac{81}{4}$
$b^2-9b+\frac{81}{4};\quad \left(b-\frac{9}{2}\right)^2$

13) 1) $\frac{1}{2}\left(\frac{1}{3}\right)=\frac{1}{6}$ 2) $\left(\frac{1}{6}\right)^2=\frac{1}{36}$
$x^2+\frac{1}{3}x+\frac{1}{36};\quad \left(x+\frac{1}{6}\right)^2$

15) To solve $ax^2+bx+c=0$ by completing the square,
   i) divide both sides of the equation by a to obtain a leading coeffecient of 1.

ii) Get the variables on one side of the equal sign and the constant on the other side.
iii) Complete the square.
iv) Factor.
v) Solve using the square root property.

17) $x^2+6x+8=0$

$$x^2+6x=-8$$
$$x^2+6x+9=-8+9$$
$$(x+3)^2=1$$
$$x+3=\pm\sqrt{1}$$
$$x+3=\pm1$$
$$x+3=1 \text{ or } x+3=-1$$
$$x=-2 \qquad x=-4 \qquad \{-4,-2\}$$

19) $k^2-8k+15=0$

$$k^2-8k=-15$$
$$k^2-8k+16=-15+16$$
$$(k-4)^2=1$$
$$k-4=\pm\sqrt{1}$$
$$k-4=\pm1$$
$$k-4=1 \text{ or } k-4=-1$$
$$k=5 \qquad k=3 \qquad \{3, 5\}$$

21) $s^2+10=-10s$

$$s^2+10s=-10$$
$$s^2+10s+25=-10+25$$
$$(s+5)^2=15$$
$$s+5=\pm\sqrt{15}$$
$$s=-5\pm\sqrt{15}$$
$$\{-5-\sqrt{15},-5+\sqrt{15}\}$$

23) $t^2=2t-9$

$$t^2-2t=-9$$
$$t^2-2t+1=-9+1$$
$$(t-1)^2=-8$$
$$t-1=\pm\sqrt{-8}$$
$$t=1\pm 2i\sqrt{2}$$
$$\{1-2i\sqrt{2},\ 1+2i\sqrt{2}\}$$

25) $v^2+4v+8=0$

$$v^2+4v=-8$$
$$v^2+4v+4=-8+4$$
$$(v+2)^2=-4$$
$$v+2=\pm\sqrt{-4}$$
$$v=-2\pm 2i$$
$$\{-2-2i,-2+2i\}$$

27) $d^2+3=12d$

$$d^2-12d=-3$$
$$d^2-12d+36=-3+36$$
$$(d-6)^2=33$$
$$d-6=\pm\sqrt{33}$$
$$d=6\pm\sqrt{33}$$
$$\{6-\sqrt{33},6+\sqrt{33}\}$$

29) $m^2+3m-40=0$

$$m^2+3m=40$$
$$m^2+3m+\frac{9}{4}=40+\frac{9}{4}$$
$$\left(m+\frac{3}{2}\right)^2=\frac{169}{4}$$
$$m+\frac{3}{2}=\pm\sqrt{\frac{169}{4}}$$
$$m+\frac{3}{2}=\pm\frac{13}{2}$$

$$m+\frac{3}{2}=\frac{13}{2} \quad \text{or} \quad m+\frac{3}{2}=-\frac{13}{2}$$
$$m=\frac{10}{2} \qquad m=-\frac{16}{2}$$
$$m=5 \qquad m=-8$$
$$\{-8,5\}$$

31) $x^2-7x+12=0$
$$x^2-7x=-12$$
$$x^2-7x+\frac{49}{4}=-12+\frac{49}{4}$$
$$\left(x-\frac{7}{2}\right)^2=\frac{1}{4}$$
$$x-\frac{7}{2}=\pm\sqrt{\frac{1}{4}}=\pm\frac{1}{2}$$
$$x-\frac{7}{2}=\frac{1}{2} \quad \text{or} \quad x-\frac{7}{2}=-\frac{1}{2}$$
$$x=\frac{8}{2} \qquad x=\frac{6}{2}$$
$$x=4 \qquad x=3 \qquad \{3,4\}$$

33) $r^2-r=3$
$$r^2-r+\frac{1}{4}=3+\frac{1}{4}$$
$$\left(r-\frac{1}{2}\right)^2=\frac{13}{4}$$
$$r-\frac{1}{2}=\pm\sqrt{\frac{13}{4}}$$
$$r=\frac{1}{2}\pm\frac{\sqrt{13}}{2}$$
$$\left\{\frac{1}{2}-\frac{\sqrt{13}}{2},\ \frac{1}{2}+\frac{\sqrt{13}}{2}\right\}$$

35) $c^2+5c+7=0$
$$c^2+5c=-7$$
$$c^2+5c+\frac{25}{4}=-7+\frac{25}{4}$$
$$\left(c+\frac{5}{2}\right)^2=-\frac{3}{4}$$
$$c+\frac{5}{2}=\pm\sqrt{-\frac{3}{4}}$$
$$c=-\frac{5}{2}\pm\frac{\sqrt{3}}{2}i$$
$$\left\{-\frac{5}{2}-\frac{\sqrt{3}}{2}i,\ -\frac{5}{2}+\frac{\sqrt{3}}{2}i\right\}$$

37) $3k^2-6k+12=0$
$$\frac{3k^2}{3}-\frac{6k}{3}+\frac{12}{3}=\frac{0}{3}$$
$$k^2-2k+4=0$$
$$k^2-2k=-4$$
$$k^2-2k+1=-4+1$$
$$(k-1)^2=-3$$
$$k-1=\pm\sqrt{-3}$$
$$k=1\pm i\sqrt{3}$$
$$\{1-i\sqrt{3},\ 1+i\sqrt{3}\}$$

39) $4r^2+24r=8$
$$\frac{4r^2}{4}+\frac{24r}{4}=\frac{8}{4}$$
$$r^2+6r=2$$
$$r^2+6r+9=2+9$$
$$(r+3)^2=11$$
$$r+3=\pm\sqrt{11}$$
$$r=-3\pm\sqrt{11}$$
$$\{-3-\sqrt{11},\ -3+\sqrt{11}\}$$

41)
$$10d = 2d^2 + 12$$
$$\frac{10d}{2} = \frac{2d^2}{2} + \frac{12}{2}$$
$$5d = d^2 + 6$$
$$-6 = d^2 - 5d$$
$$-6 + \frac{25}{4} = d^2 - 5d + \frac{25}{4}$$
$$\frac{1}{4} = \left(d - \frac{5}{2}\right)^2$$
$$\pm\sqrt{\frac{1}{4}} = d - \frac{5}{2}$$
$$\pm\frac{1}{2} = d - \frac{5}{2}$$
$$d - \frac{5}{2} = \frac{1}{2} \text{ or } d - \frac{5}{2} = -\frac{1}{2}$$
$$d = \frac{6}{2} \qquad d = \frac{4}{2}$$
$$d = 3 \qquad d = 2 \qquad \{2,3\}$$

43)
$$2n^2 + 8 = 5n$$
$$\frac{2n^2}{2} + \frac{8}{2} = \frac{5n}{2}$$
$$n^2 + 4 = \frac{5}{2}n$$
$$n^2 - \frac{5}{2}n = -4$$
$$n^2 - \frac{5}{2}n + \frac{25}{16} = -4 + \frac{25}{16}$$
$$\left(n - \frac{5}{4}\right)^2 = -\frac{39}{16}$$
$$n - \frac{5}{4} = \pm\sqrt{-\frac{39}{16}}$$
$$x = \frac{5}{4} \pm \frac{\sqrt{39}}{4}i$$
$$\left\{\frac{5}{4} - \frac{\sqrt{39}}{4}i, \frac{5}{4} + \frac{\sqrt{39}}{4}i\right\}$$

45)
$$4a^2 - 7a + 3 = 0$$
$$\frac{4a^2}{4} - \frac{7a}{4} + \frac{3}{4} = 0$$
$$a^2 - \frac{7}{4}t + \frac{3}{4} = 0$$
$$a^2 - \frac{7}{4}t = -\frac{3}{4}$$
$$a^2 - \frac{7}{4}t + \frac{49}{64} = -\frac{3}{4} + \frac{49}{64}$$
$$\left(a - \frac{7}{8}\right)^2 = \frac{1}{64}$$
$$a - \frac{7}{8} = \pm\sqrt{\frac{1}{64}}$$
$$a - \frac{7}{8} = \pm\frac{1}{8}$$
$$a - \frac{7}{8} = \frac{1}{8} \text{ or } a - \frac{7}{8} = -\frac{1}{8}$$
$$a = 1 \qquad a = \frac{3}{4} \qquad \left\{\frac{3}{4}, 1\right\}$$

47)
$$(y+5)(y-3) = 5$$
$$y^2 + 2y - 15 = 5$$
$$y^2 + 2y = 20$$
$$y^2 + 2y + 1 = 20 + 1$$
$$(y+1)^2 = 21$$
$$y + 1 = \pm\sqrt{21}$$
$$y = -1 \pm \sqrt{21}$$
$$\left\{-1 - \sqrt{21}, -1 + \sqrt{21}\right\}$$

49) $(2m+1)(m-3)=-7$

$$2m^2-5m-3=-7$$
$$2m^2-5m=-4$$
$$\frac{2m^2}{2}-\frac{5m}{2}=\frac{-4}{2}$$
$$m^2-\frac{5}{2}m=-2$$
$$m^2-\frac{5}{2}m+\frac{25}{16}=-2+\frac{25}{16}$$
$$\left(m-\frac{5}{4}\right)^2=-\frac{7}{16}$$
$$m-\frac{5}{4}=\pm\sqrt{-\frac{7}{16}}$$
$$m=\frac{5}{4}\pm\frac{\sqrt{7}}{4}i$$
$$\left\{\frac{5}{4}-\frac{\sqrt{7}}{4}i,\ \frac{5}{4}+\frac{\sqrt{7}}{4}i\right\}$$

51) $w=\text{width}$

$w+8=\text{length}$

$\text{Area}=(\text{length})(\text{width})$

$$153=(w+8)w$$
$$153=w^2+8w$$
$$153+16=w^2+8w+16$$
$$169=(w+4)^2$$
$$\pm\sqrt{169}=w+4$$
$$\pm13=w+4$$

$w+4=13$ or $w+4=-13$

$\boxed{w=9}$ $\qquad w=-17$

Reject $w=-17$ as a solution.

width $=9$ in.

length $=9+8=17$ in.

Section 11.5: Exercises

1) The fraction bar should also be under $-b$: $x=\dfrac{-b\pm\sqrt{b^2-4ac}}{2a}$

3) You cannot divide only the -2 by 2

$$\frac{-2\pm6\sqrt{11}}{2}=\frac{2\left(-1\pm3\sqrt{11}\right)}{2}$$
$$=-1\pm3\sqrt{11}$$

5) $x^2+4x+3=0$

$a=1$, $b=4$ and $c=3$

$$x=\frac{-4\pm\sqrt{4^2-4(1)(3)}}{2(1)}$$
$$=\frac{-4\pm\sqrt{16-12}}{2}$$
$$=\frac{-4\pm\sqrt{4}}{2}=\frac{-4\pm2}{2}$$
$$\frac{-4+2}{2}=\frac{-2}{2}=-1,$$
$$\frac{-4-2}{2}=\frac{-6}{2}=-3 \qquad \{-3, -1\}$$

7) $3t^2+t-10=0$

$a=3$, $b=1$ and $c=-10$

$$t=\frac{-1\pm\sqrt{1^2-4(3)(-10)}}{2(3)}$$
$$=\frac{-1\pm\sqrt{1+120}}{6}$$
$$=\frac{-1\pm\sqrt{121}}{6}=\frac{-1\pm11}{6}$$
$$\frac{-1+11}{6}=\frac{10}{6}=\frac{5}{3},$$
$$\frac{-1-11}{6}=\frac{-12}{6}=-2 \qquad \left\{-2, \frac{5}{3}\right\}$$

9) $k^2+2=5k$

$k^2-5k+2=0$

$a=1,\ b=-5$ and $c=2$

$$k=\frac{-(-5)\pm\sqrt{(-5)^2-4(1)(2)}}{2(1)}$$

$$=\frac{5\pm\sqrt{25-8}}{2}=\frac{5\pm\sqrt{17}}{2}$$

$$\left\{\frac{5-\sqrt{17}}{2},\ \frac{5+\sqrt{17}}{2}\right\}$$

11) $y^2=8y-25$

$y^2-8y+25=0$

$a=1,\ b=-8$ and $c=25$

$$y=\frac{-(-8)\pm\sqrt{(-8)^2-4(1)(25)}}{2(1)}$$

$$=\frac{8\pm\sqrt{64-100}}{2}=\frac{8\pm\sqrt{-36}}{2}$$

$$=\frac{8\pm 6i}{2}=\frac{8}{2}\pm\frac{6}{2}i=4\pm 3i$$

$\{4-3i,\ 4+3i\}$

13) $3-2w=-5w^2$

$5w^2-2w+3=0$

$a=5,\ b=-2$ and $c=3$

$$w=\frac{-(-2)\pm\sqrt{(-2)^2-4(5)(3)}}{2(5)}$$

$$=\frac{2\pm\sqrt{4-60}}{10}=\frac{2\pm\sqrt{-56}}{10}$$

$$=\frac{2\pm 2i\sqrt{14}}{10}=\frac{2}{10}\pm\frac{2i\sqrt{14}}{10}$$

$$=\frac{1}{5}\pm\frac{\sqrt{14}}{5}i$$

$$\left\{\frac{1}{5}-\frac{\sqrt{14}}{5}i,\ \frac{1}{5}+\frac{\sqrt{14}}{5}i\right\}$$

15) $r^2+7r=0$

$a=1,\ b=7$ and $c=0$

$$r=\frac{-7\pm\sqrt{7^2-4(1)(0)}}{2(1)}$$

$$=\frac{-7\pm\sqrt{49}}{2}=\frac{-7\pm 7}{2}$$

$$\frac{-7+7}{2}=\frac{0}{2}=0,$$

$$\frac{-7-7}{2}=\frac{-14}{2}=-7$$

$\{-7,\ 0\}$

17) $3v(v+3)=7v+4$

$3v^2+9v=7v+4$

$3v^2+2v-4=0$

$a=3,\ b=2$ and $c=-4$

$$v=\frac{-2\pm\sqrt{2^2-4(3)(-4)}}{2(3)}$$

$$=\frac{-2\pm\sqrt{4+48}}{6}=\frac{-2\pm\sqrt{52}}{6}$$

$$=\frac{-2\pm 2\sqrt{13}}{6}=\frac{2\left(-1\pm\sqrt{13}\right)}{6}$$

$$=\frac{-1\pm\sqrt{13}}{3}$$

$$\left\{\frac{-1-\sqrt{13}}{3},\ \frac{-1+\sqrt{13}}{3}\right\}$$

19) $(2c-5)(c-5)=-3$

$2c^2-15c+25=-3$

$2c^2-15c+28=0$

$a=2,\ b=-15$ and $c=28$

$$c=\frac{-(-15)\pm\sqrt{(-15)^2-4(2)(28)}}{2(2)}$$

$$=\frac{15\pm\sqrt{225-224}}{4}=\frac{15\pm\sqrt{1}}{4}$$

$$=\frac{15\pm 1}{4}$$

$$\frac{15+1}{4}=\frac{16}{4}=4,$$
$$\frac{15-1}{4}=\frac{14}{4}=\frac{7}{2} \qquad \left\{\frac{7}{2}, 4\right\}$$

21) $\frac{1}{6}u^2+\frac{4}{3}u=\frac{5}{2}$

$$6\left(\frac{1}{6}u^2+\frac{4}{3}u\right)=6\left(\frac{5}{2}\right)$$
$$u^2+8u=15$$
$$u^2+8u-15=0$$
$a=1$, $b=8$ and $c=-15$
$$u=\frac{-8\pm\sqrt{8^2-4(1)(-15)}}{2(1)}$$
$$=\frac{-8\pm\sqrt{64+60}}{2}=\frac{-8\pm\sqrt{124}}{2}$$
$$=\frac{-8\pm2\sqrt{31}}{2}=\frac{2\left(-4\pm\sqrt{31}\right)}{2}$$
$$=-4\pm\sqrt{31}$$
$\left\{-4-\sqrt{31}, -4+\sqrt{31}\right\}$

23) $m^2+\frac{4}{3}m+\frac{5}{9}=0$

$$9\left(m^2+\frac{4}{3}m+\frac{5}{9}\right)=9(0)$$
$$9m^2+12m+5=0$$
$a=9$, $b=12$ and $c=5$
$$k=\frac{-12\pm\sqrt{(12)^2-4(9)(5)}}{2(9)}$$
$$=\frac{-12\pm\sqrt{144-180}}{18}=\frac{-12\pm\sqrt{-36}}{18}$$
$$=\frac{-12\pm6i}{18}=-\frac{12}{18}\pm\frac{6}{18}i=-\frac{2}{3}\pm\frac{1}{3}i$$
$\left\{-\frac{2}{3}-\frac{1}{3}i, -\frac{2}{3}+\frac{1}{3}i\right\}$

25) $2(p+10)=(p+10)(p-2)$

$$2p+20=p^2+8p-20$$
$$0=p^2+6p-40$$
$a=1$, $b=6$ and $c=-40$
$$p=\frac{-6\pm\sqrt{6^2-4(1)(-40)}}{2(1)}$$
$$=\frac{-6\pm\sqrt{36+160}}{2}=\frac{-6\pm\sqrt{196}}{2}$$
$$=\frac{-6\pm14}{2}$$
$$\frac{-6+14}{2}=\frac{8}{2}=4,$$
$$\frac{-6-14}{2}=\frac{-20}{2}=-10 \qquad \{-10, 4\}$$

27) $4g^2+9=0$

$a=4$, $b=0$ and $c=9$
$$g=\frac{-0\pm\sqrt{0^2-4(4)(9)}}{2(4)}$$
$$=\frac{\pm\sqrt{-144}}{8}=\frac{\pm12i}{8}=\pm\frac{3}{2}i$$
$\left\{-\frac{3}{2}i, \frac{3}{2}i\right\}$

29) $x(x+6)=-34$

$$x^2+6x+34=0$$
$a=1$, $b=6$ and $c=34$
$$x=\frac{-6\pm\sqrt{6^2-4(1)(34)}}{2(1)}$$
$$=\frac{-6\pm\sqrt{36-136}}{2}=\frac{-6\pm\sqrt{-100}}{2}$$
$$=\frac{-6\pm10i}{2}=-\frac{6}{2}\pm\frac{10i}{2}=-3\pm5i$$
$\{-3-5i, -3+5i\}$

31) $(2s+3)(s-1)=s^2-s+6$

$$2s^2+s-3=s^2-s+6$$
$$s^2+2s-9=0$$
$a=1$, $b=2$ and $c=-9$

$$s=\frac{-2\pm\sqrt{2^2-4(1)(-9)}}{2(1)}$$
$$=\frac{-2\pm\sqrt{4+36}}{2}=\frac{-2\pm\sqrt{40}}{2}$$
$$=\frac{-2\pm2\sqrt{10}}{2}=\frac{2\left(-1\pm\sqrt{10}\right)}{2}$$
$$=-1\pm\sqrt{10}$$
$$\left\{-1-\sqrt{10},\ -1+\sqrt{10}\right\}$$

33) $3(3-4y)=-4y^2$

$$9-12y=-4y^2$$
$$4y^2-12y+9=0$$
$a=4$, $b=-12$ and $c=9$

$$y=\frac{-(-12)\pm\sqrt{(-12)^2-4(4)(9)}}{2(4)}$$
$$=\frac{12\pm\sqrt{144-144}}{8}=\frac{12\pm\sqrt{0}}{8}$$
$$=\frac{12}{8}=\frac{3}{2} \qquad \left\{\frac{3}{2}\right\}$$

35) $-\frac{1}{6}=\frac{2}{3}p^2+\frac{1}{2}p$

$$6\left(-\frac{1}{6}\right)=6\left(\frac{2}{3}p^2+\frac{1}{2}p\right)$$
$$-1=4p^2+3p$$
$$0=4p^2+3p+1$$
$a=4$, $b=3$ and $c=1$

$$p=\frac{-3\pm\sqrt{3^2-4(4)(1)}}{2(4)}$$
$$=\frac{-3\pm\sqrt{9-16}}{8}=\frac{-3\pm\sqrt{-7}}{8}$$
$$=\frac{-3\pm i\sqrt{7}}{8}=-\frac{3}{8}\pm\frac{\sqrt{7}}{8}i$$
$$\left\{-\frac{3}{8}-\frac{\sqrt{7}}{8}i,\ -\frac{3}{8}+\frac{\sqrt{7}}{8}i\right\}$$

37) $4q^2+6=20q$

$$\frac{4q^2}{2}+\frac{6}{2}=\frac{20q}{2}$$
$$2q^2+3q=10q$$
$$2q^2-10q+3q=0$$
$a=2$, $b=-10$ and $c=3$

$$q=\frac{-(-10)\pm\sqrt{(-10)^2-4(2)(3)}}{2(2)}$$
$$=\frac{10\pm\sqrt{100-24}}{4}=\frac{10\pm\sqrt{76}}{4}$$
$$=\frac{10\pm2\sqrt{19}}{4}=\frac{2\left(5\pm\sqrt{19}\right)}{4}$$
$$=\frac{5\pm\sqrt{19}}{2} \qquad \left\{\frac{5-\sqrt{19}}{2},\ \frac{5+\sqrt{19}}{2}\right\}$$

39) $a=10$, $b=-9$ and $c=3$

$$b^2-4ac=(-9)^2-4(10)(3)$$
$$=81-120=-39$$
two complex solutions

41) $4y^2-49=-28y$

$$4y^2+28y-49=0$$
$a=4$, $b=28$ and $c=49$

$$b^2-4ac=28^2-4(4)(49)$$
$$=784-784=0$$
one rational solution

43) $-5 = u(u+6)$

$-5 = u^2 + 6u$

$0 = u^2 + 6u + 5$

$a = 1,\ b = 6$ and $c = 5$

$b^2 - 4ac = 6^2 - 4(1)(5)$

$= 36 - 20 = 16$

two rational solutions

45) $a = 2,\ b = -4$ and $c = -5$

$b^2 - 4ac = (-4)^2 - 4(2)(-5)$

$= 16 + 40 = 56$

two irrational solutions

47) $a = 1,\ b = b$ and $c = 16$

$b^2 - 4ac = 0$

$b^2 - 4(1)(16) = 0$

$b^2 - 64 = 0$

$b^2 = 64$

$b = \pm\sqrt{64} = \pm 8$

-8 or 8

49) $a = 4,\ b = -12$ and $c = c$

$b^2 - 4ac = 0$

$(-12)^2 - 4(4)c = 0$

$144 - 16c = 0$

$144 = 16c$

$9 = c$

51) $a = a,\ b = 12$ and $c = 9$

$b^2 - 4ac = 0$

$(12)^2 - 4a(9) = 0$

$144 - 36a = 0$

$144 = 36a$

$4 = a$

53) $x =$ length of one leg

$2x + 1 =$ length of other leg

$\sqrt{29} =$ length of hypotenuse

$x^2 + (2x+1)^2 = \left(\sqrt{29}\right)^2$

$x^2 + 4x^2 + 4x + 1 = 29$

$5x^2 + 4x - 28 = 0$

$(5x+14)(x-2) = 0$

$5x + 14 = 0$ or $x - 2 = 0$

$5x = -14$ $\quad \boxed{x = 2}$

$x = -\frac{14}{5}$

$2x + 1 = 2(2) + 1 = 5$

The lengths of the legs are 2 in. and 5 in.

55) a) Let $h = 8$ and solve for t.

$8 = -16t^2 + 44t + 24$

$0 = -16t^2 + 44t + 16$

$0 = 2t^2 - 3t - 2$

$0 = (2t+1)(t-2)$

$2t + 1 = 0$ or $t - 2 = 0$

$2t = -1$ $\quad \boxed{t = 2}$

$t = -\frac{1}{2}$

The ball reaches 8 feet after 2 sec.

b) Let $h = 0$ and solve for t.

$0 = -16t^2 + 44t + 24$

$0 = 2t^2 - 3t - 3$

$a = 2,\ b = -3$ and $c = -3$

$$t = \frac{-(-3) \pm \sqrt{(-3)^2 - 4(2)(-3)}}{2(2)}$$

$$= \frac{3 \pm \sqrt{9 + 24}}{4} = \frac{3 \pm \sqrt{33}}{4}$$

Reject $t = \dfrac{3-\sqrt{33}}{4}$ because it is negative. The ball will hit the ground after $\dfrac{3+\sqrt{33}}{4}$ sec ≈ 2.2 sec.

Mid-Chapter Summary

1) $z^2 - 50 = 0$

$$z^2 = 50$$
$$z = \pm\sqrt{50} = \pm 5\sqrt{2}$$
$$\{-5\sqrt{2}, 5\sqrt{2}\}$$

3) $a(a+1) = 20$

$$a^2 + a = 20$$
$$a^2 + a - 20 = 0$$
$$(a+5)(a-4) = 0$$
$$a+5=0 \text{ or } a-4=0$$
$$a=-5 \qquad a=4 \qquad \{-5, 4\}$$

5) $u^2 + 7u + 9 = 0$

$$a=1,\ b=7 \text{ and } c=9$$
$$u = \frac{-7 \pm \sqrt{7^2 - 4(1)(9)}}{2(1)}$$
$$= \frac{-7 \pm \sqrt{49-36}}{2} = \frac{-7 \pm \sqrt{13}}{2}$$
$$\left\{\frac{-7-\sqrt{13}}{2}, \frac{-7+\sqrt{13}}{2}\right\}$$

7) $2k(2k+7) = 3(k+1)$

$$4k^2 + 14k = 3k + 3$$
$$4k^2 + 11k - 3 = 0$$
$$(4k-1)(k+3) = 0$$
$$4k-1=0 \text{ or } k+3=0$$
$$4k=1 \qquad k=-3$$
$$k = \frac{1}{4} \qquad \left\{-3, \frac{1}{4}\right\}$$

9) $m^2 + 14m + 60 = 0$

$$m^2 + 14m = -60$$
$$m^2 + 14m + 49 = -60 + 49$$
$$(m+7)^2 = -11$$
$$m+7 = \pm\sqrt{-11}$$
$$m+7 = \pm i\sqrt{11}$$
$$m = -7 \pm i\sqrt{11}$$
$$\{-7 - i\sqrt{11}, -7 + i\sqrt{11}\}$$

11) $10 + (3b-1)^2 = 4$

$$(3b-1)^2 = -6$$
$$3b-1 = \pm\sqrt{-6}$$
$$3b-1 = \pm i\sqrt{6}$$
$$3b = 1 \pm i\sqrt{6}$$
$$b = \frac{1 \pm i\sqrt{6}}{3}$$
$$\left\{\frac{1}{3} - \frac{\sqrt{6}}{3}i, \frac{1}{3} + \frac{\sqrt{6}}{3}i\right\}$$

13) $1=\frac{x^2}{12}-\frac{x}{3}$

$12(1)=12\left(\frac{x^2}{12}-\frac{x}{3}\right)$

$12=x^2-4x$

$0=x^2-4x-12$

$0=(x-6)(x+2)$

$x-6=0$ or $x+2=0$

$x=6$ $\quad x=-2$ $\quad \{-2, 6\}$

15) $r^2-4r=3$

$r^2-4r+4=3+4$

$(r-2)^2=7$

$r-2=\pm\sqrt{7}$

$r=2\pm\sqrt{7}$

$\{2-\sqrt{7}, 2+\sqrt{7}\}$

17) $p(p+8)=3(p^2+2)+p$

$p^2+8p=3p^2+6+p$

$0=2p^2-7p+6$

$0=(2p-3)(p-2)$

$2p-3=0$ or $p-2=0$

$2p=3$ $\quad p=2$

$p=\frac{3}{2}$ $\quad \left\{\frac{3}{2}, 2\right\}$

19) $\frac{10}{z}=1+\frac{21}{z^2}$

$z^2\left(\frac{10}{z}\right)=z^2\left(1+\frac{21}{z^2}\right)$

$10z=z^2+21$

$0=z^2-10z+21$

$0=(z-7)(z-3)$

$z-7=0$ or $z-3=0$

$z=7$ $\quad z=3$ $\quad \{3, 7\}$

21) $(3v+4)(v-2)=-9$

$3v^2-2v-8=-9$

$3v^2-2v+1=0$

$a=3$, $b=-2$ and $c=1$

$$v=\frac{-(-2)\pm\sqrt{(-2)^2-4(3)(1)}}{2(3)}$$

$$=\frac{2\pm\sqrt{4-12}}{6}=\frac{2\pm\sqrt{-8}}{6}$$

$$=\frac{2\pm 2i\sqrt{2}}{6}=\frac{1\pm i\sqrt{2}}{3}$$

$\left\{\frac{1}{3}-\frac{\sqrt{2}}{3}i, \frac{1}{3}+\frac{\sqrt{2}}{3}i\right\}$

23) $(c-5)^2+16=0$

$(c-5)^2=-16$

$c-5=\pm\sqrt{-16}$

$c-5=\pm 4i$

$c=5\pm 4i$

$\{5-4i, 5+4i\}$

25) $3g=g^2$

$0=g^2-3g$

$0=g(g-3)$

$g-3=0$ or $g=0$

$g=3$ $\quad \{0, 3\}$

27)
$$4m^3 = 9m$$
$$4m^3 - 9m = 0$$
$$m\left(4m^2 - 9\right) = 0$$
$$m(2m+3)(2m-3) = 0$$
$$2m+3=0 \text{ or } 2m-3=0 \text{ or } m=0$$
$$2m=-3 \qquad 2m=3$$
$$m=-\frac{3}{2} \qquad m=\frac{3}{2}$$
$$\left\{-\frac{3}{2}, 0, \frac{3}{2}\right\}$$

29)
$$\frac{1}{3}q^2 + \frac{5}{6}q + \frac{4}{3} = 0$$
$$6\left(\frac{1}{3}q^2 + \frac{5}{6}q + \frac{4}{3}\right) = 6(0)$$
$$2q^2 + 5q + 8 = 0$$
$$a=2,\ b=5 \text{ and } c=8$$
$$q = \frac{-5 \pm \sqrt{5^2 - 4(2)(8)}}{2(2)}$$
$$= \frac{-5 \pm \sqrt{25-64}}{4} = \frac{-5 \pm \sqrt{-39}}{4}$$
$$= \frac{-5 \pm i\sqrt{39}}{4}$$
$$\left\{\frac{-5 - i\sqrt{39}}{4}, \frac{-5 - i\sqrt{39}}{4}\right\}$$

Section 11.6: Exercises

1)
$$t - \frac{48}{t} = 8$$
$$t\left(t - \frac{48}{t}\right) = t(8)$$
$$t^2 - 48 = 8t$$
$$t^2 - 8t - 48 = 0$$
$$(t-12)(t+4) = 0$$
$$t-12=0 \text{ or } t+4=0$$
$$t=12 \qquad t=-4 \qquad \{-4,\ 12\}$$

3)
$$\frac{2}{x} + \frac{6}{x-2} = -\frac{5}{2}$$
$$2x(x-2)\left(\frac{2}{x} + \frac{6}{x-2}\right) = 2x(x-2)\left(-\frac{5}{2}\right)$$
$$4(x-2) + 12x = -5x(x-2)$$
$$4x - 8 + 12x = -5x^2 + 10x$$
$$5x^2 + 6x - 8 = 0$$
$$(5x-4)(x+2) = 0$$
$$5x-4=0 \text{ or } x+2=0$$
$$5x=4 \qquad x=-2$$
$$x=\frac{4}{5} \qquad \left\{-2,\ \frac{4}{5}\right\}$$

5)
$$1 = \frac{2}{c} + \frac{1}{c-5}$$
$$c(c-5)(1) = c(c-5)\left(\frac{2}{c} + \frac{1}{c-5}\right)$$
$$c^2 - 5c = 2(c-5) + c$$
$$c^2 - 5c = 2c - 10 + c$$
$$c^2 - 5c = 3c - 10$$
$$c^2 - 8c = -10$$
$$c^2 - 8c + 16 = -10 + 16$$
$$(c-4)^2 = 6$$
$$c - 4 = \pm\sqrt{6}$$
$$c = 4 \pm \sqrt{6}$$
$$\left\{4-\sqrt{6},\ 4+\sqrt{6}\right\}$$

7)
$$\frac{3}{2v+2} + \frac{1}{v} = \frac{3}{2}$$
$$\frac{3}{2(v+1)} + \frac{1}{v} = \frac{3}{2}$$
$$2v(v+1)\left(\frac{3}{2(v+1)} + \frac{1}{v}\right) = 2v(v+1)\left(\frac{3}{2}\right)$$
$$3v + 2v + 2 = 3v^2 + 3v$$
$$0 = 3v^2 - 2v - 2$$

$$v=\frac{-(-2)\pm\sqrt{(-2)^2-4(3)(-2)}}{2(3)}$$
$$=\frac{2\pm\sqrt{4+24}}{6}=\frac{2\pm\sqrt{28}}{6}$$
$$=\frac{2\pm 2\sqrt{7}}{6}=\frac{1\pm\sqrt{7}}{3}$$
$$\left\{\frac{1-\sqrt{7}}{3},\frac{1+\sqrt{7}}{3}\right\}$$

9) $$\frac{9}{n^2}=5+\frac{4}{n}$$
$$n^2\left(\frac{9}{n^2}\right)=n^2\left(5+\frac{4}{n}\right)$$
$$9=5n^2+4n$$
$$0=5n^2+4n-9$$
$$0=(5n+9)(n-1)$$
$$5n+9=0 \text{ or } n-1=0$$
$$5n=-9 \qquad n=1$$
$$n=-\frac{9}{5} \qquad \left\{-\frac{9}{5},1\right\}$$

11) $$\frac{5}{6r}=1-\frac{r}{6r-6}$$
$$6r(r-1)\left(\frac{5}{6r}\right)=6r(r-1)\left(1-\frac{r}{6(r-1)}\right)$$
$$5r-5=6r^2-6r-r^2$$
$$0=5r^2-11r+5$$

$$r=\frac{-(-11)\pm\sqrt{(-11)^2-4(5)(5)}}{2(5)}$$
$$=\frac{11\pm\sqrt{121-100}}{10}=\frac{11\pm\sqrt{21}}{10}$$
$$\left\{\frac{11-\sqrt{21}}{10},\frac{11+\sqrt{21}}{10}\right\}$$

13) $$g=\sqrt{g+20}$$
$$g^2=g+20$$
$$g^2-g-20=0$$
$$(g-5)(g+4)=0$$
$$g-5=0 \text{ or } g+4=0$$
$$g=5 \qquad g=-4$$
Only one solution satisfies the original equation. $\{5\}$

15) $$a=\sqrt{\frac{14a-8}{5}}$$
$$a^2=\frac{14a-8}{5}$$
$$5a^2=14a-8$$
$$5a^2-14a+8=0$$
$$(5a-4)(a-2)=0$$
$$5a-4=0 \text{ or } a-2=0$$
$$5a=4 \qquad a=2$$
$$a=\frac{4}{5} \qquad \left\{\frac{4}{5},2\right\}$$

17) $$p-\sqrt{p}=6$$
$$p-6=\sqrt{p}$$
$$(p-6)^2=\left(\sqrt{p}\right)^2$$
$$p^2-12p+36=p$$
$$p^2-13p+36=0$$
$$(p-9)(p-4)=0$$
$$p-9=0 \text{ or } p-4=0$$
$$p=9 \qquad p=4$$
Only one solution satisfies the original equation. $\{9\}$

19)
$$x = 5\sqrt{x} - 4$$
$$x + 4 = 5\sqrt{x}$$
$$(x+4)^2 = \left(5\sqrt{x}\right)^2$$
$$x^2 + 8x + 16 = 25x$$
$$x^2 - 17x + 16 = 0$$
$$(x-16)(x-1) = 0$$
$$x - 16 = 0 \text{ or } x - 1 = 0$$
$$x = 16 \qquad x = 1 \qquad \{1, 16\}$$

21)
$$2 + \sqrt{2y-1} = y$$
$$\sqrt{2y-1} = y - 2$$
$$\left(\sqrt{2y-1}\right)^2 = (y-2)^2$$
$$2y - 1 = y^2 - 4y + 4$$
$$0 = y^2 - 6y + 5$$
$$0 = (y-5)(y-1)$$
$$y - 5 = 0 \text{ or } y - 1 = 0$$
$$y = 5 \qquad y = 1$$
Only one solution satisfies the original equation. $\{5\}$

23)
$$2 = \sqrt{6k+4} - k$$
$$k + 2 = \sqrt{6k+4}$$
$$(k+2)^2 = \left(\sqrt{6k+4}\right)^2$$
$$k^2 + 4k + 4 = 6k + 4$$
$$k^2 - 2k = 0$$
$$k(k-2) = 0$$
$$k - 2 = 0 \text{ or } k = 0$$
$$k = 2 \qquad \{0, 2\}$$

25) yes 27) yes 29) no

31) yes 33) no

35)
$$x^4 - 10x^2 + 9 = 0$$
$$\left(x^2 - 9\right)\left(x^2 - 1\right) = 0$$
$$x^2 - 9 = 0 \text{ or } x^2 - 1 = 0$$
$$x^2 = 9 \qquad x^2 = 1$$
$$x = \pm\sqrt{9} \qquad x = \pm\sqrt{1}$$
$$x = \pm 3 \qquad x = \pm 1$$
$$\{-3, -1, 1, 3\}$$

37)
$$p^4 - 11p^2 + 28 = 0$$
$$\left(p^2 - 7\right)\left(p^2 - 4\right) = 0$$
$$p^2 - 7 = 0 \text{ or } p^2 - 4 = 0$$
$$p^2 = 7 \qquad p^2 = 4$$
$$p = \pm\sqrt{7} \qquad p = \pm 2$$
$$\left\{-\sqrt{7}, -2, 2, \sqrt{7}\right\}$$

39)
$$a^4 + 12a^2 = -35$$
$$a^4 + 12a^2 + 35 = 0$$
$$\left(a^2 + 7\right)\left(a^2 + 5\right) = 0$$
$$a^2 + 7 = 0 \text{ or } a^2 + 5 = 0$$
$$a^2 = -7 \qquad a^2 = -5$$
$$a = \pm i\sqrt{7} \qquad a = \pm i\sqrt{5}$$
$$\left\{-i\sqrt{7}, -i\sqrt{5}, i\sqrt{5}, i\sqrt{7}\right\}$$

41)
$$b^{2/3} + 3b^{1/3} + 2 = 0$$
$$\left(b^{1/3} + 2\right)\left(b^{1/3} + 1\right) = 0$$
$$b^{1/3} + 2 = 0 \quad \text{or} \quad b^{1/3} + 1 = 0$$
$$b^{1/3} = -2 \qquad b^{1/3} = -1$$
$$\left(\sqrt[3]{b}\right)^3 = (-2)^3 \qquad \left(\sqrt[3]{b}\right)^3 = (-1)^3$$
$$b = -8 \qquad b = -1$$
$$\{-8, -1\}$$

43) $t^{2/3} - 6t^{1/3} = 40$

$t^{2/3} - 6t^{1/3} - 40 = 0$

$(t^{1/3} + 4)(t^{1/3} - 10) = 0$

$t^{1/3} + 4 = 0$ or $t^{1/3} - 10 = 0$

$t^{1/3} = -4$ $\quad$ $t^{1/3} = 10$

$(\sqrt[3]{t})^3 = (-4)^3$ $\quad$ $(\sqrt[3]{t})^3 = 10^3$

$t = -64$ $\quad$ $t = 1000$

$\{-64, 1000\}$

45) $2n^{2/3} = 7n^{1/3} + 15$

$2n^{2/3} - 7n^{1/3} - 15 = 0$

$(2n^{1/3} + 3)(n^{1/3} - 5) = 0$

$2n^{1/3} + 3 = 0$ or $n^{1/3} - 5 = 0$

$2n^{1/3} = -3$ $\quad$ $n^{1/3} = 5$

$n^{1/3} = -\dfrac{3}{2}$ $\quad$ $(\sqrt[3]{n})^3 = 5^3$

$(\sqrt[3]{n})^3 = \left(-\dfrac{3}{2}\right)^3$ $\quad$ $n = 125$

$n = -\dfrac{27}{8}$

$\left\{-\dfrac{27}{8}, 125\right\}$

47) $v - 8v^{1/2} + 12 = 0$

$(v^{1/2} - 2)(v^{1/2} - 6) = 0$

$v^{1/2} - 2 = 0$ or $v^{1/2} - 6 = 0$

$v^{1/2} = 2$ $\quad$ $v^{1/2} = 6$

$(\sqrt{v})^2 = 2^2$ $\quad$ $(\sqrt{v})^2 = 6^2$

$v = 4$ $\quad$ $v = 36$ $\quad$ $\{4, 36\}$

49) $4h^{1/2} + 21 = h$

$0 = h - 4h^{1/2} - 21$

$0 = (h^{1/2} + 3)(h^{1/2} - 7)$

$h^{1/2} + 3 = 0$ or $h^{1/2} - 7 = 0$

$h^{1/2} = -3$ $\quad$ $h^{1/2} = 7$

$(\sqrt{h})^2 = (-3)^2$ $\quad$ $(\sqrt{h})^2 = 7^2$

$h = 9$ $\quad$ $h = 49$

Only one solution satisfies the original equation. $\{49\}$

51) $2a - 5a^{1/2} - 12 = 0$

$(2a^{1/2} + 3)(a^{1/2} - 4) = 0$

$2a^{1/2} + 3 = 0$ or $a^{1/2} - 4 = 0$

$2a^{1/2} = -3$ $\quad$ $a^{1/2} = 4$

$a^{1/2} = -\dfrac{3}{2}$ $\quad$ $(\sqrt{a})^2 = 4^2$

$(\sqrt{a})^2 = \left(-\dfrac{3}{2}\right)^2$ $\quad$ $a = 16$

$a = \dfrac{9}{4}$

Only one solution satisfies the original equation. $\{16\}$

53) $9n^4 = -15n^2 - 4$

$9n^4 + 15n^2 + 4 = 0$

$(3x^2 + 4)(3x^2 + 1) = 0$

$3x^2 + 4 = 0$ or $3x^2 + 1 = 0$

$3x^2 = -4$ $\quad$ $3x^2 = -1$

$x^2 = -\dfrac{4}{3}$ $\quad$ $x^2 = -\dfrac{1}{3}$

$x = \pm\dfrac{2}{\sqrt{3}}i$ $\quad$ $x = \pm\dfrac{1}{\sqrt{3}}i$

$$x=\pm\frac{2\sqrt{3}}{3}i \quad \text{or} \quad x=\pm\frac{\sqrt{3}}{3}i$$

$$\left\{-\frac{2\sqrt{3}}{3}i,\ -\frac{\sqrt{3}}{3}i,\ \frac{\sqrt{3}}{3}i,\ \frac{2\sqrt{3}}{3}i\right\}$$

55) $z^4-2z^2=15$

$$z^4-2z^2-15=0$$
$$(z^2-5)(z^2+3)=0$$
$$z^2-5=0 \text{ or } z^2+3=0$$
$$z^2=5 \qquad z^2=-3$$
$$z=\pm\sqrt{5} \qquad z=\pm i\sqrt{3}$$
$$\{-\sqrt{5},\ \sqrt{5},\ -i\sqrt{3},\ i\sqrt{3}\}$$

57) $w^4-6w^2+2=0$

Let $u=w^2$ and $u^2=w^4$

$$u^2-6u+2=0$$
$$u=\frac{-(-6)\pm\sqrt{(-6)^2-4(1)(2)}}{2(1)}$$
$$=\frac{6\pm\sqrt{28}}{2}=\frac{6\pm2\sqrt{7}}{2}=3\pm\sqrt{7}$$
$$u=3+\sqrt{7} \quad \text{or} \quad u=3-\sqrt{7}$$
$$u=w^2 \qquad u=w^2$$
$$w^2=3+\sqrt{7} \qquad w^2=3-\sqrt{7}$$
$$w=\pm\sqrt{3+\sqrt{7}} \qquad w=\pm\sqrt{3-\sqrt{7}}$$
$$\left\{-\sqrt{3+\sqrt{7}},\sqrt{3+\sqrt{7}},-\sqrt{3-\sqrt{7}},\sqrt{3-\sqrt{7}}\right\}$$

59) $2m^4+1=7m^2$

$$2m^4-7m^2+1=0$$

Let $u=m^2$ and $u^2=m^4$

$$2u^2-7u+1=0$$
$$u=\frac{-(-7)\pm\sqrt{(-7)^2-4(2)(1)}}{2(2)}.$$
$$=\frac{7\pm\sqrt{41}}{4}$$
$$u=\frac{7+\sqrt{41}}{2} \quad \text{or} \quad u=\frac{7-\sqrt{41}}{2}$$
$$u=m^2 \qquad u=m^2$$
$$m^2=\frac{7+\sqrt{41}}{4} \qquad m^2=3-\sqrt{7}$$
$$m=\pm\frac{\sqrt{7+\sqrt{41}}}{2} \qquad m=\pm\frac{\sqrt{7-\sqrt{41}}}{2}$$
$$\left\{-\frac{\sqrt{7+\sqrt{41}}}{2},\frac{\sqrt{7+\sqrt{41}}}{2},-\frac{\sqrt{7-\sqrt{41}}}{2},\frac{\sqrt{7-\sqrt{41}}}{2}\right\}$$

61) $t^{-2}-4t^{-1}-12=0$

$$(t^{-1}+2)(t^{-1}-6)=0$$
$$t^{-1}+2=0 \text{ or } t^{-1}-6=0$$
$$t^{-1}=-2 \qquad t^{-1}=6$$
$$t=-\frac{1}{2} \qquad t=\frac{1}{6} \qquad \left\{-\frac{1}{2},\frac{1}{6}\right\}$$

63) $4=13y^{-1}-3y^{-2}$

$$3y^{-2}-13y^{-1}+4=0$$
$$(3y^{-1}+1)(y^{-1}-4)=0$$
$$3y^{-1}-1=0 \text{ or } y^{-1}-4=0$$
$$3y^{-1}=1 \qquad y^{-1}=4$$
$$y^{-1}=\frac{1}{3} \qquad y=\frac{1}{4}$$
$$y=3$$
$$\left\{\frac{1}{4},3\right\}$$

65) $(x-2)^2+11(x-2)+24=0$

Let $u=x-2$

$$u^2+11u+24=0$$
$$(u+8)(u+3)=0$$
$$u+8=0 \text{ or } u+3=0$$
$$u=-8 \qquad u=-3$$

Solve for x using $u = x - 2$.

$-8 = x - 2 \qquad -3 = x - 2$

$-6 = x \qquad -1 = x$

$\{-6, -1\}$

67) $2(3q+4)^2 - 13(3q+4) + 20 = 0$

Let $u = 3q + 4$

$2u^2 - 13u + 20 = 0$

$(2u-5)(u-4) = 0$

$2u - 5 = 0$ or $u - 4 = 0$

$2u = 5$

$u = \frac{5}{2} \qquad u = 4$

Solve for q using $u = 3q + 4$.

$\frac{5}{2} = 3q + 4 \qquad 4 = 3q + 4$

$-\frac{3}{2} = 3q \qquad 0 = 3q$

$-\frac{1}{2} = q \qquad 0 = q$

$\left\{-\frac{1}{2}, 0\right\}$

69) $(5a-3)^2 + 6(5a-3) = -5$

Let $u = 5a - 3$

$u^2 + 6u + 5 = 0$

$(u-1)(u-5) = 0$

$u + 1 = 0$ or $u + 5 = 0$

$u = -1 \qquad u = -5$

Solve for a using $u = 5a - 3$.

$-1 = 5a - 3 \qquad -5 = 5a - 3$

$2 = 5a \qquad -2 = 5a$

$\frac{2}{5} = a \qquad -\frac{2}{5} = a$

$\left\{-\frac{2}{5}, \frac{2}{5}\right\}$

71) $3(k+8)^2 + 5(k+8) = 12$

Let $u = k + 8$

$3u^2 + 5u - 12 = 0$

$(3u-4)(u+3) = 0$

$3u - 4 = 0$ or $u + 3 = 0$

$3u = 4 \qquad u = -3$

$u = \frac{4}{3}$

Solve for k using $u = k + 8$.

$\frac{4}{3} = k + 8 \qquad -3 = k + 8$

$-\frac{20}{3} = k \qquad -11 = k$

$\left\{-11, -\frac{20}{3}\right\}$

73) $1 - \frac{8}{2w+1} = -\frac{16}{(2w+1)^2}$

Let $u = 2w + 1$

$1 - \frac{8}{u} = -\frac{16}{u^2}$

$u^2\left(1 - \frac{8}{u}\right) = u^2\left(-\frac{16}{u^2}\right)$

$u^2 - 8u = -16$

$u^2 - 8u + 16 = 0$

$(u-4)^2 = 0$

$u - 4 = 0$

$u = 4$

Solve for w using $u = 2w + 1$.

$4 = 2w + 1$

$3 = 2w$

$\frac{3}{2} = w \qquad \left\{\frac{3}{2}\right\}$

75) $1+\frac{2}{h-3}=\frac{1}{(h-3)^2}$

Let $u=h-3$

$$1+\frac{2}{u}=\frac{1}{u^2}$$
$$u^2\left(1+\frac{2}{u}\right)=u^2\left(\frac{1}{u^2}\right)$$
$$u^2+2u=1$$
$$u^2+2u+1=1+1$$
$$(u+1)^2=2$$
$$u+1=\pm\sqrt{2}$$
$$u=-1\pm\sqrt{2}$$

Solve for h using $u=h-3$.

$-1-\sqrt{2}=h-3$ $\quad$ $-1+\sqrt{2}=h-3$

$2-\sqrt{2}=h$ $\quad$ $2+\sqrt{2}=h$

$\{2-\sqrt{2},\ 2+\sqrt{2}\}$

Section 11.7: Exercises

1) $A=\pi r^2$

$$\frac{A}{\pi}=r^2$$
$$\pm\sqrt{\frac{A}{\pi}}=r$$
$$\pm\frac{\sqrt{A}}{\sqrt{\pi}}\cdot\frac{\sqrt{\pi}}{\sqrt{\pi}}=r$$
$$\frac{\pm\sqrt{A\pi}}{\pi}=r$$

3) $a=\frac{v^2}{r}$

$$ar=v^2$$
$$\pm\sqrt{ar}=v$$

5) $E=\frac{I}{d^2}$

$$d^2E=I$$
$$d^2=\frac{I}{E}$$
$$d=\pm\sqrt{\frac{I}{E}}$$
$$d=\pm\frac{\sqrt{I}}{\sqrt{E}}\cdot\frac{\sqrt{E}}{\sqrt{E}}=\frac{\pm\sqrt{IE}}{E}$$

7) $F=\frac{kq_1q_2}{r^2}$

$$r^2F=kq_1q_2$$
$$r^2=\frac{kq_1q_2}{F}$$
$$r=\pm\sqrt{\frac{kq_1q_2}{F}}$$
$$r=\frac{\pm\sqrt{kq_1q_2}}{\sqrt{F}}\cdot\frac{\sqrt{F}}{\sqrt{F}}=\frac{\pm\sqrt{kq_1q_2F}}{F}$$

9) $d=\sqrt{\frac{4A}{\pi}}$

$$d^2=\frac{4A}{\pi}$$
$$\pi d^2=4A$$
$$\frac{1}{4}\pi d^2=A$$

11) $T_p=2\pi\sqrt{\frac{l}{g}}$

$$\frac{T_p}{2\pi}=\sqrt{\frac{l}{g}}$$
$$\frac{T_p^2}{4\pi^2}=\frac{l}{g}$$
$$\frac{gT_p^2}{4\pi^2}=l$$

13) $T_p=2\pi\sqrt{\frac{l}{g}}$

$$T_p^2=4\pi^2\left(\frac{l}{g}\right)$$
$$gT_p^2=4\pi^2 l$$
$$g=\frac{4\pi^2 l}{T_p^2}$$

15) a) Both are written in the standard form for a quadratic equation, $ax^2+bx+c=0$

b) Use the quadratic formula.

17) $x=\dfrac{-(-5)\pm\sqrt{(-5)^2-4rs}}{2r}$

$=\dfrac{5\pm\sqrt{25-4rs}}{2r}$

19) $z=\dfrac{-r\pm\sqrt{r^2-4p(-q)}}{2p}$

$=\dfrac{-r\pm\sqrt{r^2+4pq}}{2p}$

21) $da^2-ha=k$

$da^2-ha-k=0$

$a=\dfrac{-(-h)\pm\sqrt{(-h)^2-4d(-k)}}{2d}$

$=\dfrac{h\pm\sqrt{h^2+4dk}}{2d}$

23) $s=\dfrac{1}{2}gt^2+vt$

$0=\dfrac{1}{2}gt^2+vt-s$

$t=\dfrac{-v\pm\sqrt{v^2+2\left(\frac{1}{2}g\right)s}}{2\left(\frac{1}{2}g\right)}$

$\dfrac{-v\pm\sqrt{v^2+gs}}{g}$

25) $x=$ width of sheet metal

$x+5=$ length of sheet metal

length of box $=x+3-1-1$

$=x+1$

width of box $=x-1-1$

$=x-2$

height of box $=1$

Volume $=(\text{length})(\text{width})(\text{height})$

$70=(x+1)(x-2)(1)$

$70=x^2-x-2$

$0=x^2-x-72$

$0=(x+8)(x-9)$

$x+8=0$ or $x-9=0$

$x=-8$ $\quad \boxed{x=9}$

width $=9$ in.

length $=9+3=12$ in.

27) $x = \text{width of non-skid surface}$

$80+2x = \text{length of pool plus two strips of non-skid surface}$

$60+2x = \text{width of pool plus two strips of non-skid surface}$

Area of Pool plus strips − Area of Pool = Area of Strips

$$(80+2x)(60+2x) \quad - \quad 80(60) \quad = \quad 576$$

$$4800+280x+4x^2-480=576$$

$$4x^2+280x-576=0$$

$$x^2+70x-144=0$$

$$(x+72)(x-2)=0$$

$$x+72=0 \text{ or } x-2=0$$

$$x=-72 \qquad \boxed{x=2}$$ The width of non-skid surface is 2 ft.

29) $x = \text{base of the sail}$

$2x+1 = \text{height of the sail}$

$$\text{Area} = \frac{1}{2}(\text{base})(\text{height})$$

$$60 = \frac{1}{2}(x)(2x-1)$$

$$120 = 2x^2 - x$$

$$0 = 2x^2 - x - 120$$

$$0 = (2x+15)(x-8)$$

$$2x+15=0 \text{ or } x-8=0$$

$$2x=-15 \qquad \boxed{x=8}$$

$$x = -\frac{15}{2}$$

base $= 8$ in. height $= 2(8)-1 = 15$ in.

31) $x = \text{height of the ramp}$

$2x+4 = \text{base of the ramp}$

$3x-4 = \text{hypotenuse of the ramp}$

$$a^2+b^2=c^2$$

$$x^2+(2x+4)^2=(3x-4)^2$$

$$x^2+4x^2+16x+16=9x^2-24x+16$$

$$5x^2+16x+16=9x^2-24x+16$$

$$0=4x^2-40x$$

$$0=4x(x-10)$$

$$4x=0 \text{ or } x-10=0$$

$$x=0 \qquad \boxed{x=10}$$

The height of the ramp is 10 in.

33) a) $h=40$

$$h=-16t^2+60t+4$$

$$40=-16t^2+60t+4$$

$$0=-16t^2+60t-36$$

$$0=4t^2-15t+9$$

$$0=(4t-3)(t-3)$$

$$4t-3=0 \text{ or } t-3=0$$

$$4t=3$$

$$t=\frac{3}{4} \qquad t=3$$

0.75 sec on the way up,

3 sec on the way down.

b) $h=0$

$$h=-16t^2+60t+4$$

$$0=-16t^2+60t+4$$

$$0=4t^2-15t-1$$

$$t=\frac{-(-15)\pm\sqrt{(-15)^2-4(4)(-1)}}{2(4)}$$

$$=\frac{15\pm\sqrt{225-16}}{8}=\frac{15\pm\sqrt{241}}{8}$$

Reject $\dfrac{15-\sqrt{241}}{8}$ as a solution since this is a negative number.

$\dfrac{15+\sqrt{241}}{8}$ sec or about 3.8 sec

35) a) $x=0$

$$y=-0.25x^2+1.5x+9.5$$
$$y=-0.25(0)^2+1.5(0)+9.5$$
$$y=9.5$$

9.5 million

b) $y=11.75$

$$y=-0.25x^2+1.5x+9.5$$
$$11.75=-0.25x^2+1.5x+9.5$$
$$0=-0.25x^2+1.5x-2.25$$
$$0=25x^2-150x+225$$
$$0=x^2-6x+9$$
$$0=(x-3)^2$$
$$0=x-3$$
$$3=x$$

11.75 million saw a Broadway play in 1999.

37) $D=\dfrac{65}{P}$; $S=10p+3$

$$D=S$$
$$\frac{65}{P}=10P+3$$
$$65=P(10P+3)$$
$$65=10P^2+3P$$
$$0=10P^2+3P-65$$

$$P=\frac{-3\pm\sqrt{3^2-4(10)(-65)}}{2(10)}$$
$$=\frac{-3\pm\sqrt{9+2600}}{20}=\frac{-3\pm\sqrt{2609}}{20}$$
$$=\frac{-3\pm\sqrt{2609}}{20}\approx\frac{-3\pm51.08}{20}$$
$$P\approx\frac{-3-51.08}{20} \quad\text{or}\quad P\approx\frac{-3+51.08}{20}$$
$$P\approx-2.70 \qquad \boxed{P\approx\$2.40}$$

Chapter 11 Review

1) $a^2-3a-54=0$

$$(a-9)(a+6)=0$$
$$a-9=0 \text{ or } a+6=0$$
$$a=9 \qquad a=-6 \qquad \{-6, 9\}$$

3)

$$\frac{2}{3}c^2=\frac{2}{3}c+\frac{1}{2}$$
$$6\left(\frac{2}{3}c^2\right)=6\left(\frac{2}{3}c+\frac{1}{2}\right)$$
$$4c^2=4c+3$$
$$4c^2-4c-3=0$$
$$(2c+1)(2c-3)=0$$
$$2c+1=0 \text{ or } 2c-3=0$$
$$2c=-1 \qquad 2c=3$$
$$c=-\frac{1}{2} \qquad c=\frac{3}{2} \qquad \left\{-\frac{1}{2}, \frac{3}{2}\right\}$$

5) $x^3+3x^2-16x-48=0$

$$x^2(x+3)-16(x+3)=0$$
$$(x^2-16)(x+3)=0$$
$$(x-4)(x+4)(x+3)=0$$
$$x-4=0 \text{ or } x+4=0 \text{ or } x+3=0$$
$$x=4 \qquad x=-4 \qquad x=-3$$

$\{-4, -3, 4\}$

7) $l = \text{length}$

$l - 4 = \text{width}$

$\text{Area} = (\text{width})(\text{length})$

$96 = (l-4)l$

$96 = l^2 - 4l$

$0 = l^2 - 4l - 96$

$0 = (l-12)(l+8)$

$l - 12 = 0$ or $l + 8 = 0$

$\boxed{l = 12}$ $\quad l = -8$

$\text{length} = 12$ cm

$\text{width} = 12 - 4 = 8$ cm

9) $d^2 = 144$

$d = \pm\sqrt{144}$

$d = \pm 12$ $\quad \{-12, 12\}$

11) $v^2 + 4 = 0$

$v^2 = -4$

$v = \pm\sqrt{-4}$

$v = \pm 2i$ $\quad \{-2i, 2i\}$

13) $(b-3)^2 = 49$

$b - 3 = \pm\sqrt{49}$

$b - 3 = \pm 7$

$b - 3 = 7$ or $b - 3 = -7$

$b = 10$ $\quad b = -4$ $\quad \{-4, 10\}$

15) $27k^2 - 30 = 0$

$27k^2 = 30$

$k^2 = \dfrac{30}{27}$

$k^2 = \dfrac{10}{9}$

$k = \pm\sqrt{\dfrac{10}{9}} k = \pm\dfrac{\sqrt{10}}{3}$

$\left\{-\dfrac{\sqrt{10}}{3}, \dfrac{\sqrt{10}}{3}\right\}$

17) Let a = length of one side

$a^2 + 3^2 = (3\sqrt{2})^2$

$a^2 + 9 = 9(2)$

$a^2 + 9 = 18$

$a^2 = 9$

$a = \pm 3$

Reject -3 for the length of the side. The length of the side is 3 units.

19) $d = \sqrt{(x_2 - x_1)^2 + (y_2 - y_1)^2}$

$d = \sqrt{(7-2)^2 + (5-3)^2}$

$d = \sqrt{5^2 + 2^2} = \sqrt{25 + 4} = \sqrt{29}$

21) $d = \sqrt{(x_2 - x_1)^2 + (y_2 - y_1)^2}$

$d = \sqrt{(0-3)^2 + [3-(-1)]^2}$

$d = \sqrt{(-3)^2 + 4^2} = \sqrt{9+16}$

$d = \sqrt{25} = 5$

23) $\sqrt{-49} = i\sqrt{49} = 7i$

25) $\sqrt{-2} \cdot \sqrt{-8} = i\sqrt{2} \cdot i\sqrt{8}$

$= i^2\sqrt{16} = -1 \cdot 4 = -4$

27) $(2+i) + (10-4i) = 12 - 3i$

29) $\left(\frac{4}{5}-\frac{1}{3}i\right)-\left(\frac{1}{2}+i\right)$
$=\left(\frac{4}{5}-\frac{1}{2}\right)+\left(-\frac{1}{3}i-i\right)$
$=\left(\frac{8}{10}-\frac{5}{10}\right)+\left(-\frac{1}{3}i-\frac{3}{3}i\right)$
$=\frac{3}{10}-\frac{4}{3}i$

31) $5(-6+7i)=-30+35i$

33) $3i(-7+12i)=-21i+36i^2$
$=-21i+36(-1)$
$=-36-21i$

35) $(4-6i)(3-6i)$
$=12-24i-18i+36i^2$
$=12-42i+36(-1)$
$=12-42i-36=-24-42i$

37) $(2-7i)(2+7i)=(2)^2+(7)^2$
$=4+49=53$

39) $\frac{6}{2+5i}=\frac{6}{2+5i}\cdot\frac{2-5i}{2-5i}$
$=\frac{12-30i}{(2)^2+(5)^2}=\frac{12-30i}{4+25}$
$=\frac{12-30i}{29}=\frac{12}{29}-\frac{30}{29}i$

41) $\frac{8}{i}=\frac{8}{i}\cdot\frac{-i}{-i}=\frac{-8i}{(1)^2}=-8i$

43) $\frac{9-4i}{6-i}=\frac{9-4i}{6-i}\cdot\frac{6+i}{6+i}$
$=\frac{54+9i-24i-4i^2}{(6)^2+(1)^2}$
$=\frac{54-15i-4(-1)}{36+1}$
$=\frac{54-15i+4}{37}=\frac{58-15i}{37}$
$=\frac{58}{37}-\frac{15}{37}i$

45) 1) $\frac{1}{2}(10)=5$ 2) $5^2=25$
$r^2+10r+25;\quad (r+5)^2$

47) 1) $\frac{1}{2}(-5)=-\frac{5}{2}$ 2) $\left(-\frac{5}{2}\right)^2=\frac{25}{4}$
$c^2-5c+\frac{25}{4};\quad \left(c-\frac{5}{2}\right)^2$

49) 1) $\frac{1}{2}\left(\frac{2}{3}\right)=\frac{1}{3}$ 2) $\left(\frac{1}{3}\right)^2=\frac{1}{9}$
$a^2+\frac{2}{3}a+\frac{1}{9};\quad \left(a+\frac{1}{3}\right)^2$

51) $p^2-6p-16=0$
$p^2-6p=16$
$p^2-6p+9=16+9$
$(p-3)^2=25$
$p-3=\pm\sqrt{25}$
$p-3=\pm5$
$p-3=5$ or $p-3=-5$
$p=8$ $\quad p=-2 \quad \{-2, 8\}$

53) $$n^2+10n=6$$
$$n^2+10n+25=6+25$$
$$(n+5)^2=31$$
$$n+5=\pm\sqrt{31}$$
$$n=-5\pm\sqrt{31}$$
$$\{-5-\sqrt{31},-5+\sqrt{31}\}$$

55) $$f^2+3f+1=0$$
$$f^2+3f=-1$$
$$f^2+3f+\frac{9}{4}=-1+\frac{9}{4}$$
$$\left(f+\frac{3}{2}\right)^2=\frac{5}{4}$$
$$f+\frac{3}{2}=\pm\sqrt{\frac{5}{4}}$$
$$f+\frac{3}{2}=\pm\frac{\sqrt{5}}{2}$$
$$f=-\frac{3}{2}\pm\frac{\sqrt{5}}{2}$$
$$\left\{-\frac{3}{2}-\frac{\sqrt{5}}{2},-\frac{3}{2}+\frac{\sqrt{5}}{2}\right\}$$

57) $$-3q^2+7q=12$$
$$q^2-\frac{7}{3}q=-4$$
$$q^2-\frac{7}{3}q+\frac{49}{36}=-4+\frac{49}{36}$$
$$\left(q-\frac{7}{6}\right)^2=-\frac{95}{36}$$
$$q-\frac{7}{6}=\pm\sqrt{-\frac{95}{36}}$$
$$q-\frac{7}{6}=\pm\frac{\sqrt{95}}{6}i$$
$$q=\frac{7}{6}\pm\frac{\sqrt{95}}{6}i$$
$$\left\{\frac{7}{6}-\frac{\sqrt{95}}{6}i,\frac{7}{6}+\frac{\sqrt{95}}{6}i\right\}$$

59) $$m^2+4m-12=0$$
$a=1$, $b=4$ and $c=-12$
$$m=\frac{-4\pm\sqrt{4^2-4(1)(-12)}}{2(1)}$$
$$=\frac{-4\pm\sqrt{16+48}}{2}$$
$$=\frac{-4\pm\sqrt{64}}{2}=\frac{-4\pm 8}{2}$$
$$\frac{-4+8}{2}=\frac{4}{2}=2,$$
$$\frac{-4-8}{2}=\frac{-12}{2}=-6 \qquad \{-6, 2\}$$

61) $$10g-5=2g^2$$
$$0=2g^2-10g+5$$
$a=2$, $b=-10$ and $c=5$
$$g=\frac{-(-10)\pm\sqrt{(-10)^2-4(2)(5)}}{2(2)}$$
$$=\frac{10\pm\sqrt{100-40}}{4}$$
$$=\frac{10\pm\sqrt{60}}{4}=\frac{10\pm 2\sqrt{15}}{4}=\frac{5\pm\sqrt{15}}{2}$$
$$\left\{\frac{5-\sqrt{15}}{2},\frac{5+\sqrt{15}}{2}\right\}$$

63) $$\frac{1}{6}t^2-\frac{1}{3}t+\frac{2}{3}=0$$
$$6\left(\frac{1}{6}t^2-\frac{1}{3}t+\frac{2}{3}\right)=6(0)$$
$$t^2-2t+4=0$$
$a=1$, $b=-2$ and $c=4$

$$t=\frac{-(-2)\pm\sqrt{(-2)^2-4(1)(4)}}{2(1)}$$
$$=\frac{2\pm\sqrt{4-16}}{2}=\frac{2\pm\sqrt{-12}}{2}$$
$$=\frac{2\pm 2i\sqrt{3}}{2}=1\pm i\sqrt{3}$$
$$\left\{1-i\sqrt{3},\ 1+i\sqrt{3}\right\}$$

65) $(6r+1)(r-4)=-2(12r+1)$
$$6r^2-23r-4=-24r-2$$
$$6r^2+r-2=0$$
$a=6,\ b=1$ and $c=-2$
$$r=\frac{-1\pm\sqrt{1^2-4(6)(-2)}}{2(6)}$$
$$=\frac{-1\pm\sqrt{1+48}}{12}=\frac{-1\pm\sqrt{49}}{12}$$
$$=\frac{-1\pm 7}{12}$$
$$\frac{-1+7}{12}=\frac{6}{12}=\frac{1}{2},$$
$$\frac{-1-7}{12}=\frac{-8}{12}=-\frac{2}{3} \qquad \left\{-\frac{2}{3},\frac{1}{2}\right\}$$

67) $a=3,\ b=-2$ and $c=-5$
$$b^2-4ac=(-2)^2-4(3)(-5)$$
$$=4-60=64$$
two rational solutions

69) $t^2=-3(t+2)$
$$t^2=-3t-6$$
$$t^2+3t+6=0$$
$a=1,\ b=3$ and $c=6$
$$b^2-4ac=3^2-4(1)(6)$$
$$=9-24=-15$$
two irrational solutions

71) $4k^2+bk+9=0$
$a=4,\ b=b$ and $c=9$
$$b^2-4ac=0$$
$$b^2-4(4)(9)=0$$
$$b^2-144=0$$
$$b^2=144$$
$$b=\pm 12$$

73) $z+2=\frac{15}{z}$
$$z(z+2)=z\left(\frac{15}{z}\right)$$
$$z^2+2z=15$$
$$z^2+2z-15=0$$
$$(z+5)(z-3)=0$$
$$z+5=0 \text{ or } z-3=0$$
$$z=-5 \qquad z=3 \qquad \{-5,3\}$$

75) $\frac{10}{m}=3+\frac{8}{m^2}$
$$m^2\left(\frac{10}{m}\right)=m^2\left(3+\frac{8}{m^2}\right)$$
$$10m=3m^2+8$$
$$0=3m^2-10+8$$
$$0=(3m-4)(m-2)$$
$$3m-4=0 \text{ or } m-2=0$$
$$3m=4 \qquad m=2$$
$$m=\frac{4}{3} \qquad \left\{\frac{4}{3},2\right\}$$

77) $x-4\sqrt{x}=5$

$x-5=4\sqrt{x}$

$(x-5)^2=(4\sqrt{x})^2$

$x^2-10x+25=16x$

$x^2-26x+25=0$

$(x-25)(x-1)=0$

$x-25=0$ or $x-1=0$

$x=25 \quad x=1$

Only one solution satisfies the original equation. $\{25\}$

79) $b^4+5b^2-14=0$

$(b^2+7)(b^2-2)=0$

$b^2+7=0$ or $b^2-2=0$

$b^2=-7 \quad b^2=2$

$b=\pm i\sqrt{7} \quad b=\pm\sqrt{2}$

$\{-\sqrt{2},\sqrt{2},-i\sqrt{7},i\sqrt{7}\}$

81) $y+2=3y^{1/2}$

$y-3y^{1/2}+2=0$

$(y^{1/2}-2)(y^{1/2}-1)=0$

$y^{1/2}-2=0$ or $y^{1/2}-1=0$

$y^{1/2}=2 \quad y^{1/2}=1$

$(\sqrt{y})^2=2^2 \quad (\sqrt{y})^2=1^2$

$y=4 \quad y=1 \quad \{1,4\}$

83) $2(v+2)^2+(v+2)-3=0$

Let $u=v+2$

$2u^2+u-3=0$

$(2u+3)(u-1)=0$

$2u+3=0$ or $u-1=0$

$2u=-3 \quad u=1$

$u=-\frac{3}{2}$

Solve for v using $u=v+2$.

$-\frac{3}{2}=v+2 \quad 1=v+2$

$-\frac{7}{2}=v \quad -1=v$

$\left\{-\frac{7}{2},-1\right\}$

85) $F=\frac{mv^2}{r}$

$Fr=mv^2$

$\frac{Fr}{m}=v^2$

$\pm\sqrt{\frac{Fr}{m}}=v$

$\pm\frac{\sqrt{Fr}}{\sqrt{m}}\cdot\frac{\sqrt{m}}{\sqrt{m}}=v$

$\pm\frac{\sqrt{Frm}}{r}=v$

87) $r=\sqrt{\frac{A}{\pi}}$

$r^2=\frac{A}{\pi}$

$\pi r^2=A$

89) $n=\dfrac{-(-l)\pm\sqrt{(-l)^2-4(k)(-m)}}{2k}$

$=\dfrac{l\pm\sqrt{l^2+4km}}{2k}$

91) $x=$ width of border

$18+2x=$ length of case plus borders

$27+2x=$ width of case plus borders

Area of case plus border $=792$

$(18+2x)(27+2x)=792$

$486+45x+4x^2=792$

$4x^2+90x-306=0$

$(4x+102)(x-3)=0$

$4x+102=0$ or $x-3=0$

$4x=-102$ $\boxed{x=3}$

$x=-\dfrac{51}{2}$

The width of the border is 3 in.

93) $D=\dfrac{240}{P}$; $S=4P-2$

$D=S$

$\dfrac{240}{P}=4P-2$

$240=P(4P-2)$

$240=4P^2-2P$

$0=4P^2-2P-240$

$0=2P^2-P-120$

$0=(2P+15)(P-8)$

$2P+15=0$ or $P-8=0$

$2P=-15$ $\boxed{P=\$8.00}$

$P=-\dfrac{15}{2}$

Chapter 11 Test

1) $d=\sqrt{(x_2-x_1)^2+(y_2-y_1)^2}$

$d=\sqrt{[4-(-6)]^2+(3-2)^2}$

$d=\sqrt{10^2+1^2}=\sqrt{100+1}=\sqrt{101}$

3) $\sqrt{-18}$

$=i\sqrt{18}$

$=3i\sqrt{2}$

5) $(-3+7i)-(-8+4i)$

$=5+3i$

7) $\left(\dfrac{1}{2}+3i\right)\left(\dfrac{2}{3}-i\right)=\dfrac{1}{3}-\dfrac{1}{2}i+2i-3i^2$

$=\dfrac{1}{3}+\dfrac{3}{2}i+3$

$=\dfrac{10}{3}+\dfrac{3}{2}i$

9) $\dfrac{8}{6+i}=\dfrac{8}{6+i}\cdot\dfrac{6-i}{6-i}=\dfrac{48-8i}{6^2+1^2}$

$=\dfrac{48-8i}{36+1}=\dfrac{48-8i}{37}=\dfrac{48}{37}-\dfrac{8}{37}i$

11) $b^2+4b-7=0$

$b^2+4b=7$

$b^2+4b+4=7+4$

$(b+2)^2=11$

$b+2=\pm\sqrt{11}$

$b=-2\pm\sqrt{11}$

$\{-2-\sqrt{11},-2+\sqrt{11}\}$

13) $a=1,\ b=-8$ and $c=17$

$$x=\frac{-(-8)\pm\sqrt{(-8)^2-4(1)(17)}}{2(1)}$$

$$=\frac{8\pm\sqrt{64-68}}{2}=\frac{8\pm\sqrt{-4}}{2}$$

$$=\frac{8\pm 2i}{2}=4\pm i$$

$\{4-i,4+i\}$

15) $3q^2+2q=8$

$3q^2+2q-8=0$

$(3q-4)(q+2)=0$

$3q-4=0$ or $q-2=0$

$3q=4 \qquad q=-2$

$q=\frac{4}{3} \qquad \left\{-2,\frac{4}{3}\right\}$

17) $(2t-3)(t-2)=2$

$2t^2-7t+6=2$

$2t^2-7t+4=0$

$a=2,\ b=-7$ and $c=4$

$$t=\frac{-(-7)\pm\sqrt{(-7)^2-4(2)(4)}}{2(2)}$$

$$=\frac{7\pm\sqrt{49-32}}{4}=\frac{7\pm\sqrt{17}}{4}$$

$\left\{\frac{7-\sqrt{17}}{4},\frac{7+\sqrt{17}}{4}\right\}$

19) $\frac{3}{10x}=\frac{x}{x-1}-\frac{4}{5}$

$$10x(x-1)\left(\frac{3}{10x}\right)=10x(x-1)\left(\frac{x}{x-1}-\frac{4}{5}\right)$$

$3x-3=10x^2-8x^2+8x$

$3x-3=2x^2+8x$

$0=2x^2+5x+3$

$0=(2x+3)(x+1)$

$2x+3=0$ or $x+1=0$

$2x=-3 \qquad x=-1$

$x=-\frac{3}{2}$

$\left\{-\frac{3}{2},-1\right\}$

21) Let a = length of one side

$a^2+\left(3\sqrt{5}\right)^2=8^2$

$a^2+9(5)=64$

$a^2+45=64$

$a^2=19$

$a=\pm\sqrt{19}$

Reject $-\sqrt{19}$ for the length of the side. The length of the side is $\sqrt{19}$ units.

23) $r=\sqrt{\frac{3V}{\pi h}}$

$r^2=\frac{3V}{\pi h}$

$\pi r^2 h=3V$

$\frac{1}{3}\pi r^2 h=V$

25) $x = \text{width of sheet metal}$

$x+6 = \text{length of sheet metal}$

$\text{length of box} = x+6-3-3$

$= x$

$\text{width of box} = x-3-3$

$= x-6$

$\text{height of box} = 3$

$\text{Volume} = (\text{length})(\text{width})(\text{height})$

$273 = x(x-6)(3)$

$273 = 3x^2 - 18x$

$0 = 3x^2 - 18x - 273$

$0 = x^2 - 6x - 91$

$0 = (x+7)(x-13)$

$x+7=0$ or $x-13=0$

$x=-7$ $\quad \boxed{x=13}$

$\text{width} = 13$ in.

$\text{length} = 13+6 = 19$ in.

Cumulative Review: Chapters 1-11

1) $\dfrac{4}{15}+\dfrac{1}{6}-\dfrac{3}{5}=\dfrac{8}{30}+\dfrac{5}{30}-\dfrac{18}{30}$

$= -\dfrac{5}{30} = -\dfrac{1}{6}$

3) $A = (20)(21)-(21-8)(20-6)$

$= 420-(13)(14) = 420-182$

$= 238 \text{ cm}^2$

$P = 2(20)+2(21) = 40+42 = 82$ cm

5) $\left(5x^4y^{-10}\right)\left(3xy^3\right)^2$

$= 5x^4y^{-10}3^2x^2y^6$

$= 5\cdot 9\cdot x^4\cdot x^2\cdot y^{-10}\cdot y^6$

$= 45x^{4+2}y^{-10+6} = 5x^6y^{-4} = \dfrac{5x^6}{y^4}$

7) $x = \text{cameras sold in December 2006}$

$108 = x + x(.20)$

$108 = 1.20x$

$90 = x$

9) $2x-5y=8$

x-int: Let $y=0$, and solve for x.

$2x-5(0)=8$

$2x=8$

$x=4$ $\quad (4, 0)$

y-int: Let $x=0$, and solve for y.

$x(0)-5y=8$

$-5y=8$

$y=-\dfrac{8}{5}$ $\quad \left(0, -\dfrac{8}{5}\right)$

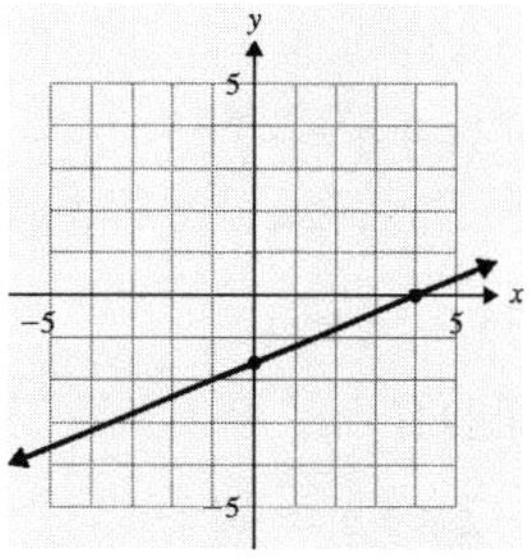

11) $x =$ cost for a bag of chips
$y =$ cost for a can of soda
$2x+3y=3.85$
$x+2y=2.30$
Use substitution. $x=2.30-2y$

$$\begin{aligned} 2(2.30-2y)+3y&=3.85 & x&=2.30-2y;\ y=0.75 \\ 4.60-4y+3y&=3.85 & x&=2.30-2(0.75) \\ -y&=-0.75 & x&=2.30-1.50 \\ y&=0.75 & x&=0.80 \end{aligned}$$

chips: \$0.80; soda: \$0.75

13) $3(r-5)^2 = 3(r^2-10r+25)$
$= 3r^2-30r+75$

15) $a^3+125=(a+5)(a^2-5a+25)$

17) $\dfrac{2+\frac{6}{c}}{\frac{2}{c^2}-\frac{8}{c}} = \dfrac{c^2\left(2+\frac{6}{c}\right)}{c^2\left(\frac{2}{c^2}-\frac{8}{c}\right)} = \dfrac{2c^2+6c}{2-8c}$
$= \dfrac{2c(c+3)}{2(1-4c)} = \dfrac{c(c+3)}{1-4c}$

19) I $4x-2y+z=-7$
II $-3x+y-2z=5$
III $2x+3y+5z=4$
Add: $\mathrm{I}+(2)\mathrm{II}$

$$\begin{aligned} 4x-2y+z&=-7 \\ -6x+2y-4z&=10 \\ \hline \text{A } -10x \quad -3z&=3 \end{aligned}$$

Add: $\mathrm{III}+(-3)\mathrm{II}$

$$\begin{aligned} 2x+3y+5z&=4 \\ 9x-3y+6z&=-15 \\ \hline 11x \quad +11z&=-11 \end{aligned}$$

B $x+z=-1;\ x=-z-1$
Substitute $x=-z-1$ into A.

$$\begin{aligned} -10(-z-1)-3z&=3 \\ 10z+10-3z&=3 \\ -7z&=-7 \\ z&=-1 \end{aligned}$$

Substitute $z=-1$ into A.

$$\begin{aligned} -10x-3(-1)&=3 \\ -10x+3&=3 \\ -10x&=0 \\ x&=0 \end{aligned}$$

Substitute $x=0$ and $z=-1$ into II.

$$\begin{aligned} -3(0)+y-2(-1)&=5 \\ y+2&=5 \\ y&=3 \qquad (0, 3, -1) \end{aligned}$$

21) $\sqrt[3]{40}=\sqrt[3]{8}\cdot\sqrt[3]{5}=2\sqrt[3]{5}$

23) $64^{2/3}=\left(\sqrt[3]{64}\right)^2=(4)^2=16$

25) $(10+3i)(1-8i)=10-80i+3i-24i^2$
$=10-77i-24(-1)$
$=10-77i+24$
$=34-77i$

27) $\frac{3}{5}x^2+\frac{1}{5}=\frac{1}{5}x$

$$5\left(\frac{3}{5}x^2+\frac{1}{5}\right)=5\left(\frac{1}{5}x\right)$$

$$3x^2+1=x$$

$$3x^2-x+1=0$$

$a=3,\ b=-1$ and $c=1$

$$x=\frac{-(-1)\pm\sqrt{(-1)^2-4(3)(1)}}{2(3)}$$

$$=\frac{1\pm\sqrt{1-12}}{6}=\frac{1\pm\sqrt{-11}}{6}$$

$$=\frac{1\pm i\sqrt{11}}{6}=\frac{1}{6}\pm\frac{\sqrt{11}}{6}i$$

$$\left\{\frac{1}{6}-\frac{\sqrt{11}}{6}i,\frac{1}{6}+\frac{\sqrt{11}}{6}i\right\}$$

29) $p^2+6p=27$

$$p^2+6p-27=0$$

$$(p+9)(p-3)=0$$

$p+9=0$ or $p-3=0$

$p=-9$ $\quad p=3$ $\qquad \{-9,3\}$

Section 12.1: Exercises

1) It is a special type of relation in which each element of the domain corresponds to exactly one element in the range.

3) Domain: $\{5,6,14\}$
Range: $\{-3,0,1,3\}$
Not a function

5) Domain: $\{-2,2,5,8\}$
Range: $\{4,25,64\}$
Is a function

7) Domain: $(-\infty,\infty)$ Range: $[-4,\infty)$
Is a function

9) yes 11) yes

13) no 15) no

17) False. It is read as "f of x."

19)

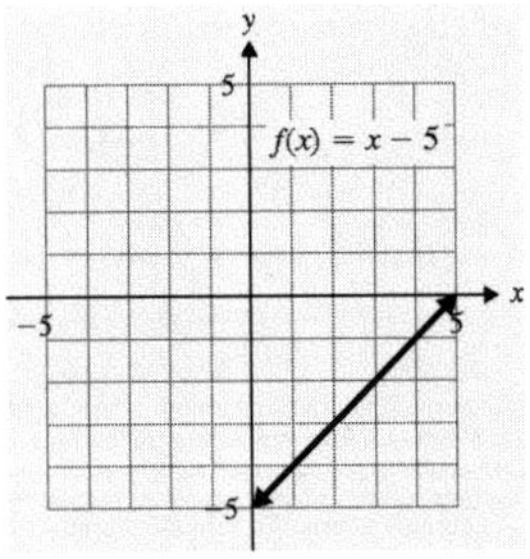

21)

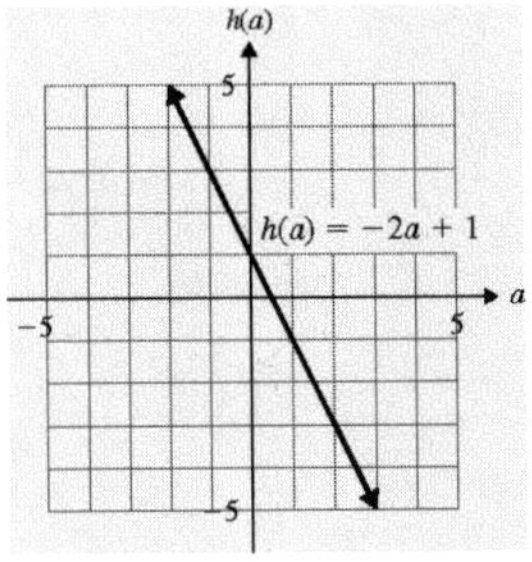

23)

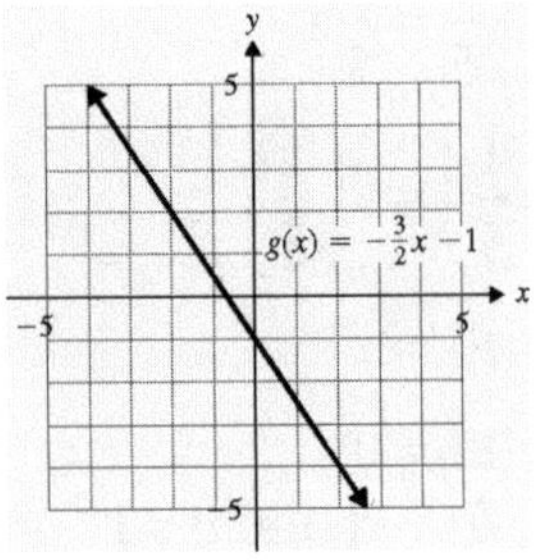

25)

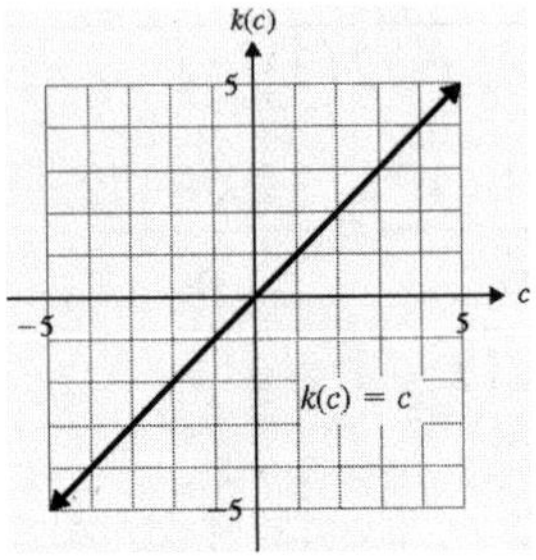

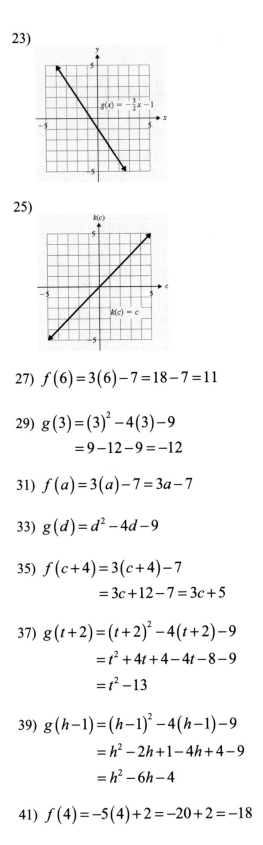

27) $f(6)=3(6)-7=18-7=11$

29) $g(3)=(3)^2-4(3)-9$
$=9-12-9=-12$

31) $f(a)=3(a)-7=3a-7$

33) $g(d)=d^2-4d-9$

35) $f(c+4)=3(c+4)-7$
$=3c+12-7=3c+5$

37) $g(t+2)=(t+2)^2-4(t+2)-9$
$=t^2+4t+4-4t-8-9$
$=t^2-13$

39) $g(h-1)=(h-1)^2-4(h-1)-9$
$=h^2-2h+1-4h+4-9$
$=h^2-6h-4$

41) $f(4)=-5(4)+2=-20+2=-18$

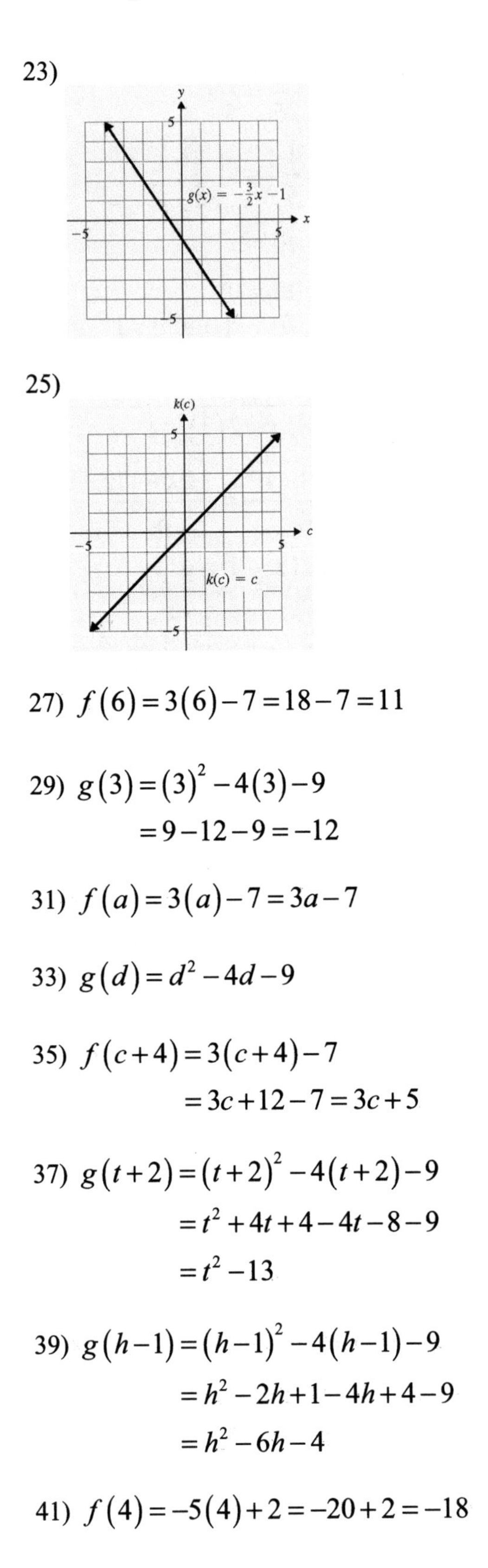

43) $g(-6) = (-6)^2 + 7(-6) + 2$
$= 36 - 42 + 2 = -4$

45) $f(-3k) = -5(-3k) + 2 = 15k + 2$

47) $g(5t) = (5t)^2 + 7(5t) + 2$
$= 25t^2 + 35t + 2$

49) $f(b+1) = -5(b+1) + 2$
$= -5b - 5 + 2 = -5b - 3$

51) $g(r+4) = (r+4)^2 + 7(r+4) + 2$
$= r^2 + 8r + 16 + 7r + 28 + 2$
$= r^2 + 15r + 46$

53) $f(x) = 4x + 3$
$23 = 4x + 3$
$20 = 4x$
$5 = x$

55) $h(x) = -2x - 5$
$0 = -2x - 5$
$5 = -2x$
$-\frac{5}{2} = x$

57) $p(x) = x^2 - 6x - 16$
$0 = x^2 - 6x - 16$
$0 = (x+2)(x-8)$
$x + 2 = 0$ or $x - 8 = 0$
$x = -2 \qquad x = 8$

59) $(-\infty, \infty)$

61) Set the denominator equal to 0 and solve for the variable. The domain consists of all real numbers except the values that make the denominator equal to 0.

63) $(-\infty, \infty)$ 65) $(-\infty, \infty)$

67) Solve: $x + 8 = 0$
$x = -8$
$(-\infty, -8) \cup (-8, \infty)$

69) Solve: $x = 0$
$(-\infty, 0) \cup (0, \infty)$

71) Solve: $2c - 1 = 0$
$2c = 1$
$c = \frac{1}{2}$
$\left(-\infty, \frac{1}{2}\right) \cup \left(\frac{1}{2}, \infty\right)$

73) Solve: $7t + 3 = 0$
$7t = -3$
$t = -\frac{3}{7}$
$\left(-\infty, -\frac{3}{7}\right) \cup \left(-\frac{3}{7}, \infty\right)$

75) $(-\infty, \infty)$

77) Solve: $x^2 + 11z + 24 = 0$
$(x+8)(x+3) = 0$
$x + 8 = 0$ or $x + 3 = 0$
$x = -8 \qquad x = -3$
$(-\infty, -8) \cup (-8, -3) \cup (-3, \infty)$

79) Solve: $c^2 - 5c - 36 = 0$
$(c-9)(c+4) = 0$
$c - 9 = 0$ or $c + 4 = 0$
$c = 9 \qquad c = -4$
$(-\infty, -4) \cup (-4, 9) \cup (9, \infty)$

81) Solve: $x \geq 0 \qquad [0, \infty)$

83) Solve: $n+2\geq 0$

$n\geq -2 \qquad [-2,\infty)$

85) Solve: $a-8\geq 0$

$a\geq 8 \qquad [8,\infty)$

87) Solve: $2x-5\geq 0$

$2x\geq 5$

$x\geq \frac{5}{2} \qquad \left[\frac{5}{2},\infty\right)$

89) Solve: $-t\geq 0$

$t\leq 0 \qquad (-\infty,0]$

91) Solve: $9-a\geq 0$

$-a\geq -9$

$a\leq 9 \qquad (-\infty,9]$

93) $(-\infty,\infty)$

95) a) $C(20)=22(20)=440$

$440

b) $C(56)=22(56)=1232$

$1232

c) $C(y)=770$

$770=22y$

$35=y$ 35 sq. yards

d)

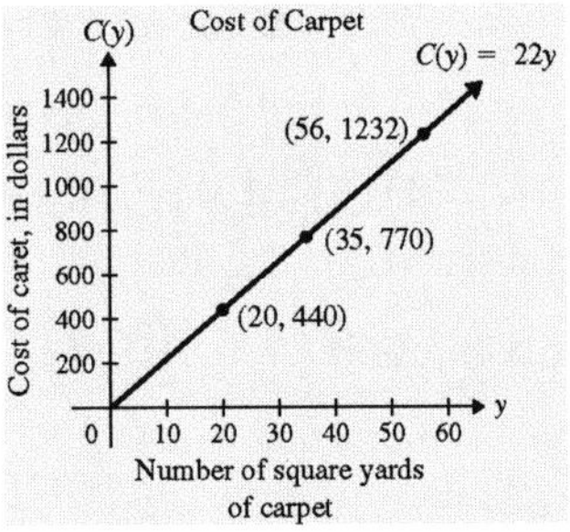

97) a) $L(1)=50(1)+40=50+40=90$

The labor charge for a 1-hour repair job is $90.

b) $L(1.5)=50(1.5)+40$

$=75+40=115$

The labor charge for a 1.5-hour repair job is $115.

c) $L(h)=165$

$165=50h+40$

$125=50h$

$2.5=h$

If the labor charge is $165, the repair job took 2.5 hours.

99) a) $A(r)=\pi r^2$

b) $A(3)=\pi(3)^2=9\pi$

When the radius of the circle is 3 cm, the area of the circle is 9π sq. cm.

c) $A(5)=\pi(5)^2=25\pi$

When the radius of the circle is 5 in, the area of the circle is 25π sq. in.

d) $A(r)=64\pi$

$64\pi=\pi r^2$

$64=r^2$

$\pm 8=r \qquad r=8$ inches

Section 12.2: Exercises

1) Domain: $(-\infty, \infty)$; Range: $[3, \infty)$

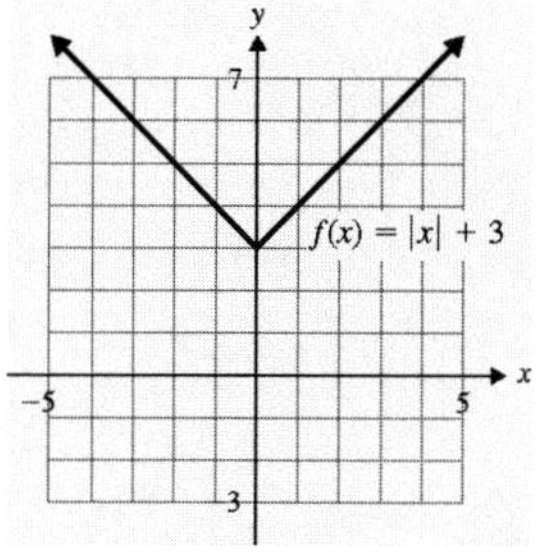

3) Domain: $(-\infty, \infty)$; Range: $[0, \infty)$

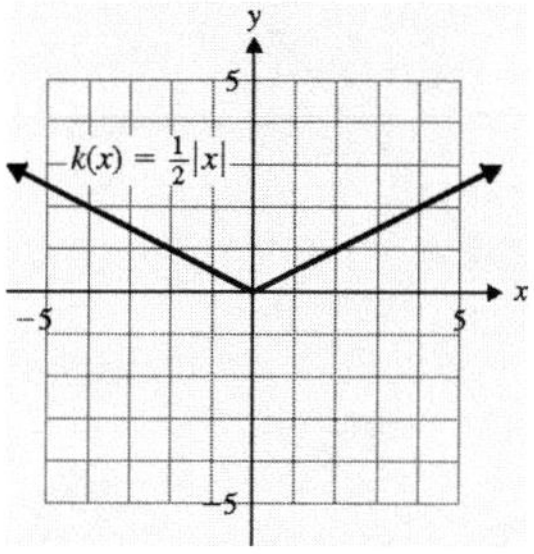

5) Domain: $(-\infty, \infty)$; Range: $[-4, \infty)$

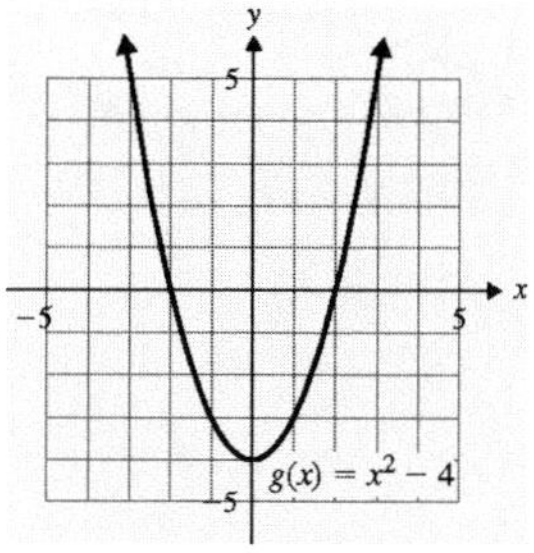

7) Domain: $(-\infty, \infty)$; Range: $(-\infty, -1]$

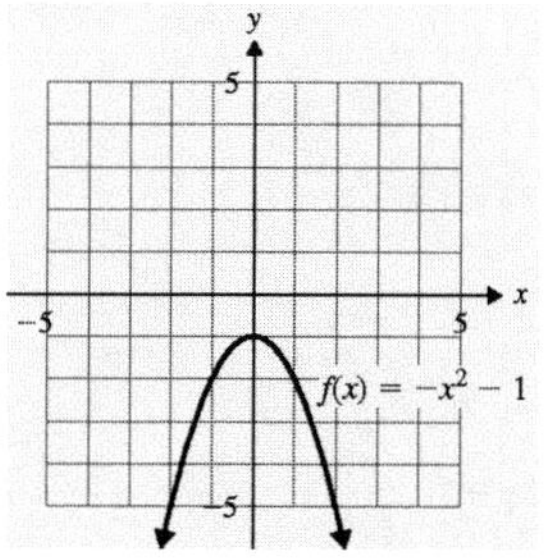

9) Domain: $[-3, \infty)$; Range: $[0, \infty)$

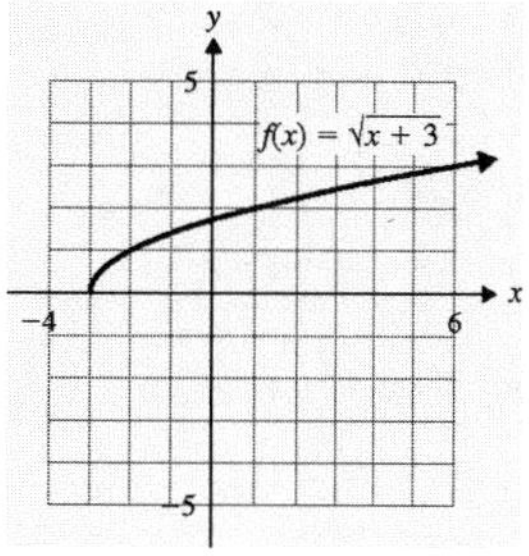

11) Domain: $[0, \infty)$; Range: $[0, \infty)$

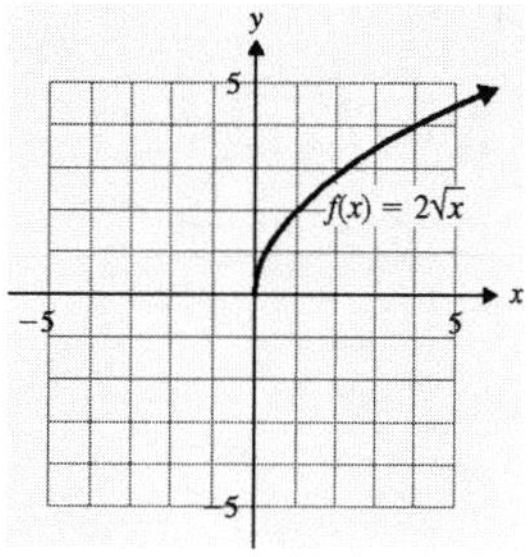

13) The graph of $g(x)$ is the same shape as $f(x)$, but g is shifted down 2 units.

15) The graph of $g(x)$ is the same shape as $f(x)$, but g is shifted left 2 units.

17) The graph of $g(x)$ is the reflection of $f(x)$ about the x-axis.

19)

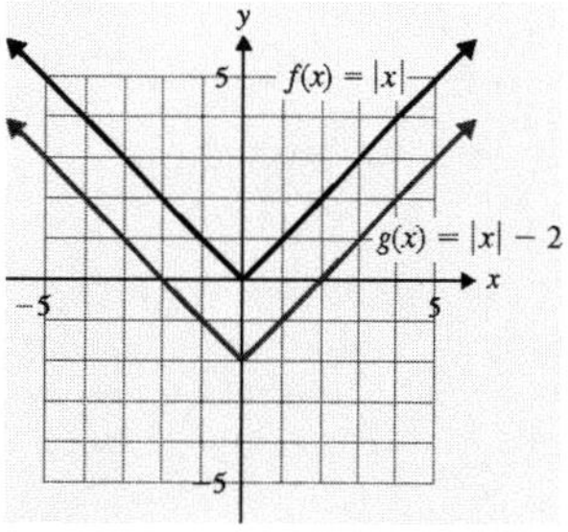

21)

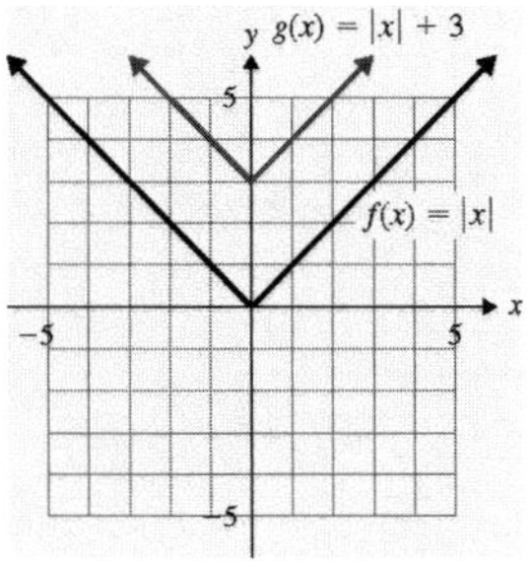

23)

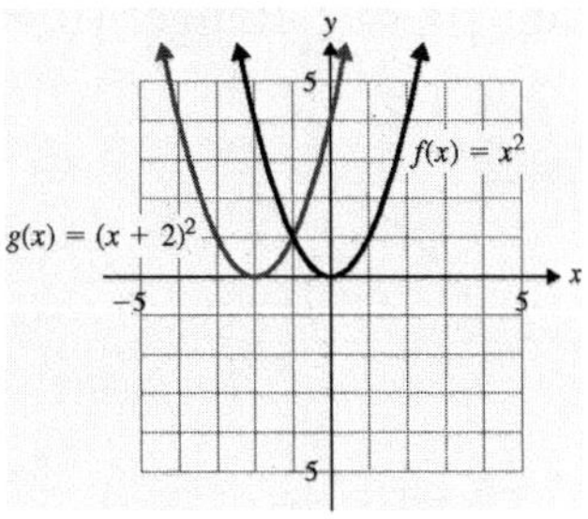

25)

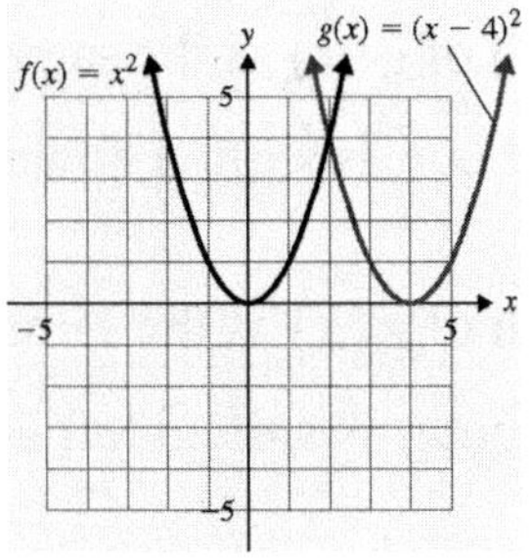

27)

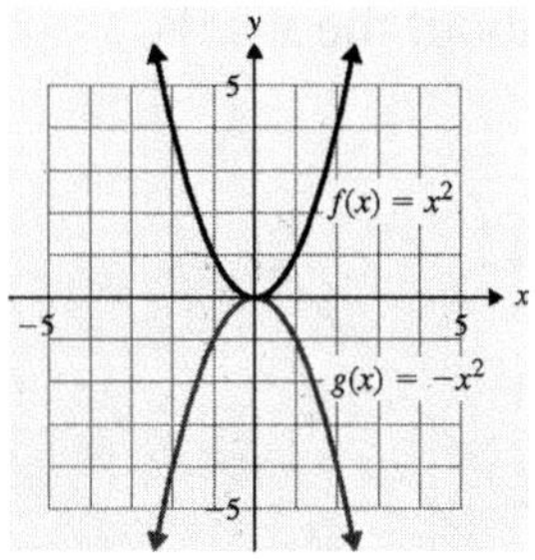

29)

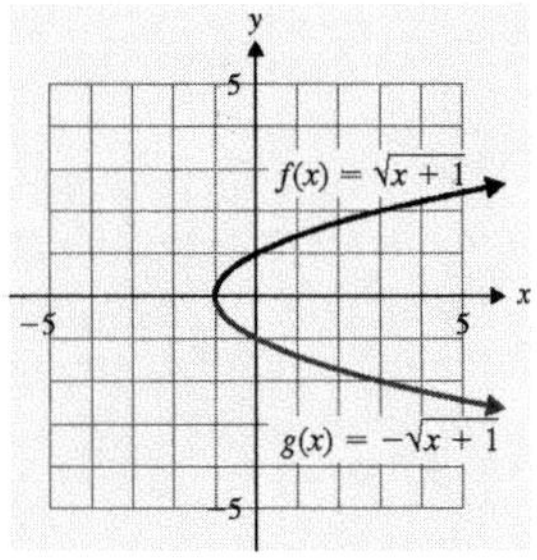

31)

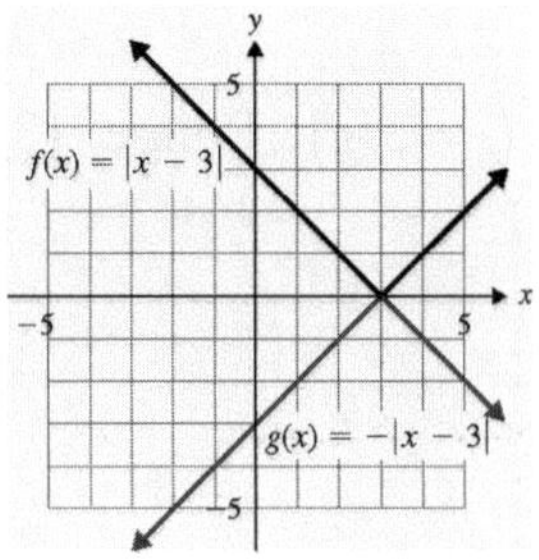

33)

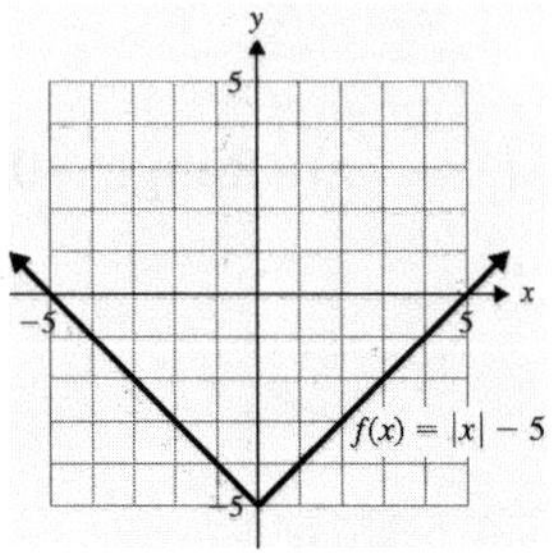

35)

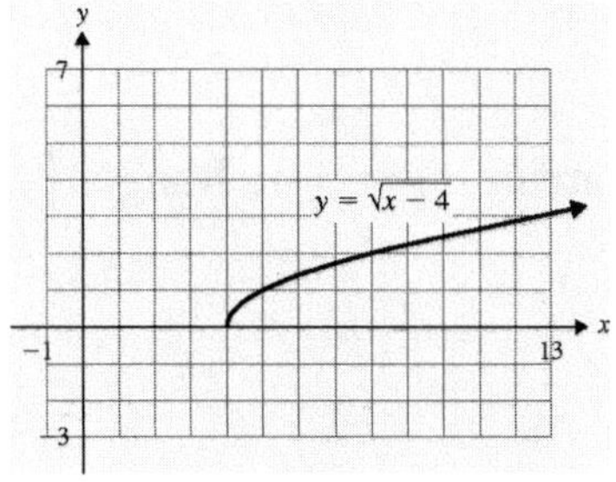

37)

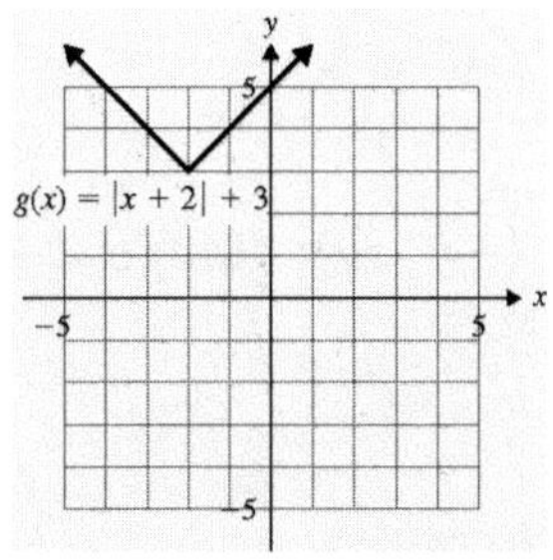

39)

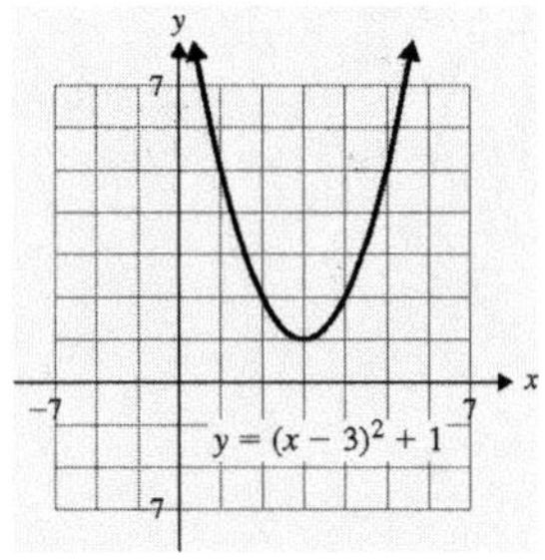

41)

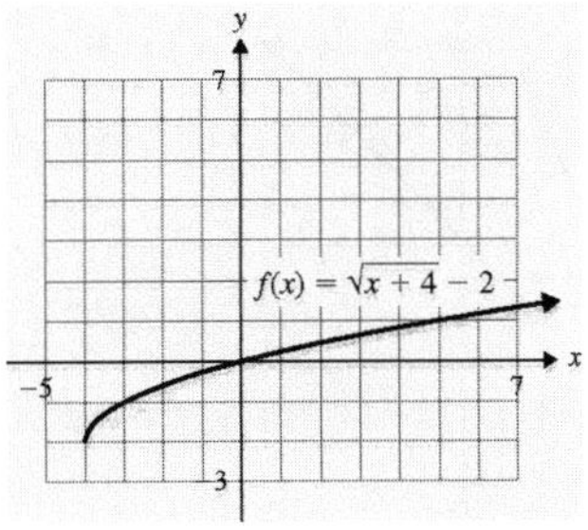

43)

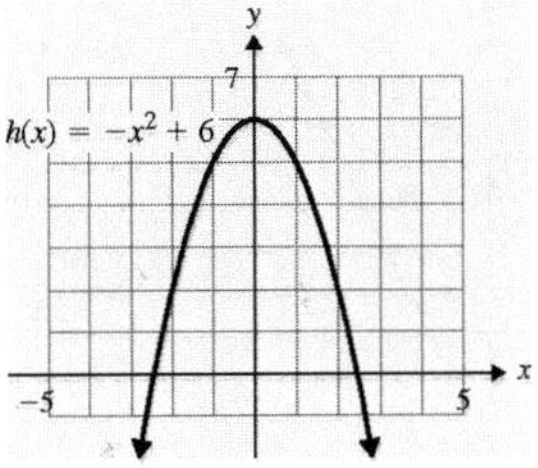

45)

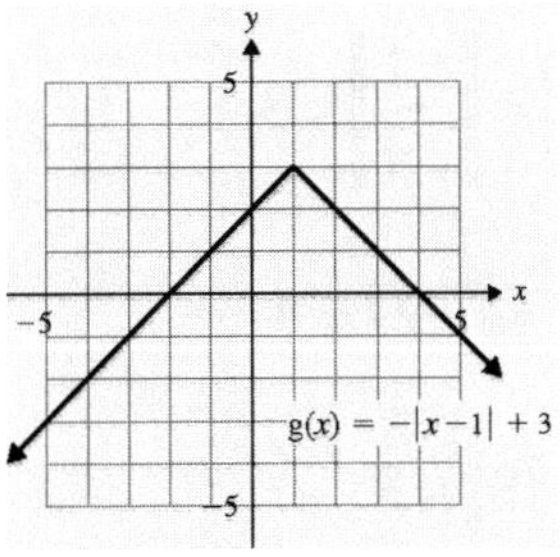

47)

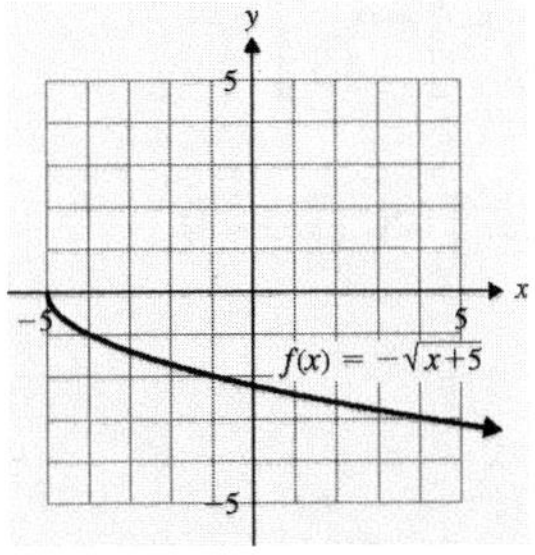

49) a) $h(x)$ b) $f(x)$

c) $g(x)$ d) $k(x)$

51) $g(x)=\sqrt{x+5}$

53) $g(x)=|x+2|-1$

55) $g(x)=(x+3)^2+\dfrac{1}{2}$

57) $g(x)=-x^2$

59)

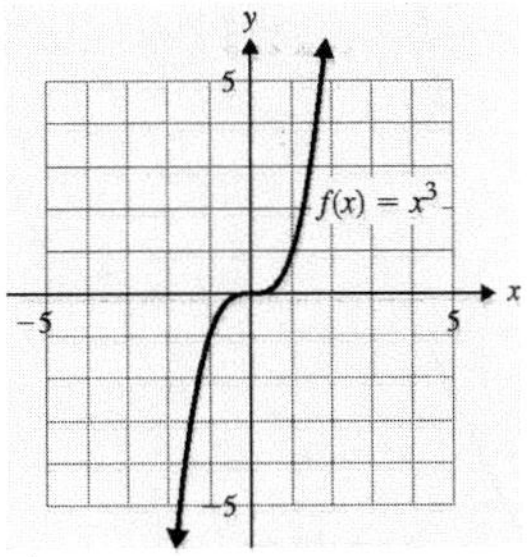

a)

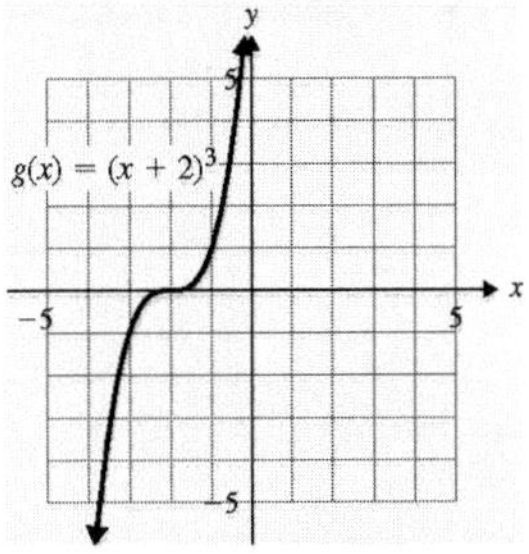

b)

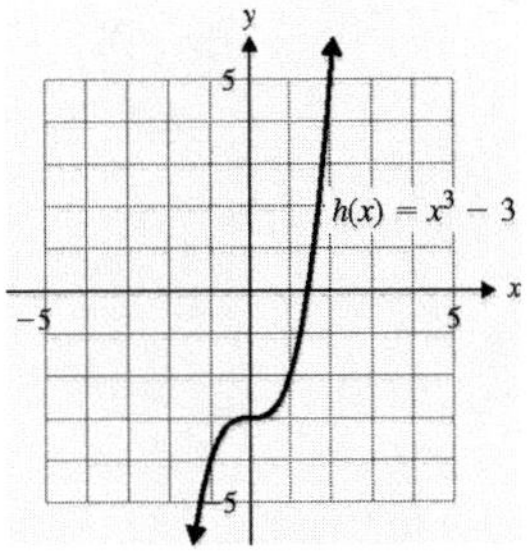

c)

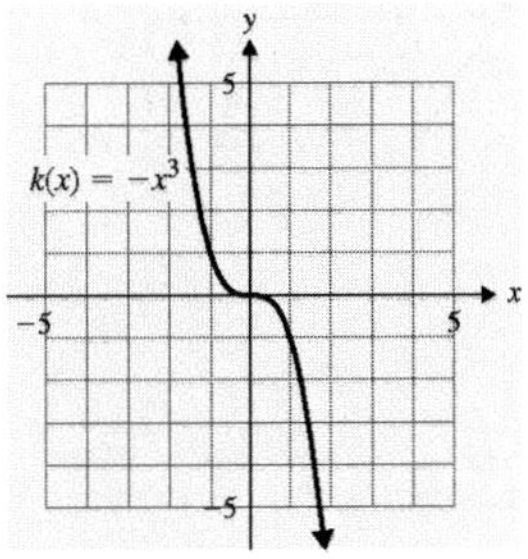

d)

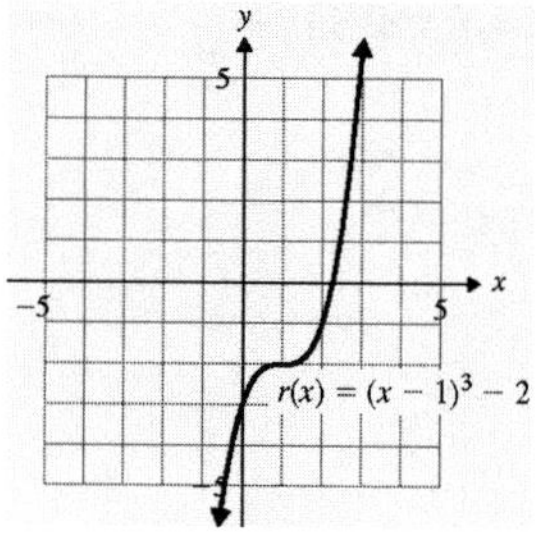

61)

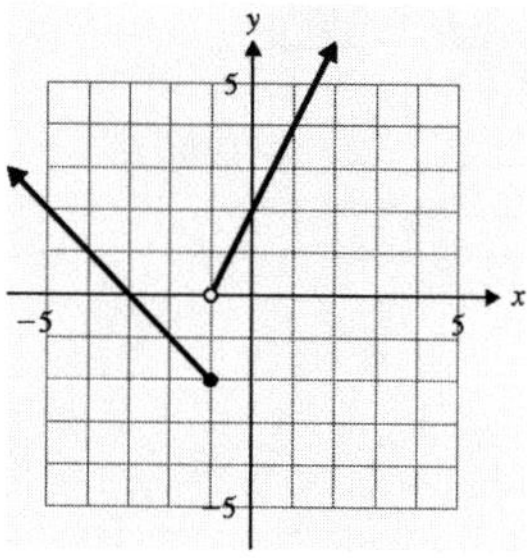

63)

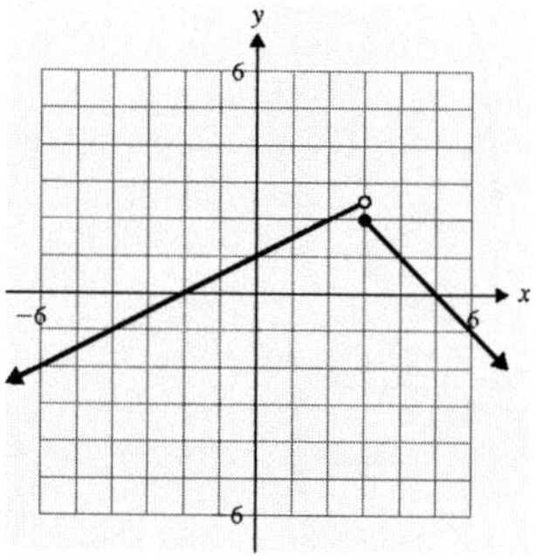

65)

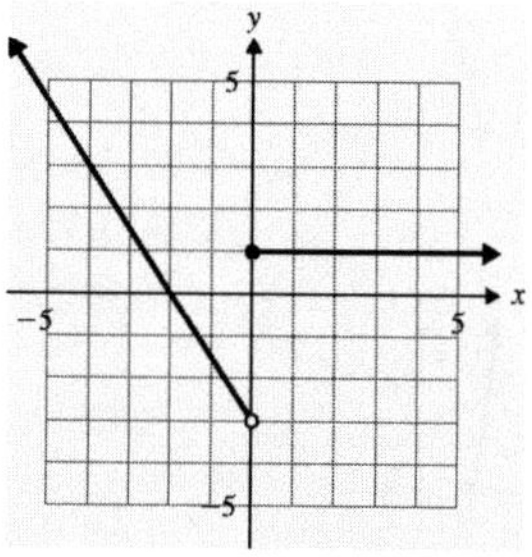

67)

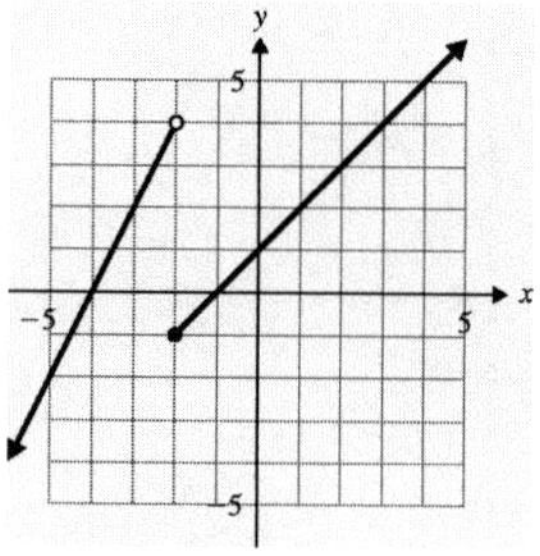

69)

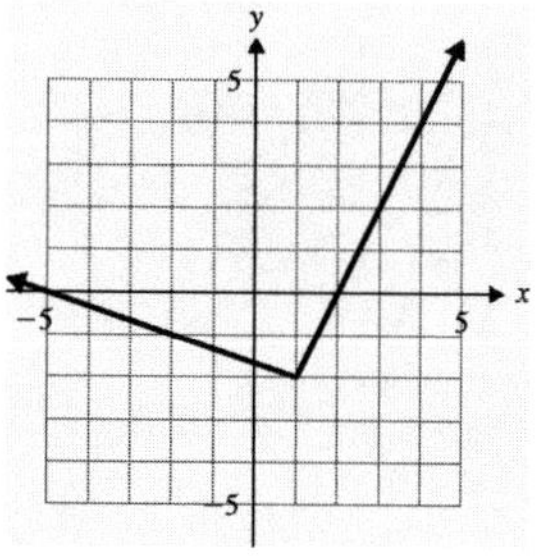

71) $f\left(3\frac{1}{4}\right)=\left[\!\left[3\frac{1}{4}\right]\!\right]=3$

73) $f(9.2)=[\![9.2]\!]=9$

75) $f(8)=[\![8]\!]=8$

77) $f\left(-6\frac{2}{5}\right)=\left[\!\left[-6\frac{2}{5}\right]\!\right]=-7$

79) $f(-8.1)=[\![-8.1]\!]=-9$

81)

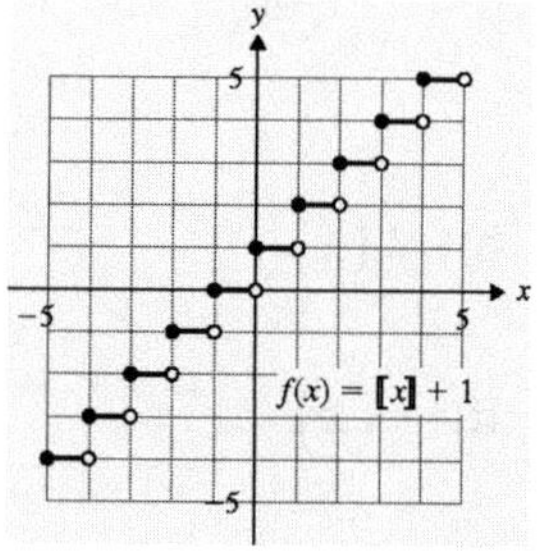

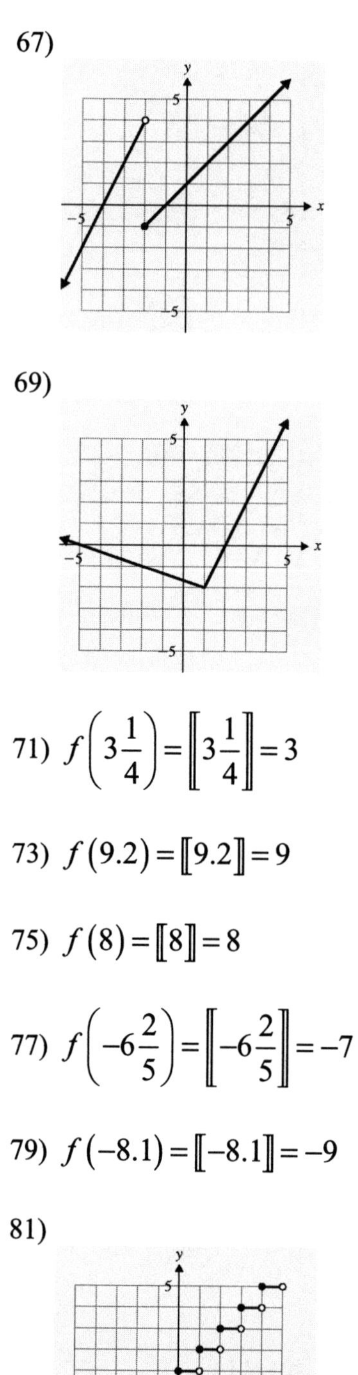

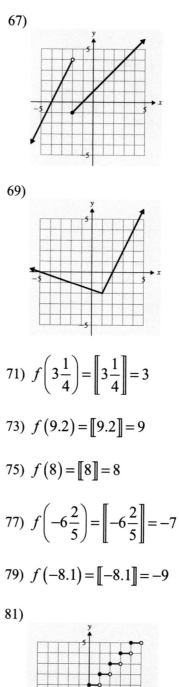

83)

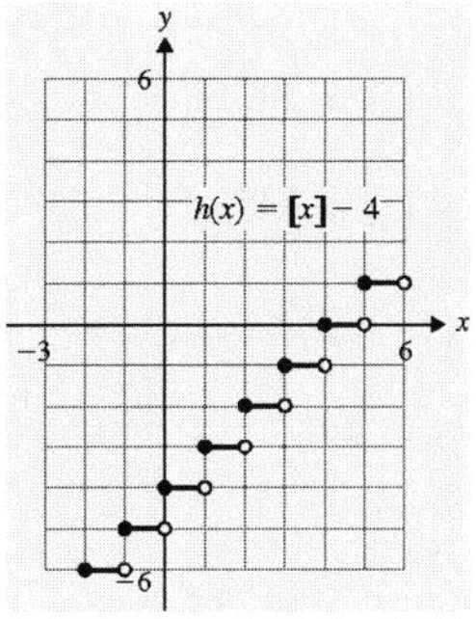

85)

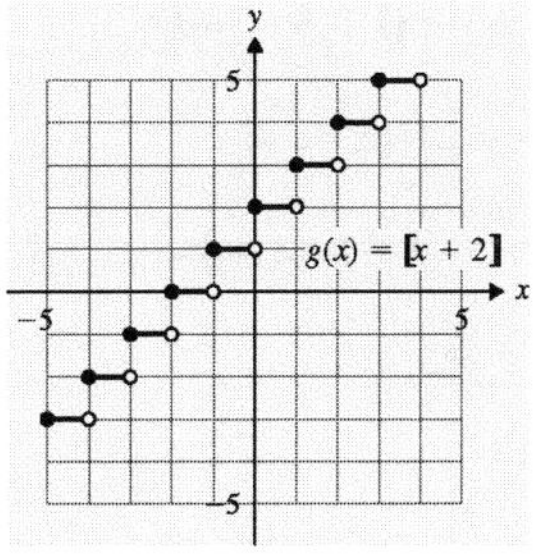

87)

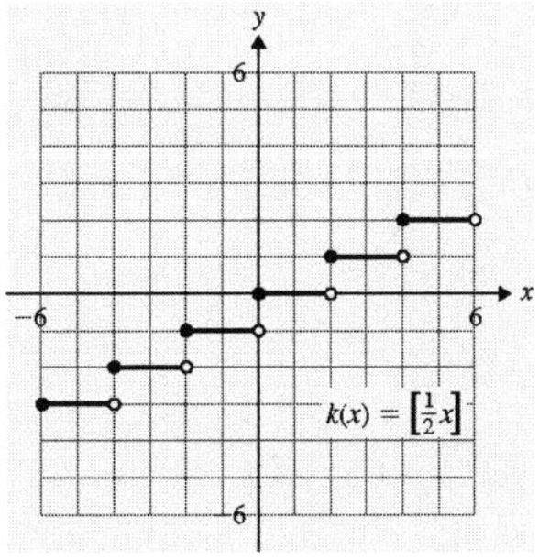

89)

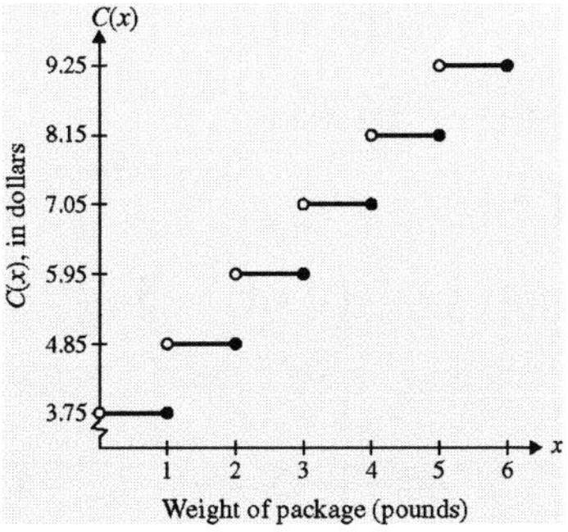

91)

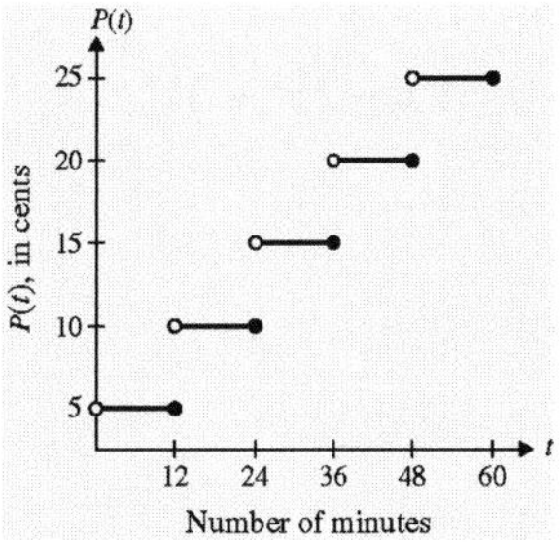

Section 12.3: Exercises

1) (h,k) 3) a is positive

5) $a>1$ or $a<-1$

7) $V(-1,-4); x=-1$

x-int $0=(x+1)^2-4$

$4=(x+1)^2$

$\pm\sqrt{4}=x+1$

$\pm 2=x+1$

$-1\pm 2=x$

$x=-3$ or $x=1$ $\quad (-3,0),(1,0)$

y-int $y=(0+1)^2-4$

$y=(1)^2-4$

$y=1-4=-3$ $\quad (0,-3)$

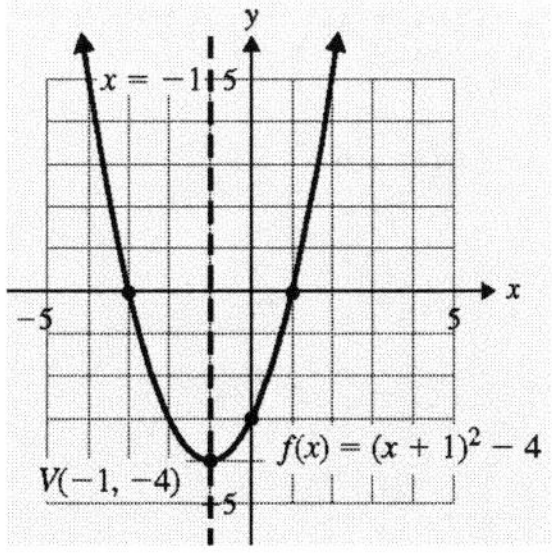

9) $V(2,3); x=2$

x-int $0=(x-2)^2+3$

$-3=(x-2)^2$

$\pm\sqrt{-3}=x-2$

$\pm i\sqrt{3}=x-2$

$2\pm i\sqrt{3}=x$ none

y-int $y=(0-2)^2+3$

$y=(-2)^2+3$

$y=4+3=7$ $(0,7)$

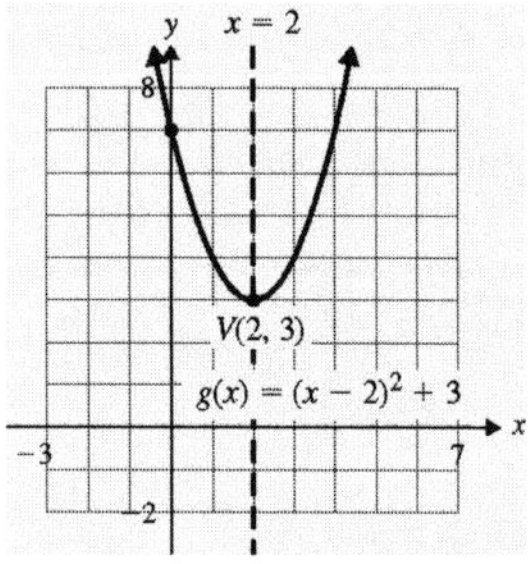

11) $V(4,-2); x=4$

x-int $0=(x-4)^2-2$

$2=(x-4)^2$

$\pm\sqrt{2}=x-4$

$4\pm\sqrt{2}=x$

$(4-\sqrt{2},0),(4+\sqrt{2},0)$

y-int $y=(0-4)^2-2$

$y=(-4)^2-2$

$y=16-2=14$ $(0,14)$

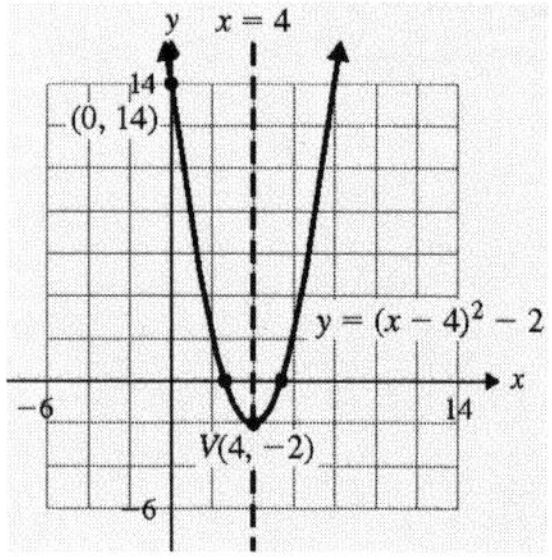

13) $V(-3,6); x=-3$

x-int $0=-(x+3)^2+6$

$-6=-(x+3)^2$

$6=(x+3)^2$

$\pm\sqrt{6}=x+3$

$-3\pm\sqrt{6}=x$

$(-3+\sqrt{6},0),(-3-\sqrt{6},0)$

y-int $y=-(0+3)^2+6$

$y=-(3)^2+6$

$y=-9+6=-3$ $(0,-3)$

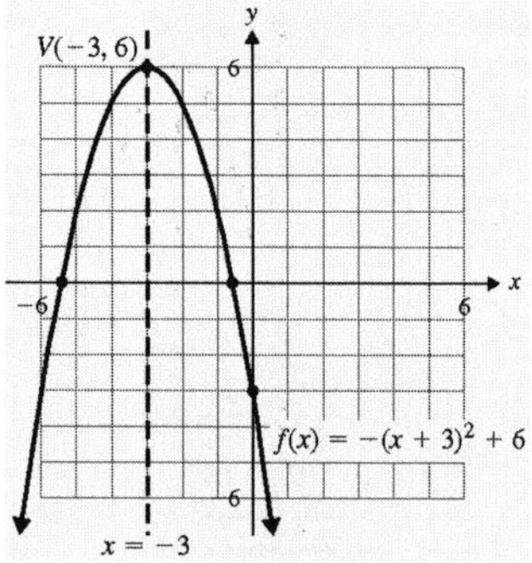

15) $V(-1,-5); x=-1$

x-int $0=-(x+1)^2-5$

$5=-(x+1)^2$

$-5=(x+1)^2$

$\pm\sqrt{-5}=x+1$

$-1\pm i\sqrt{5}=x$ none

y-int $y=-(0+1)^2-5$

$y=-(1)^2-5$

$y=-1-5=-6$ $(0,-6)$

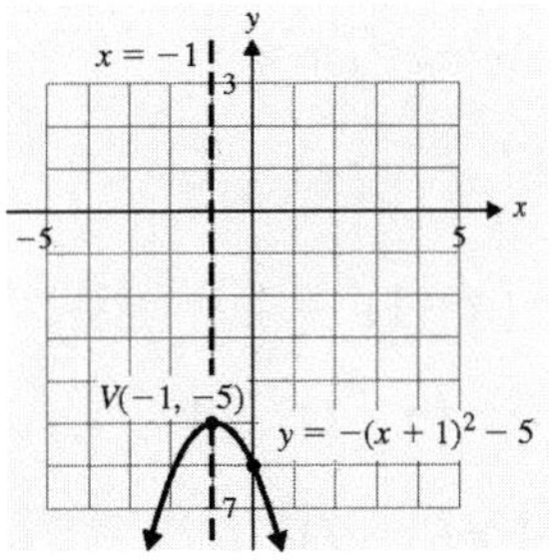

17) $V(1,-8); x=1$

$x\text{-int}\quad 0=2(x-1)^2-8$

$8=2(x-1)^2$

$4=(x-1)^2$

$\pm 2=x-1$

$1\pm 2=x$

$x=-1 \text{ or } x=3 \qquad (-1,0),(3,0)$

$y\text{-int}\quad y=2(0-1)^2-8$

$y=2(-1)^2-8$

$y=2(1)-8$

$y=2-8=-6 \qquad (0,-6)$

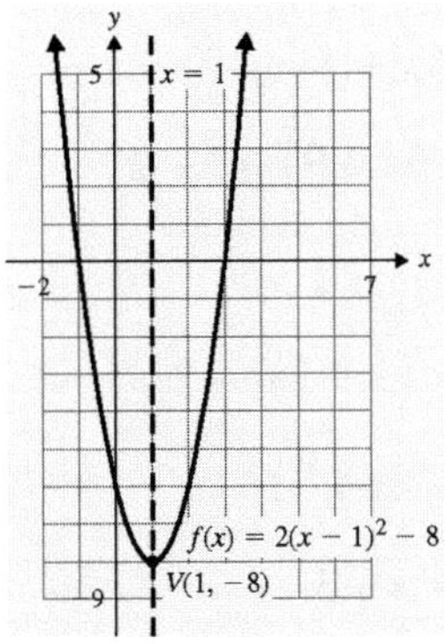

19) $V(-4,0); x=-4$

$x\text{-int}\quad 0=\frac{1}{2}(x+4)^2$

$0=(x+4)^2$

$0=x+4$

$-4=x \qquad (-4,0)$

$y\text{-int}\quad y=\frac{1}{2}(0+4)^2$

$y=\frac{1}{2}(4)^2$

$y=\frac{1}{2}(16)=8 \qquad (0,8)$

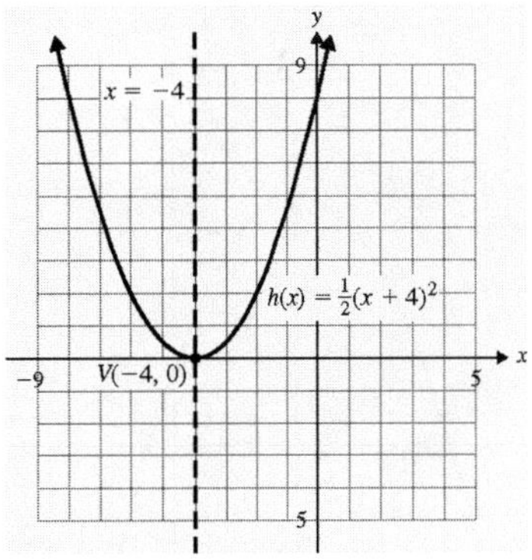

21) $V(0,5); x=0$

$x\text{-int}\quad 0=-x^2+5$

$-5=-x^2$

$5=x^2$

$\pm\sqrt{5}=x \qquad (-\sqrt{5},0),(\sqrt{5},0)$

$y\text{-int}\quad y=-(0)^2+5$

$y=5 \qquad (0,5)$

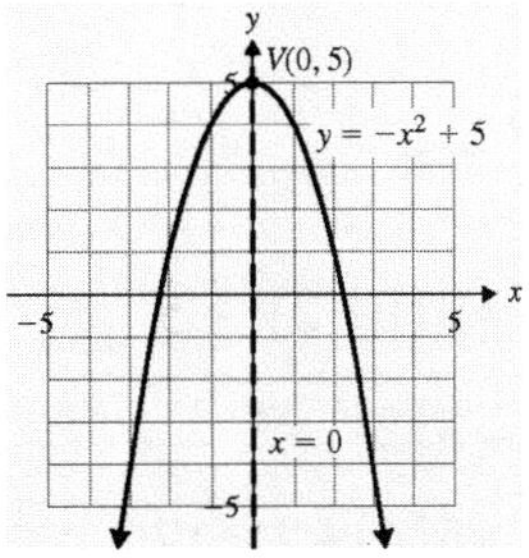

23) $V(-4,3); x=-4$

$x\text{-int}\quad 0=-\frac{1}{3}(x+4)^2+3$

$-3=-\frac{1}{3}(x+4)^2$

$9=(x+4)^2$

$\pm 3=x+4$

$-4\pm 3=x$

$x=-7$ or $x=-1$ $\qquad (-7,0),(-1,0)$

y-int $\quad y=-\dfrac{1}{3}(0+4)^2+3$

$y=-\dfrac{1}{3}(4)^2+3$

$y=-\dfrac{16}{3}+3=-\dfrac{7}{3} \qquad \left(0,-\dfrac{7}{3}\right)$

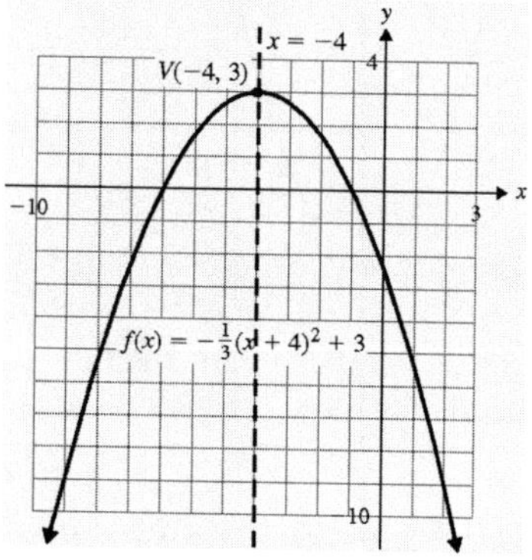

25) $V(-2,5); x=-2$

x-int $\quad 0=3(x+2)^2+5$

$-5=3(x+2)^2$

$-\dfrac{5}{3}=(x+2)^2$

$\pm\sqrt{-\dfrac{5}{3}}=x+2$

$-2\pm i\sqrt{\dfrac{5}{3}}=x \qquad$ none

y-int $\quad y=3(0+2)^2+5$

$y=3(2)^2+5$

$y=3(4)+2=17 \qquad (0,17)$

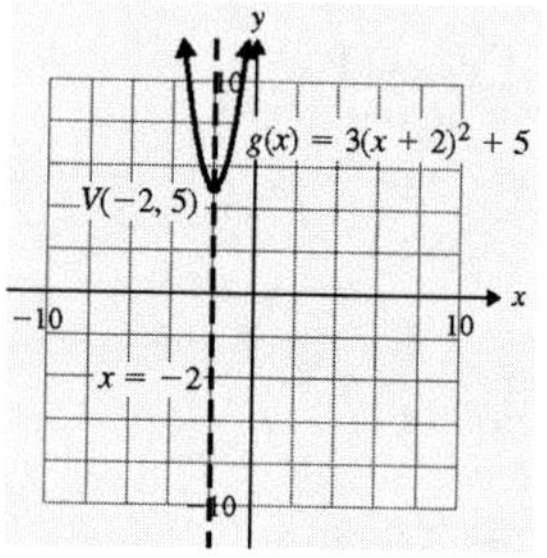

27) $f(x)=x^2-2x-3$

$f(x)=\left(x^2-2x+1\right)-3-1$

$f(x)=(x-1)^2-4$

x-int $\quad 0=(x-1)^2-4$

$4=(x-1)^2$

$\pm\sqrt{4}=x-1$

$\pm 2=x-1$

$1\pm 2=x$

$x=3$ or $x=-1 \qquad (3,0),(-1,0)$

y-int $\quad f(0)=0^2-2(0)-3$

$=-3 \qquad (0,-3)$

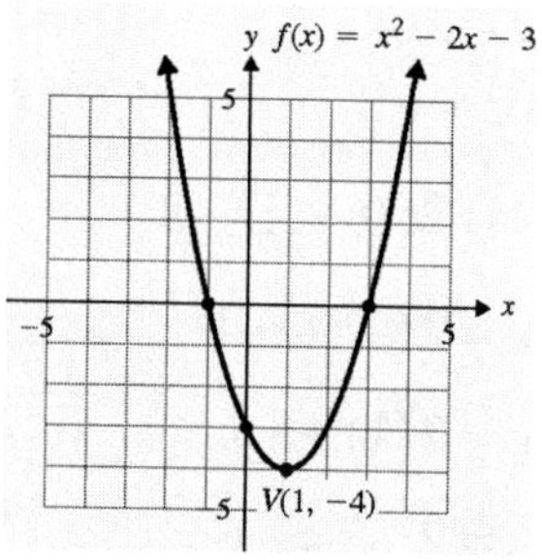

29) $y=x^2+6x+7$

$y=\left(x^2+6x+9\right)+7-9$

$y=(x+3)^2-2$

x-int $\quad 0=(x+3)^2-2$

$2=(x+3)^2$

$\pm\sqrt{2}=x+3$

$-3\pm\sqrt{2}=x$

$\left(-3-\sqrt{2},0\right),\left(-3-\sqrt{2},0\right)$

y-int $\quad y=0^2+6(0)+7$

$=7 \qquad (0,7)$

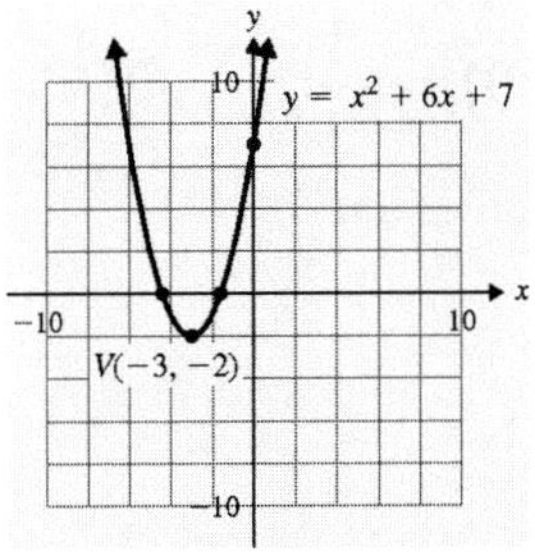

31) $g(x)=x^2+4x$

$g(x)=(x^2+4x+4)-4$

$g(x)=(x+2)^2-4$

x-int $0=(x+2)^2-4$

$4=(x+2)^2$

$\pm 2=x+2$

$-2\pm 2=x$

$x=-4$ or $x=0$ $\quad (-4,0),(0,0)$

y-int $g(0)=0^2+4(0)$ $\quad (0,0)$

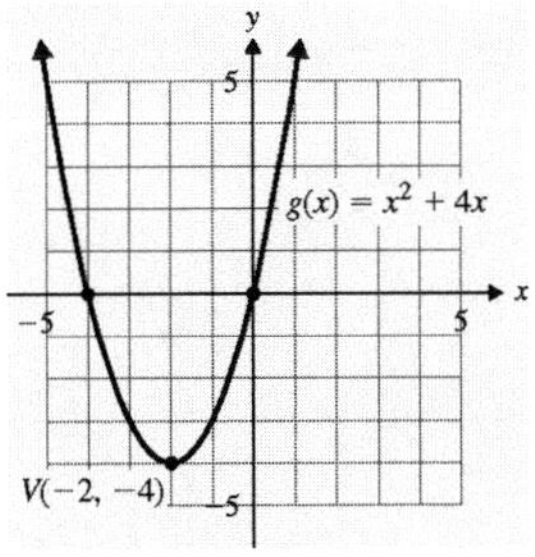

33) $h(x)=-x^2-4x+5$

$h(x)=-(x^2+4x)+5$

$h(x)=-(x^2+4x+4)+5+4$

$h(x)=-(x+2)^2+9$

x-int $0=-(x+2)^2+9$

$-9=-(x+2)^2$

$9=(x+2)^2$

$\pm 3=x+2$

$-2\pm 3=x$

$x=-5$ or $x=1$ $\quad (-5,0),(1,0)$

y-int $h(0)=-(0)^2-4(0)+5$

$=5$ $\quad (0,5)$

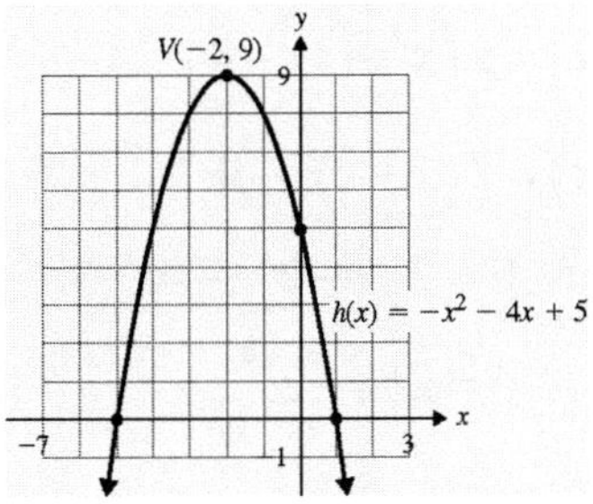

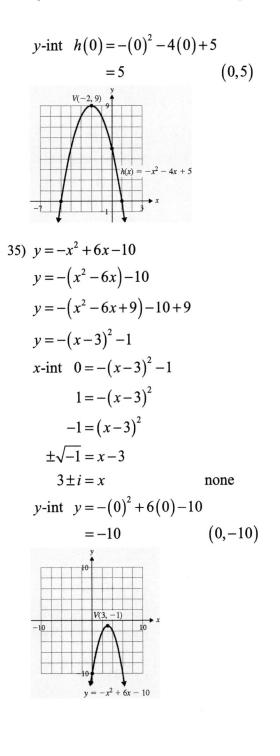

35) $y=-x^2+6x-10$

$y=-(x^2-6x)-10$

$y=-(x^2-6x+9)-10+9$

$y=-(x-3)^2-1$

x-int $0=-(x-3)^2-1$

$1=-(x-3)^2$

$-1=(x-3)^2$

$\pm\sqrt{-1}=x-3$

$3\pm i=x$ $\quad$ none

y-int $y=-(0)^2+6(0)-10$

$=-10$ $\quad (0,-10)$

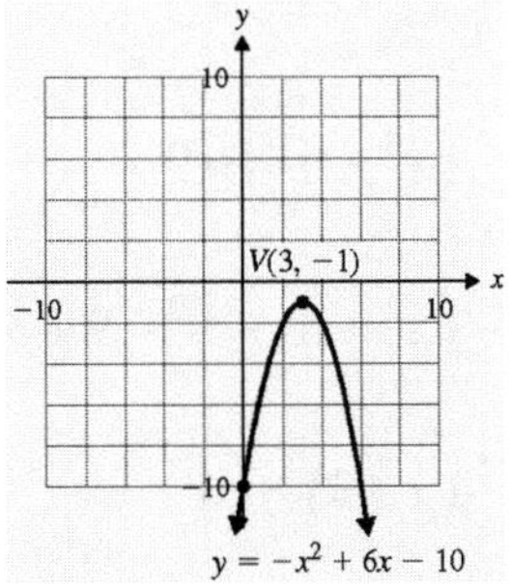

37) $f(x)=2x^2-8x+4$

$f(x)=2(x^2-4x)+4$

$f(x)=2(x^2-4x+4)+4-8$

$f(x)=2(x-2)^2-4$

x-int $0=2(x-2)^2-4$

$4=2(x-2)^2$

$2=(x-2)^2$

$\pm\sqrt{2}=x-2$

$2\pm\sqrt{2}=x$

$(2-\sqrt{2},0),(2+\sqrt{2},0)$

y-int $f(0)=2(0)^2-8(0)+4$

$=4$ $(0,4)$

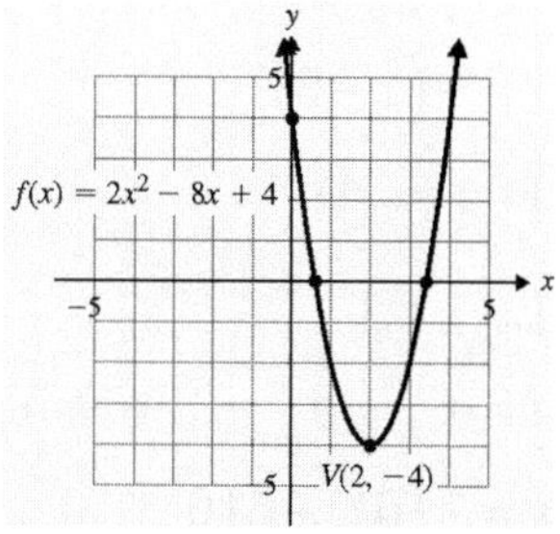

39) $g(x)=-\frac{1}{3}x^2-2x-9$

$g(x)=-\frac{1}{3}(x^2+6x)-9$

$g(x)=-\frac{1}{3}(x^2+6x+9)-9+3$

$g(x)=-\frac{1}{3}(x+3)^2-6$

x-int $0=-\frac{1}{3}(x+3)^2-6$

$6=-\frac{1}{3}(x+3)^2$

$-18=(x+3)^2$

$\pm\sqrt{-18}=x+3$

$-3\pm 3i\sqrt{2}=x$ none

y-int $g(0)=-\frac{1}{3}(0)^2-2(0)-9$

$=-9$

$(0,-9)$

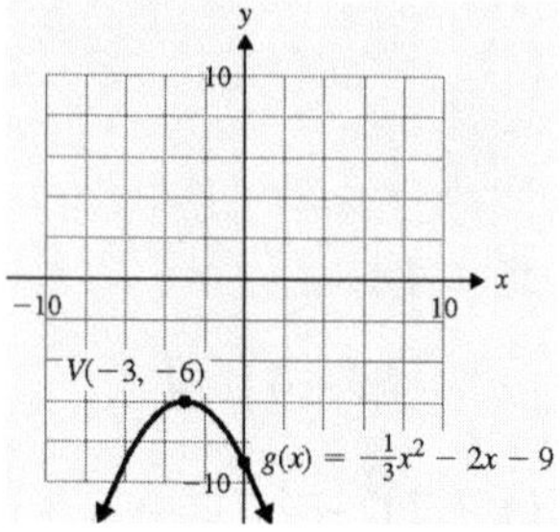

41) $y=x^2-3x+2$

$y=\left(x^2-3x+\frac{9}{4}\right)+2-\frac{9}{4}$

$y=\left(x-\frac{3}{2}\right)^2-\frac{1}{4}$

x-int $0=\left(x-\frac{3}{2}\right)^2-\frac{1}{4}$

$\frac{1}{4}=\left(x-\frac{3}{2}\right)^2$

$\pm\frac{1}{2}=x-\frac{3}{2}$

$\frac{3}{2}\pm\frac{1}{2}=x$

$x=1$ or $x=2$ $(1,0),(2,0)$

y-int $y=(0)^2-3(0)+2$

$=2$ $(0,2)$

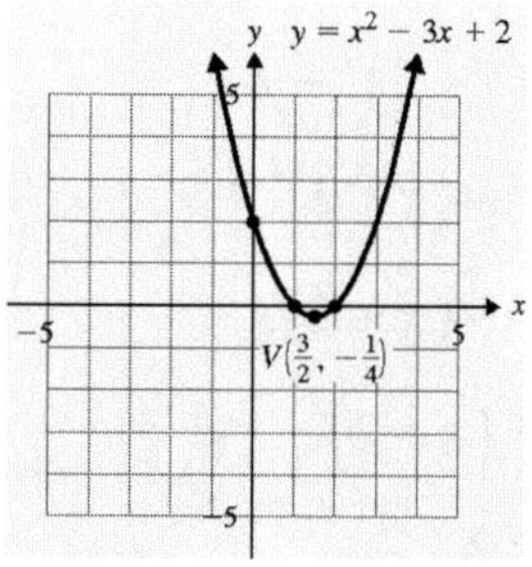

43) $h=-\dfrac{2}{2(1)}=-1$

$k=(-1)^2+2(-1)-3$

$=1-2-3=-4 \qquad V(-1,-4)$

x-int $(-3,0),(1,0) \qquad y$-int $(0,-3)$

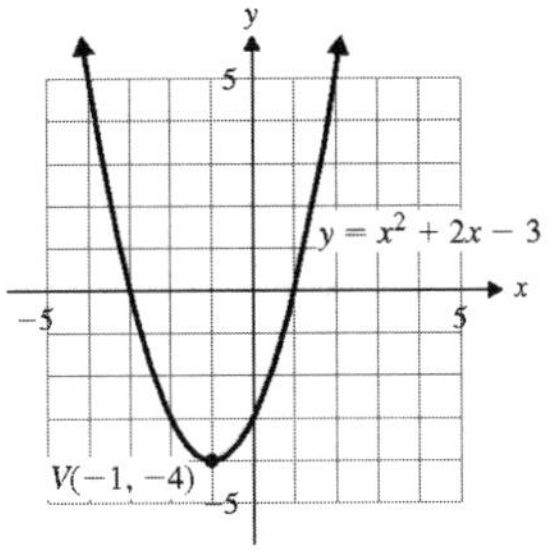

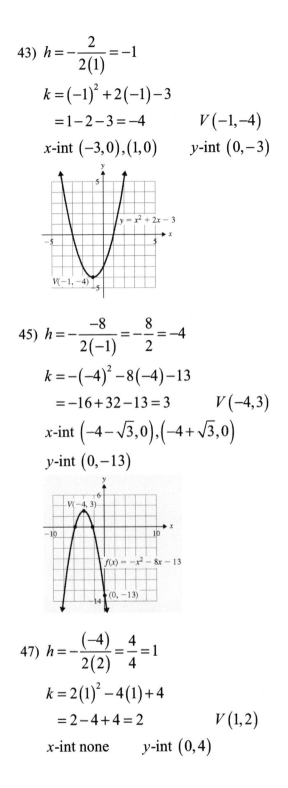

45) $h=-\dfrac{-8}{2(-1)}=-\dfrac{8}{2}=-4$

$k=-(-4)^2-8(-4)-13$

$=-16+32-13=3 \qquad V(-4,3)$

x-int $\left(-4-\sqrt{3},0\right),\left(-4+\sqrt{3},0\right)$

y-int $(0,-13)$

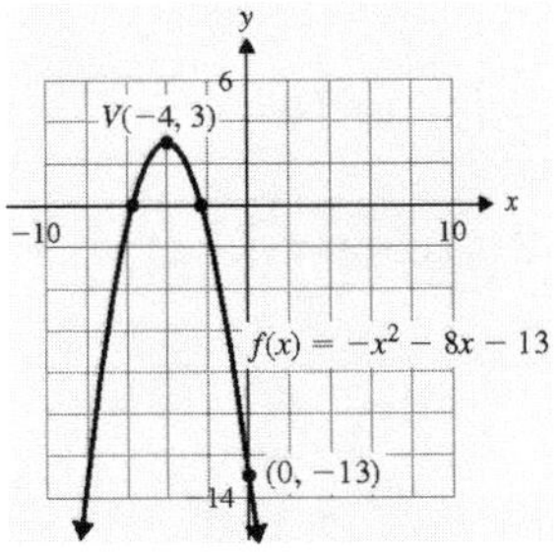

47) $h=-\dfrac{(-4)}{2(2)}=\dfrac{4}{4}=1$

$k=2(1)^2-4(1)+4$

$=2-4+4=2 \qquad V(1,2)$

x-int none $\qquad y$-int $(0,4)$

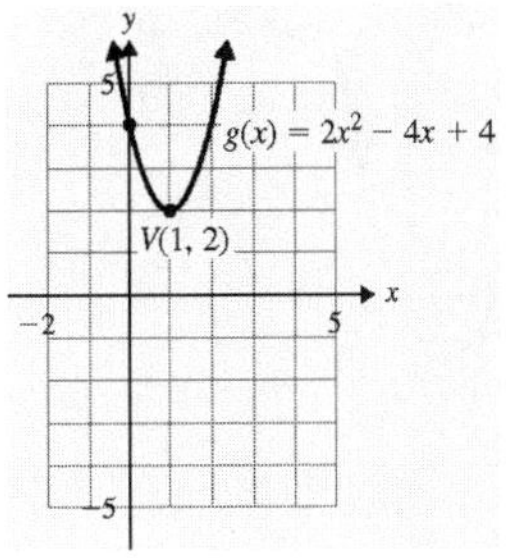

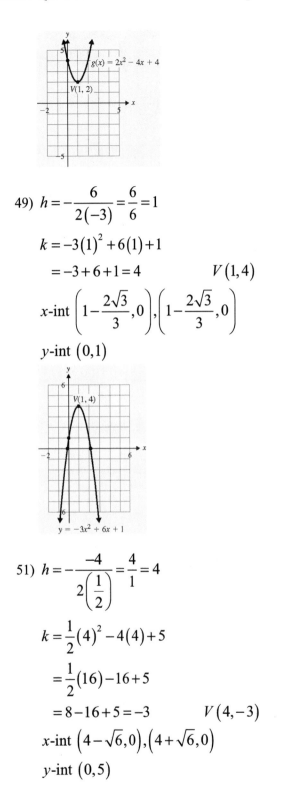

49) $h=-\dfrac{6}{2(-3)}=\dfrac{6}{6}=1$

$k=-3(1)^2+6(1)+1$

$=-3+6+1=4 \qquad V(1,4)$

x-int $\left(1-\dfrac{2\sqrt{3}}{3},0\right),\left(1-\dfrac{2\sqrt{3}}{3},0\right)$

y-int $(0,1)$

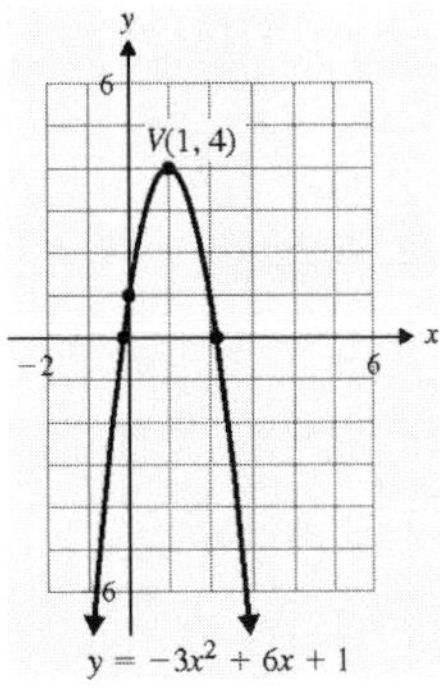

51) $h=-\dfrac{-4}{2\left(\dfrac{1}{2}\right)}=\dfrac{4}{1}=4$

$k=\dfrac{1}{2}(4)^2-4(4)+5$

$=\dfrac{1}{2}(16)-16+5$

$=8-16+5=-3 \qquad V(4,-3)$

x-int $\left(4-\sqrt{6},0\right),\left(4+\sqrt{6},0\right)$

y-int $(0,5)$

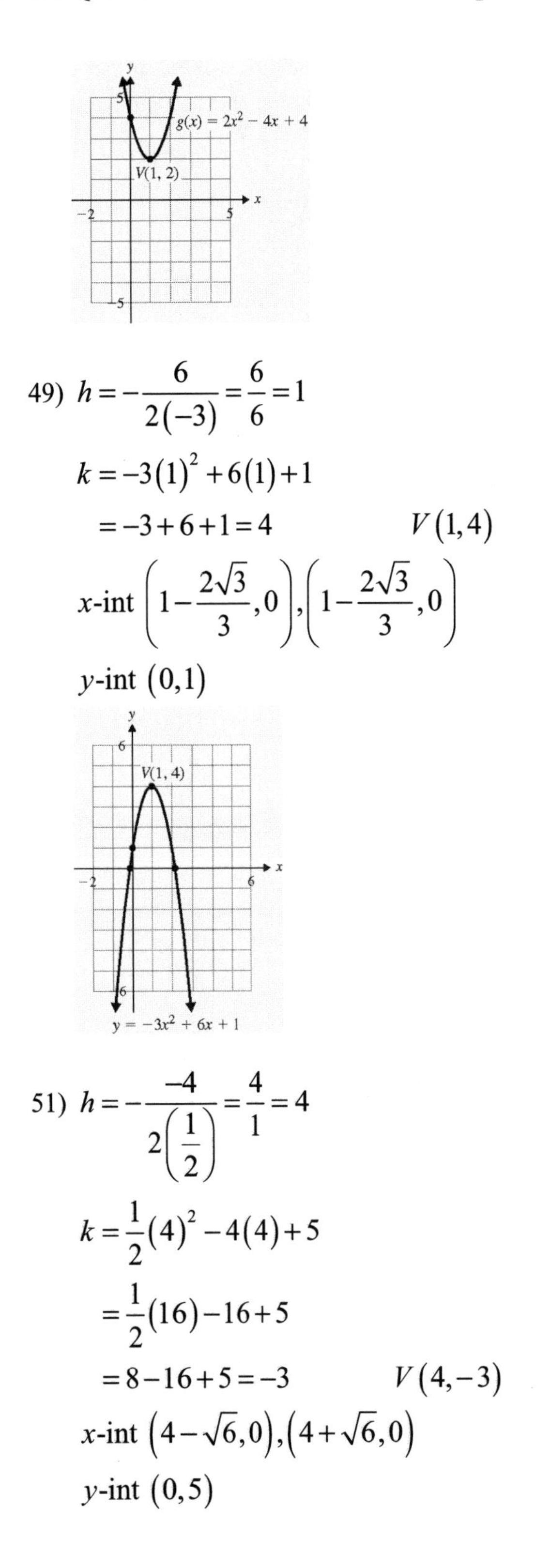

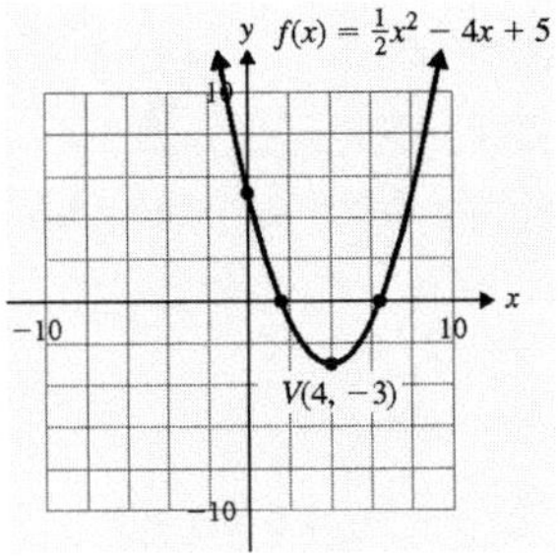

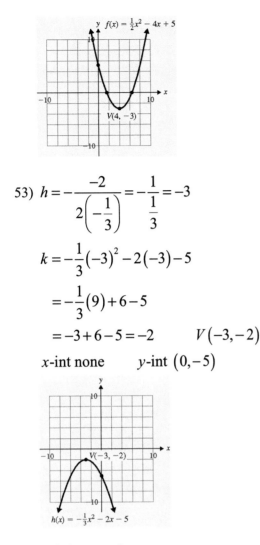

53) $h = -\dfrac{-2}{2\left(-\dfrac{1}{3}\right)} = -\dfrac{1}{\dfrac{1}{3}} = -3$

$k = -\dfrac{1}{3}(-3)^2 - 2(-3) - 5$

$= -\dfrac{1}{3}(9) + 6 - 5$

$= -3 + 6 - 5 = -2 \qquad V(-3, -2)$

x-int none $\qquad$ y-int $(0, -5)$

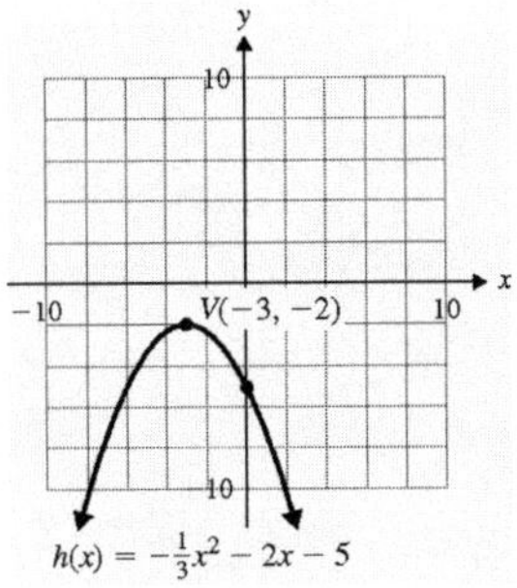

Section 12.4: Exercises

1) If a is positive the graph opens upward, so the y-coordinate of the vertex is the minimum value of the function. If a is negative the graph opens downward, so the y-coordinate of the vertex is the maximum value of the function.

3) a) minimum

b) $h = -\dfrac{6}{2(1)} = -3$

$k = (-3)^2 + 6(-3) + 9$

$= 9 - 18 + 9 = 0 \qquad V(-3, 0)$

c) 0 $\qquad$ d)

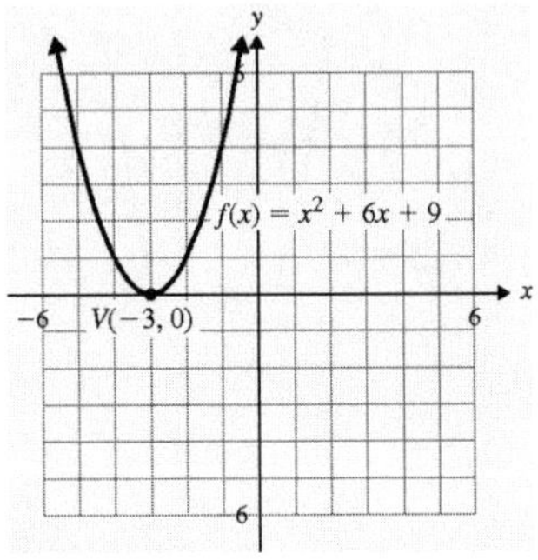

5) a) maximum

b) $h = -\dfrac{4}{2\left(-\dfrac{1}{2}\right)} = \dfrac{4}{1} = 4$

$k = -\dfrac{1}{2}(4)^2 + 4(4) - 6$

$= -\dfrac{1}{2}(16) + 16 - 6$

$= -8 + 16 - 6 = 2 \qquad V(4, 2)$

c) 2 $\qquad$ d)

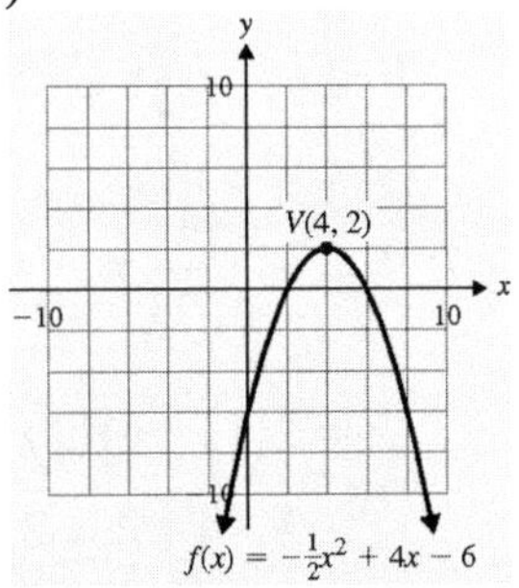

7) a) The max height occurs at the t-coordinate of the vertex.

$t = -\dfrac{320}{2(-16)} = \dfrac{320}{32} = 10$

10 seconds

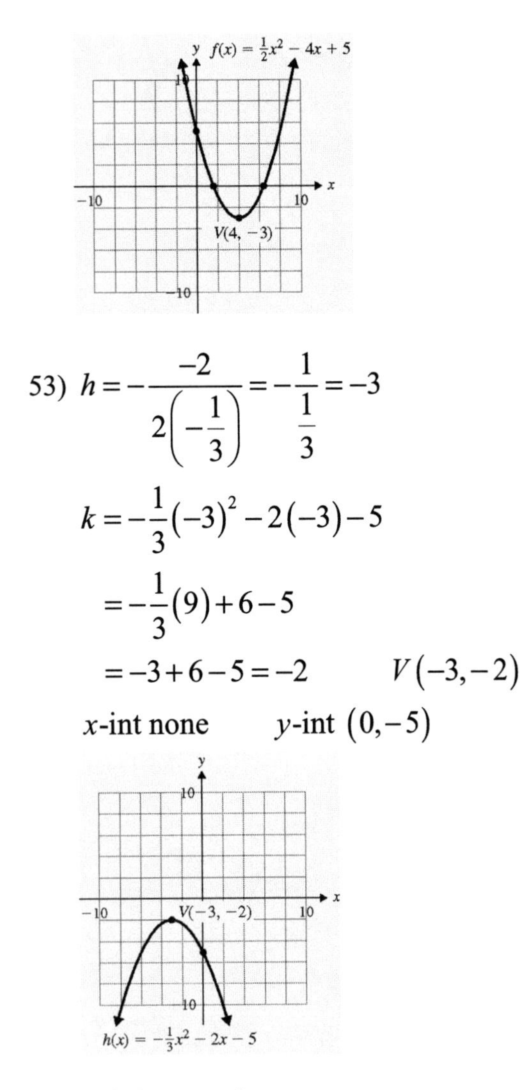

b) Since the object reaches max height $t=10$, find $h(10)$.

$h(10)=-16(10)^2+320(10)$
$=-16(100)+3200=1600$

1600 ft

c) The object hits ground when $h(t)=0$.

$0=-16t^2+320t$

$0=t^2-20t$

$0=t(t-20)$

$t=0$ or $t-20=0$

$t=20$

The object will hit the ground after 20 sec.

9) The x-coordinate of the vertex represents the number of months after January that had the greatest number of guests.

$x=-\dfrac{120}{2(-10)}=\dfrac{120}{20}=6$

The inn had the greatest number of guests in July.

The number of guests at the inn during July is $N(6)$.

$N(6)=-10(6)^2+120(6)+120$
$=-10(36)+720+120=480$

480 guests stayed at the inn during the month of July.

11) The t-coordinate of the vertex represents the number of years after 1989 in which the greatest number of babies was born to teen mothers.

$t=-\dfrac{2.75}{2(-0.721)}\approx 2$

Greatest number born in 1991.

The number of babies born in (in thousands) to teen mothers is $N(2)$.

$N(2)=-0.721(2)^2+2.75(2)+528$
$=-2.884+5.5+528\approx 531$

Approximately 531,000 babies were born to teen mothers in 1991.

13) $w=$ width of ice rink

$l=$ length of ice rink

$A=$ area of the rink

Maximize: $A=lw$

Constraint: $2l+2w=100$

$2l+2w=100$

$2l=100-2w$

$l=50-w$

$A=lw$

$A(w)=(50-w)w$

$A(w)=-w^2+50w$

Find the w-coordinate of the vertex, the value that maximizes the area.

$w=-\dfrac{50}{2(-1)}=25$

Find $A(25)$

$A(25)=-(25)^2+50(25)$
$=-625+1250=625\text{ ft}^2$

15) $w =$ width of dog pen
$l =$ length of dog pen
$A =$ area of the dog pen
Maximize: $A = lw$
Since the barn is 1 side of the pen, the fence is used for only 3 sides.
Constraint: $l + 2w = 48$

$$l + 2w = 48$$
$$l = 48 - 2w$$
$$A = lw$$
$$A(w) = (48 - 2w)w = -2w^2 + 48w$$

Find the w-coordinate of the vertex, the value that maximizes the area.

$$w = -\frac{48}{2(-2)} = 12$$

Use $l + 2w = 40$ with $w = 12$ to find the corresponding length.

$$l + 2(12) = 48$$
$$l + 24 = 48$$
$$l = 24$$

The dog pen will be 12 ft×24 ft.

17) $x =$ one integer, $y =$ other integer,
$P =$ product
Maximize: $P = xy$
Constraint: $x + y = 18$

$$x + y = 18$$
$$y = 18 - x$$
$$P = xy$$
$$P(x) = x(18 - x) = -x^2 + 18x$$
$$x = -\frac{18}{2(-1)} = 9$$
$$x + y = 18$$
$$9 + y = 18$$
$$y = 9$$

9 and 9

19) $x =$ one integer, $y =$ other integer,
$P =$ product
Maximize: $P = xy$
Constraint: $x - y = 12$

$$x - y = 12$$
$$x - 12 = y$$
$$P = xy$$
$$P(x) = x(x - 12) = x^2 - 12x$$
$$x = -\frac{-12}{2(1)} = 6$$
$$x - y = 12$$
$$6 - y = 12$$
$$-y = 6$$
$$y = -6$$

−6 and 6

21) (h, k) 23) to the left

25) $V(-4, 1); y = 1$

x-int $x = (0 - 1)^2 - 4$
$x = 1 - 4$
$x = -3$ $(-3, 0)$

y-int $0 = (y - 1)^2 - 4$
$4 = (y - 1)^2$
$\pm\sqrt{4} = y - 1$
$1 \pm 2 = y$
$y = -1$ or $y = 3$ $(0, -1), (0, 3)$

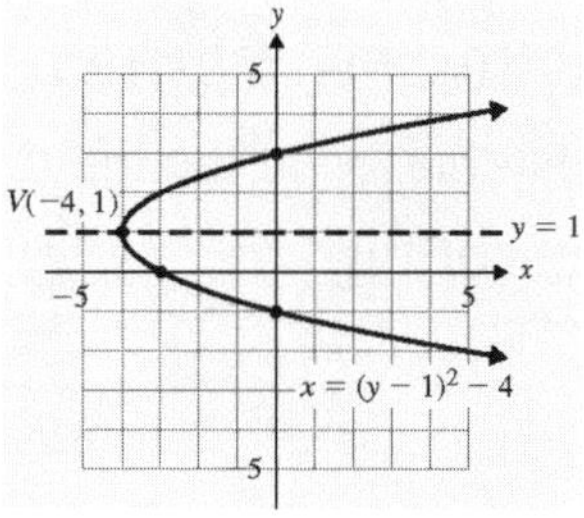

27) $V(2,0);\ y=0$

x-int $x=0^2+2$

$x=2$ $\quad (2,0)$

y-int $0=y^2+2$

$-2=y^2$

$\pm i\sqrt{2}=y$ $\quad$ none

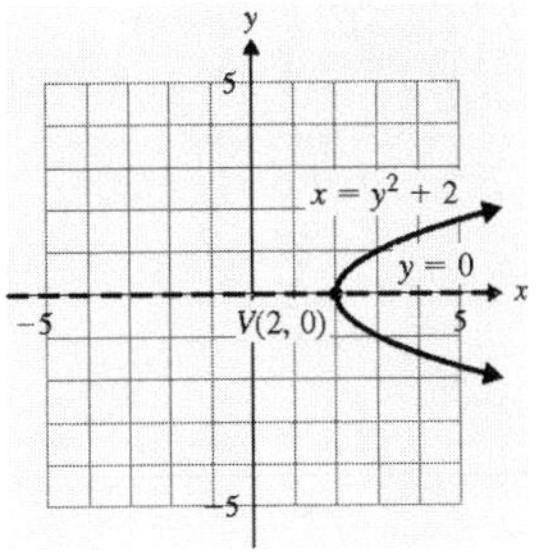

29) $V(5,4);\ y=4$

x-int $x=-(0-4)^2+5$

$x=-(-4)^2+5$

$x=-16+5$ $\quad (-11,0)$

y-int $0=-(y-4)^2+5$

$-5=-(y-4)^2$

$5=(y-4)^2$

$\pm\sqrt{5}=y-4$

$4\pm\sqrt{5}=y$

$\left(0,4-\sqrt{5}\right),\left(0,4+\sqrt{5}\right)$

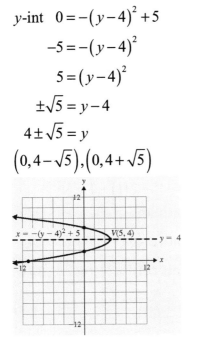

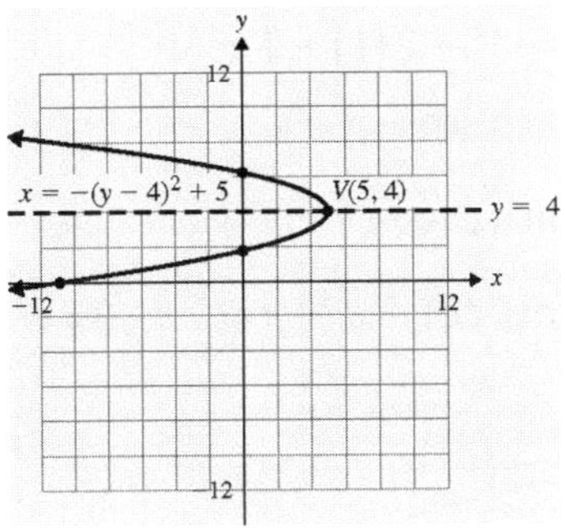

31) $V(-9,2);\ y=2$

x-int $x=-2(0-2)^2-9$

$x=-2(4)-9$

$x=-8-9=-17$ $\quad (-17,0)$

y-int $0=-2(y-2)^2-9$

$9=-2(y-2)^2$

$-\frac{9}{2}=(y-2)^2$

$\pm i\frac{3}{\sqrt{2}}=y+1$

$-1\pm i\frac{3}{\sqrt{2}}=y$ $\quad$ none

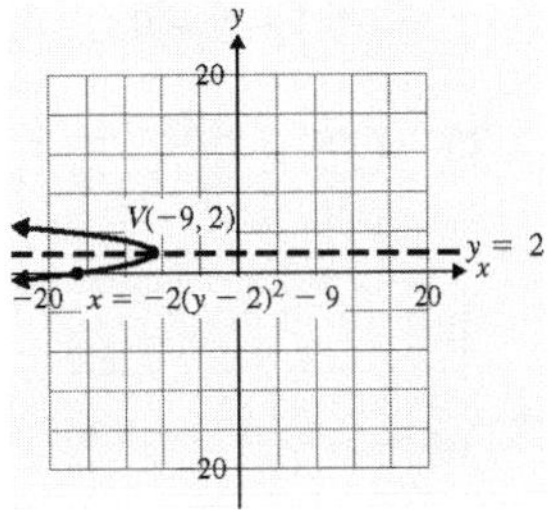

33) $V(0,-2);\ y=-2$

x-int $x=\frac{1}{4}(0+2)^2$

$x=\frac{1}{4}(4)=1$ $\quad (1,0)$

y-int $0=\frac{1}{4}(y+2)^2$

$0=(y+2)^2$

$0=y+2$

$-2=y$ $\quad (0,-2)$

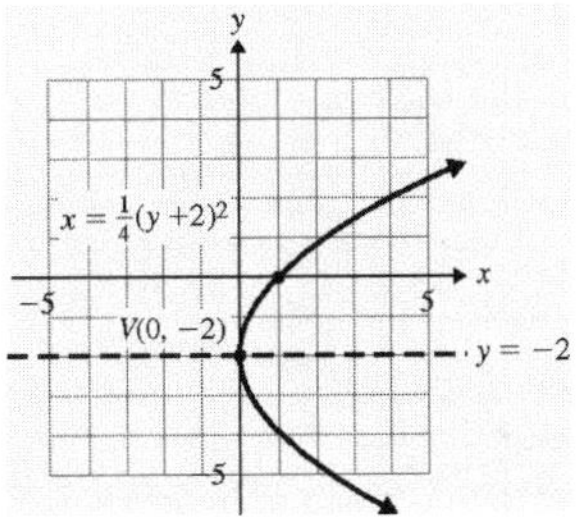

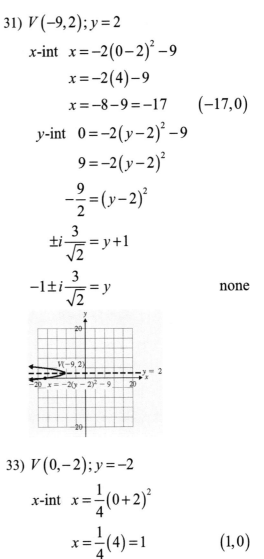

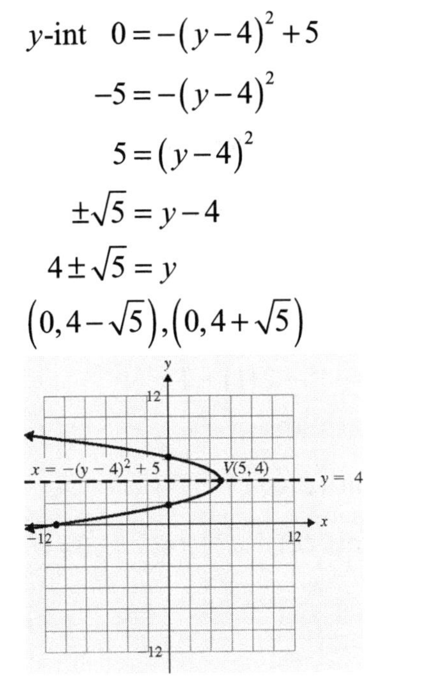

35) $x = y^2 - 4y + 5$

$x = \left(y^2 - 4y + 4\right) + 5 - 4$

$x = \left(y - 2\right)^2 + 1$

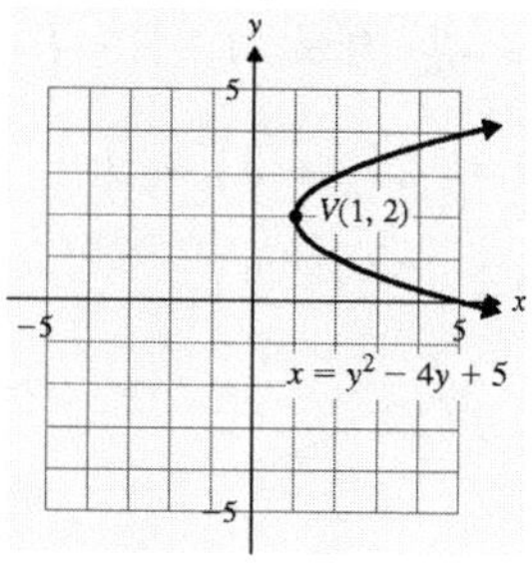

37) $x = -y^2 + 6y + 6$

$x = -\left(y^2 - 6y\right) + 6$

$x = -\left(y^2 - 6y + 9\right) + 6 + 9$

$x = -\left(y - 3\right)^2 + 15$

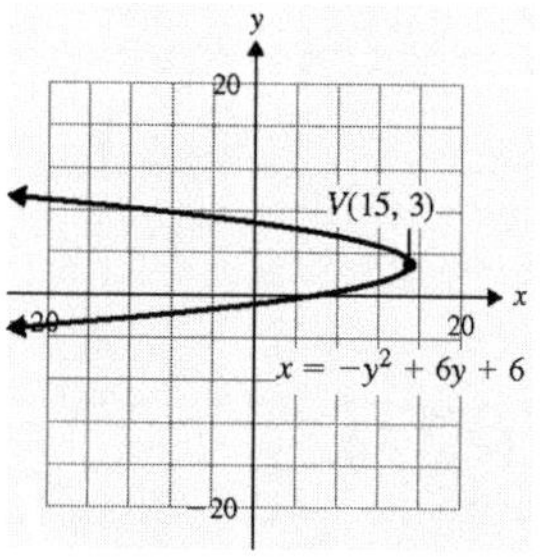

39) $x = \dfrac{1}{3}y^2 + \dfrac{8}{3}y - \dfrac{5}{3}$

$3x = y^2 + 8y - 5$

$3x = \left(y^2 + 8y + 16\right) - 5 - 16$

$3x = \left(y + 4\right)^2 - 21$

$x = \dfrac{1}{3}\left(y + 4\right)^2 - 7$

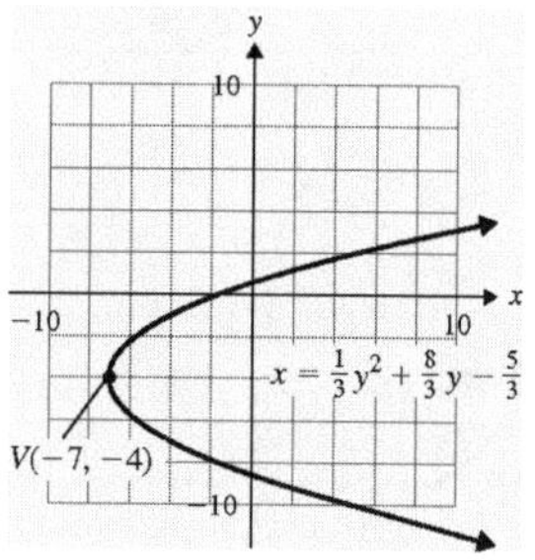

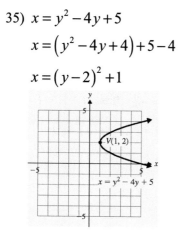

41) $x = -4y^2 - 8y - 10$

$x = -4\left(y^2 + 2y\right) - 10$

$x = -4\left(y^2 + 2y + 1\right) - 10 + 4$

$x = -4\left(y + 1\right)^2 - 6$

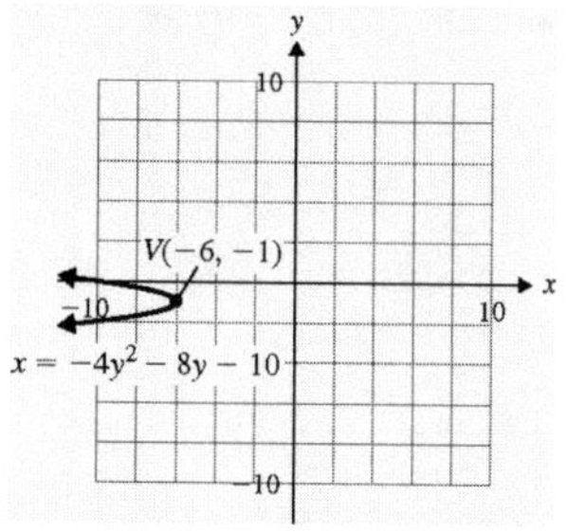

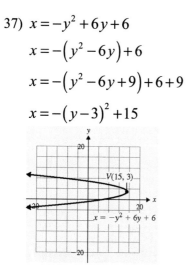

43) $y = -\dfrac{-4}{2(1)} = \dfrac{4}{2} = 2$

$x = 2^2 - 4(2) + 3$

$= 4 - 8 + 3 = -1 \qquad V(-1, 2)$

x-int $(3, 0) \qquad y$-int $(0, 1), (0, 3)$

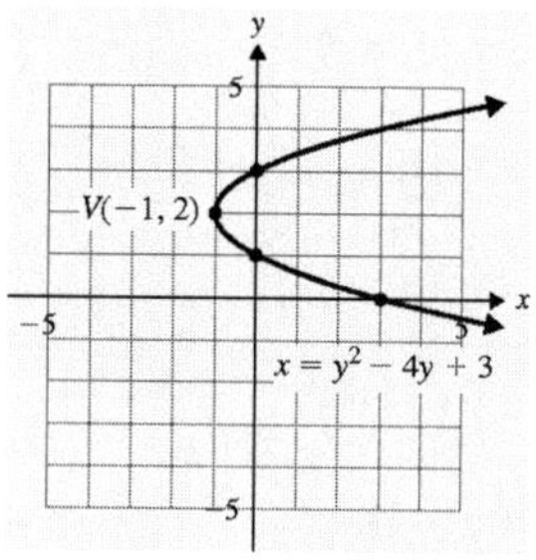

45) $y = -\dfrac{2}{2(-1)} = \dfrac{2}{2} = 1$

$x = -1^2 + 2(1) + 2$

$= -1 + 2 + 2 = 3 \qquad V(3, 1)$

x-int $(2, 0)$

y-int $\left(0, 1 - \sqrt{3}\right), \left(0, 1 + \sqrt{3}\right)$

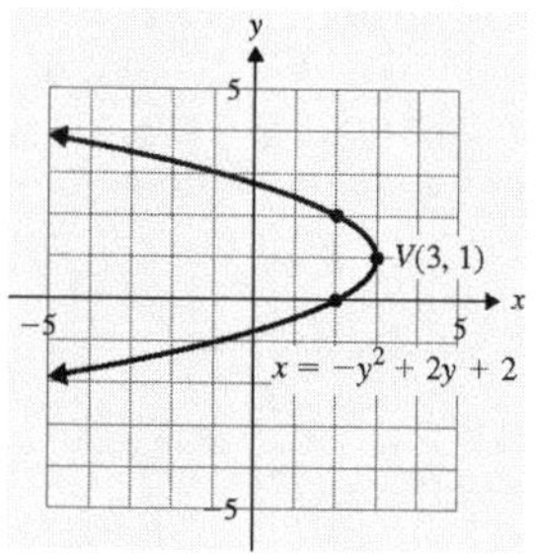

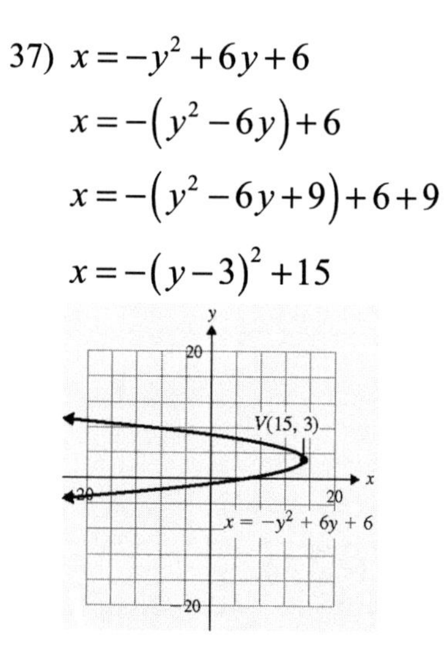

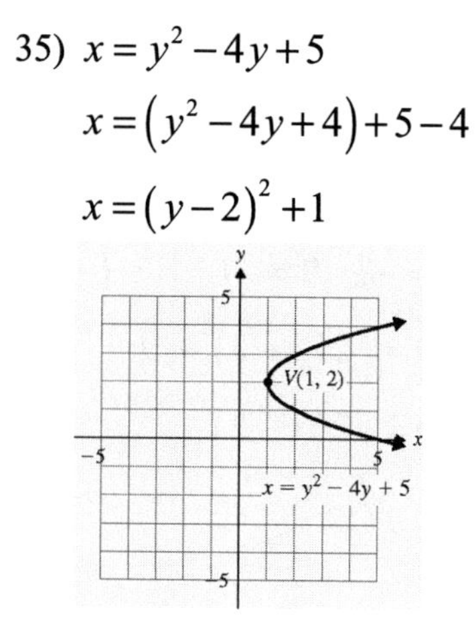

47) $y=-\dfrac{4}{2(-2)}=\dfrac{4}{4}=1$

$x=-2(1)^2+4(1)-6$

$=-2+4-6=-4 \qquad V(-4,1)$

x-int $(-6,0) \qquad y$-int none

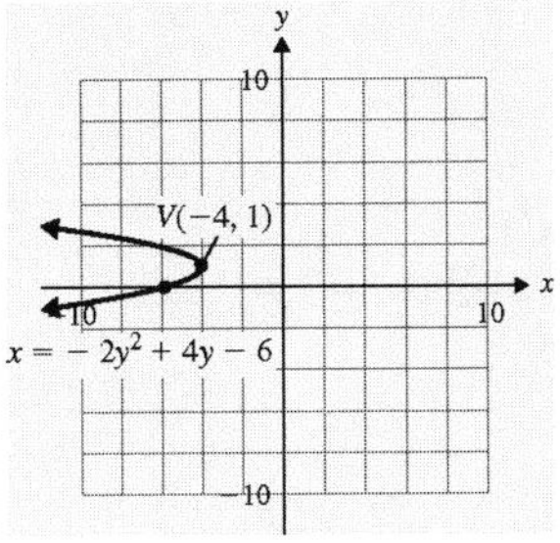

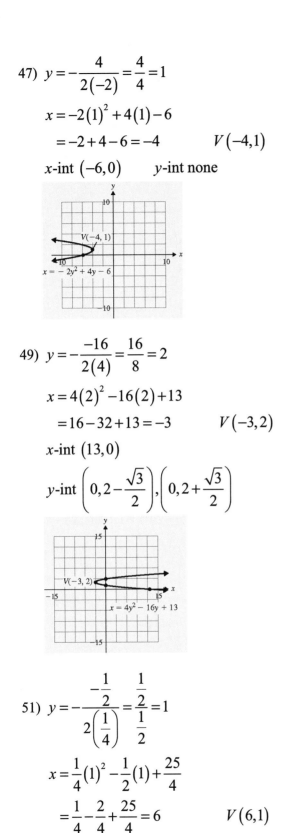

49) $y=-\dfrac{-16}{2(4)}=\dfrac{16}{8}=2$

$x=4(2)^2-16(2)+13$

$=16-32+13=-3 \qquad V(-3,2)$

x-int $(13,0)$

y-int $\left(0,2-\dfrac{\sqrt{3}}{2}\right),\left(0,2+\dfrac{\sqrt{3}}{2}\right)$

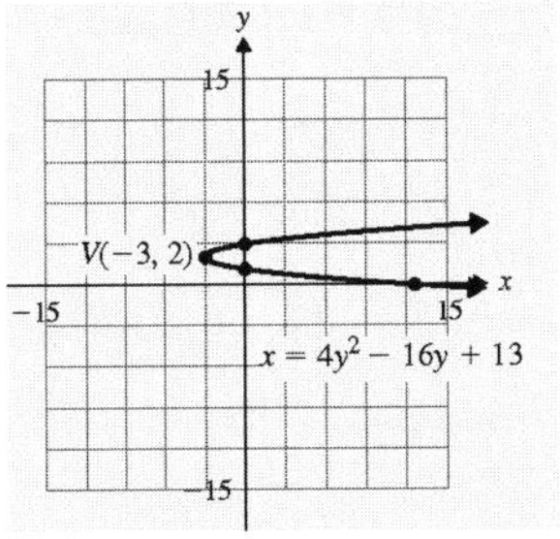

51) $y=-\dfrac{-\frac{1}{2}}{2\left(\frac{1}{4}\right)}=\dfrac{\frac{1}{2}}{\frac{1}{2}}=1$

$x=\dfrac{1}{4}(1)^2-\dfrac{1}{2}(1)+\dfrac{25}{4}$

$=\dfrac{1}{4}-\dfrac{2}{4}+\dfrac{25}{4}=6 \qquad V(6,1)$

x-int $\left(\dfrac{25}{4},0\right) \qquad y$-int none

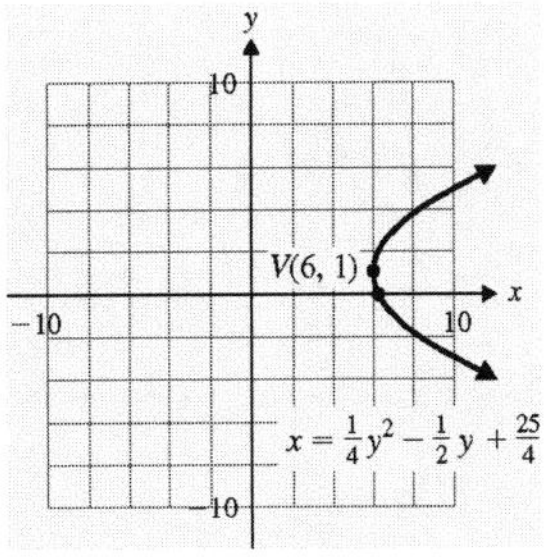

Section 12.5: Exercises

1) a) $(f+g)(x)=f(x)+g(x)$

$=(-3x+1)+(2x-11)$

$=-x-10$

b) $(f+g)(5)=-5-10=-15$

c) $(f-g)(x)=f(x)-g(x)$

$=(-3x+1)-(2x-11)$

$=-3x+1-2x+11$

$=-5x+12$

d) $(f-g)(2)=-5(2)+12$

$=-10+12=2$

3) a) $(f+g)(x)$

$=f(x)+g(x)$

$=(4x^2-7x-1)+(x^2+3x-6)$

$=5x^2-4x-7$

b) $(f+g)(5)=5(5)^2-4(5)-7$

$=5(25)-20-7$

$=125-20-7$

$=98$

c) $(f-g)(x)$
$= f(x)-g(x)$
$= (4x^2-7x-1)-(x^2+3x-6)$
$= 4x^2-7x-1-x^2-3x+6$
$= 3x^2-10x+5$

d) $(f-g)(2)=3(2)^2-10(2)+5$
$= 12-20+5=-3$

5) a) $(fg)(x)=f(x)\cdot g(x)$
$= (x)(-x+5)$
$= -x^2+5x$

b) $(fg)(-3)=-(-3)^2+5(-3)$
$= -9-15=-24$

7) a) $(fg)(x)=f(x)\cdot g(x)$
$= (2x+3)(3x+1)$
$= 6x^2+11x+3$

b) $(fg)(-3)=6(-3)^2+11(-3)+3$
$= 54-33+3=24$

9) a) $\left(\frac{f}{g}\right)(x)=\frac{f(x)}{g(x)}=\frac{6x+9}{x+4},\ x\neq 4$

b) $\left(\frac{f}{g}\right)(-2)=\frac{6(-2)+9}{-2+4}$
$= \frac{-12+9}{2}=-\frac{3}{2}$

11) a) $\left(\frac{f}{g}\right)(x)=\frac{f(x)}{g(x)}$
$= \frac{x^2-5x-24}{x-8}$
$= \frac{(x-8)(x+3)}{x-8}$
$= x+3, \qquad x\neq 8$

b) $\left(\frac{f}{g}\right)(-2)=-2+3=1$

13) a) $\left(\frac{f}{g}\right)(x)=\frac{f(x)}{g(x)}$
$= \frac{3x^2+14x+8}{3x+2}$
$= \frac{(3x+2)(x+4)}{3x+2}$
$= x+4, \qquad x\neq -\frac{2}{3}$

b) $\left(\frac{f}{g}\right)(-2)=-2+4=2$

15) a) $P(x)=R(x)-C(x)$
$= 12x-(8x+2000)$
$= 4x-2000$

b) $P(1500)=4(1500)-2000$
$= 6000-2000=\$4000$

17) a) $P(x)=R(x)-C(x)$
$= 18x-(15x+2400)$
$= 3x-2400$

b) $P(800)=3(800)-2400$
$= 2400-2400=0$

19) a) $(f\circ g)(x)=f(g(x))$
$=5(x+7)-4$
$=5x+35-4$
$=5x+31$

b) $(g\circ f)(x)=g(f(x))$
$=(5x-4)+7$
$=5x+3$

c) $(f\circ g)(3)=5(3)+31$
$=15+31=46$

21) a) $(k\circ h)(x)=k(h(x))$
$=3(-2x+9)-1$
$=-6x+27-1$
$=-6x+26$

b) $(h\circ k)(x)=h(k(x))$
$=-2(3x-1)+9$
$=-6x+2+9$
$=-6x+11$

c) $(k\circ h)(-1)=-6(-1)+26$
$=6+26=32$

23) a) $(h\circ g)(x)=h(g(x))$
$=(x^2-6x+11)-4$
$=x^2-6x+7$

b) $(g\circ h)(x)$
$=g(h(x))$
$=(x-4)^2-6(x-4)+11$
$=x^2-8x+16-6x+24+11$
$=x^2-14x+51$

c) $(g\circ h)(4)=(4)^2-14(4)+51$
$=16-56+51=11$

25) a) $(f\circ g)(x)$
$=f(g(x))$
$=2(3x-5)^2+3(3x-5)-10$
$=2(9x^2-30x+25)+9x-15-10$
$=18x^2-60x+50+9x-25$
$=18x^2-51x+25$

b) $(g\circ f)(x)=g(f(x))$
$=3(2x^2+3x-10)-5$
$=6x^2+9x-30-5$
$=6x^2+9x-35$

c) $(f\circ g)(1)=18(1)^2-51(1)+25$
$=18-51+25=-8$

27) a) $(n\circ m)(x)$
$=n(m(x))$
$=-(x+8)^2+3(x+8)-8$
$=-(x^2+16x+64)+3x+24-8$
$=-x^2-16x-64+3x+16$
$=-x^2-13x-48$

b) $(m\circ n)(x)=m(n(x))$
$=(-x^2+3x-8)+8$
$=-x^2+3x$

c) $(m\circ n)(0)=-(0)^2+3(0)=0$

29) a) $r(5)=4(5)=20$ The radius of the spill 5 min. after the ship started leaking was 20 ft.

b) $A(20)=\pi(20)^2=400\pi$

The area of the oil slick is 400π ft^2 when its radius is 20 ft.

c) $A(r(t))=\pi(4t)^2=16\pi t^2$

This is the area of the oil slick in terms of t, the number of minutes after the leak began.

d) $A(r(5))=16\pi(5)^2$

$=16\pi(25)=400\pi$

The area of the oil slick 5 minutes after the ship began leaking was 400π ft^2.

Section 12.6: Exercises

1) increases 3) direct

5) inverse 7) combined

9) $M=kn$ 11) $h=\frac{k}{j}$

13) $T=\frac{k}{c^2}$ 15) $s=krt$

17) $Q=\frac{k\sqrt{z}}{m}$

19) a) $z=kx$ b) $z=9x$ c) $z=9(6)$

$63=k(7)$ $=54$

$9=k$

21) a) $N=\frac{k}{y}$ b) $N=\frac{48}{y}$

$4=\frac{k}{12}$

$48=k$

c) $N=\frac{48}{3}=16$

23) a) $Q=\frac{kr^2}{w}$ b) $Q=\frac{5r^2}{w}$

$25=\frac{k(10)^2}{20}$

$500=k(100)$

$5=k$

c) $Q=\frac{5(6)^2}{4}=\frac{5(36)}{4}=5(9)=45$

25) $B=kR$

$35=k(5)$

$7=k$

$B=7R=7(8)=56$

27) $L=\frac{k}{h^2}$

$8=\frac{k}{3^2}$

$72=k$

$L=\frac{72}{h^2}=\frac{72}{2^2}=18$

29) $y=kxz$

$60=k(4)(3)$

$60=12k$

$5=k$

$y=5xz=5(7)(2)=70$

31) $E=kh$

$437.50=k(35)$

$12.50=k$

$E=12.50h=12.50(40)=\$500$

33) $t=\dfrac{d}{r}$

$14=\dfrac{d}{60}$

$840=d$

$t=\dfrac{840}{r}=\dfrac{840}{70}=12$ hours

35) $P=kIR^2$

$100=k(4)(5)^2$

$100=k(100)$

$1=k$

$P=(1)IR^2=(5)(6)^2=180$ watts

37) $E_k=kmv^2$

$112{,}500=k(1000)(15)^2$

$112{,}500=225{,}000k$

$0.5=k$

$E_k=0.5mv^2=0.5(1000)(18)^2$

$=500(324)=162{,}000$ J

39) $f=\dfrac{k}{L}$

$100=\dfrac{k}{5}$

$500=k$

$f=\dfrac{500}{L}=\dfrac{500}{2.5}=200$ cycles/sec

41) $R=\dfrac{kl}{A}$

$2=\dfrac{k(40)}{0.05}$

$0.1=40k$

$0.0025=k$

$R=\dfrac{0.0025l}{A}=\dfrac{0.0025(60)}{0.05}$

$=\dfrac{0.15}{0.05}=3$ ohms

43) $F=kx$

$200=k5$

$40=k$

$F=40x=40(8)=320$ lb

Chapter 12 Review

1) Domain: $\{-7,-5,2,4\}$
Range: $\{-4,-1,3,5,9\}$
Not a function

3) Domain: $(-\infty,\infty)$ Range: $[0,\infty)$
Is a function

5) yes 7) no

9) yes 11) yes

13)

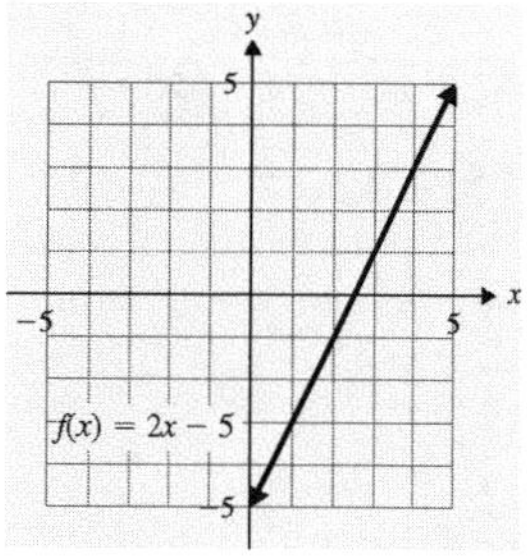

15)

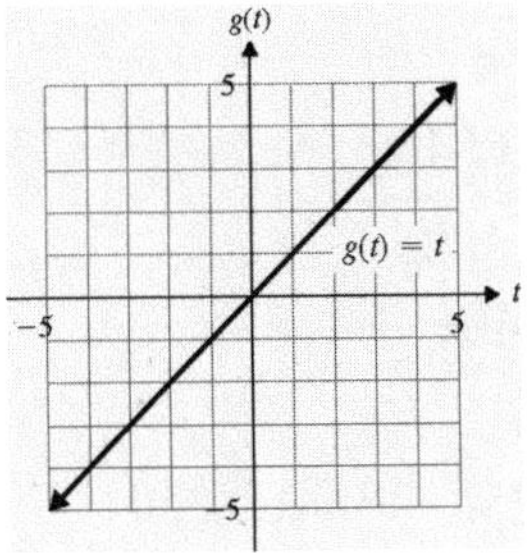

17) a) $f(5)=-8(5)+3=-40+3=-37$

b) $f(-4)=-8(-4)+3=32+3=35$

c) $g(-2)=(-2)^2+7(-2)-12$
$=4-14-12=-22$

d) $g(3)=(3)^2+7(3)-12$
$=9+21-12=18$

e) $f(c)=-8(c)+3=-8c+3$

f) $g(r)=r^2+7r-12$

g) $f(p-3)=-8(p-3)+3$
$=-8p+24+3$
$=-8p+27$

h) $g(t+4)$
$=(t+4)^2+7(t+4)-12$
$=t^2+8t+16+7t+28-12$
$=t^2+15t+32$

19) $k(x)=-\frac{2}{3}x+8$
$0=-\frac{2}{3}x+8$
$-8=-\frac{2}{3}x$
$12=x$

21) $p(x)=x^2-8x+15$
$3=x^2-8x+15$
$0=x^2-8x+12$
$0=(x-2)(x-4)$
$x-2=0$ or $x-6=0$
$x=2 \qquad x=6$

23) Solve: $x-5=0$
$x=5 \qquad (-\infty,5)\cup(5,\infty)$

25) $(-\infty,\infty)$

27) Solve: $5t-7\geq 0$
$5t\geq 7$
$t\geq \frac{7}{5} \qquad \left[\frac{7}{5},\infty\right)$

29) Solve: $x=0 \qquad (-\infty,0)\cup(0,\infty)$

31) $(-\infty,\infty)$

33) Solve: $a^2-7a-8=0$
$(a+1)(a-8)=0$
$a+1=0$ or $a-8=0$
$a=-1 \qquad a=8$
$(-\infty,-1)\cup(-1,8)\cup(8,\infty)$

35) a) $C(30)=0.20(30)+26=32$
\$32

b) $C(100)=0.20(100)+26=46$
\$46

c) $C(m)=0.20m+26$
$56=0.20m+26$
$30=0.20m$
$150=m$ 150 miles

d) $C(m) = 0.20m + 26$

$42 = 0.20m + 26$

$16 = 0.20m$

$80 = m$ 80 miles

37) Domain: $[0, \infty)$; Range: $[0, \infty)$

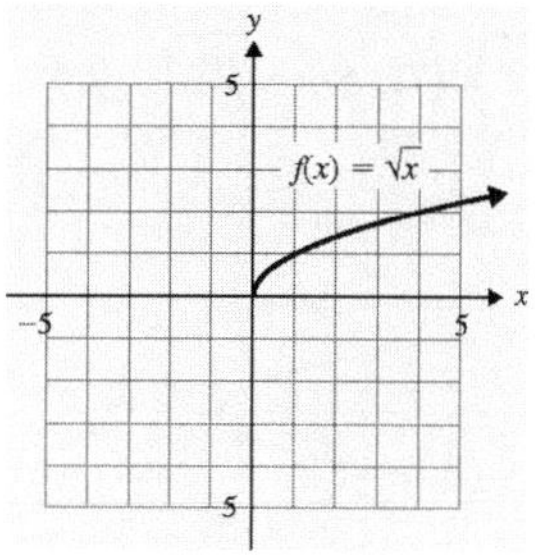

39) Domain: $(-\infty, \infty)$; Range: $[0, \infty)$

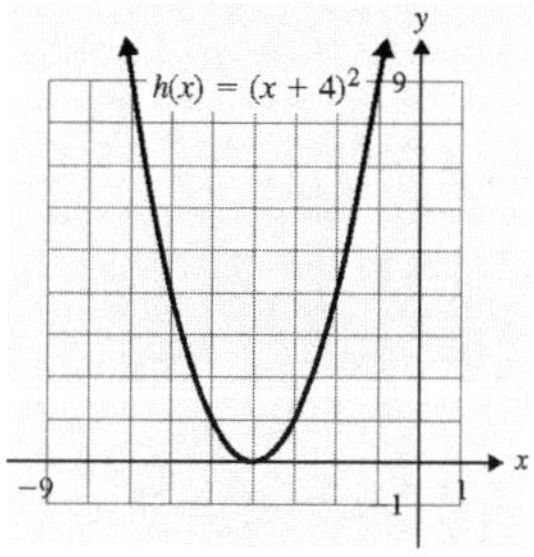

41) Domain: $(-\infty, \infty)$; Range: $(-\infty, 5]$

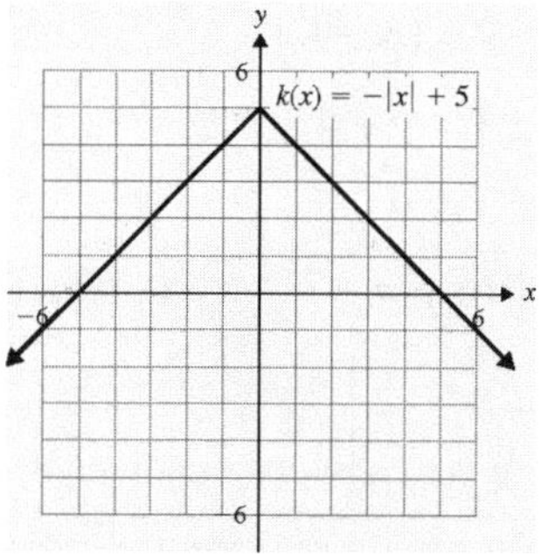

43) Domain: $[2, \infty)$; Range: $[-1, \infty)$

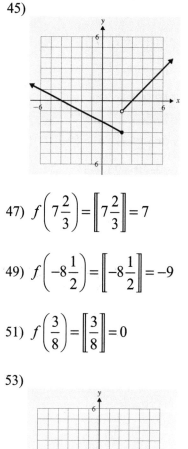

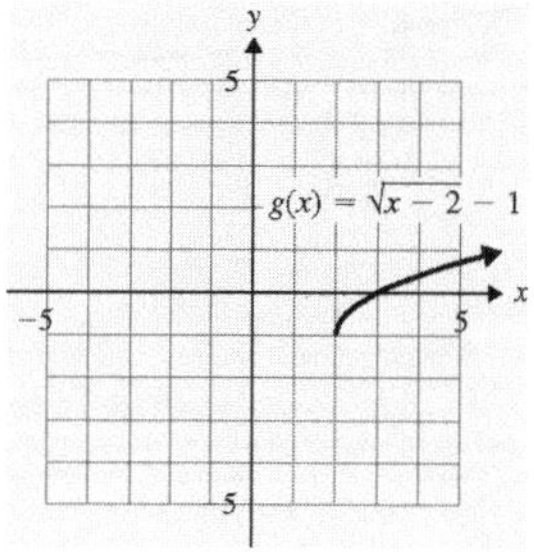

45)

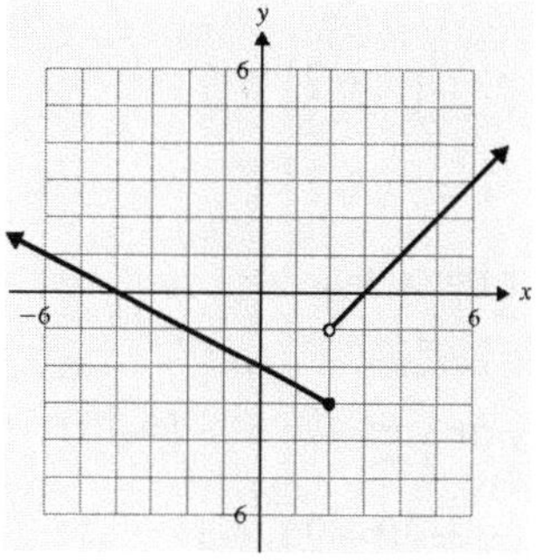

47) $f\left(7\frac{2}{3}\right) = \left[\!\left[7\frac{2}{3}\right]\!\right] = 7$

49) $f\left(-8\frac{1}{2}\right) = \left[\!\left[-8\frac{1}{2}\right]\!\right] = -9$

51) $f\left(\frac{3}{8}\right) = \left[\!\left[\frac{3}{8}\right]\!\right] = 0$

53)

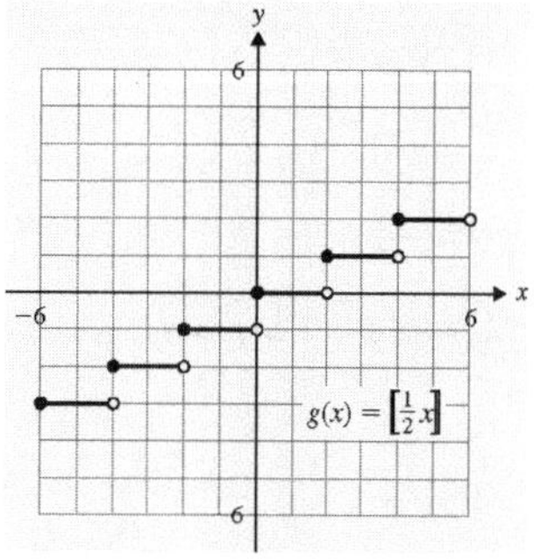

55) $g(x) = |x - 5|$ 57) a) (h, k) b) $x = h$

c) If a is positive, the parabola opens upward. If a is negative, the parabola opens downward.

59) a) (h, k) b) $y = k$

c) If a is positive, the parabola opens to the right. If a is negative, the parabola opens to the left.

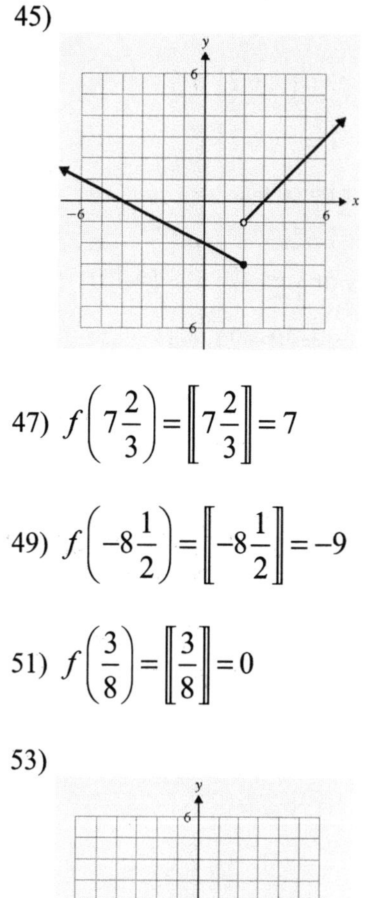

61) $V(-2,-1); x=-2$

x-int $\quad 0=(x+2)^2-1$

$1=(x+2)^2$

$\pm 1 = x+2$

$-2\pm 1 = x$

$x=-3$ or $x=-1 \qquad (-3,0),(-1,0)$

y-int $\quad y=(0+2)^2-1$

$y=(2)^2-1$

$y=4-1=3 \qquad (0,3)$

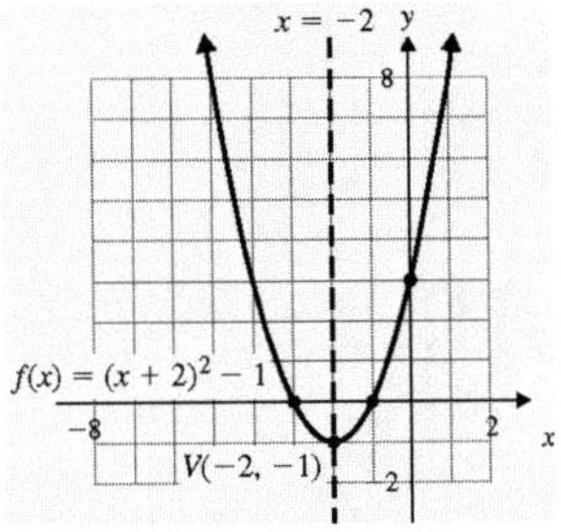

63) $V(-1,0); y=0$

x-int $\quad x=-0^2-1$

$x=-1 \qquad (-1,0)$

y-int $\quad 0=-y^2-1$

$1=-y^2$

$-1=y^2$

$\pm i = y \qquad$ none

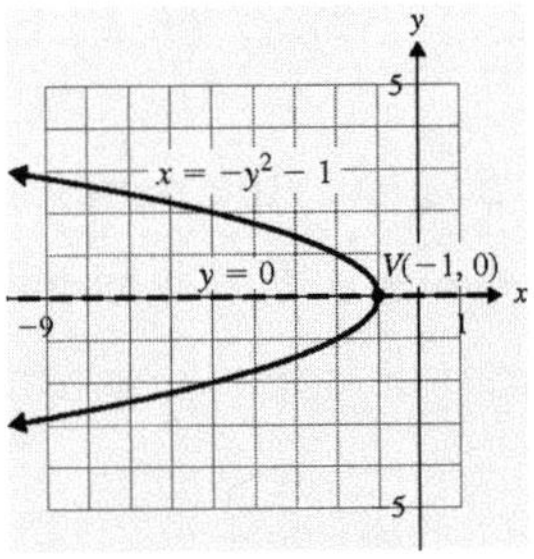

65) $V(11,3); y=3$

x-int $\quad x=-(0-3)^2+11$

$x=-(-3)^2+11$

$x=-9+11=2 \qquad (2,0)$

y-int $\quad 0=-(y-3)^2+11$

$-11=-(y-3)^2$

$11=(y-3)^2$

$\pm\sqrt{11}=y-3$

$3\pm\sqrt{11}=y$

$\left(0,3-\sqrt{11}\right),\left(0,3+\sqrt{11}\right)$

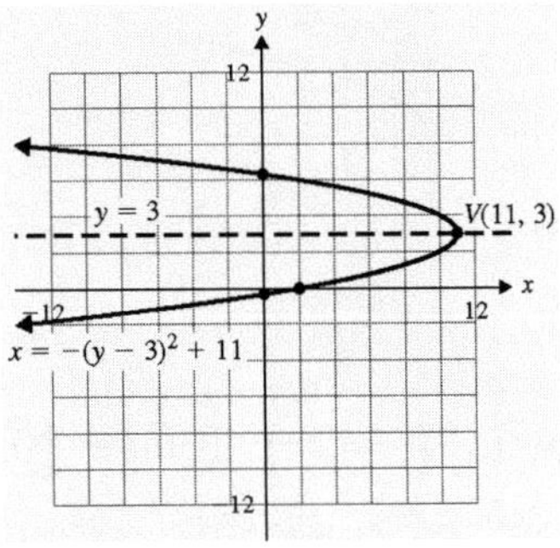

67) $x=y^2+8y+7$

$x=\left(y^2+8y+16\right)+7-16$

$x=(y+4)^2-9$

x-int $\quad x=(0)^2+8(0)+7=7 \qquad (7,0)$

y-int $\quad 0=(y+4)^2-9$

$9=(y+4)^2$

$\pm 3 = y+4$

$-4\pm 3 = y$

$y=-1$ or $y=-7 \qquad (0,-1),(0,-7)$

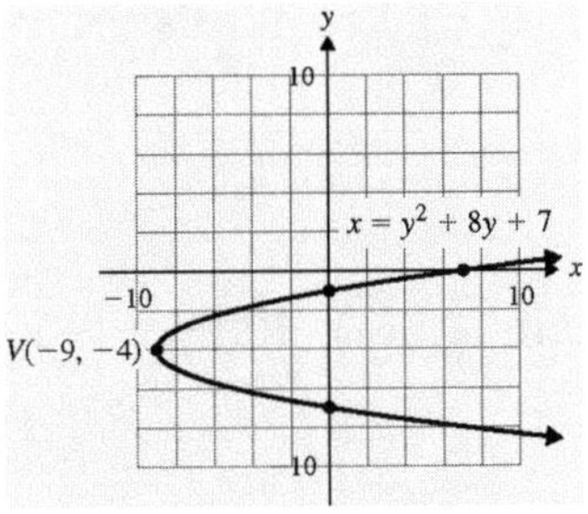

69) $y=\frac{1}{2}x^2-4x+9$

$y=\frac{1}{2}\left(x^2-8x\right)+9$

$y=\frac{1}{2}\left(x^2-8x+16\right)+9-8$

$y=\frac{1}{2}(x-4)^2+1$

x-int $0=\frac{1}{2}(x-4)^2+1$

$-1=\frac{1}{2}(x-4)^2$

$-2=(x-4)^2$

$\pm i\sqrt{2}=x-4$

$4\pm i\sqrt{2}=x$ none

y-int $y=\frac{1}{2}(0)^2-4(0)+9=9$ $(0,9)$

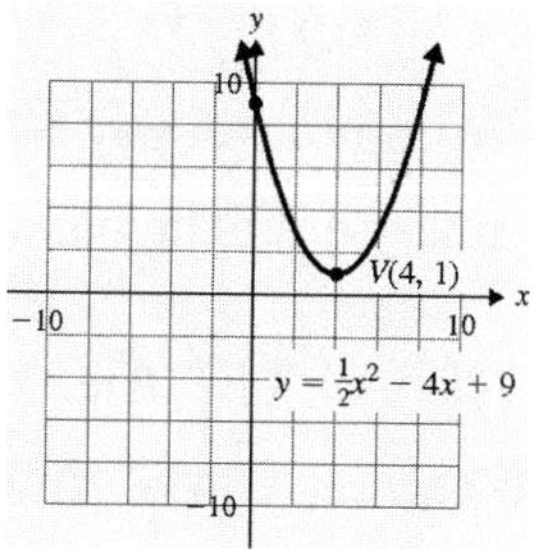

71) $h=-\frac{-2}{2(1)}=1$

$k=1^2-2(1)-4$

$=1-2-4=-5$ $V(1,-5)$

x-int $\left(1-\sqrt{5},0\right),\left(1+\sqrt{5},0\right)$

y-int $(0,-4)$

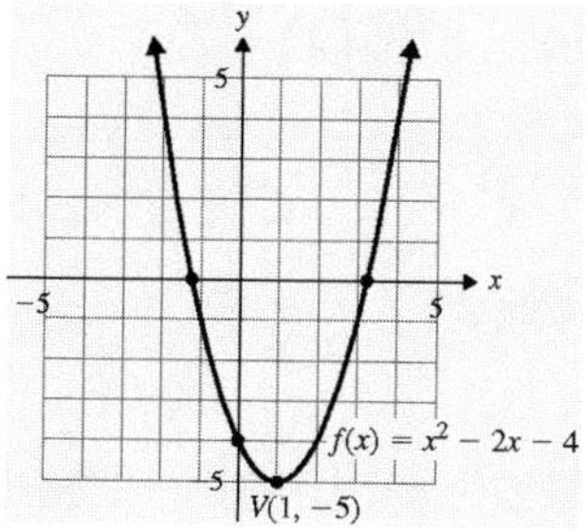

73) $h=-\frac{-3}{2\left(-\frac{1}{2}\right)}=-\frac{3}{1}=-3$

$k=-\frac{1}{2}(-3)^2-3(-3)-\frac{5}{2}$

$=-\frac{9}{2}+9-\frac{5}{2}=2$ $V(2,-3)$

x-int $\left(-\frac{5}{2},0\right)$

y-int $(0,-5),(0,-1)$

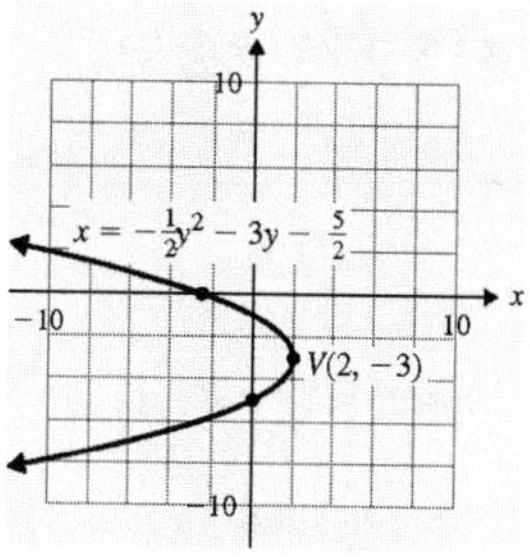

75) a) The max height occurs at the t-coordinate of the vertex.

$t=-\frac{32}{2(-16)}=\frac{32}{32}=1$

1 second

b) Since the object reaches max height $t=1$, find $h(1)$.

$h(1)=-16(1)^2+32(1)+240$

$=-16+32+240=256$

256 ft

c) The object hits ground

when $h(t)=0$.

$$0=-16t^2+32t+240$$
$$0=t^2-2t-15$$
$$0=(t-5)(t+3)$$
$$t-5=0 \text{ or } t+3=0$$
$$\boxed{t=5} \qquad t=-3$$

The object will hit the ground after 5 sec.

77) $(f+g)(x)=f(x)+g(x)$
$$=(5x+2)+(-x+4)$$
$$=4x+6$$

79) $(g-h)(2)=g(2)-h(2)$
$$=(-2+4)-\left(3(2)^2-7\right)$$
$$=2-(12-7)$$
$$=2-5=-3$$

81) $(fg)(x)=f(x)\cdot g(x)$
$$=(5x+2)(-x+4)$$
$$=-5x^2+18x+8$$

83) a) $\left(\frac{f}{g}\right)(x)=\frac{f(x)}{g(x)}=\frac{6x-5}{x+4}, x\neq -4$

b) $\left(\frac{f}{g}\right)(3)=\frac{6(3)-5}{3+4}=\frac{13}{7}$

85) a) $P(x)=R(x)-C(x)$
$$=20x-(14x+400)$$
$$=6x-400$$

b) $P(200)=6(200)-400$
$$=1200-400=\$800$$

87) a) $(g\circ h)(x)=g(h(x))$
$$=4(x^2+5)-3$$
$$=4x^2+20-3$$
$$=4x^2+17$$

b) $(h\circ g)(x)=h(g(x))$
$$=(4x-3)^2+5$$
$$=16x^2-24x+9+5$$
$$=16x^2-24x+14$$

c) $(h\circ g)(-2)$
$$=16(-2)^2-24(-2)+14$$
$$16(4)+48+14=64+62=126$$

89) a) $(N\circ G)(h)=N(G(x))$
$$=0.8(12h)=9.6h$$

This is Antoine's net pay in terms of how many hours he has worked.

b) $(N\circ G)(30)=9.6(30)=288$

When Antoine works 30 hours, his net pay is \$288.

c) $(N\circ G)(40)=9.6(40)=\$384$

91) $A=ktr$
$$15=k\left(\frac{1}{2}\right)(5)$$
$$6=k$$
$$A=6tr=6(3)(4)=72$$

93) $w=kr^3$
$$0.96=k(2)^3$$
$$0.12=k$$
$$w=0.12r^3=0.12(3)^3=3.24 \text{ lb}$$

Chapter 12 Test

1) It is a special type of relation in which each element of the domain corresponds to exactly one element of the range.

3) a) $\left[-\frac{7}{3},\infty\right)$ b) yes

5) $\left(-\infty,\frac{8}{7}\right)\cup\left(\frac{8}{7},\infty\right)$

7) $f(c)=4c+3$

9) $g(k+5)$

$=(k+5)^2-6(k+5)+10$

$=k^2+10k+25-6k-30+10$

$=k^2+4k+5$

11) a) $C(3)=50(3)+60$

$=150+60=210$

The cost of delivering 3 cubic yards of cedar mulch is \$210.

b) $C(m)=50m+60$

$360=50m+60$

$300=50m$

$6=m$ 6 cubic yards

13) Domain: $[-3,\infty)$; Range: $[0,\infty)$

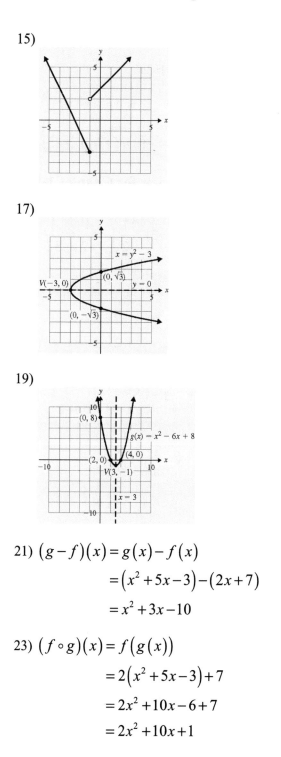

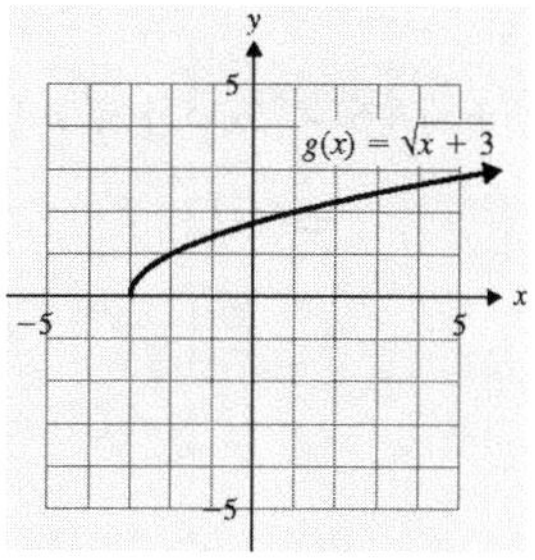

15)

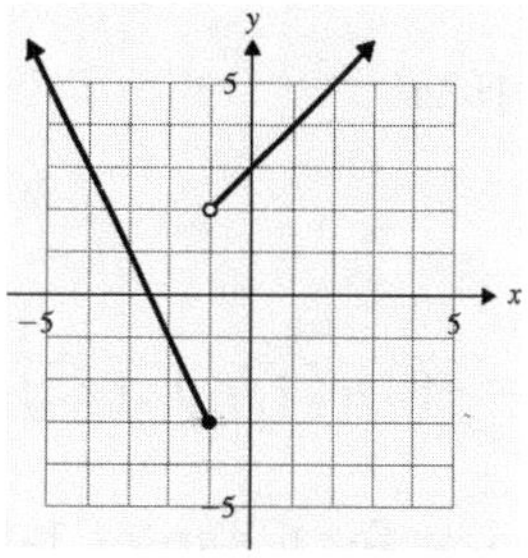

17)

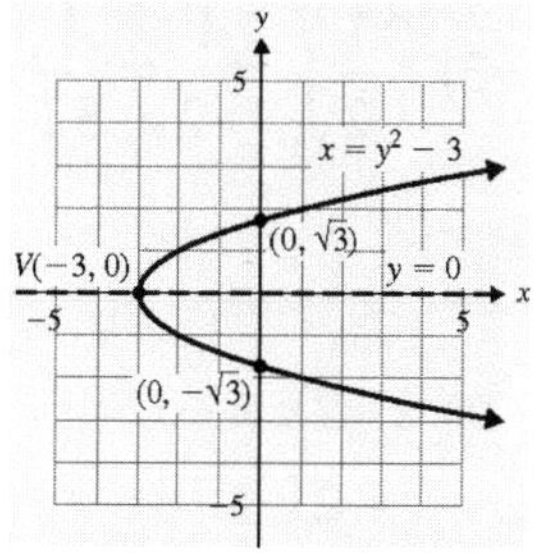

19)

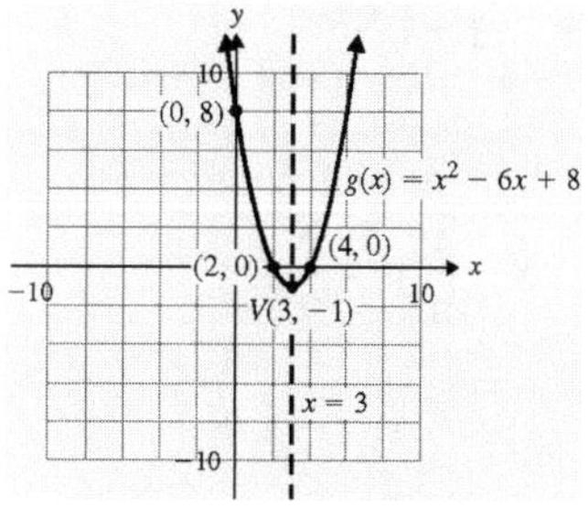

21) $(g-f)(x)=g(x)-f(x)$

$=\left(x^2+5x-3\right)-(2x+7)$

$=x^2+3x-10$

23) $(f\circ g)(x)=f(g(x))$

$=2\left(x^2+5x-3\right)+7$

$=2x^2+10x-6+7$

$=2x^2+10x+1$

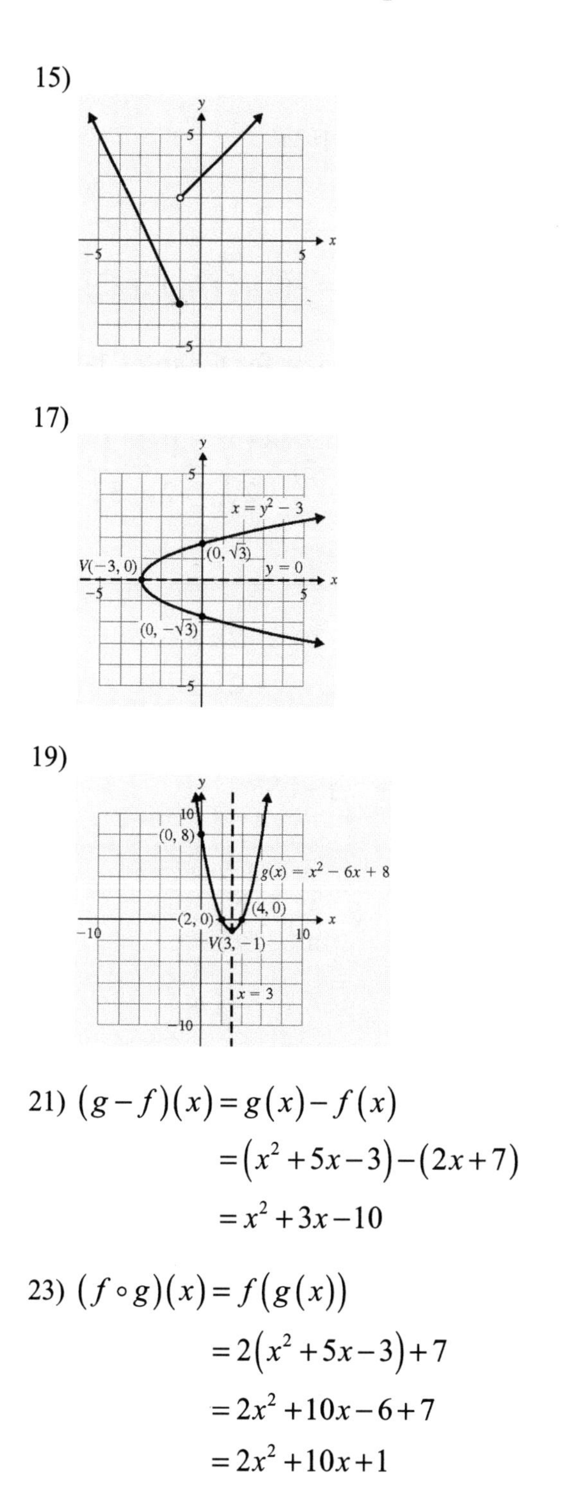

25) $n = krs^2$

$$72 = k(2)(3)^2$$
$$72 = 18k$$
$$4 = k$$
$$n = 4rs^2 = 4(3)(5)^2 = 12(25) = 300$$

Cumulative Review for Chapter 1-12

1) inverse

3) $\left(\frac{1}{2}\right)^5 = \frac{1^5}{2} = \frac{1}{32}$

5) $10^0 = 1$

7) $x + 8 \le 6$ or $1 - 2x \le -5$

$$x \le -2 \qquad -2x \le -6$$
$$x \ge 3$$
$$(-\infty, -2] \cup [3, \infty)$$

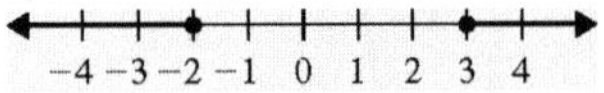

9) $x - \frac{1}{4}y = \frac{5}{2}$

$$x = \frac{1}{4}y + \frac{5}{2}$$
$$6\left(\frac{1}{2}x + \frac{1}{3}y\right) = 6\left(\frac{13}{6}\right)$$
$$3x + 2y = 13$$

Use substitution.

$$3\left(\frac{1}{4}y + \frac{5}{2}\right) + 2y = 13$$
$$4\left(\frac{3}{4}y + \frac{15}{2} + 2y\right) = 4(13)$$
$$3y + 30 + 8y = 52$$
$$11y = 22$$
$$y = 2$$
$$x - \frac{1}{4}(2) = \frac{5}{2}$$
$$x - \frac{1}{2} = \frac{5}{2}$$
$$x = \frac{6}{2} = 3 \qquad (3,\ 2)$$

11) $\dfrac{12r - 40r^2 + 6r^3 + 4}{4r^2}$

$$= \frac{6r^3}{4r^2} - \frac{40r^2}{4r^2} + \frac{12r}{4r^2} + \frac{4}{4r^2}$$
$$= \frac{3}{2}r - 10 + \frac{3}{r} + \frac{1}{r^2}$$

13) $100 - 9m^2 = (10 + 3m)(10 - 3m)$

15) $\dfrac{c-8}{2c^2 - 5c - 12} \div \dfrac{3c - 24}{c^2 - 16}$

$$= \frac{c-8}{2c^2 - 5c - 12} \cdot \frac{c^2 - 16}{3c - 24}$$
$$= \frac{\cancel{c-8}}{(2c+3)\cancel{(c-4)}} \cdot \frac{(c+4)\cancel{(c-4)}}{3\cancel{(c-8)}}$$
$$= \frac{c+4}{3(2c+3)}$$

17) $|7y + 6| \le -8 \qquad \varnothing$

19) $\sqrt{60} = \sqrt{4} \cdot \sqrt{15} = 2\sqrt{15}$

21) $\sqrt{18c^6 d^{11}} = 3\sqrt{2} \cdot c^3 \cdot d^5 \sqrt{d}$

$$= 3c^3 d^5 \sqrt{2d}$$

23) $\sqrt{12} + \sqrt{3} + \sqrt{48} = 2\sqrt{3} + \sqrt{3} + 4\sqrt{3}$

$$= 7\sqrt{3}$$

25) $\dfrac{4-2i}{2+3i}=\dfrac{4-2i}{2+3i}\cdot\dfrac{2-3i}{2-3i}$

$=\dfrac{8-12i-4i+6i^2}{2^2+3^2}$

$=\dfrac{8-16i+6(-1)}{4+9}$

$=\dfrac{2-16i}{13}=\dfrac{2}{13}-\dfrac{16}{13}i$

27) $4(y^2+2y)=5$

$4y^2+8y=5$

$4y^2+8y-5=0$

$(2y+5)(2y-1)=0$

$2y+5=0$ or $2y-1=0$

$2y=-5 \qquad 2y=1$

$y=-\dfrac{5}{2} \qquad y=\dfrac{1}{2}$

$\left\{-\dfrac{5}{2},\dfrac{1}{2}\right\}$

29) a) $g(7)=7+1=8$

b) $\left(\dfrac{h}{g}\right)(x)=\dfrac{h(x)}{g(x)}$

$=\dfrac{x^2+4x+3}{x+1}$

$=\dfrac{(x+3)(x+1)}{x+1}$

$=x+3, \qquad x\neq -1$

c) $g(x)=x+1$

$5=x+1$

$4=x$

d) $(g\circ h)(x)=g(h(x))$

$=(x^2+4x+3)+1$

$=x^2+4x+4$

31)

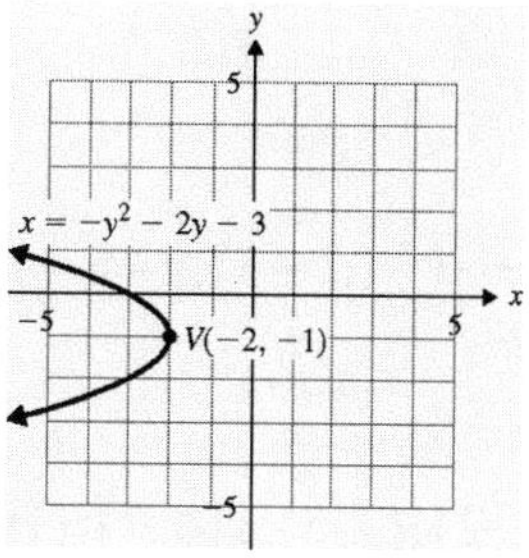

Section 13.1: Exercises

1) no

3) yes; $h^{-1} = \{(-16,-5),(-4,-1),(8,3)\}$

5) yes;
$g^{-1} = \{(1,2),(2,5),(14,7),(19,10)\}$

7) yes

9) No; only one-to-one functions have inverses.

11) False; it is read "f inverse of x."

13) True

15) False; they are symmetric with respect to $y = x$

17) a) yes b)

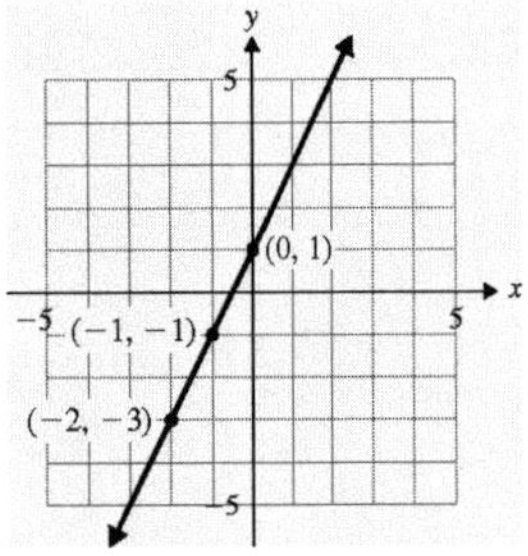

19) no

21) a) yes b)

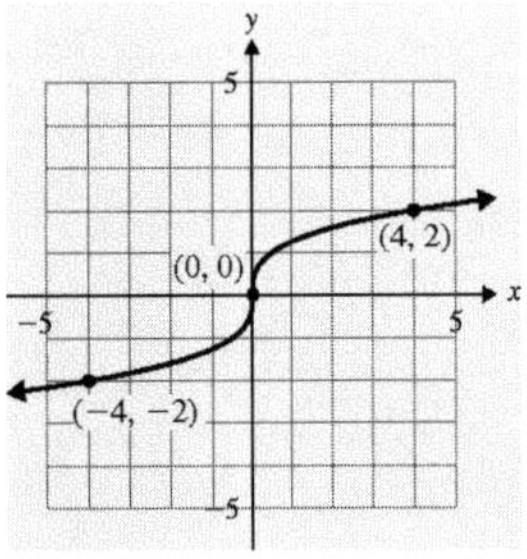

23)
$$y = x - 6$$
$$x = y - 6$$
$$x + 6 = y$$
$$g^{-1}(x) = x + 6$$

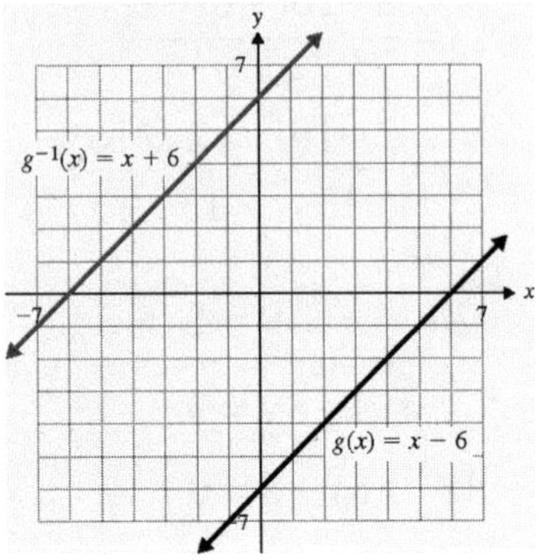

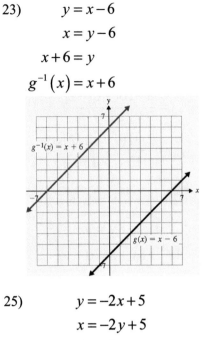

25)
$$y = -2x + 5$$
$$x = -2y + 5$$
$$x - 5 = -2y$$
$$-\frac{1}{2}x + \frac{5}{2} = y$$
$$f^{-1}(x) = -\frac{1}{2}x + \frac{5}{2}$$

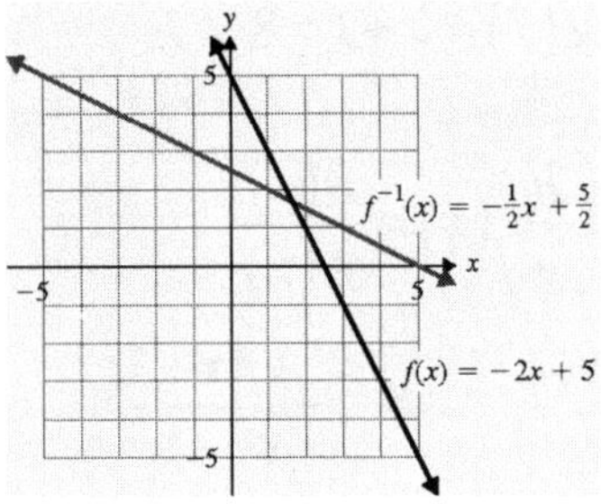

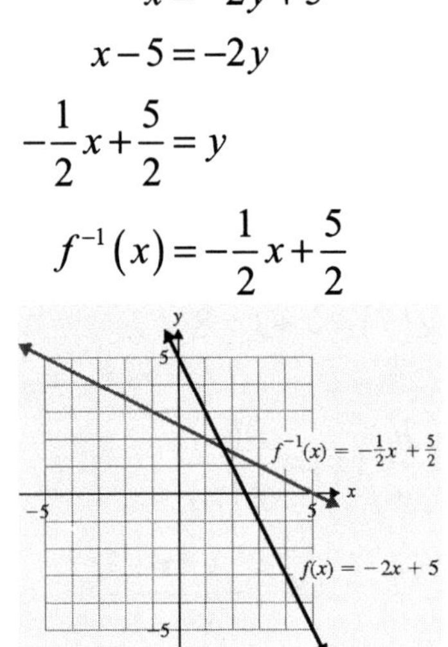

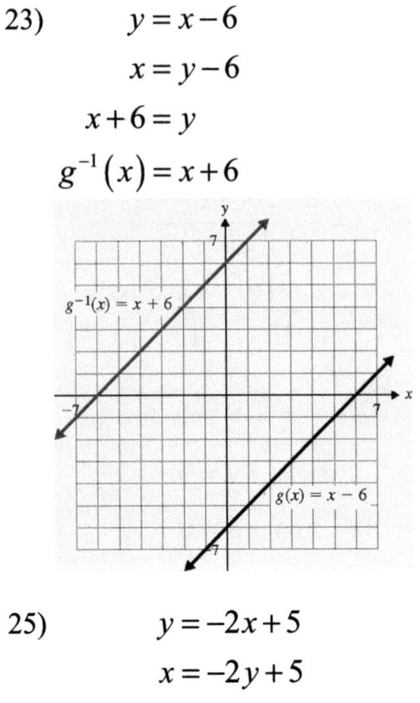

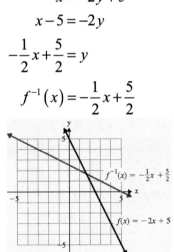

27)
$$y=\frac{1}{2}x$$
$$x=\frac{1}{2}y$$
$$2x=y$$
$$g^{-1}(x)=2x$$

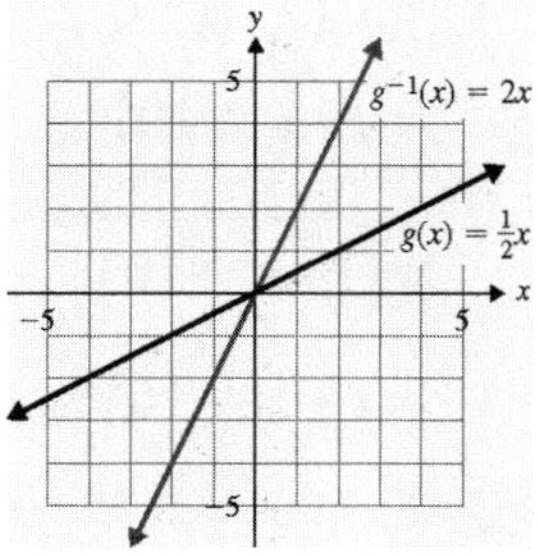

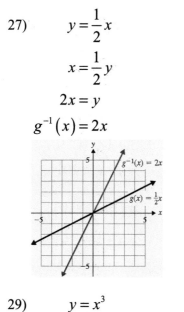

29)
$$y=x^3$$
$$x=y^3$$
$$\sqrt[3]{x}=y$$
$$f^{-1}(x)=\sqrt[3]{x}$$

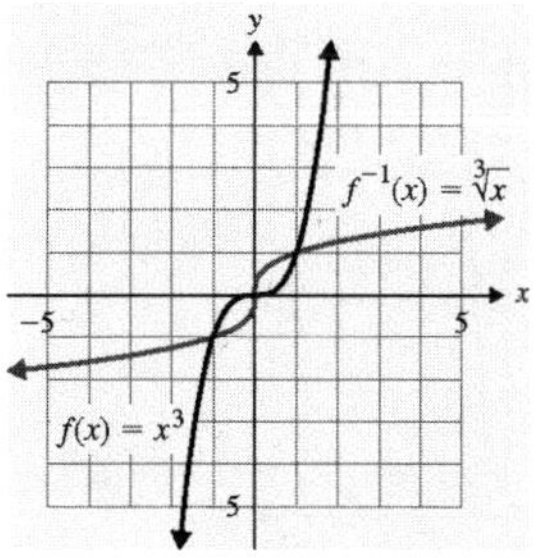

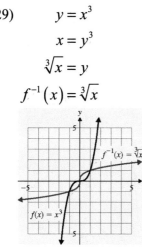

31)
$$y=2x-6$$
$$x=2y-6$$
$$x+6=2y$$
$$\frac{1}{2}x+3=y$$
$$f^{-1}(x)=\frac{1}{2}x+3$$

33)
$$y=-\frac{3}{2}x+4$$
$$x=-\frac{3}{2}y+4$$
$$x-4=-\frac{3}{2}y$$
$$-\frac{2}{3}x+\frac{8}{3}=y$$
$$h^{-1}(x)=-\frac{2}{3}x+\frac{8}{3}$$

35)
$$y=\sqrt[3]{x+2}$$
$$x=\sqrt[3]{y+2}$$
$$x^3=y+2$$
$$x^3-2=y$$
$$g^{-1}(x)=x^3-2$$

37)
$$y=\sqrt{x}$$
$$x=\sqrt{y}$$
$$x^2=y$$
$$f^{-1}(x)=x^2,\ x\geq 0$$

39) a) $f(1)=5(1)-2=3$ b) $f^{-1}(3)=1$

41) a) $f(9)=-\frac{1}{3}(9)+5=-3+5=2$

b) $f^{-1}(2)=9$

43) a) $f(-7)=-(-7)+3=7+3=10$

b) $f^{-1}(10)=-7$

45) a) $f(3)=2^3=8$

b) $f^{-1}(8)=3$

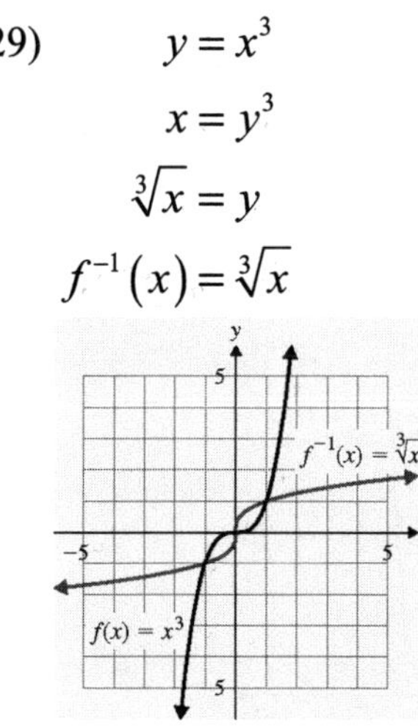

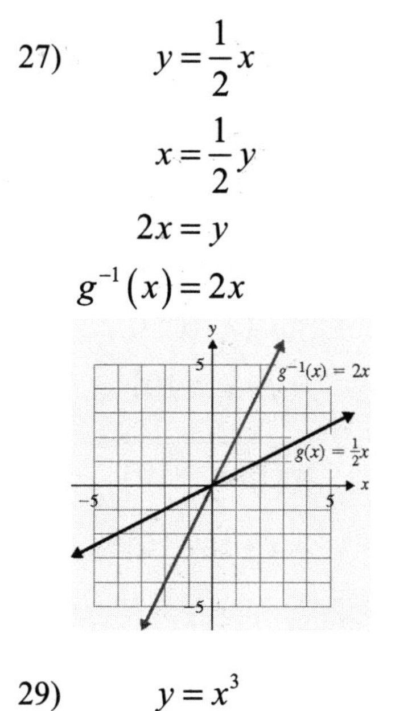

47) $(f^{-1} \circ f)(x) = f^{-1}(f(x))$
$= f^{-1}(x+9)$
$= (x+9) - 9$
$= x$
$(f \circ f^{-1})(x) = f(f^{-1}(x))$
$= f(x-9)$
$= (x-9) + 9 = x$

49) $(f^{-1} \circ f)(x) = f^{-1}(f(x))$
$= f^{-1}(-6x+4)$
$= -\frac{1}{6}(-6x+4) + \frac{2}{3}$
$= x - \frac{4}{6} + \frac{2}{3} = x$
$(f \circ f^{-1})(x) = f(f^{-1}(x))$
$= f\left(-\frac{1}{6}x + \frac{2}{3}\right)$
$= -6\left(-\frac{1}{6}x + \frac{2}{3}\right) + 4$
$= x - 4 + 4 = x$

51) $(f^{-1} \circ f)(x) = f^{-1}(f(x))$
$= f^{-1}\left(\frac{3}{2}x - 9\right)$
$= \frac{2}{3}\left(\frac{3}{2}x - 9\right) + 6$
$= x - 6 + 6 = x$
$(f \circ f^{-1})(x) = f(f^{-1}(x))$
$= f\left(\frac{2}{3}x + 6\right)$
$= \frac{3}{2}\left(\frac{2}{3}x + 6\right) - 9$
$= x + 9 - 9 = x$

53) $(f^{-1} \circ f)(x) = f^{-1}(f(x))$
$= f^{-1}\left(\sqrt[3]{x-10}\right)$
$= \left(\sqrt[3]{x-10}\right)^3 + 10$
$= x - 10 + 10 = x$
$(f \circ f^{-1})(x) = f(f^{-1}(x))$
$= f(x^3 + 10)$
$= \left(\sqrt[3]{(x^3+10)}\right)^3 - 10$
$= x + 10 - 10 = x$

Section 13.2: Exercises

1) Choose values for the variable that will give positive numbers, negative numbers, and zero in the exponent.

3)

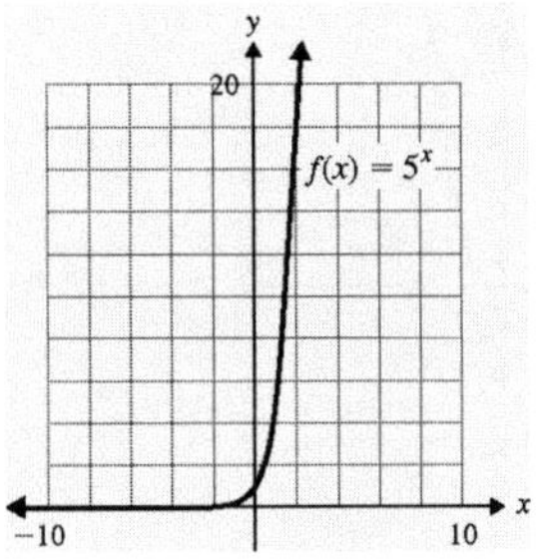

5)

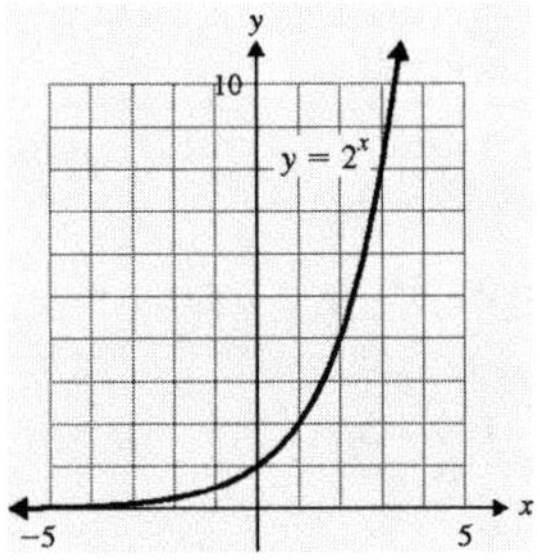

7)

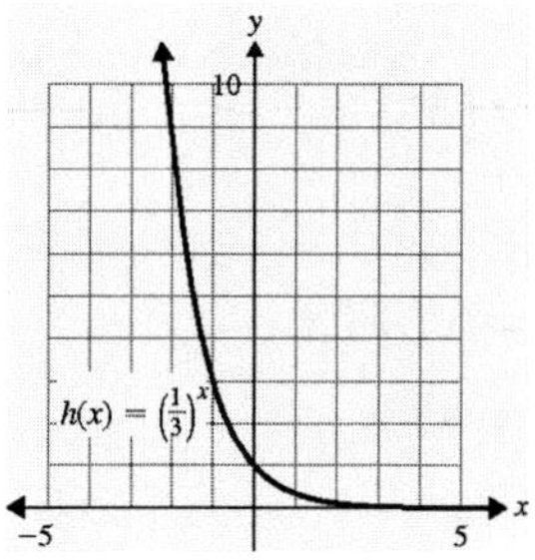

9) $(-\infty, \infty)$

11)

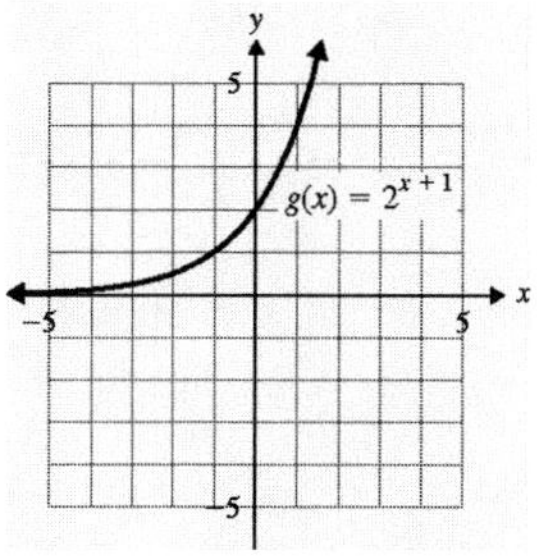

13)

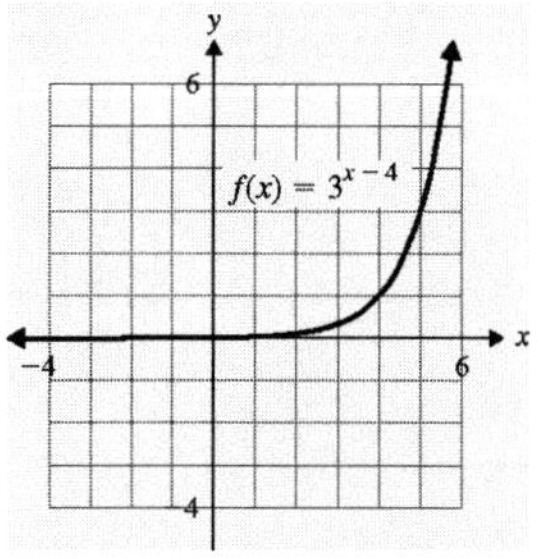

15)

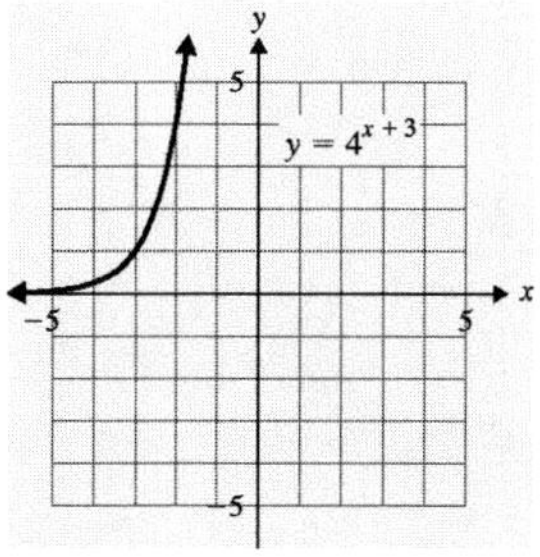

17)

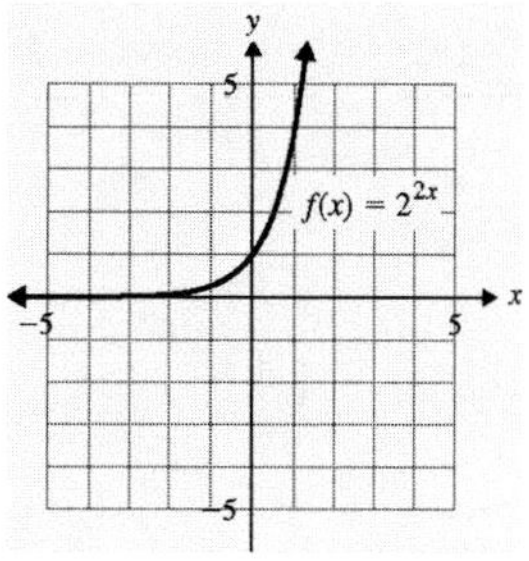

19)

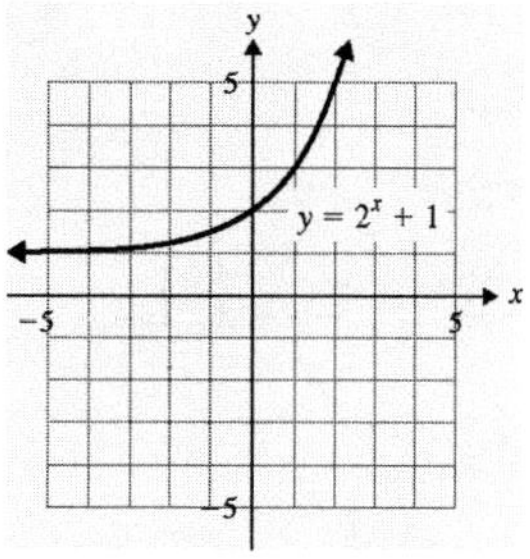

21)

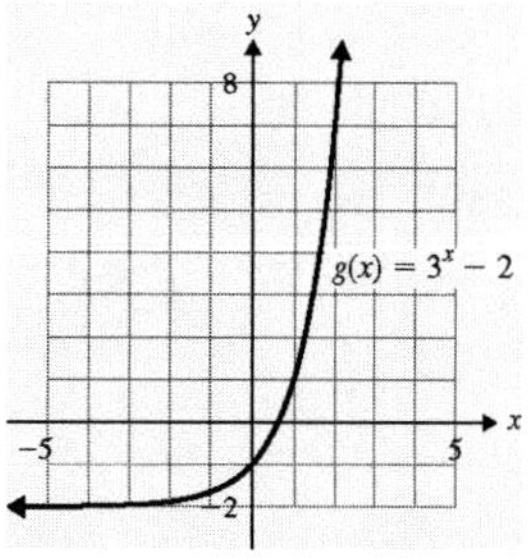

23)

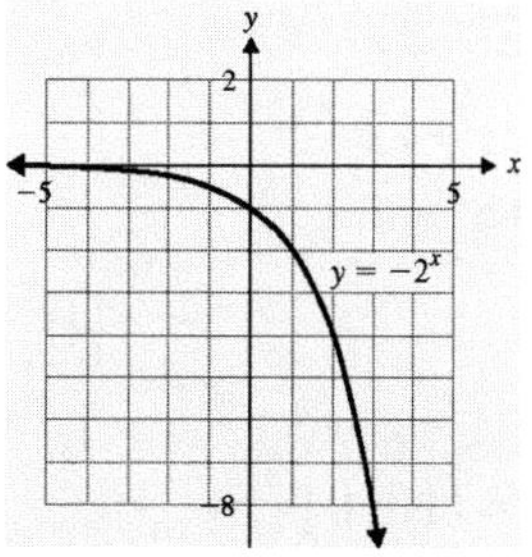

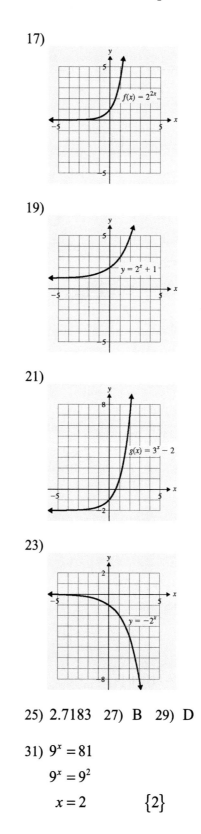

25) 2.7183 27) B 29) D

31) $9^x = 81$

$9^x = 9^2$

$x = 2$ $\{2\}$

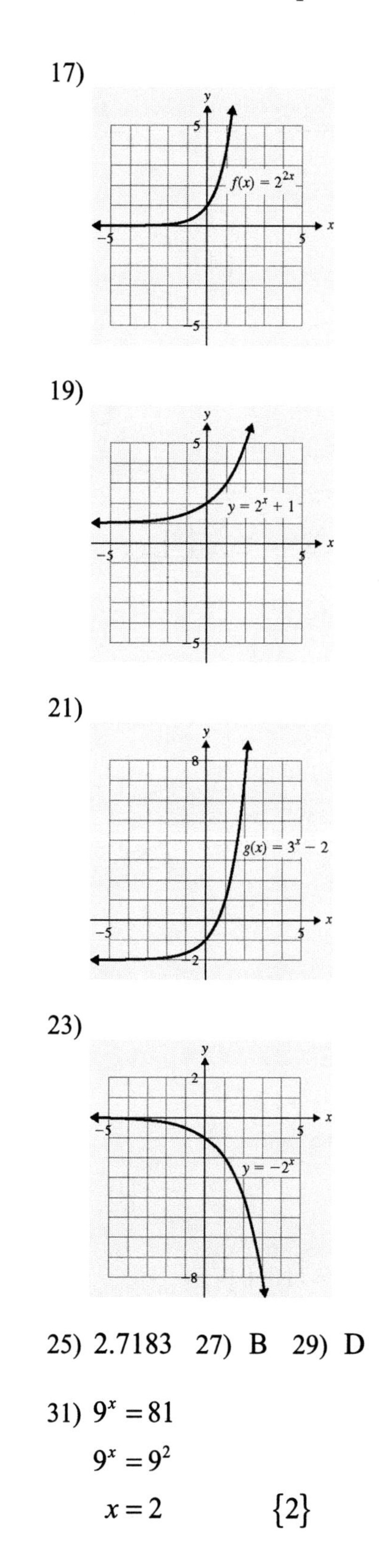

33) $5^{4d} = 125$

$5^{4d} = 5^3$

$4d = 3$

$d = \frac{3}{4}$ $\left\{\frac{3}{4}\right\}$

35) $16^{m-2} = 2^{3m}$

$\left(2^4\right)^{m-2} = 2^{3m}$

$2^{4m-8} = 2^{3m}$

$4m - 8 = 3m$

$m = 8$ $\{8\}$

37) $7^{2k-6} = 49^{3k+1}$

$7^{2k-6} = \left(7^2\right)^{3k+1}$

$7^{2k-6} = 7^{6k+2}$

$2k - 6 = 6k + 2$

$-8 = 4k$

$-2 = k$ $\{-2\}$

39) $32^{3c} = 8^{c+4}$

$\left(2^5\right)^{3c} = \left(2^3\right)^{c+4}$

$2^{15c} = 2^{3c+12}$

$15c = 3c + 12$

$12c = 12$

$c = 1$ $\{1\}$

41) $100^{5z-1} = (1000)^{2z+7}$

$\left(10^2\right)^{5z-1} = \left(10^3\right)^{2z+7}$

$10^{10z-2} = 10^{6z+21}$

$10z - 2 = 6z + 21$

$4z = 23$

$z = \frac{23}{4}$ $\left\{\frac{23}{4}\right\}$

43) $81^{3n+9} = 27^{2n+6}$

$\left(3^4\right)^{3n+9} = \left(3^3\right)^{2n+6}$

$3^{12n+36} = 3^{6n+18}$

$12n + 36 = 6n + 18$

$6n = -18$

$n = -3$ $\{-3\}$

45) $6^x = \frac{1}{36}$

$6^x = \left(\frac{1}{6}\right)^2$

$6^x = 6^{-2}$

$x = -2$ $\{-2\}$

47) $2^a = \frac{1}{8}$

$2^a = \left(\frac{1}{2}\right)^3$

$2^a = 2^{-3}$

$a = -3$ $\{-3\}$

49) $9^r = \frac{1}{27}$

$\left(3^2\right)^r = \left(\frac{1}{3}\right)^3$

$3^{2r} = 3^{-3}$

$2r = -3$

$r = -\frac{3}{2}$ $\left\{-\frac{3}{2}\right\}$

51) $\left(\frac{3}{4}\right)^{5k}=\left(\frac{27}{64}\right)^{k+1}$

$\left(\frac{3}{4}\right)^{5k}=\left[\left(\frac{3}{4}\right)^{3}\right]^{k+1}$

$\left(\frac{3}{4}\right)^{5k}=\left(\frac{3}{4}\right)^{3k+3}$

$5k=3k+3$

$2k=3$

$k=\frac{3}{2}$ $\quad\left\{\frac{3}{2}\right\}$

53) $\left(\frac{5}{6}\right)^{3x+7}=\left(\frac{36}{25}\right)^{2x}$

$\left(\frac{5}{6}\right)^{3x+7}=\left[\left(\frac{6}{5}\right)^{2}\right]^{2x}$

$\left(\frac{5}{6}\right)^{3x+7}=\left[\left(\frac{5}{6}\right)^{-2}\right]^{2x}$

$\left(\frac{5}{6}\right)^{3x+7}=\left(\frac{5}{6}\right)^{-4x}$

$3x+7=-4x$

$7=-7x$

$-1=x$ $\quad\{-1\}$

55) a) When the SUV was purchased $t=0$. Find $V(0)$.

$V(0)=32,700(0.812)^{0}$

$V(0)=32,700$

$32,700

b) Find $V(3)$.

$V(3)=32,700(0.812)^{3}$

$V(3)\approx 17,507.17$

$17,507.17

57) a) When the minivan was purchased $t=0$. Find $V(0)$.

$V(0)=16,800(0.803)^{0}$

$V(0)=16,800$

$16,800

b) Find $V(6)$.

$V(6)=16,800(0.803)^{6}$

$V(6)\approx 4504.04$

$4504.04

59) a) The value of the house in 1995 was $V(0)$.

$V(0)=185,200(1.03)^{0}$

$V(0)=185,200$

$185,200

b) Find $V(7)$.

$V(7)=185,200(1.03)^{7}$

$V(7)\approx 21,973.12$

$227,772.64

61) $c=2000,\ t=18,\ r=0.09$

$$A=2000\left[\frac{(1+0.09)^{18}-1}{0.09}\right](1+0.09)$$

$$=2000\left[\frac{(1.09)^{18}-1}{0.09}\right](1.09)$$

$\approx \$90,036.92$

63) $c = 4000,\ t = 10,\ r = 0.07$

$$A = 4000\left[\frac{(1+0.07)^{10}-1}{0.07}\right](1+0.07)$$

$$= 4000\left[\frac{(1.07)^{10}-1}{0.07}\right](1.07)$$

$$\approx \$59{,}134.40$$

65) $A(6) = 1000e^{-0.5332(6)}$

≈ 40.8 mg

Section 13.3: Exercises

1) a must be a positive real number that is not equal to 1.

3) 10 5) $7^2 = 49$ 7) $2^3 = 8$

9) $9^{-2} = \frac{1}{81}$ 11) $10^6 = 1{,}000{,}000$

13) $25^{1/2} = 5$ 15) $13^1 = 13$

17) $\log_9 81 = 2$ 19) $\log 100 = 2$

21) $\log_3 \frac{1}{81} = -4$ 23) $\log_{10} 1 = 0$

25) $\log_{169} 13 = \frac{1}{2}$ 27) $\sqrt{9} = 3$

$9^{1/2} = 3$

$\log_9 3 = \frac{1}{2}$

29) $\sqrt[3]{64} = 4$

$64^{1/3} = 4$

$\log_{64} 4 = \frac{1}{3}$

31) Write the equation in exponential form, then solve for the variable.

33) $\log_{11} x = 2$

$11^2 = x$

$121 = x$ $\{121\}$

35) $\log_4 r = 3$

$4^3 = r$

$64 = k$ $\{64\}$

37) $\log p = 5$

$10^5 = p$

$100{,}000 = p$ $\{100{,}000\}$

39) $\log_m 49 = 2$

$m^2 = 49$

$m = \pm 7$

the base must be positive $\{7\}$

41) $\log_6 h = -2$

$6^{-2} = h$

$\frac{1}{36} = h$ $\left\{\frac{1}{36}\right\}$

43) $\log_2(a+2) = 4$

$2^4 = a+2$

$16 = a+2$

$14 = a$ $\{14\}$

45) $\log_3(4t-3) = 3$

$3^3 = 4t-3$

$27 = 4t-3$

$30 = 4t$

$\frac{30}{4} = t$

$\frac{15}{2} = t$ $\left\{\frac{15}{2}\right\}$

47) $\log_{81}\sqrt[4]{9}=x$

$$81^x=\sqrt[4]{9}$$
$$\left(9^2\right)^x=9^{1/4}$$
$$9^{2x}=9^{1/4}$$
$$2x=\frac{1}{4}$$
$$x=\frac{1}{8} \qquad \left\{\frac{1}{8}\right\}$$

49) $\log_{125}\sqrt{5}=c$

$$125^c=\sqrt{5}$$
$$\left(5^3\right)^c=5^{1/2}$$
$$5^{3c}=5^{1/2}$$
$$3c=\frac{1}{2}$$
$$c=\frac{1}{6} \qquad \left\{\frac{1}{6}\right\}$$

51) $\log_{144}w=\frac{1}{2}$

$$144^{1/2}=w$$
$$12=w \qquad \{12\}$$

53) $\log_8 x=\frac{2}{3}$

$$8^{2/3}=x$$
$$\left(\sqrt[3]{8}\right)^2=x$$
$$2^2=x$$
$$4=x \qquad \{4\}$$

55) $\log_{(3m-1)}25=2$

$$(3m-1)^2=25$$
$$3m-1=\pm 5$$

$$3m-1=5 \text{ or } 3m-1=-5$$
$$3m=6 \qquad 3m=-4$$
$$m=2 \qquad m=-\frac{4}{3}$$

the base must be positive $\{2\}$

57) $\log_5 25=2$ since $5^{\boxed{2}}=25$

59) $\log_2 32=5$ since $2^{\boxed{5}}=32$

61) $\log 100=2$ since $10^{\boxed{2}}=100$

63) Let $\log_{49}7=x$

$$49^x=7$$
$$\left(7^2\right)^x=7^1$$
$$2x=1$$
$$x=\frac{1}{2}$$
$$\log_{49}7=\frac{1}{2}$$

65) Let $\log_8\frac{1}{8}=x$

$$8^x=\frac{1}{8}$$
$$8^x=8^{-1}$$
$$x=-1$$
$$\log_8\frac{1}{8}=-1$$

67) $\log_5 5=1$

69) Let $\log_{1/4}16=x$

$$\left(\frac{1}{4}\right)^x=16$$
$$\left(4^{-1}\right)^x=4^2$$
$$-x=2$$
$$x=-2$$
$$\log_{1/4}16=-2$$

71) Replace $f(x)$ with y, write $y = \log_a x$ in exponential form, make a table of values, then plot the points and draw the curve.

73)

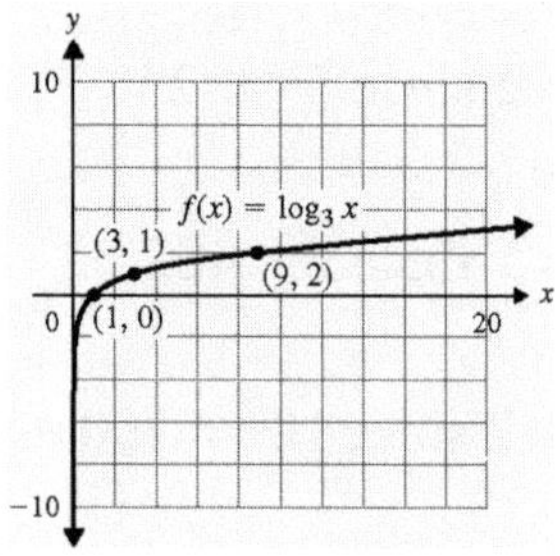

75)

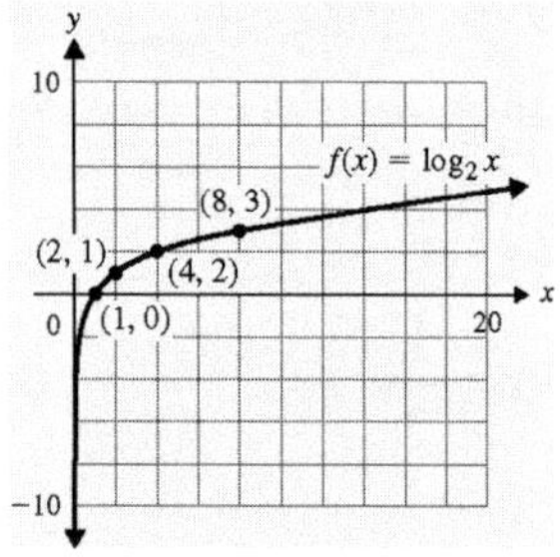

77)

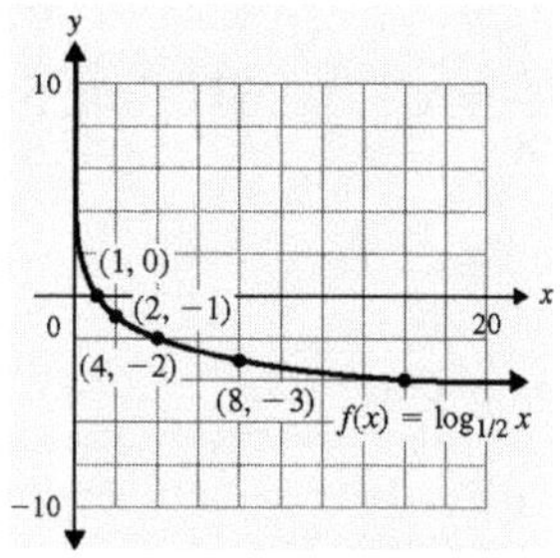

79)

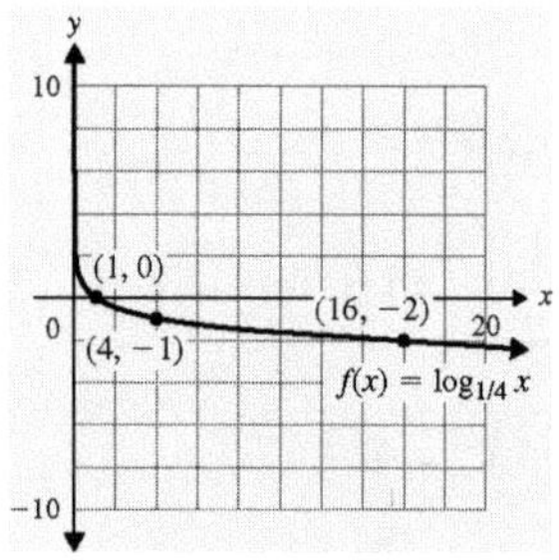

81) $y = 3^x$

$x = 3^y$

$y = \log_3 x$

$f^{-1}(x) = \log_3 x$

83) $y = \log_2 x$

$x = \log_2 y$

$y = 2^x$

$f^{-1}(x) = 2^x$

85) a) Find $L(0)$.

$$L(0) = 1800 + 68\log_3(0+3)$$
$$= 1800 + 68\log_3 3$$
$$= 1800 + 68(1) = 1868$$

b) Find $L(24)$.

$$L(24) = 1800 + 68\log_3(24+3)$$
$$= 1800 + 68\log_3 27$$
$$= 1800 + 68(3)$$
$$= 1800 + 204 = 2004$$

c) Let $L(t) = 2072$ and solve for t.

$$2072 = 1800 + 68\log_3(t+3)$$
$$272 = 68\log_3(t+3)$$
$$4 = \log_3(t+3)$$
$$3^4 = t+3$$
$$81 = t+3$$
$$78 = t$$

78 years after 1980 is the year 2058.

87) a) Find $S(1)$.

$$S(1) = 14\log_3[2(1)+1]$$
$$= 14\log_3 3$$
$$= 14(1) = 14 \qquad 14{,}000$$

b) Find $S(4)$.

$$S(4) = 14\log_3[2(4)+1]$$
$$= 14\log_3 9$$
$$= 14(2) = 28 \qquad 28{,}000$$

c) Find $S(13)$ and compare this value to the actual value.

$S(13) = 14\log_3[2(13)+1]$
$= 14\log_3 27$
$= 14(3) = 42$ 42,000

The actual number sold was 1000 more boxes than what was predicted by the formula.

Section 13.4: Exercises

1) True 3) False

5) False 7) True

9) $\log_8(3\cdot 10) = \log_8 3 + \log_8 10$

11) $\log_7 5d = \log_7 5 + \log_7 d$

13) $\log_9 \frac{4}{7} = \log_9 4 - \log_9 7$

15) $\log_5 2^3 = 3\log_5 2$

17) $\log p^8 = 8\log p$

19) $\log_3 \sqrt{7} = \log_3 7^{1/2} = \frac{1}{2}\log_3 7$

21) $\log_5 25t = \log_5 25 + \log_5 t$
$= 2 + \log_5 t$

23) $\log_2 \frac{8}{k} = \log_2 8 - \log_2 k = 3 - \log_2 k$

25) $\log_7 49^3 = 3\log_7 49 = 3(2) = 6$

27) $\log 1000b = \log 1000 + \log b$
$= 3 + \log b$

29) $\log_2 32^7 = 7\log_2 32 = 7(5) = 35$

31) $\log_5 \sqrt{5} = \log_5 5^{1/2} = \frac{1}{2}\log_5 5 = \frac{1}{2}$

33) $\log \sqrt[3]{100} = \log 100^{1/3} = \frac{1}{3}\log 100$
$= \frac{1}{3}(2) = \frac{2}{3}$

35) $\log_6 w^4 z^3 = \log_6 w^4 + \log_6 z^3$
$= 4\log_6 w + 3\log_6 z$

37) $\log_7 \frac{a^2}{b^5} = \log_7 a^2 - \log_7 b^5$
$= 2\log_7 a - 5\log_7 b$

39) $\log \frac{\sqrt[5]{11}}{y^2} = \log \sqrt[5]{11} - \log y^2$
$= \log 11^{1/5} - 2\log y$
$= \frac{1}{5}\log 11 - 2\log y$

41) $\log_2 \frac{4\sqrt{n}}{m^3}$
$= \log_2 4\sqrt{n} - \log_2 m^3$
$= \log_2 4 + \log_2 \sqrt{n} - 3\log_2 m$
$= 2 + \log_2 n^{1/2} - 3\log_2 m$
$= 2 + \frac{1}{2}\log_2 n - 3\log_2 m$

43) $\log_4 \frac{x^3}{yz^2} = \log_4 x^3 - \log_4 yz^2$
$= 3\log_4 x - (\log_4 y + \log_4 z^2)$
$= 3\log_4 x - (\log_4 y + 2\log_4 z)$
$= 3\log_4 x - \log_4 y - 2\log_4 z$

45) $\log_5 \sqrt{5c} = \log_5 (5c)^{1/2} = \frac{1}{2}\log_5 5c$
$= \frac{1}{2}(\log_5 5 + \log_5 c)$
$= \frac{1}{2}(1 + \log_5 c)$
$= \frac{1}{2} + \frac{1}{2}\log_5 c$

47) $\log k(k-6) = \log k + \log(k-6)$

49) $\log_a m + \log_a n = \log_a mn$

51) $\log_7 d - \log_7 3 = \log_7 \frac{d}{3}$

53) $4\log_3 f + \log_3 g = \log_3 f^4 + \log_3 g$
$= \log_3 f^4 g$

55) $\log_8 t + 2\log_8 u - 3\log_8 v$
$= \log_8 t + \log_8 u^2 - \log_8 v^3$
$= \log_8 tu^2 - \log_8 v^3 = \log_8 \frac{tu^2}{v^3}$

57) $\log(r^2+3) - 2\log(r^2-3)$
$= \log(r^2+3) - \log(r^2-3)^2$
$= \log\frac{r^2+3}{(r^2-3)^2}$

59) $3\log_n 2 + \frac{1}{2}\log_n k$
$= \log_n 2^3 + \log_n k^{1/2}$
$= \log_n 8 + \log_n \sqrt{k} = \log_n 8\sqrt{k}$

61) $\frac{1}{3}\log_d 5 - 2\log_d z$
$= \log_d 5^{1/3} - \log_d z^2$
$= \log_d \sqrt[3]{5} - \log_d z^2 = \log_d \frac{\sqrt[3]{5}}{z^2}$

63) $\log_6 y - \log_6 3 - 3\log_6 z$
$= \log_6 y - \log_6 3 - \log_6 z^3$
$= \log_6 y - (\log_6 3 + \log_6 z^3)$
$= \log_6 y - \log_6 3z^3 = \log_6 \frac{y}{3z^3}$

65) $4\log_3 t - 2\log_3 6 - 2\log_3 u$
$= \log_3 t^4 - \log_3 6^2 - \log_3 u^2$
$= \log_3 t^4 - (\log_3 36 + \log_3 u^2)$
$= \log_3 t^4 - (\log_3 36u^2) = \log_3 \frac{t^4}{36u^2}$

67) $\frac{1}{2}\log_b(c+4) - 2\log_b(c+3)$
$= \log_b (c+4)^{1/2} - \log_b (c+3)^2$
$= \log_b \sqrt{c+4} - \log_b (c+3)^2$
$= \log_b \frac{\sqrt{c+4}}{(c+3)^2}$

69) $\log 45 = \log(5\cdot 9) = \log 5 + \log 9$
$= 0.6990 + 0.9542 = 1.6532$

71) $\log 81 = \log 9^2 = 2\log 9$
$= 2(0.9542) = 1.9084$

73) $\log\frac{5}{9} = \log 5 - \log 9$
$= 0.6990 - 0.9542 = -0.2552$

75) $\log 3 = \log\sqrt{9} = \log 9^{1/2} = \frac{1}{2}\log 9$
$= \frac{1}{2}(0.9542) = 0.4771$

77) $\log\frac{1}{5} = \log 5^{-1} = -1\log 5$
$= -(0.6990) = -0.6990$

79) $\log\frac{1}{81} = \log 1 - \log 81$
$= 0 - \log 9^2 = -2\log 9$
$= -2(0.9542) = -1.9084$

81) $\log 50 = \log(10\cdot 5) = \log 10 + \log 5$
$= 1 + 0.6990 = 1.6990$

Section 13.5: Exercises

1) e

3) $\log 100 = 2$ since $10^{\boxed{2}} = 100$

5) $\log\frac{1}{1000} = \log 10^{-3}$
$= -3\log 10 = -3\cdot 1 = -3$

7) $\log 0.1 = \log\frac{1}{10} = \log 10^{-1}$
$= -1\log 10 = -1\cdot 1 = -1$

9) $\log 10^9 = 9\log 10 = 9\log 10 = 9\cdot 1 = 9$

11) $\log\sqrt[4]{10} = \log 10^{1/4}$
$= \frac{1}{4}\log 10 = \frac{1}{4}\cdot 1 = \frac{1}{4}$

13) $\ln e^6 = 6\ln e = 6\cdot 1 = 6$

15) $\ln\sqrt{e} = \ln e^{1/2} = \frac{1}{2}\ln e = \frac{1}{2}\cdot 1 = \frac{1}{2}$

17) $\ln\frac{1}{e^5} = \ln e^{-5} = -5\ln e = -5\cdot 1 = -5$

19) $\ln 1 = 0$ 21) $\log 16 \approx 1.2041$

23) $\log 0.5 = -0.3010$

25) $\ln 3 \approx 1.0986$ 27) $\ln 1.31 \approx 0.2700$

29) $\log x = 3$
$10^3 = x$
$1000 = x$ $\{1000\}$

31) $\log k = -1$
$10^{-1} = k$
$\frac{1}{10} = k$ $\left\{\frac{1}{10}\right\}$

33) $\log(4a) = 2$
$10^2 = 4a$
$100 = 4a$
$25 = a$ $\{25\}$

35) $\log(3t+4) = 1$
$10^1 = 3t+4$
$10 = 3t+4$
$6 = 3t$
$2 = t$ $\{2\}$

37) $\log a = 1.5$
$10^{1.5} = a$ $\{10^{1.5}\};\{31.6228\}$

39) $\log r = 0.8$
$10^{0.8} = r$ $\{10^{0.8}\};\{6.3096\}$

41) $\ln x = 1.6$
$e^{1.6} = x$ $\{e^{1.6}\};\{4.9530\}$

43) $\ln t = -2$

$e^{-2} = t$

$\frac{1}{e^2} = t \qquad \left\{\frac{1}{e^2}\right\};\{0.1353\}$

45) $\ln(3q) = 2.1$

$e^{2.1} = 3q$

$\frac{e^{2.1}}{3} = q \qquad \left\{\frac{e^{2.1}}{3}\right\};\{2.7221\}$

47) $\log\left(\frac{1}{2}c\right) = 0.47$

$10^{0.47} = \frac{1}{2}c$

$2(10)^{0.47} = c$

$\left\{2(10)^{0.47}\right\};\{5.9024\}$

49) $\log(5y-3) = 3.8$

$10^{3.8} = 5y-3$

$3+10^{3.8} = 5y$

$\frac{3+10^{3.8}}{5} = y$

$\left\{\frac{3+10^{3.8}}{5}\right\};\{1262.5147\}$

51) $\ln(10w+19) = 1.85$

$e^{1.85} = 10w+19$

$e^{1.85} - 19 = 10w$

$\frac{e^{1.85}-19}{10} = w$

$\left\{\frac{e^{1.85}-19}{10}\right\};\{-1.2640\}$

53) $\ln(2d-5) = 0$

$e^0 = 2d-5$

$5+1 = 2d$

$6 = 2d$

$3 = d \qquad \{3\}$

55) $\log_2 13 = \frac{\ln 13}{\ln 2} \approx 3.7004$

57) $\log_9 70 = \frac{\ln 70}{\ln 9} \approx 1.9336$

59) $\log_{1/3} 16 = \frac{\ln 16}{\ln \frac{1}{3}} \approx -2.5237$

61) $\log_5 3 = \frac{\ln 3}{\ln 5} \approx 0.6826$

63) $L(0.1) = 10\log\frac{0.1}{10^{-12}}$

$= 10\log\frac{10^{-1}}{10^{-12}}$

$= 10\log 10^{11}$

$= 110\log 10 = 110 \text{ dB}$

65) $L(0.00000001) = 10\log\frac{0.00000001}{10^{-12}}$

$= 10\log\frac{10^{-8}}{10^{-12}}$

$= 10\log 10^4$

$= 40\log 10$

$= 40 \text{ dB}$

67) $A = 3000\left(1+\frac{.05}{12}\right)^{12(3)}$

$\approx 3000(1.004)^{36} \approx \$3484.42$

69) $A=4000\left(1+\frac{.065}{4}\right)^{4(5)}$

$\approx 4000(1.0163)^{20}\approx \$5521.68$

71) $A=3000e^{.05(3)}=3000e^{.15}\approx \$3485.50$

73) $A=10{,}000e^{.075(6)}=10{,}000e^{.45}$

$\approx \$15{,}683.12$

75) a) $N(0)=5000e^{0.0617(0)}$

$=5000e^{0}$

$=5000(1)=5000$ bacteria

b) $N(8)=5000e^{0.0617(8)}$

$=5000e^{0.4936}$

≈ 8191 bacteria

77) $N(24)=10{,}000e^{0.0492(24)}$

$=10{,}000e^{1.1808}$

$=32{,}570$ bacteria

79) $\text{pH}=-\log\left[2\times 10^{-3}\right]$

$=-\log(0.002)\approx 2.7$; acidic

81) $\text{pH}=-\log\left[6\times 10^{-12}\right]$

$=-\log(0.000000000006)$

≈ 11.2; basic

83) $y=\ln x$

$x=\ln y$

$y=e^{x}$

Section 13.6: Exercises

1) $7^{x}=49$

$7^{x}=7^{2}$

$x=2\quad \{2\}$

3) $7^{n}=15$

$\ln 7^{n}=\ln 15$

$n\ln 7=\ln 15$

$n=\frac{\ln 15}{\ln 7}$

$\left\{\frac{\ln 15}{\ln 7}\right\};\{1.3917\}$

5) $8^{z}=3$

$\ln 8^{z}=\ln 3$

$z\ln 8=\ln 3$

$z=\frac{\ln 3}{\ln 8}$

$\left\{\frac{\ln 3}{\ln 8}\right\};\{0.5283\}$

7) $6^{5p}=36$

$6^{5p}=6^{2}$

$5p=2$

$p=\frac{2}{5}\qquad \left\{\frac{2}{5}\right\}$

9) $4^{6k}=2.7$

$\ln 4^{6k}=\ln 2.7$

$6k\ln 4=\ln 2.7$

$k=\frac{\ln 2.7}{6\ln 4}$

$\left\{\frac{\ln 2.7}{6\ln 4}\right\};\{0.1194\}$

11) $2^{4n+1}=5$

$\ln 2^{4n+1}=\ln 5$

$(4n+1)\ln 2=\ln 5$

$4n\ln 2+\ln 2=\ln 5$

$4n\ln 2=\ln 5-\ln 2$

$n=\frac{\ln 5-\ln 2}{4\ln 2}$

$\left\{\frac{\ln 5-\ln 2}{4\ln 2}\right\},\ \{0.3305\}$

13)
$$5^{3a-2}=8$$
$$\ln 5^{3a-2}=\ln 8$$
$$(3a-2)\ln 5=\ln 8$$
$$3a\ln 5-2\ln 5=\ln 8$$
$$3a\ln 5=\ln 8+2\ln 5$$
$$a=\frac{\ln 8+2\ln 5}{3\ln 5}$$
$$\left\{\frac{\ln 8+2\ln 5}{3\ln 5}\right\},\ \{1.0973\}$$

15)
$$4^{2c+7}=64^{3c-1}$$
$$4^{2c+7}=\left(4^3\right)^{3c-1}$$
$$4^{2c+7}=4^{9c-3}$$
$$10=7c$$
$$\frac{10}{7}=c \qquad \left\{\frac{10}{7}\right\}$$

17)
$$9^{5d-2}=4^{3d}$$
$$\ln 9^{5d-2}=\ln 4^{3d}$$
$$(5d-2)\ln 9=3d\ln 4$$
$$5d\ln 9-2\ln 9=3d\ln 4$$
$$5d\ln 9-3d\ln 4=2\ln 9$$
$$d(5\ln 9-3\ln 4)=2\ln 9$$
$$d=\frac{2\ln 9}{5\ln 9-3\ln 4}$$
$$\left\{\frac{2\ln 9}{5\ln 9-3\ln 4}\right\};\{0.6437\}$$

19)
$$e^y=12.5$$
$$\ln e^y=\ln 12.5$$
$$y\ln e=\ln 12.5$$
$$y(1)=\ln 12.5$$
$$y=\ln 12.5$$
$$\{\ln 12.5\};\{2.5257\}$$

21)
$$e^{-4x}=9$$
$$\ln e^{-4x}=\ln 9$$
$$-4x\ln e=\ln 9$$
$$-4x(1)=\ln 9$$
$$x=-\frac{\ln 9}{4}$$
$$\left\{-\frac{\ln 9}{4}\right\};\{-0.5493\}$$

23)
$$e^{0.01r}=2$$
$$\ln e^{0.01r}=\ln 2$$
$$0.01r\ln e=\ln 2$$
$$0.01r(1)=\ln 2$$
$$r=\frac{\ln 2}{0.01}$$
$$\left\{\frac{\ln 2}{0.01}\right\},\ \{69.3147\}$$

25)
$$e^{0.006t}=3$$
$$\ln e^{0.006t}=\ln 3$$
$$0.006t\ln e=\ln 3$$
$$0.006t(1)=\ln 3$$
$$t=\frac{\ln 3}{0.006}$$
$$\left\{\frac{\ln 3}{0.006}\right\},\ \{183.1021\}$$

27)
$$e^{-0.4y}=5$$
$$\ln e^{-0.4y}=\ln 5$$
$$-0.4y\ln e=\ln 5$$
$$-0.4y(1)=\ln 5$$
$$y=-\frac{\ln 5}{0.4}$$
$$\left\{-\frac{\ln 5}{0.4}\right\},\ \{-4.0236\}$$

29) $\log_6(k+9)=\log_6 11$

$k+9=11$

$k=2$ $\{2\}$

31) $\log_7(3p-1)=\log_7 9$

$3p-1=9$

$3p=10$

$p=\dfrac{10}{3}$ $\left\{\dfrac{10}{3}\right\}$

33) $\log x+\log(x-2)=\log 15$

$\log x(x-2)=\log 15$

$x(x-2)=15$

$x^2-2x=15$

$x^2-2x-15=0$

$(x-5)(x+3)=0$

$x-5=0$ or $x+3=0$

$x=5$ $x=-3$

Only one solution satisfies the original equation. $\{5\}$

35) $\log_3 n+\log_3(12-n)=\log_3 20$

$\log_3 n(12-n)=\log_3 20$

$n(12-n)=20$

$12n-n^2=20$

$n^2-12n+20=0$

$(n-10)(n-2)=0$

$n-10=0$ or $n-2=0$

$n=10$ $n=2$ $\{2,10\}$

37) $\log_2(-z)+\log_2(z-8)=\log_2 15$

$\log_2[-z(z-8)]=\log_2 15$

$-z(z-8)=15$

$-z^2+8z=15$

$z^2-8z+15=0$

$(z-5)(z-3)=0$

$z-5=0$ or $z-3=0$

$z=5$ $z=3$

Neither solution satisfies the original equation. $\varnothing$

39) $\log_6(5b-4)=2$

$6^2=5b-4$

$36=5b-4$

$40=5b$

$8=b$ $\{8\}$

41) $\log(3p+4)=1$

$10^1=3p+4$

$6=3p$

$2=p$ $\{2\}$

43) $\log_3 y+\log_3(y-8)=2$

$\log_3 y(y-8)=2$

$3^2=y(y-8)$

$9=y^2-8y$

$0=y^2-8y-9$

$0=(y-9)(y+1)$

$y-9=0$ or $y+1=0$

$y=9$ $y=-1$

Only one solution satisfies the original equation. $\{9\}$

45) $\log_2 r + \log_2 (r+2) = 3$

$$\log_2 r(r+2) = 3$$
$$2^3 = r(r+2)$$
$$8 = r^2 + 2r$$
$$0 = r^2 + 2r - 8$$
$$0 = (r+4)(r-2)$$
$$r+4=0 \text{ or } r-2=0$$
$$r=-4 \qquad r=2$$

Only one solution satisfies the original equation. $\{2\}$

47) $\log_4 20c - \log_4 (c+1) = 2$

$$\log_4 \frac{20c}{c+1} = 2$$
$$4^2 = \frac{20c}{c+1}$$
$$16 = \frac{20c}{c+1}$$
$$16(c+1) = 20c$$
$$16c + 16 = 20c$$
$$16 = 4c$$
$$4 = c \qquad \{4\}$$

49) $\log_2 8d - \log_2 (2d-1) = 4$

$$\log_2 \frac{8d}{2d-1} = 4$$
$$2^4 = \frac{8d}{2d-1}$$
$$16 = \frac{8d}{2d-1}$$
$$16(2d-1) = 8d$$
$$32d - 16 = 8d$$
$$-16 = -24d$$
$$\frac{-16}{-24} = d$$
$$\frac{2}{3} = d \qquad \left\{\frac{2}{3}\right\}$$

51) a) $2500 = 2000e^{0.06t}$

$$\frac{5}{4} = e^{0.06t}$$
$$\ln\frac{5}{4} = \ln e^{0.06t}$$
$$\ln\frac{5}{4} = 0.06t \ln e$$
$$\ln\frac{5}{4} = 0.06t(1)$$
$$\frac{\ln\frac{5}{4}}{0.06} = t$$
$$3.72 \text{ yr} \approx t$$

b) $4000 = 2000e^{0.06t}$

$$2 = e^{0.06t}$$
$$\ln 2 = \ln e^{0.06t}$$
$$\ln 2 = 0.06t \ln e$$
$$\ln 2 = 0.06t(1)$$
$$\frac{\ln 2}{0.06} = t$$
$$11.55 \text{ yr} \approx t$$

53) $7800 = 7000e^{0.075t}$

$$\frac{7800}{7000} = e^{0.075t}$$

$$\ln\frac{39}{35} = \ln e^{0.075t}$$

$$\ln\frac{39}{35} = 0.075t \ln e$$

$$\ln\frac{39}{35} = 0.075t(1)$$

$$\frac{\ln\frac{39}{35}}{0.075} = t$$

$$1.44 \text{ yr} \approx t$$

55) $5000 = Pe^{0.08(10)}$

$$5000 = Pe^{0.80}$$

$$\frac{5000}{e^{0.40}} = P$$

$$\$2246.64 \approx P$$

57) $4000 = 3000e^{r(4)}$

$$\frac{4}{3} = e^{4r}$$

$$\ln\frac{4}{3} = \ln e^{4r}$$

$$\ln\frac{4}{3} = 4r \ln e$$

$$\ln\frac{4}{3} = 4r(1)$$

$$\frac{\ln\frac{4}{3}}{4} = r$$

$0.072 \approx r$ 7.2%

59) a) $5000 = 4000e^{0.0374t}$

$$\frac{5}{4} = e^{0.0374t}$$

$$\ln\frac{5}{4} = \ln e^{0.0374t}$$

$$\ln\frac{5}{4} = 0.0374t \ln e$$

$$\ln\frac{5}{4} = 0.0374t(1)$$

$$\frac{\ln\frac{5}{4}}{0.0374} = t$$

$$6 \text{ hr} \approx t$$

b) $8000 = 4000e^{0.0374t}$

$$2 = e^{0.0374t}$$

$$\ln 2 = \ln e^{0.0374t}$$

$$\ln 2 = 0.0374t \ln e$$

$$\ln 2 = 0.0374t(1)$$

$$\frac{\ln 2}{0.0374} = t$$

$$18.5 \text{ hr} \approx t$$

61) Let $t = 8$, $y_0 = 21,000$

$$y = 21,000e^{0.036(8)}$$

$$y = 21,000e^{0.288}$$

$$y \approx 28,009 \text{ people}$$

63) a) Let $t = 15, y_0 = 2470$

$$y = 2470e^{-0.013(15)}$$

$$y = 2470e^{-0.195}$$

$$y \approx 2032 \text{ people}$$

b) Let $y = 1600, y_0 = 2470$

$$1600 = 2470e^{-0.013t}$$
$$\frac{1600}{2470} = e^{-0.013t}$$
$$\ln\frac{1600}{2470} = \ln e^{-0.013t}$$
$$\ln\frac{1600}{2470} = -0.013t \ln e$$
$$\ln\frac{1600}{2470} = -0.013t(1)$$
$$\frac{\ln\frac{1600}{2470}}{-0.013} = t$$
$$33 \approx t \qquad 2023$$

65) a) Let $t = 2000, y_0 = 15$

$$y = 15e^{-0.000121(2000)}$$
$$y = 15e^{-0.242}$$
$$y \approx 11.78 \text{ g}$$

b) Let $y = 10, y_0 = 15$

$$10 = 15e^{-0.000121t}$$
$$\frac{2}{3} = e^{-0.000121t}$$
$$\ln\frac{2}{3} = \ln e^{-0.000121t}$$
$$\ln\frac{2}{3} = -0.000121t \ln e$$
$$\ln\frac{2}{3} = -0.000121t(1)$$
$$\frac{\ln\frac{2}{3}}{-0.000121} = t$$
$$3351 \text{ yr} \approx t$$

c) Let $y = 7.5, y_0 = 15$

$$7.5 = 15e^{-0.000121t}$$
$$\frac{1}{2} = e^{-0.000121t}$$
$$\ln\frac{1}{2} = \ln e^{-0.000121t}$$
$$\ln\frac{1}{2} = -0.000121t \ln e$$
$$\ln\frac{1}{2} = -0.000121t(1)$$
$$\frac{\ln\frac{1}{2}}{-0.000121} = t$$
$$5728 \text{ yr} \approx t$$

67) a) $y = 0.4e^{-0.086(0)} = 0.4e^{0} = 0.4$ units

b) $y = 0.4e^{-0.086(7)} = 0.4e^{-0.602}$
$= 0.22$ units

Chapter 13 Review

1) yes; $\{(-4,-7),(1,-2),(5,1),(11,6)\}$

3) yes

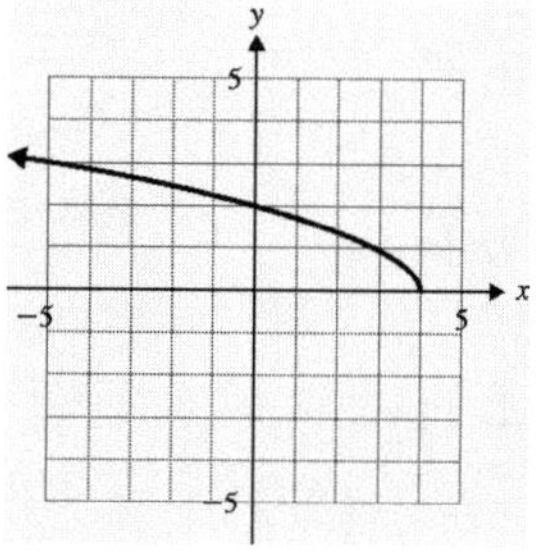

5) $y = x + 4$

$x = y + 4$

$x - 4 = y$

$f^{-1}(x) = x - 4$

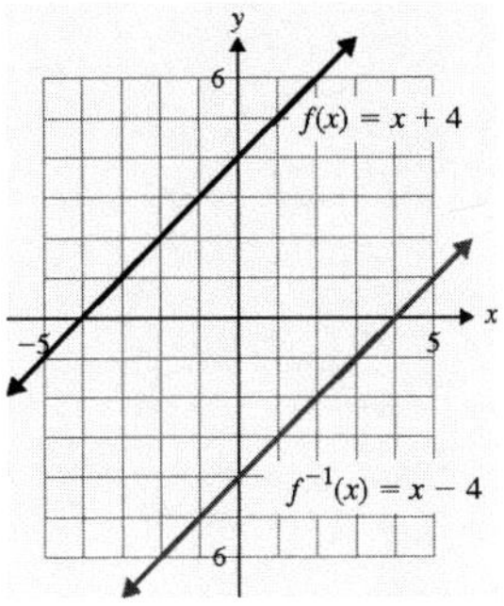

7) $y = \frac{1}{3}x - 1$

$x = \frac{1}{3}y - 1$

$x + 1 = \frac{1}{3}y$

$3x + 3 = y$

$h^{-1}(x) = 3x + 3$

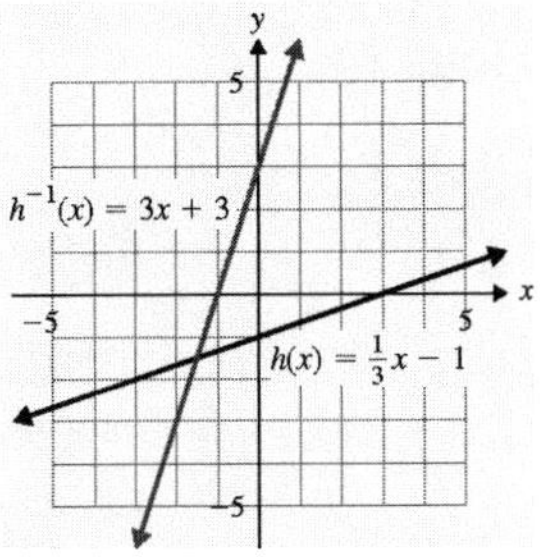

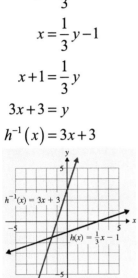

9) a) $f(2) = 6(2) - 1 = 12 - 1 = 11$

b) $f^{-1}(11) = 2$

11)

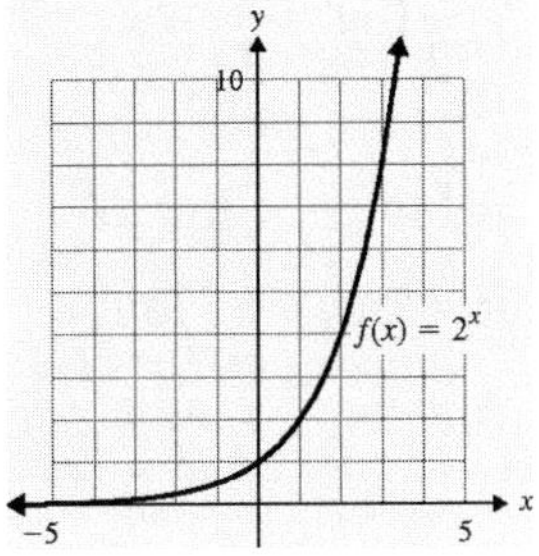

13)

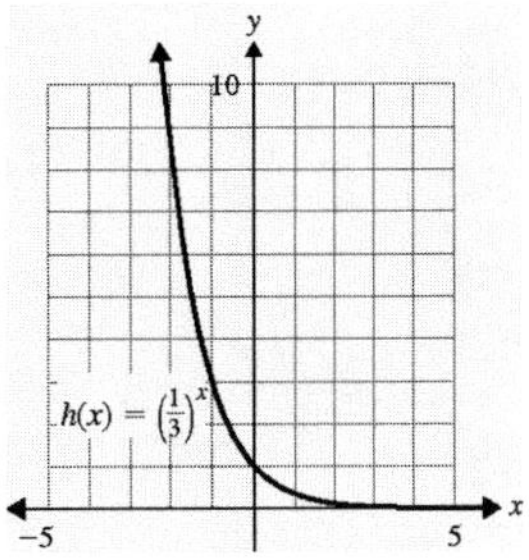

15) $2^c = 64$

$2^c = 2^6$

$c = 6$ $\qquad \{6\}$

17) $16^{3z} = 32^{2z-1}$

$\left(2^4\right)^{3z} = \left(2^5\right)^{2z-1}$

$2^{12z} = 2^{10z-5}$

$12z = 10z - 5$

$2z = -5$

$z = -\frac{5}{2}$ $\qquad \left\{-\frac{5}{2}\right\}$

19) $\left(\frac{3}{2}\right)^{x+4} = \left(\frac{4}{9}\right)^{x-3}$

$\left(\frac{3}{2}\right)^{x+4} = \left[\left(\frac{3}{2}\right)^{-2}\right]^{x-3}$

$\left(\frac{3}{2}\right)^{x+4} = \left(\frac{3}{2}\right)^{-2x+6}$

$x + 4 = -2x + 6$

$3x = 2$

$x = \frac{2}{3}$ $\qquad \left\{\frac{2}{3}\right\}$

21) x must be a positive number.

23) $5^3 = 125$ $\qquad$ 25) $10^2 = 100$

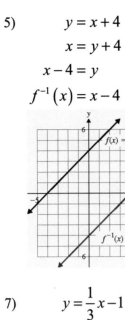

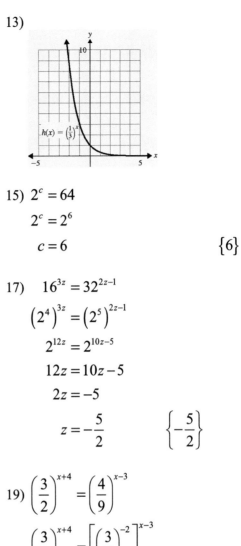

27) $\log_3 81 = 4$ $\qquad$ 29) $\log 1000 = 3$

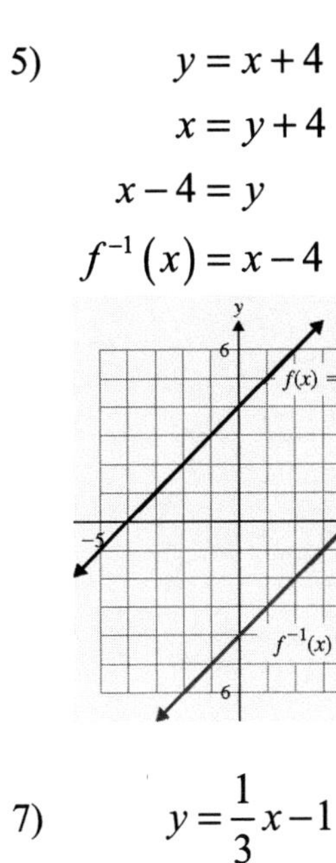

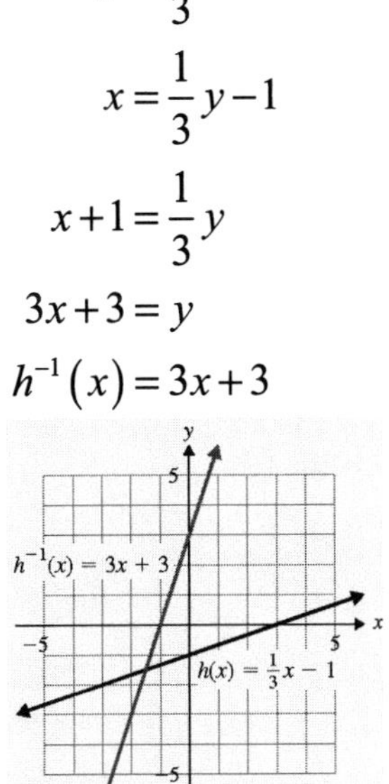

31) $\log_2 x = 3$

$2^3 = x$

$8 = x$ $\{8\}$

33) $\log_{32} 16 = x$

$32^x = 16$

$(2^5)^x = 2^4$

$2^{5x} = 2^4$

$5x = 4$

$x = \frac{4}{5}$ $\left\{\frac{4}{5}\right\}$

35) Let $\log_8 64 = x$

$8^x = 64$

$8^x = 8^2$

$x = 2$

$\log_8 64 = 2$

37) Let $\log 1000 = x$

$10^x = 1000$

$10^x = 10^3$

$x = 3$

$\log 1000 = 3$

39) Let $\log_{1/2} 16 = x$

$\left(\frac{1}{2}\right)^x = 16$

$\left(\frac{1}{2}\right)^x = \left(\frac{1}{2}\right)^{-4}$

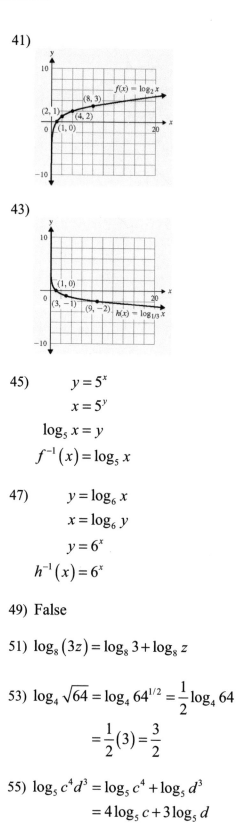

$x = -4$

$\log_{1/2} 16 = -4$

41)

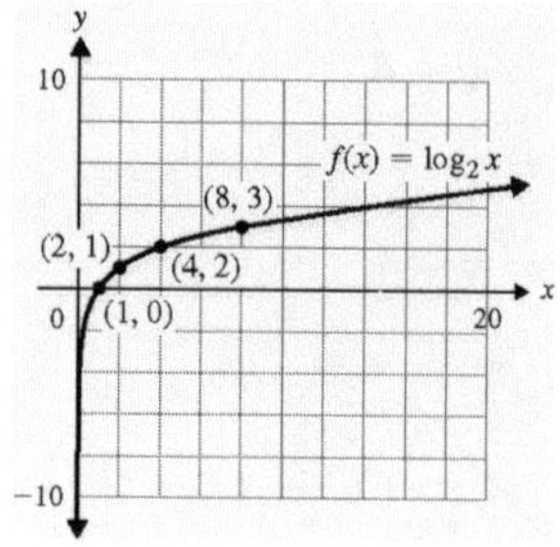

43)

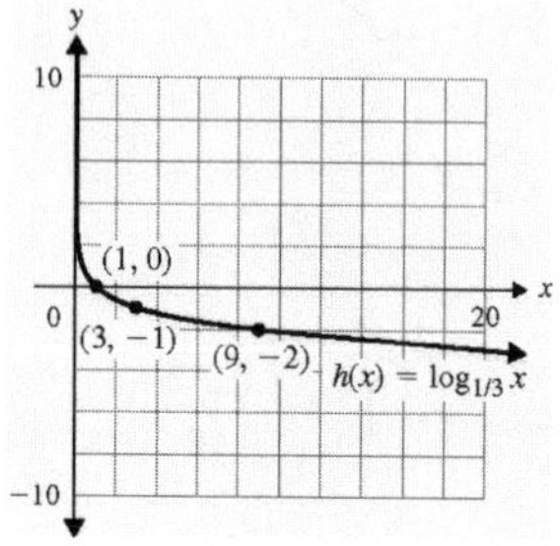

45) $y = 5^x$

$x = 5^y$

$\log_5 x = y$

$f^{-1}(x) = \log_5 x$

47) $y = \log_6 x$

$x = \log_6 y$

$y = 6^x$

$h^{-1}(x) = 6^x$

49) False

51) $\log_8(3z) = \log_8 3 + \log_8 z$

53) $\log_4 \sqrt{64} = \log_4 64^{1/2} = \frac{1}{2}\log_4 64$

$= \frac{1}{2}(3) = \frac{3}{2}$

55) $\log_5 c^4 d^3 = \log_5 c^4 + \log_5 d^3$

$= 4\log_5 c + 3\log_5 d$

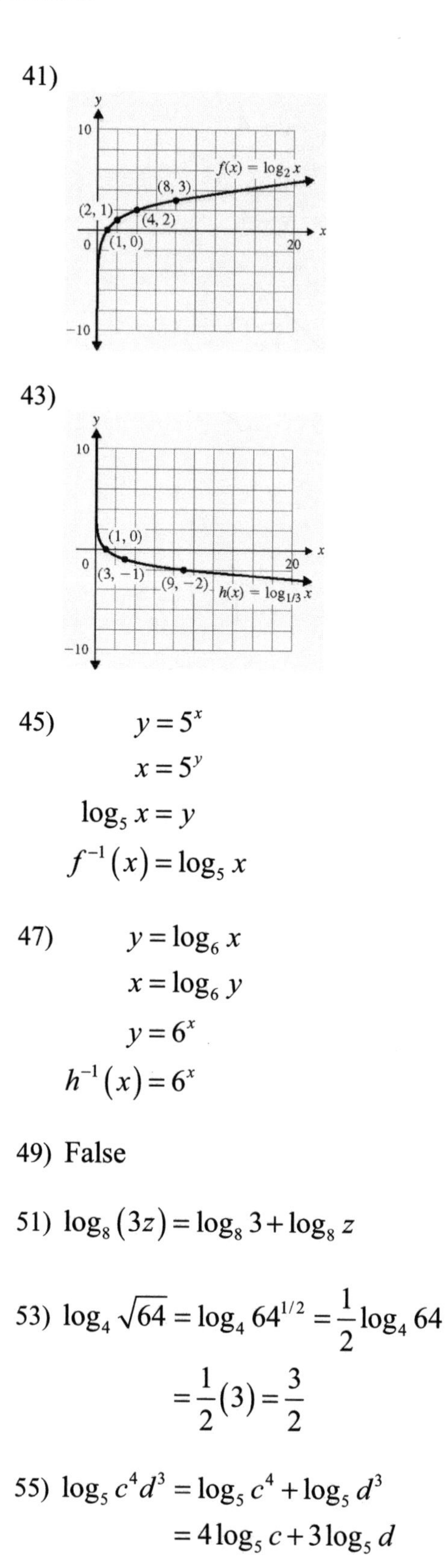

57) $\log_a \frac{xy}{z^3} = \log_a xy - \log_a z^3$

$= \log_a x + \log_a y - 3\log_a z$

59) $\log p(p+8) = \log p + \log(p+8)$

61) $\log c + \log d = \log(cd)$

63) $9\log_2 a + 3\log_2 b = \log_2 a^9 + \log_2 b^3$

$= \log_2 a^9 b^3$

65) $\log_3 5 + 4\log_3 m - 2\log_3 n$

$= \log_3 5 + \log_3 m^4 - \log_3 n^2$

$= \log_3 5m^4 - \log_3 n^2 = \log_3 \frac{5m^4}{n^2}$

67) $3\log_5 c - \log_5 d - 2\log_5 f$

$= \log_5 c^3 - (\log_5 d + \log_5 f^2)$

$= \log_5 c^3 - \log_5 df^2 = \log_5 \frac{c^3}{df^2}$

69) $\log 49 = \log 7^2 = 2\log 7$

$\approx 2(0.8451) \approx 1.6902$

71) $\log \frac{7}{9} = \log 7 - \log 9$

$\approx 0.8451 - 0.9542 \approx -0.1091$

73) e

75) $\log 10 = 1$ since $10^{1} = 10$

77) $\log\sqrt{10} = \frac{1}{2}\log 10 = \frac{1}{2}\cdot 1 = \frac{1}{2}$

79) $\log 0.001 = \log 10^{-3}$

$= -3\log 10 = -3\cdot 1 = -3$

81) $\ln 1 = 0$

83) $\log 8 \approx 0.9031$

85) $\ln 1.75 \approx 0.5596$

87) $\log p = 2$

$10^2 = p$

$100 = p \qquad \{100\}$

89) $\log\left(\frac{1}{2}c\right) = -1$

$10^{-1} = \frac{1}{2}c$

$\frac{1}{10} = \frac{1}{2}c$

$\frac{1}{5} = c \qquad \left\{\frac{1}{5}\right\}$

91) $\log x = 2.1$

$10^{2.1} = x \qquad \{10^{2.1}\};\{125.8925\}$

93) $\ln y = 2$

$e^2 = y \qquad \{e^2\};\{7.3891\}$

95) $\log(4t) = 1.75$

$10^{1.75} = 4t$

$\frac{10^{1.75}}{4} = t$

$\left\{\frac{10^{1.75}}{4}\right\};\{14.0585\}$

97) $\log_4 19 = \frac{\log 19}{\log 4} \approx 2.1240$

99) $\log_{1/2} 38 = \frac{\log 38}{\log \frac{1}{2}} \approx -5.2479$

101) $L(0.1) = 10\log\frac{0.1}{10^{-12}} = 10\log\frac{10^{-1}}{10^{-12}}$
$= 10\log 10^{11} = 110\log 10$
$= 110(1) = 110 \text{ dB}$

103) $A = 2500\left(1+\frac{.06}{4}\right)^{4(5)}$
$= 2500(1.015)^{20} \approx \$3367.14$

105) $A = 9000e^{.062(4)}$
$= 9000e^{.248} \approx \$11,533.14$

107) a) $N(0) = 6000e^{0.0514(0)}$
$= 6000(1)$
$= 6000$ bacteria

b) $N(12) = 6000e^{0.0514(12)}$
$= 6000e^{0.6168}$
$\approx 11,118$ bacteria

109) $2^y = 16$
$2^y = 2^4$
$y = 4 \qquad \{4\}$

111) $9^{4k} = 2$
$\ln 9^{4k} = \ln 2$
$4k\ln 9 = \ln 2$
$k = \frac{\ln 2}{4\ln 9}$
$\left\{\frac{\ln 2}{4\ln 9}\right\}; \{0.0789\}$

113) $6^{2c} = 8^{c-5}$
$\ln 6^{2c} = \ln 8^{c-5}$
$2c\ln 6 = (c-5)\ln 8$
$2c\ln 6 = c\ln 8 - 5\ln 8$
$5\ln 8 = c\ln 8 - 2c\ln 6$
$5\ln 8 = c(\ln 8 - 2\ln 6)$
$\frac{5\ln 8}{\ln 8 - 2\ln 6} = c$
$\left\{\frac{5\ln 8}{\ln 8 - 2\ln 6}\right\}; \{-6.9127\}$

115) $e^{5p} = 8$
$\ln e^{5p} = \ln 8$
$5p\ln e = \ln 8$
$5p = \ln 8$
$p = \frac{\ln 8}{5} \qquad \left\{\frac{\ln 8}{5}\right\}; \{0.4159\}$

117) $\log_3(5w+3) = 2$
$3^2 = 5w+3$
$9 = 5w+3$
$6 = 5w$
$\frac{6}{5} = w \qquad \left\{\frac{6}{5}\right\}$

119) $\log_2 x + \log_2(x+2) = \log_2 24$
$\log_2 x(x+2) = \log_2 24$
$x(x+2) = 24$
$x^2 + 2x = 24$
$x^2 + 2x - 24 = 0$
$(x+6)(x-4) = 0$
$x+6=0$ or $x-4=0$
$x=-6 \qquad x=4$
Only one solution satisfies the original equation. $\{4\}$

121) $\log 5r - \log(r+6) = \log 2$

$$\log \frac{5r}{r+6} = \log 2$$
$$\frac{5r}{r+6} = 2$$
$$5r = 2(r+6)$$
$$5r = 2r + 12$$
$$3r = 12$$
$$r = 4 \qquad \{4\}$$

123) $\log_4 k + \log_4 (k-12) = 3$

$$\log_4 k(k-12) = 3$$
$$4^3 = k(k-12)$$
$$64 = k^2 - 12k$$
$$0 = k^2 - 12k - 64$$
$$0 = (k-16)(k+4)$$

$k - 16 = 0$ or $k + 4 = 0$

$k = 16 \qquad k = -4$

Only one solution satisfies the original equation. $\{16\}$

125) $$10,000 = Pe^{.065(6)}$$
$$10,000 = Pe^{0.39}$$
$$\frac{10,000}{e^{0.39}} = P$$
$$\$6770.57 \approx P$$

127) a) $y = 16,416e^{0.016(5)}$

$= 16,416e^{0.08} \approx 17,777$ people

b) $$23,000 = 16,416e^{0.016t}$$
$$\frac{23,000}{16,416} = e^{0.016t}$$
$$\ln\frac{23,000}{16,416} = \ln e^{0.016t}$$
$$\ln\frac{23,000}{16,416} = 0.016t \ln e$$
$$\frac{\ln\frac{23,000}{16,416}}{0.016} = t$$
$$21 \approx t$$

The year 2011

Chapter 13 Test

1) no

3) yes

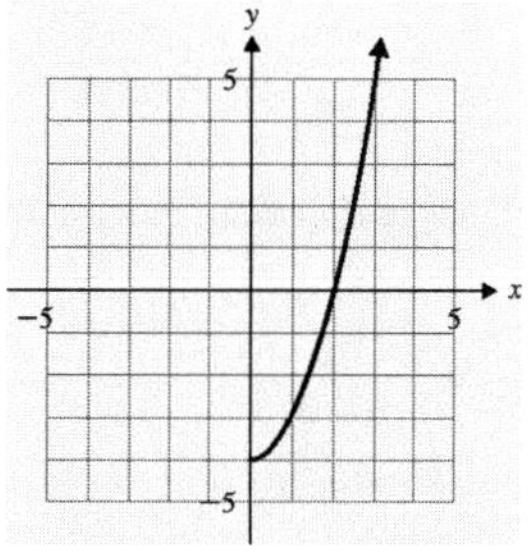

5)

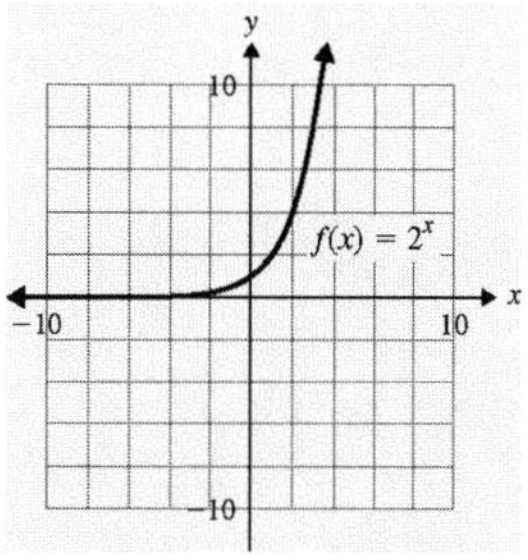

7) a) $(0, \infty)$ b) $(-\infty, \infty)$

9) $\log_3 \frac{1}{9} = -2$

11) $125^{2c} = 25^{c-4}$

$$\left(5^3\right)^{2c} = \left(5^2\right)^{c-4}$$
$$5^{6c} = 5^{2c-8}$$
$$6c = 2c - 8$$
$$4c = -8$$
$$c = -2 \qquad \{-2\}$$

13) $\log(3r+13) = 2$

$$10^2 = 3r + 13$$
$$100 = 3r + 13$$
$$87 = 3r$$
$$29 = r \qquad \{29\}$$

15) a) let $\log_2 16 = x$

$$2^x = 16$$
$$2^x = 2^4$$
$$x = 4$$

b) let $\log_7 \sqrt{7} = x$

$$7^x = \sqrt{7}$$
$$7^x = 7^{1/2}$$
$$x = \frac{1}{2}$$

17) $\log_8(5n) = \log_8 5 + \log_8 n$

19) $2\log x - 3\log(x+1)$

$$= \log x^2 - \log(x+1)^3 = \log\frac{x^2}{(x+1)^3}$$

21) $e^{0.3t} = 5$

$$\ln e^{0.3t} = \ln 5$$
$$0.3t \ln e = \ln 5$$
$$t = \frac{\ln 5}{0.3} \qquad \left\{\frac{\ln 5}{0.3}\right\};\{5.3648\}$$

23) $4^{4a+3} = 9$

$$\ln 4^{4a+3} = \ln 9$$
$$(4a+3)\ln 4 = \ln 9$$
$$4a\ln 4 + 3\ln 4 = \ln 9$$
$$4a\ln 4 = \ln 9 - 3\ln 4$$
$$a = \frac{\ln 9 - 3\ln 4}{4\ln 4}$$
$$\left\{\frac{\ln 9 - 3\ln 4}{4\ln 4}\right\};\{-0.3538\}$$

25) $A = 6000e^{0.074(5)} = 6000e^{0.37}$

$\approx \$8686.41$

Cumulative Review: Chapters 1-13

1) $40 + 8 \div 2 - 3^2 = 40 + 8 \div 2 - 9$

$$= 40 + 4 - 9$$
$$= 44 - 9 = 35$$

3) $(-5a^2)(3a^4) = -15a^{2+4} = -15a^6$

5) $\left(\frac{2c^{10}}{d^3}\right)^{-3} = \left(\frac{d^3}{2c^{10}}\right)^3 = \frac{d^9}{8c^{30}}$

7) $x =$ regular price of watch

$$38.40 = x - .20x$$
$$3840 = 100x - 20x$$
$$3840 = 80x$$
$$48 = x \qquad \$48.00$$

9) $x+4y=-2$

$2(x+4y)=2(-2)$

$2x+8y=-4$

Add the equations.

$$\begin{array}{r} 2x+8y=-4 \\ +\ -2x+3y=15 \\ \hline 11y=11 \\ y=1 \end{array}$$

Substitute $y=1$ into

$x+4y=-2$

$x+4(1)=-2$

$x+4=-2$

$x=-6$ $\quad (-6,1)$

11) $m=\dfrac{-1-5}{2-(-2)}=\dfrac{-6}{4}=-\dfrac{3}{2}$

$(x_1, y_1)=(-2,5)$

$y-y_1=m(x-x_1)$

$y-5=-\dfrac{3}{2}(x-(-2))$

$y-5=-\dfrac{3}{2}(x+2)$

$y-5=-\dfrac{3}{2}x-3$

$y=-\dfrac{3}{2}x+2$

13) $4w^2+w-18=(4w+9)(w-2)$

15) $x^2+14x=-48$

$x^2+14x+48=0$

$(x+6)(x+8)=0$

$x+6=0$ or $x+8=0$

$x=-6$ $\quad x=-8$ $\quad \{-8,-6\}$

17) $\dfrac{9}{y+6}+\dfrac{4}{y-6}=\dfrac{-4}{y^2-36}$

$\dfrac{9}{y+6}+\dfrac{4}{y-6}=\dfrac{-4}{(y+6)(y-6)}$

$9(y-6)+4(y+6)=-4$

$9y-54+4y+24=-4$

$13y-30=-4$

$13y=26$

$y=2$

$\{2\}$

19) $\sqrt{120}=\sqrt{4}\cdot\sqrt{30}=2\sqrt{30}$

21) $\sqrt{\dfrac{36a^5}{a^3}}=\sqrt{36a^2}=6a$

23) $\sqrt{h^2+2h-7}=h-3$

$h^2+2h-7=(h-3)^2$

$h^2+2h-7=h^2-6h+9$

$8h-16=0$

$8h=16$

$h=2$

The solution does not satisfy the original equation. $\varnothing$

25) $k^2-8k+4=0$

$k^2-8k=-4$

$k^2-8k+16=-4+16$

$(k-4)^2=12$

$k-4=\pm\sqrt{12}$

$k=4\pm2\sqrt{3}$

$\{4-2\sqrt{3},4+2\sqrt{3}\}$

27)
$$t^2 = 10t - 41$$
$$t^2 - 10t = -41$$
$$t^2 - 10t + 25 = -41 + 25$$
$$(t-5)^2 = -16$$
$$t - 5 = \pm 4i$$
$$t = 5 \pm 4i$$
$$\{5-4i, 5+4i\}$$

29) Solve: $3x - 2 = 0$
$$3x = 2$$
$$x = \frac{2}{3}$$
$$\left(-\infty, \frac{2}{3}\right) \cup \left(\frac{2}{3}, \infty\right)$$

31)

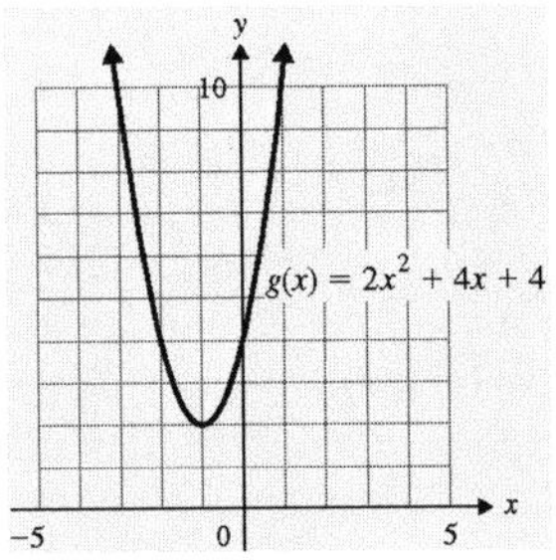

33) $16^y = \frac{1}{64}$
$$(4^2)^y = 4^{-3}$$
$$4^{2y} = 4^{-3}$$
$$2y = -3$$
$$y = -\frac{3}{2} \qquad \left\{-\frac{3}{2}\right\}$$

35) $\log a + 2\log b - 5\log c$
$$= \log a + \log b^2 - \log c^5$$
$$= \log ab^2 - \log c^5 = \log \frac{ab^2}{c^5}$$

Section 14.1: Exercises

1) No; there are values in the domain that give more than one value in the range. The graph fails the vertical line test.

3) Center: $(-2,4)$; $r=\sqrt{9}=3$

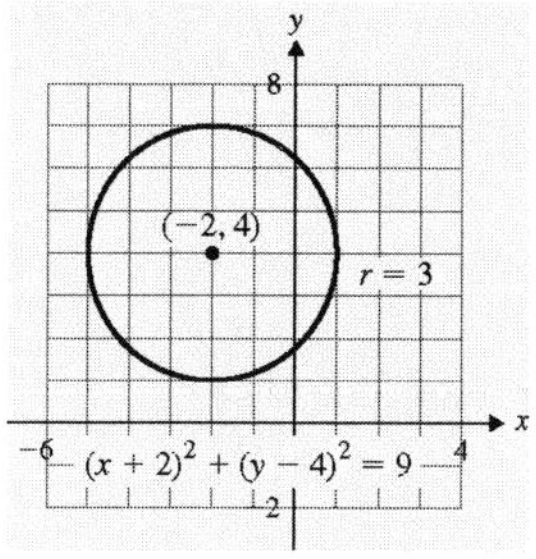

5) Center: $(5,3)$; $r=\sqrt{1}=1$

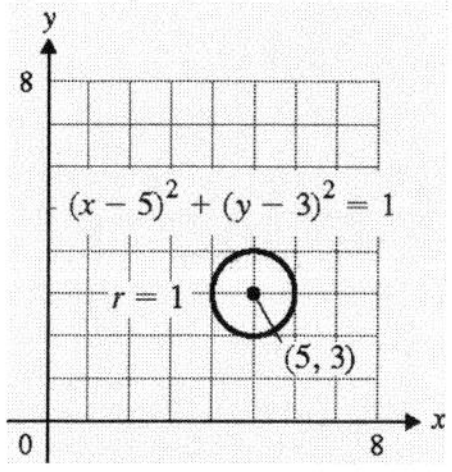

7) Center: $(-3,0)$; $r=\sqrt{4}=2$

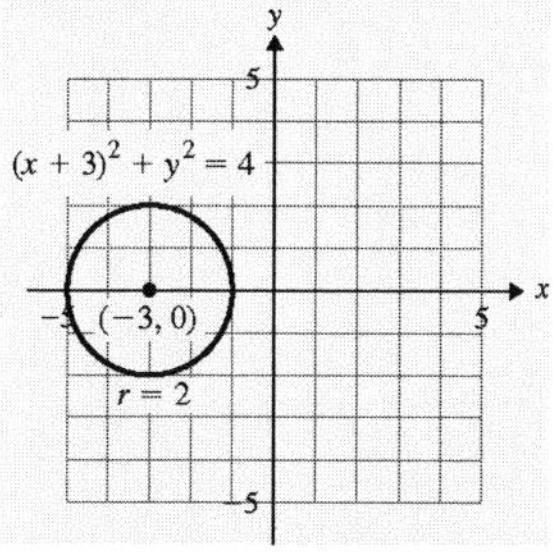

9) Center: $(6,-3)$; $r=\sqrt{16}=4$

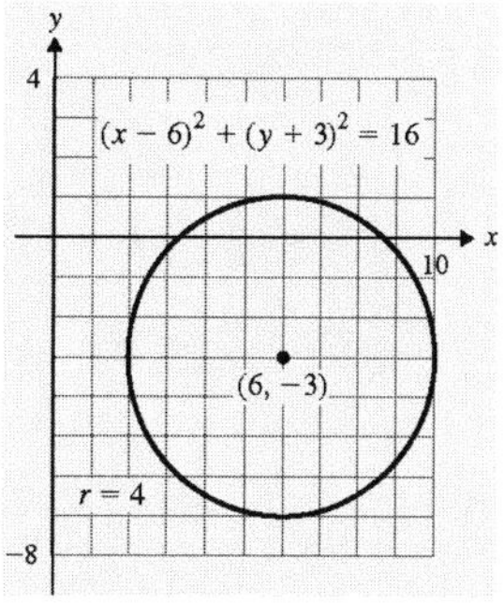

11) Center: $(0,0)$; $r=\sqrt{36}=6$

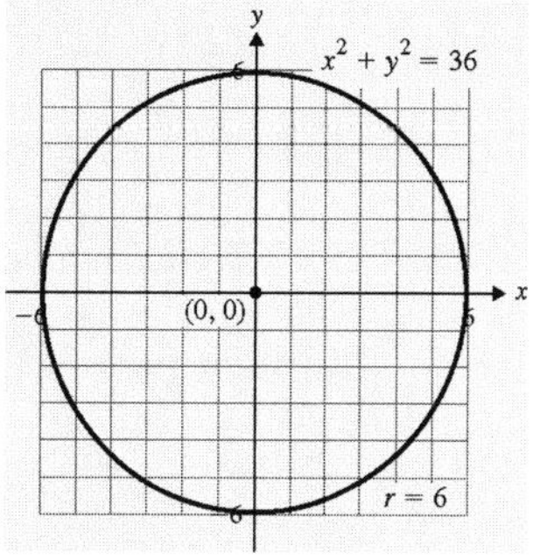

13) Center: $(0,0)$; $r=\sqrt{9}=3$

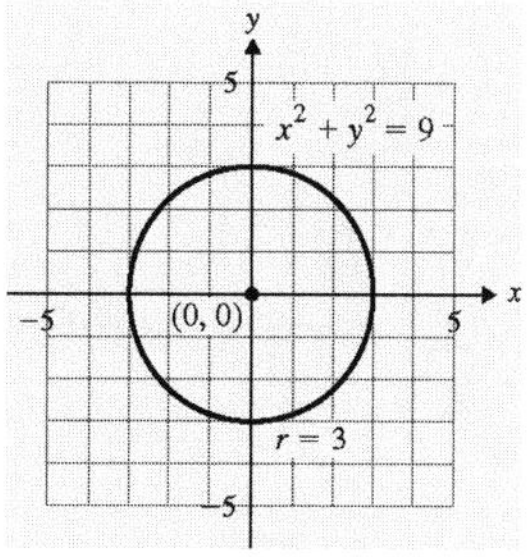

15) Center: $(0,1)$; $r=\sqrt{25}=5$

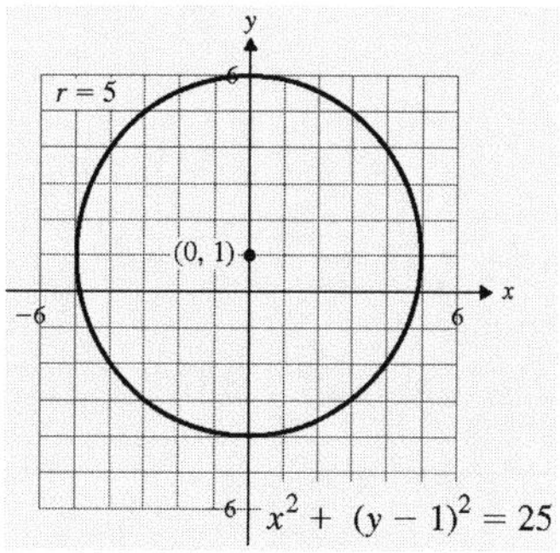

17) $(x-4)^2+(y-1)^2=5^2$

$(x-4)^2+(y-1)^2=25$

19) $(x-(-3))^2+(y-2)^2=1^2$

$(x+3)^2+(y-2)^2=1$

21) $(x-(-1))^2+(y-(-5))^2=\left(\sqrt{3}\right)^2$

$(x+1)^2+(y+5)^2=3$

23) $(x-0)^2+(y-0)^2=\left(\sqrt{10}\right)^2$

$x^2+y^2=10$

25) $(x-6)^2+(y-0)^2=(4)^2$

$(x-6)^2+y^2=16$

27) $(x-0)^2+(y-(-4))^2=\left(2\sqrt{2}\right)^2$

$x^2+(y+4)^2=8$

29) $x^2+y^2+2x+10y+17=0$

$(x^2+2x)+(y^2+10y)=-17$

$(x^2+2x+1)+(y^2+10y+25)=-17+1+25$

$(x+1)^2+(y+5)^2=9$

Center: $(-1,-5)$; $r=\sqrt{9}=3$

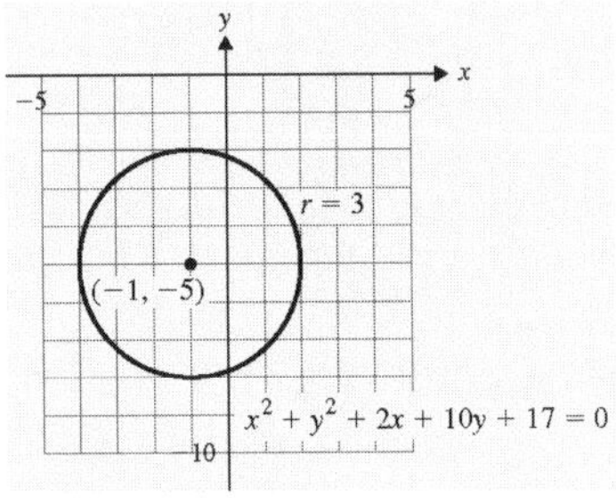

31) $x^2+y^2+8x-2y-8=0$

$(x^2+8x)+(y^2-2y)=8$

$(x^2+8x+16)+(y^2-2y+1)=8+16+1$

$(x+4)^2+(y-1)^2=25$

Center: $(-4,1)$; $r=\sqrt{25}=5$

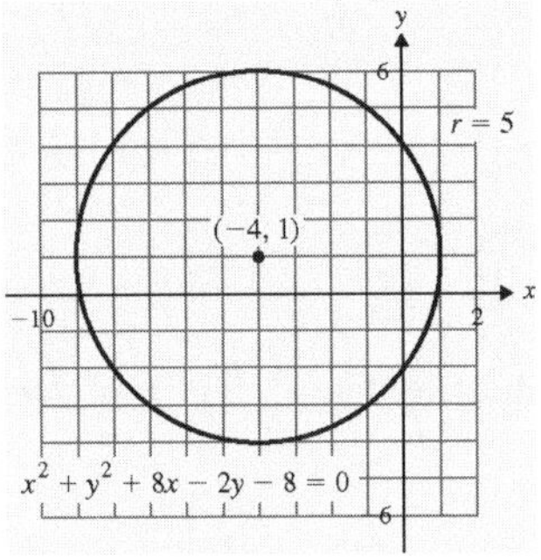

33) $x^2+y^2-10x-14y+73=0$

$(x^2-10x)+(y^2-14y)=-73$

$(x^2-10x+25)+(y^2-14y+49)=-73+25+49$

$(x-5)^2+(y-7)^2=1$

Center: $(5,7)$; $r=\sqrt{1}=1$

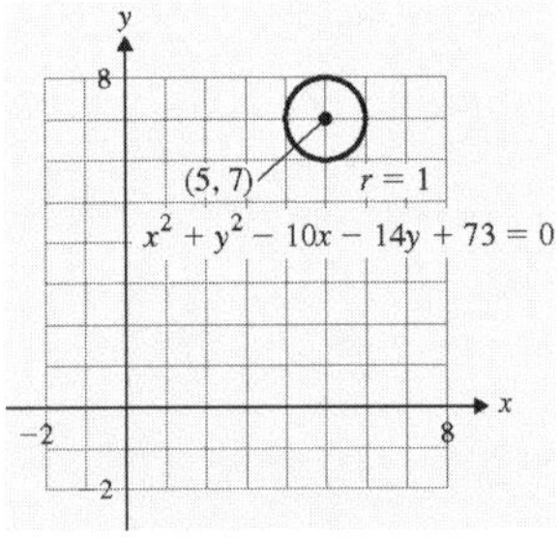

35) $x^2+y^2+6y+5=0$

$x^2+(y^2+6y)=-5$

$x^2+(y^2+6y+9)=-5+9$

$x^2+(y+3)^2=4$

Center: $(0,-3)$; $r=\sqrt{4}=2$

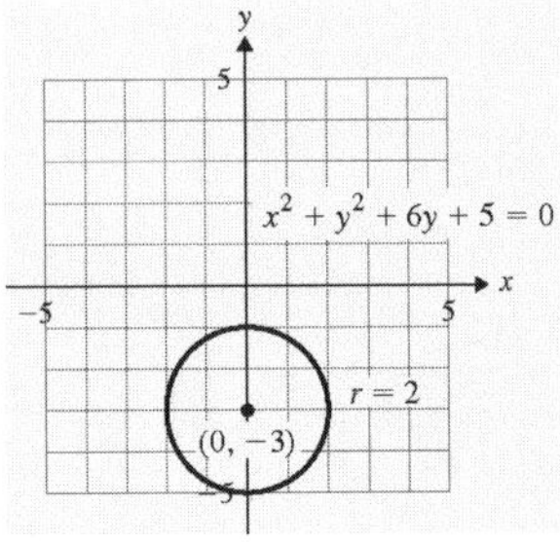

37) $x^2+y^2-4x-1=0$

$$\left(x^2-4x\right)+y^2=1$$

$$\left(x^2-4x+4\right)+y^2=1+4$$

$$\left(x-2\right)^2+y^2=5$$

Center: $(2,0)$; $r=\sqrt{5}$

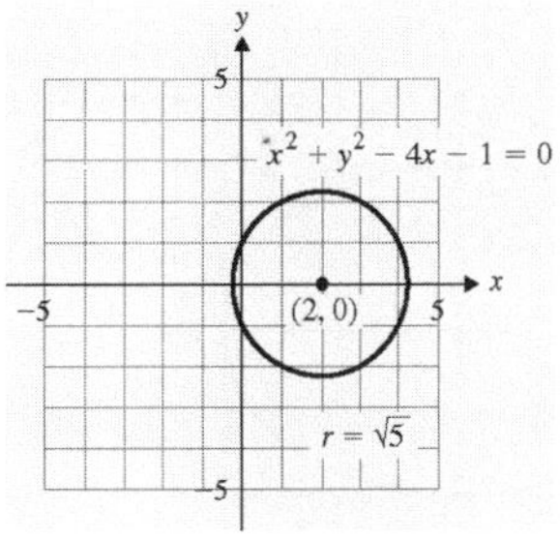

39) $x^2+y^2-8x+8y-4=0$

$$\left(x^2-8x\right)+\left(y^2+8y\right)=4$$

$$\left(x-8x+16\right)^2+\left(y^2+8y+16\right)=4+16+16$$

$$\left(x-4\right)^2+\left(y+4\right)^2=36$$

Center: $(4,-4)$; $r=\sqrt{36}=6$

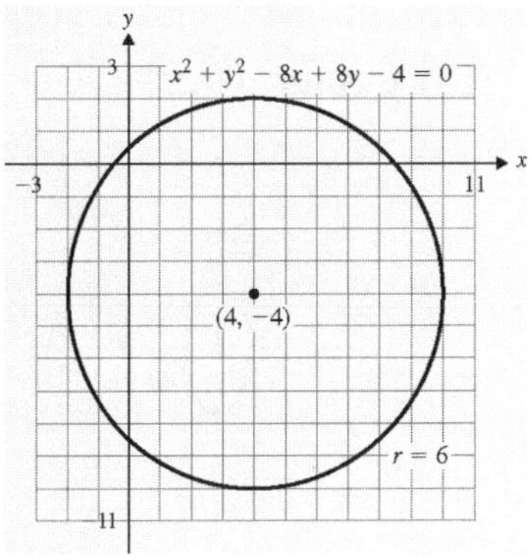

41) $4x^2+4y^2-12x-4y-6=0$

$$x^2+y^2-3x-y-\frac{3}{2}=0$$

$$\left(x^2-3x\right)+\left(y^2-y\right)=\frac{3}{2}$$

$$\left(x^2-3x+\frac{9}{4}\right)+\left(y^2-y+\frac{1}{4}\right)=\frac{3}{2}+\frac{9}{4}+\frac{1}{4}$$

$$\left(x-\frac{3}{2}\right)^2+\left(y-\frac{1}{2}\right)^2=4$$

Center: $\left(\frac{3}{2},\frac{1}{2}\right)$; $r=\sqrt{4}=2$

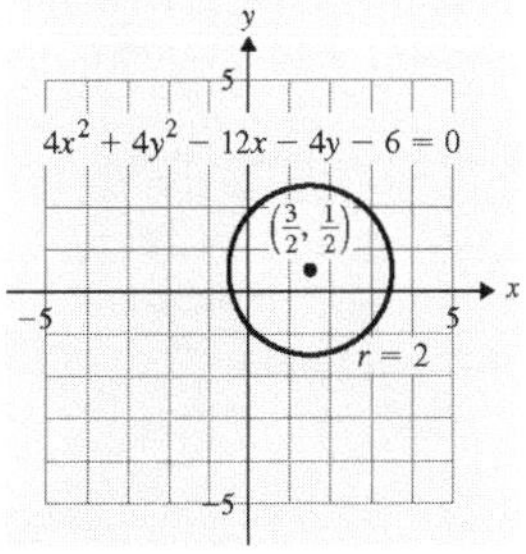

Section 14.2: Exercises

1) ellipse 3) hyperbola

5) hyperbola 7) ellipse

9) Center: $(-2,1)$

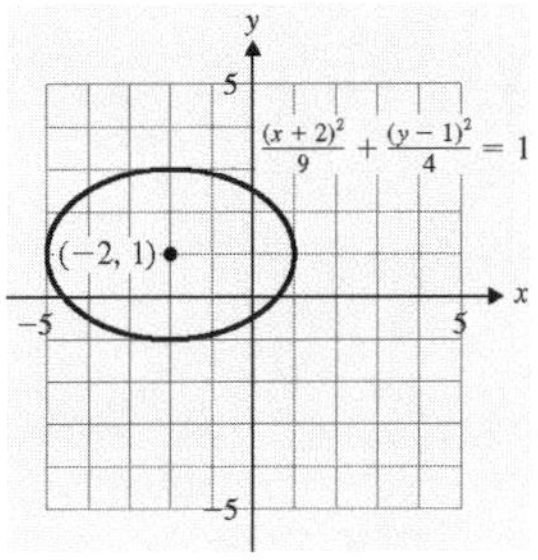

11) Center: $(3,-2)$

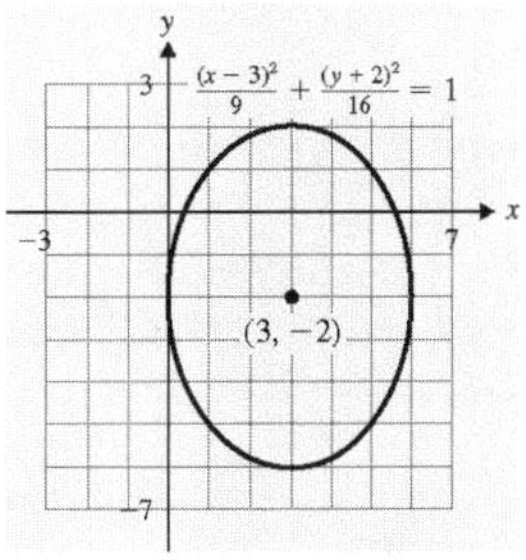

13) Center: $(0,0)$

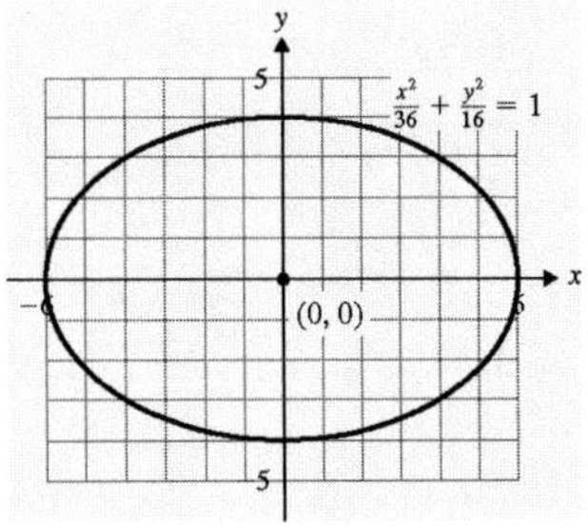

15) Center: $(0,0)$

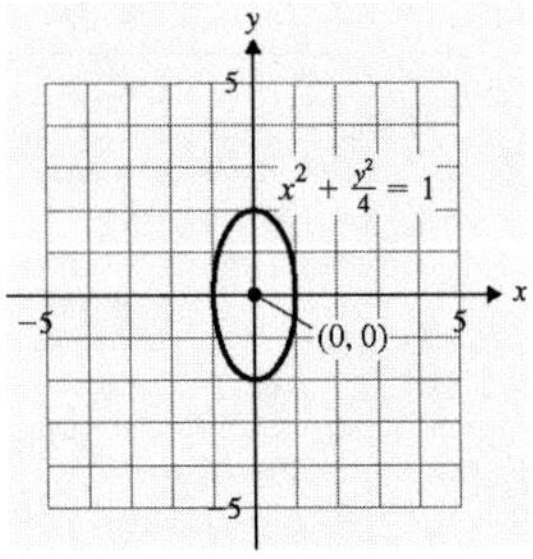

17) Center: $(0,-4)$

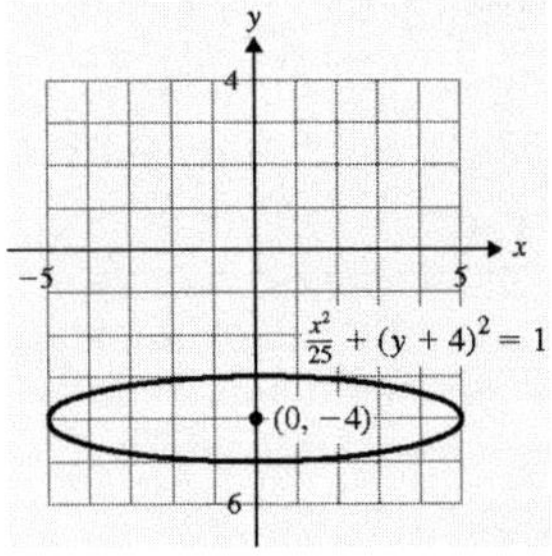

19) Center: $(-1,-3)$

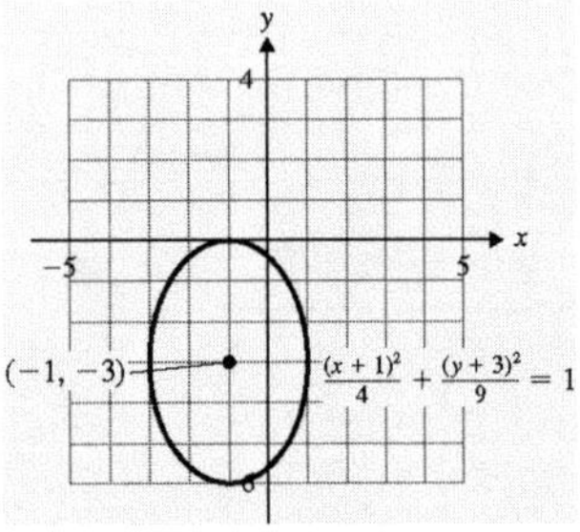

21) $4x^2+9y^2=36$

$$\frac{4x^2}{36}+\frac{9y^2}{36}=\frac{36}{36}$$

$$\frac{x^2}{9}+\frac{y^2}{4}=1$$

Center: $(0,0)$

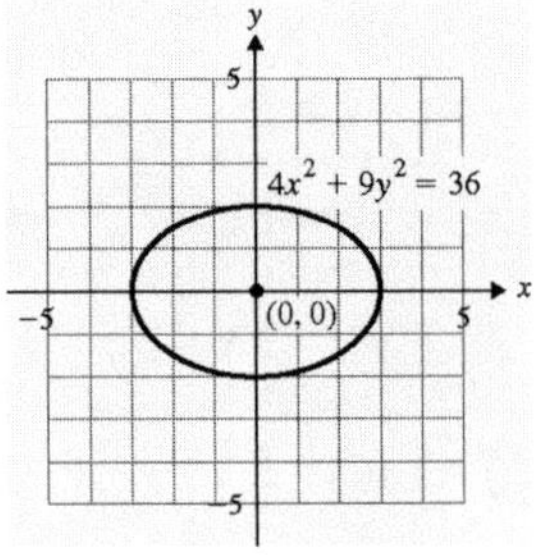

23) $25x^2+y^2=25$

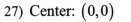

$$\frac{25x^2}{25}+\frac{y^2}{25}=\frac{25}{25}$$

$$x^2+\frac{y^2}{25}=1$$

Center: $(0,0)$

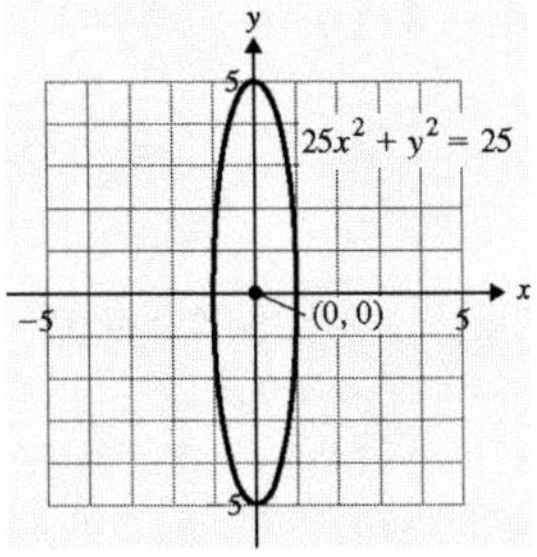

25) Center: $(0,0)$

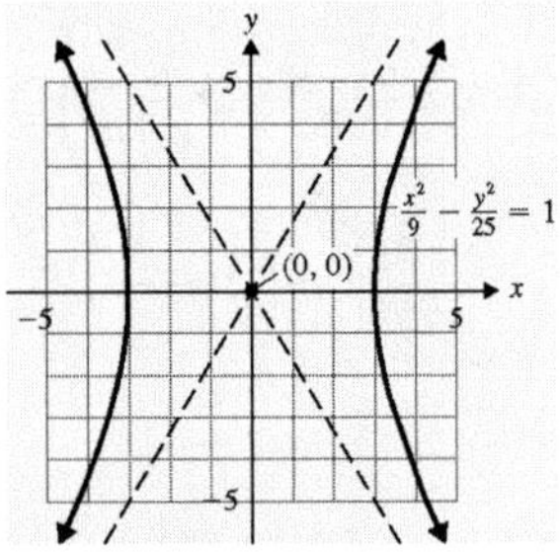

27) Center: $(0,0)$

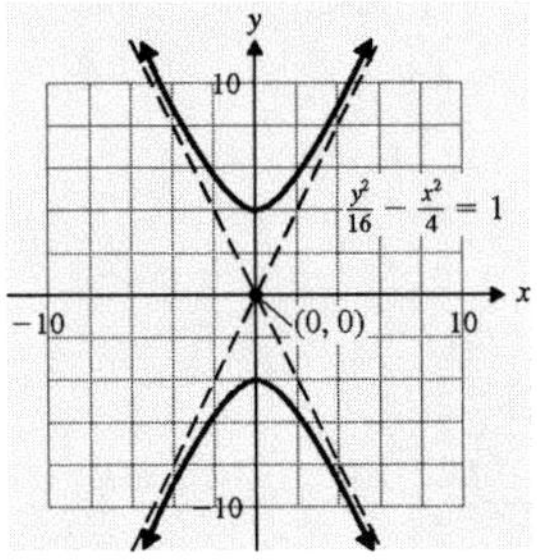

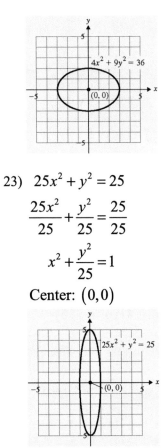

29) Center: $(2,-3)$

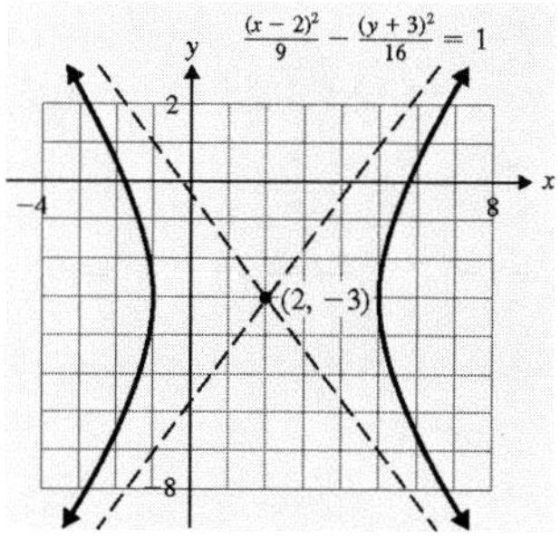

31) Center: $(-4,-1)$

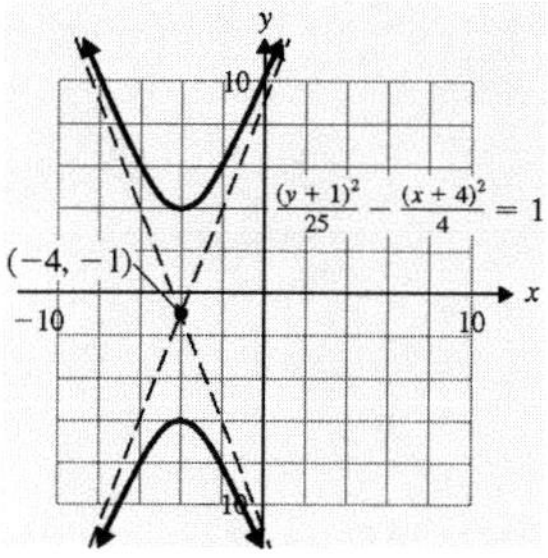

33) Center: $(1,0)$

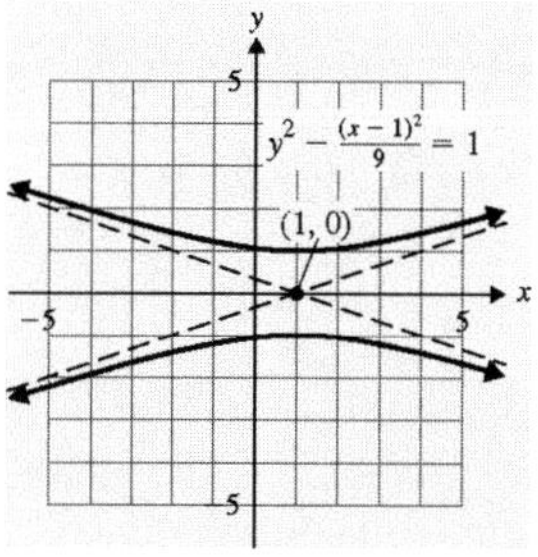

35) Center: $(1,2)$

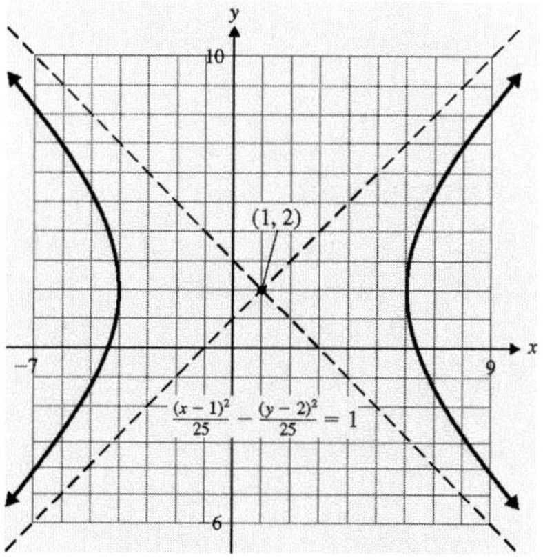

37) $9x^2 - y^2 = 36$

$$\frac{9x^2}{36} - \frac{y^2}{36} = \frac{36}{36}$$

$$\frac{x^2}{4} - \frac{y^2}{36} = 1$$

Center: $(0,0)$

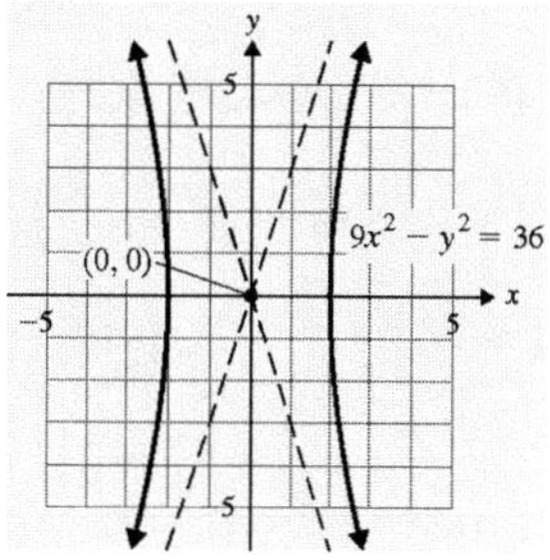

39) Center: $(0,0)$

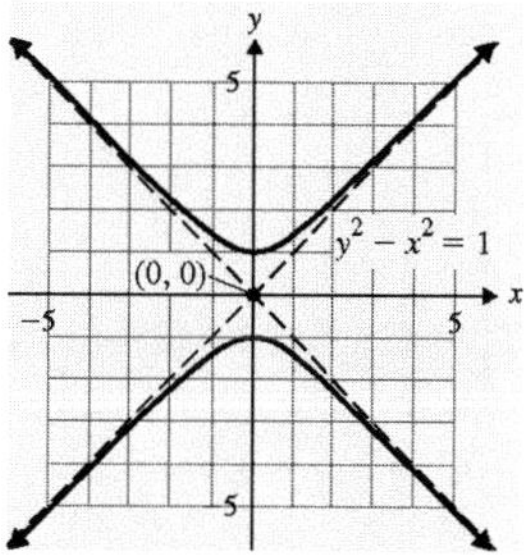

Mid-Chapter Summary

1) parabola

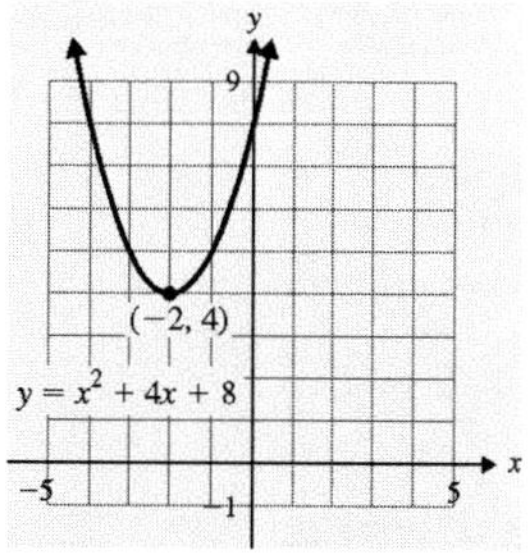

3) hyperbola

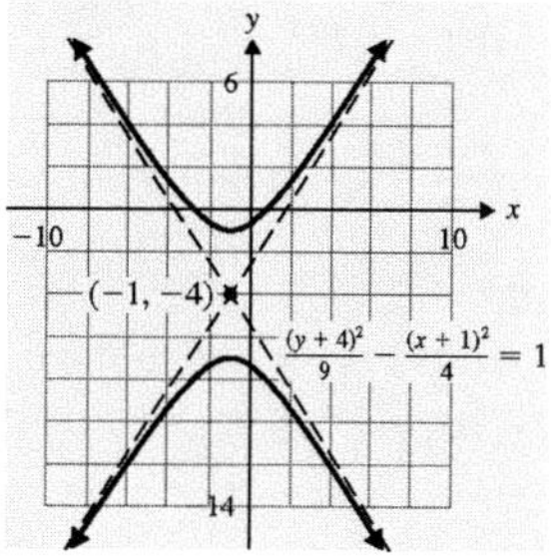

5) $16x^2 + 9y^2 = 144$

$$\frac{16x^2}{144} + \frac{9y^2}{144} = \frac{144}{144}$$

$$\frac{x^2}{9} + \frac{y^2}{16} = 1$$

ellipse

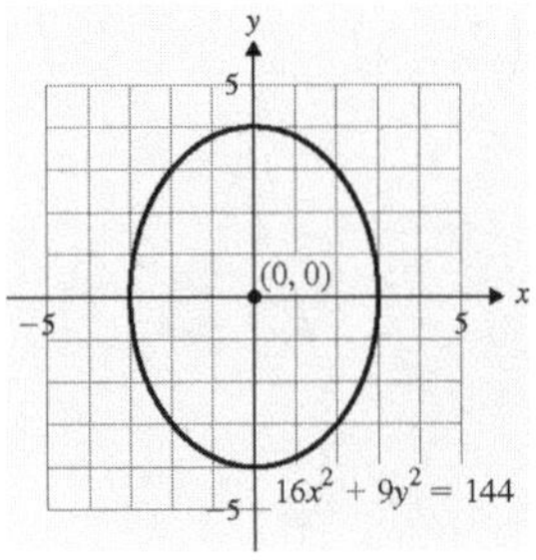

7) $x^2 + y^2 + 8x - 6y - 11 = 0$

$$\left(x^2 + 8x\right) + \left(y^2 - 6y\right) = 11$$

$$\left(x^2 + 8x + 16\right) + \left(y^2 - 6y + 9\right) = 11 + 16 + 9$$

$$(x+4)^2 + (y-3)^2 = 36$$

circle

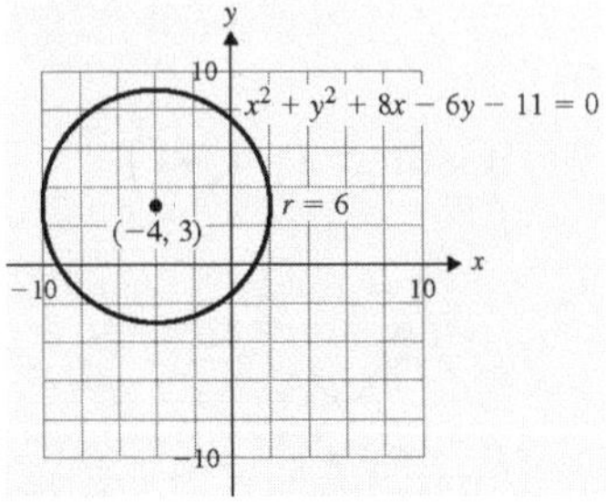

9) ellipse

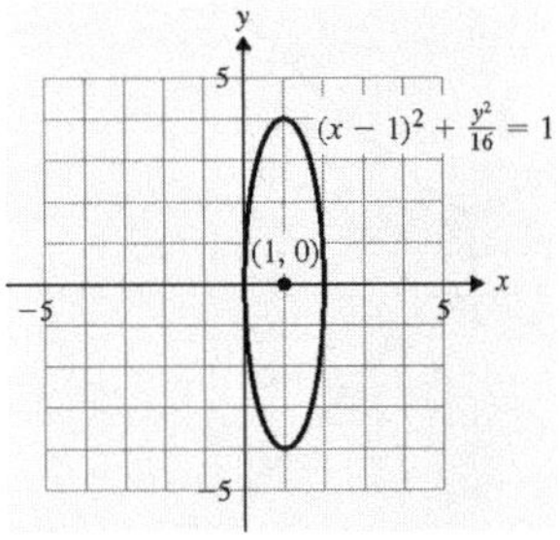

11) parabola

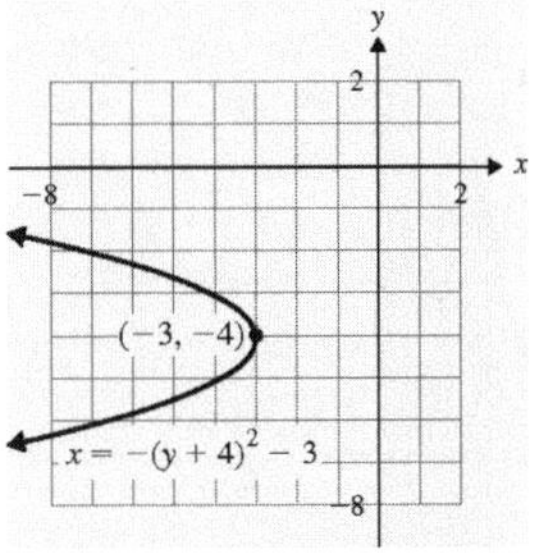

13) $25x^2 - 4y^2 = 100$

$$\frac{25x^2}{100} - \frac{4y^2}{100} = \frac{100}{100}$$

$$\frac{x^2}{4} - \frac{y^2}{25} = 1$$

hyperbola

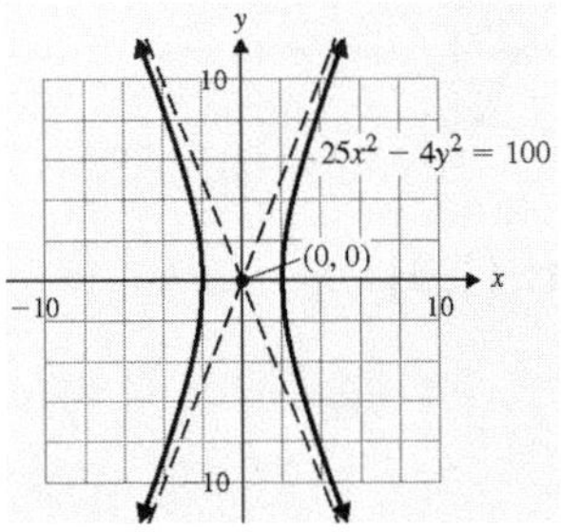

15) circle

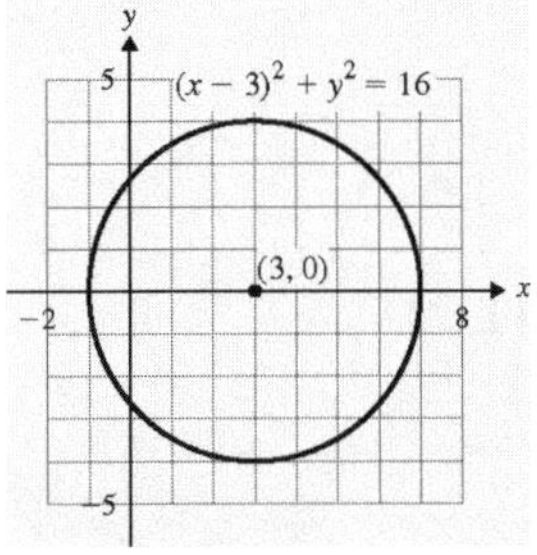

17) parabola

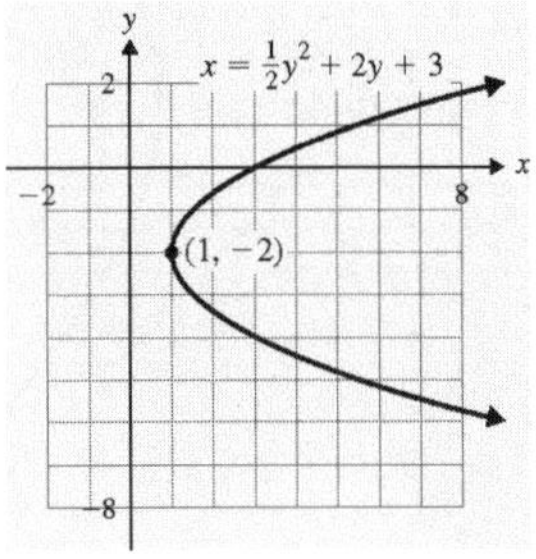

19) $(x-2)^2-(y+1)^2=9$

$$\frac{(x-2)^2}{9}-\frac{(y+1)^2}{9}=1$$

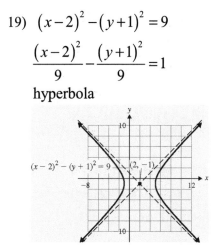

hyperbola

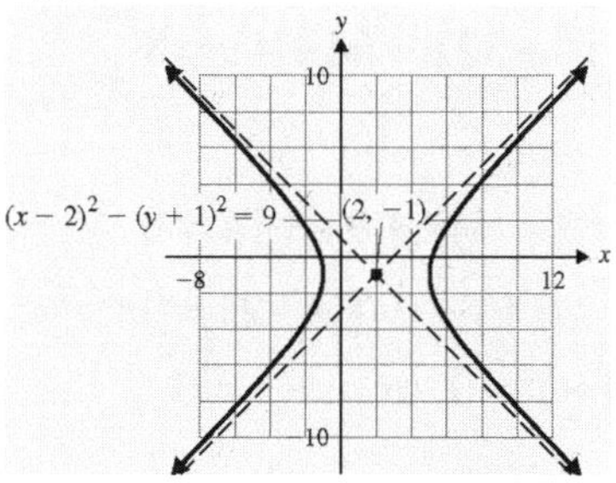

Section 14.3: Exercises

1) a)

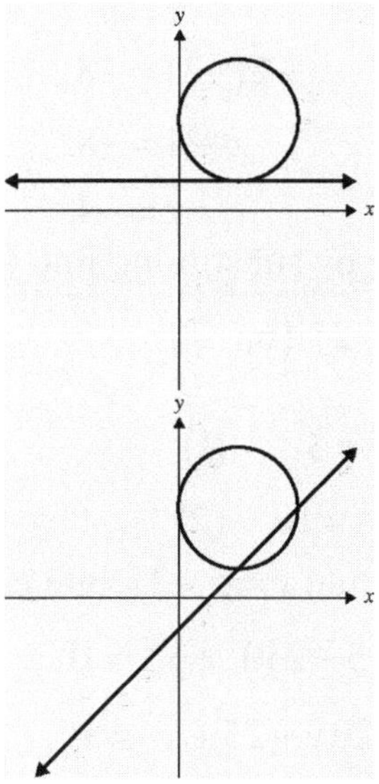

b)

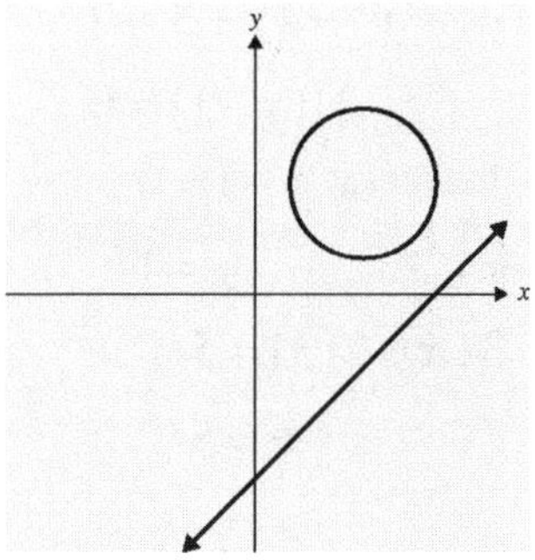

c) $0, 1, \text{or } 2$

3) a)

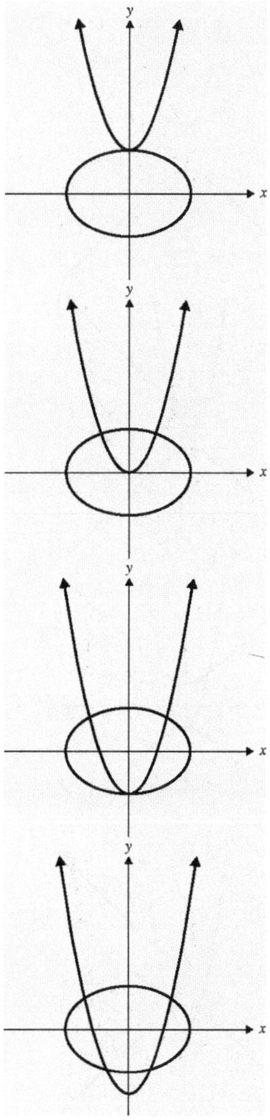

b)

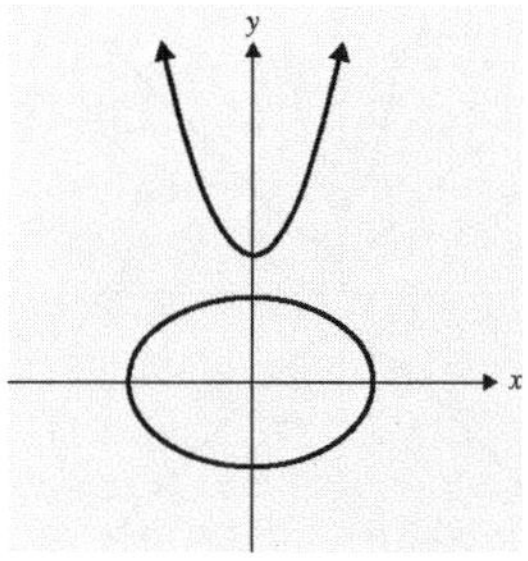

c) $0, 1, 2, 3, \text{or } 4$

5) a)

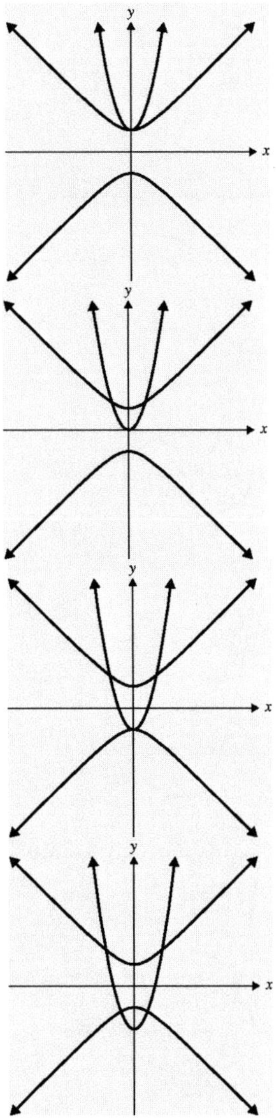

b)

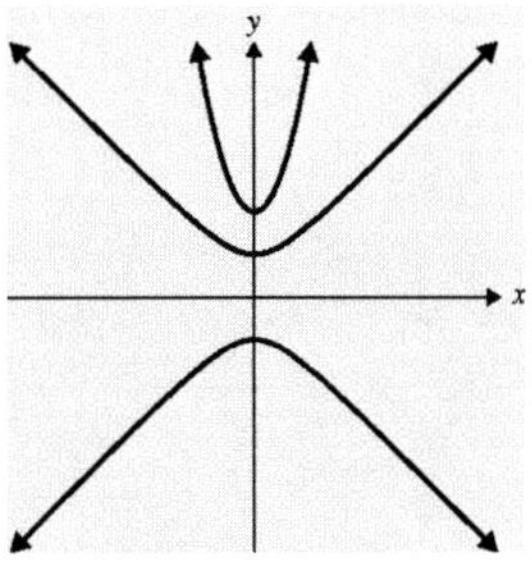

c) 0, 1, 2, 3, or 4

7) $x^2+4y=8$ (1)

$x+2y=-8$ (2)

Substitute $x=-2y-8$ into (1).

$$(-2y-8)^2+4y=8$$
$$4y^2+32y+64+4y=8$$
$$4y^2+36y+56=0$$
$$y^2+9y+14=0$$
$$(y+7)(y+2)=0$$
$$y+7=0 \text{ or } y+2=0$$
$$y=-7 \qquad y=-2$$
$$y=-7: x+2(-7)=-8$$
$$x-14=-8$$
$$x=6$$
$$y=-2: x+2(-2)=-8$$
$$x-4=-8$$
$$x=-4$$

Verify by substituing into (1).

$\{(-4,-2),(6,-7)\}$

9) $x+2y=5$ (1)

$x^2+y^2=10$ (2)

Substitute $x=5-2y$ into (2).

$$(5-2y)^2+y^2=10$$
$$4y^2-20y+25+y^2=10$$
$$5y^2-20y+15=0$$
$$y^2-4y+3=0$$
$$(y-3)(y-1)=0$$
$$y-3=0 \text{ or } y-1=0$$
$$y=3 \qquad y=1$$
$$y=3: x+2(3)=5$$
$$x+6=5$$
$$x=-1$$
$$y=1: x+2(1)=5$$
$$x+2=5$$
$$x=3$$

Verify by substituing into (2).

$\{(-1,3),(3,1)\}$

11) $y = x^2 - 6x + 10 \quad (1)$

$y = 2x - 6 \quad (2)$

Substitute (2) into (1).

$$2x - 6 = x^2 - 6x + 10$$
$$0 = x^2 - 8x + 16$$
$$0 = (x-4)^2$$
$$0 = x - 4$$
$$4 = x$$
$$x = 4: y = 2(4) - 6 = 8 - 6 = 2$$

Verify by substituing into (1).

$\{(4,2)\}$

13) $x^2 + 2y^2 = 11 \quad (1)$

$x^2 - y^2 = 8 \quad (2)$

Solve using elimination.

$$\begin{array}{rl} x^2 + 2y^2 = 11 & (1) \\ + \quad -x^2 + y^2 = -8 & -1\cdot(2) \\ \hline 3y^2 = 3 & \end{array}$$
$$y^2 = 1$$
$$y = \pm\sqrt{1} = \pm 1$$
$$y = 1: x^2 - 1^2 = 8$$
$$x^2 - 1 = 8$$
$$x^2 = 9$$
$$x = \pm 3$$
$$y = -1: x^2 - (-1)^2 = 8$$
$$x^2 - 1 = 8$$
$$x^2 = 9$$
$$x = \pm 3$$

Verify by substituing into (1).

$\{(3,1),(3,-1),(-3,1),(-3,-1)\}$

15) $x^2 + y^2 = 6 \quad (1)$

$2x^2 + 5y^2 = 18 \quad (2)$

Solve using elimination.

$$\begin{array}{rl} -2x^2 - 2y^2 = -12 & -2\cdot(1) \\ + \quad 2x^2 + 5y^2 = 18 & (2) \\ \hline 3y^2 = 6 & \end{array}$$
$$y^2 = 2$$
$$y = \pm\sqrt{2}$$
$$y = \sqrt{2}: x^2 + \left(\sqrt{2}\right)^2 = 6$$
$$x^2 + 2 = 6$$
$$x^2 = 4$$
$$x = \pm 2$$
$$y = -\sqrt{2}: x^2 + \left(-\sqrt{2}\right)^2 = 6$$
$$x^2 + 2 = 6$$
$$x^2 = 4$$
$$x = \pm 2$$

Verify by substituing into (2).

$\left\{\left(2,\sqrt{2}\right),\left(2,-\sqrt{2}\right),\left(-2,\sqrt{2}\right),\left(-2,-\sqrt{2}\right)\right\}$

17) $3x^2 + 4y = -1 \quad (1)$

$x^2 + 3y = -12 \quad (2)$

Solve using elimination.

$$\begin{array}{rl} 3x^2 + 4y = -1 & (1) \\ + \quad -3x^2 - 9y = 36 & -3\cdot(2) \\ \hline -5y = 35 & \end{array}$$
$$y = -7$$
$$y = -7: x^2 + 3(-7) = -12$$
$$x^2 - 21 = -12$$
$$x^2 = 9$$
$$x = \pm 3$$

Verify by substituing into (2).

$\{(3,-7),(-3,-7)\}$

19) $y = 6x^2 - 1$ (1)

$2x^2 + 5y = -5$ (2)

Substitute (1) into (2).

$$2x^2 + 5(6x^2 - 1) = -5$$
$$2x^2 + 30x^2 - 5 = -5$$
$$32x^2 = 0$$
$$x^2 = 0$$
$$x = 0$$

$x = 0: y = 6(0)^2 - 1 = -1$

Verify by substituing into (2).

$\{(0, -1)\}$

21) $x^2 + y^2 = 4$ (1)

$-2x^2 + 3y = 6$ (2)

Solve using elimination.

$2x^2 + 2y^2 = 8$ $\quad 2 \cdot (1)$

$+ \quad -2x^2 + 3y = 6$ $\quad (2)$

$$2y^2 + 3y = 14$$
$$2y^2 + 3y - 14 = 0$$
$$(2y+7)(y-2) = 0$$

$2y + 7 = 0$ or $y - 2 = 0$

$2y = -7$ $\quad y = 2$

$y = -\frac{7}{2}$

$$y = -\frac{7}{2}: x^2 + \left(-\frac{7}{2}\right)^2 = 4$$
$$x^2 + \frac{49}{4} = 4$$
$$x^2 = -\frac{33}{4}$$
$$x = \pm\frac{\sqrt{33}}{2}i$$

does not give real number solutions.

$$y = 2: x^2 + 2^2 = 4$$
$$x^2 + 4 = 4$$
$$x^2 = 0$$
$$x = 0$$

Verify by substituing into (2).

$\{(0, 2)\}$

23) $x^2 + y^2 = 3$ (1)

$x + y = 4$ (2)

Substitute $y = 4 - x$ into (1).

$$x^2 + (4 - x)^2 = 3$$
$$x^2 + 16 - 8x + x^2 = 3$$
$$2x^2 - 8x + 13 = 0$$
$$x = \frac{-(-8) \pm \sqrt{(-8)^2 - 4(2)(13)}}{2(2)}$$
$$= \frac{8 \pm \sqrt{64 - 104}}{4} = \frac{8 \pm \sqrt{-40}}{4}$$

No real number values for x. $\varnothing$

25) $x = \sqrt{y}$ (1)

$x^2 - 9y^2 = 9$ (2)

Substitute (1) into (2).

$$(\sqrt{y})^2 - 9y^2 = 9$$
$$y - 9y^2 = 9$$
$$0 = 9y^2 - y + 9$$
$$y = \frac{-(-1) \pm \sqrt{(-1)^2 - 4(9)(9)}}{2(9)}$$
$$= \frac{1 \pm \sqrt{1 - 324}}{18} = \frac{1 \pm \sqrt{-323}}{18}$$

No real number values for y. $\varnothing$

27) $9x^2+y^2=9 \quad (1)$

$x^2+y^2=5 \quad (2)$

Solve using elimination.

$$\begin{array}{rr} 9x^2+y^2=9 & (1) \\ + \quad -x^2-y^2=-5 & -1\cdot(2) \\ \hline 8x^2=4 & \end{array}$$

$$x^2=\frac{1}{2}$$

$$x=\pm\sqrt{\frac{1}{2}}=\pm\frac{\sqrt{2}}{2}$$

$$x=\frac{\sqrt{2}}{2}: \left(\frac{\sqrt{2}}{2}\right)^2+y^2=5$$

$$\frac{2}{4}+y^2=5$$

$$y^2=\frac{18}{4}$$

$$y=\pm\frac{3\sqrt{2}}{2}$$

$$x=-\frac{\sqrt{2}}{2}: \left(-\frac{\sqrt{2}}{2}\right)^2+y^2=5$$

$$\frac{2}{4}+y^2=5$$

$$y^2=\frac{18}{4}$$

$$y=\pm\frac{3\sqrt{2}}{2}$$

Verify by substituing into (1).

$$\left\{\left(\frac{\sqrt{2}}{2},\frac{3\sqrt{2}}{2}\right),\left(\frac{\sqrt{2}}{2},-\frac{3\sqrt{2}}{2}\right),\left(-\frac{\sqrt{2}}{2},\frac{3\sqrt{2}}{2}\right),\left(-\frac{\sqrt{2}}{2},-\frac{3\sqrt{2}}{2}\right)\right\}$$

29) $y=-x^2-2 \quad (1)$

$x^2+y^2=4 \quad (2)$

Substitute (1) into (2).

$$x^2+\left(-x^2-2\right)^2=4$$

$$x^2+x^4+4x^2+4=4$$

$$x^4+5x^2=0$$

$$x^2\left(x^2+5\right)=0$$

$$x^2+5=0 \text{ or } x^2=0$$

$$x^2=-5 \qquad \boxed{x=0}$$

$$x=\pm i\sqrt{5}$$

$x=\pm i\sqrt{5}$ does not give real number solutions.

$x=0$: $y=-(0)^2-2=-2$

Verify by substituing into (2).

$$\{(0,-2)\}$$

31) $x=$ one number

$y=$ other number

$xy=40$

$x+y=13$

Substitute $y=13-x$ into (1).

$$x(13-x)=40$$

$$13x-x^2=40$$

$$0=x^2-13x+40$$

$$0=(x-8)(x-5)$$

$$x-8=0 \text{ or } x-5=0$$

$$x=8 \qquad x=5$$

$x=8$: $y=13-8=5$

$x=5$: $y=13-5=8$

The numbers are 8 and 5.

33) $l =$ length of screen

$w =$ width of screen

$2l + 2w = 38 \quad (1)$

$lw = 88 \quad (2)$

Solve (1) for l.

$2l = 38 - 2w$

$l = 19 - w \quad (3)$

Substitute (3) into (2).

$(19 - w)w = 88$

$19w - w^2 = 88$

$0 = w^2 - 19w + 88$

$0 = (w - 11)(w - 8)$

$w - 11 = 0$ or $w - 8 = 0$

$w = 11 \qquad w = 8$

$w = 11: l = 19 - 11 = 8$

$w = 8: l = 19 - 8 = 11$

The dimensions of the screen are 8 in×11 in.

35) $15x^2 = 6x^2 + 33x + 12$

$9x^2 - 33x - 12 = 0$

$3x^2 - 11x - 4 = 0$

$(3x + 1)(x - 4) = 0$

$3x + 1 = 0$ or $x - 4 = 0$

$3x = -1 \qquad \boxed{x = 4}$

$x = -\frac{1}{3}$

$x = 4: y = 15(4)^2 = 240$

The break-even point is 4000 basketballs and $240.

Section 14.4: Exercises

1) The endpoints are included when the inequality symbol is ≤ or ≥. The endpoints are not included when the symbol is < or >.

3) a) $[-5,1]$ b) $(-\infty,-5)\cup(1,\infty)$

5) a) $[-1,3]$ b) $(-\infty,-1)\cup(3,\infty)$

7) $x^2 + 6x - 7 \ge 0$

$(x + 7)(x - 1) \ge 0$

$(x + 7)(x - 1) = 0$

$x + 7 = 0$ or $x - 1 = 0$

$x = -7 \qquad x = 1$

Interval A: $(-\infty,-7)$ Positive

Interval B: $(-7,1)$ Negative

Interval C: $(1,\infty)$ Positive

$(-\infty,-7]\cup[1,\infty)$

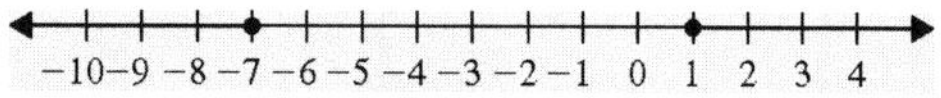

9) $c^2 + 5c < 36$

$c^2 + 5c - 36 < 0$

$(c + 9)(c - 4) < 0$

$(c + 9)(c - 4) = 0$

$c + 9 = 0$ or $c - 4 = 0$

$c = -9 \qquad c = 4$

Interval A: $(-\infty,-9)$ Positive

Interval B: $(-9,4)$ Negative

Interval C: $(4,\infty)$ Positive

$(-9,4)$

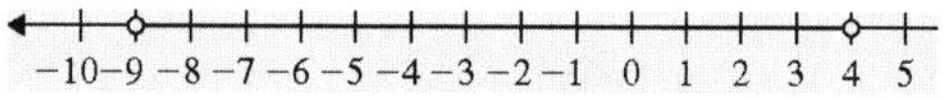

11) $r^2 - 13r > -42$

$r^2 - 13r + 42 > 0$

$(r - 6)(r - 7) > 0$

$(r - 6)(r - 7) = 0$

$r - 6 = 0$ or $r - 7 = 0$

$r = 6 \qquad r = 7$

Interval A: $(-\infty,6)$ Positive
Interval B: $(6,7)$ Negative
Interval C: $(7,\infty)$ Positive

$(-\infty,6)\cup(7,\infty)$

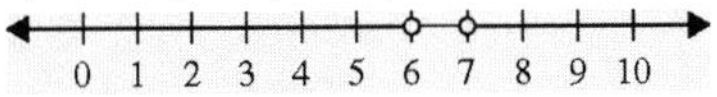

13) $3z^2+14z-24\le 0$

$(3z-4)(z+6)\le 0$

$(3z-4)(z+6)=0$

$3z-4=0$ or $z+6=0$

$3z=4 \qquad z=-6$

$z=\dfrac{4}{3}$

Interval A: $(-\infty,-6)$ Positive
Interval B: $\left(-6,\dfrac{4}{3}\right)$ Negative
Interval C: $\left(\dfrac{4}{3},\infty\right)$ Positive

$\left[-6,\dfrac{4}{3}\right]$

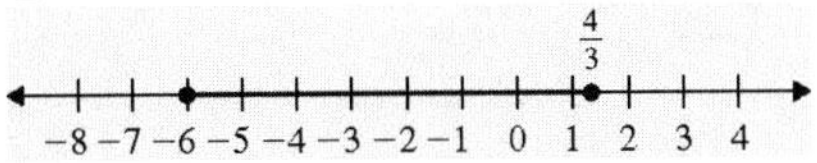

15) $7p^2-4>12p$

$7p^2-12p-4>0$

$(7p+2)(p-2)>0$

$(7p+2)(p-2)=0$

$7p+2=0$ or $p-2=0$

$7p=-2 \qquad p=2$

$p=-\dfrac{2}{7}$

Interval A: $\left(-\infty,-\dfrac{2}{7}\right)$ Positive
Interval B: $\left(-\dfrac{2}{7},2\right)$ Negative
Interval C: $(2,\infty)$ Positive

$\left(-\infty,-\dfrac{2}{7}\right)\cup(2,\infty)$

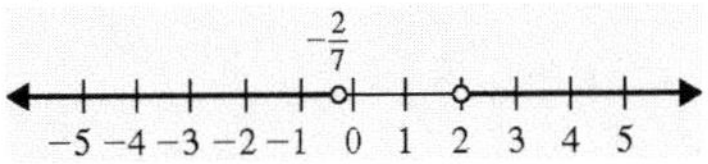

17) $b^2-9b>0$

$b(b-9)>0$

$b(b-9)=0$

$b-9=0$ or $b=0$

$b=9$

Interval A: $(-\infty,0)$ Positive
Interval B: $(0,9)$ Negative
Interval C: $(9,\infty)$ Positive

$(-\infty,0)\cup(9,\infty)$

19) $4y^2\le -5y$

$4y^2+5y\le 0$

$y(4y+5)\le 0$

$y(4y+5)=0$

$4y+5=0$ or $y=0$

$4y=-5$

$y=-\dfrac{5}{4}$

Interval A: $\left(-\infty,-\dfrac{5}{4}\right)$ Positive
Interval B: $\left(-\dfrac{5}{4},0\right)$ Negative
Interval C: $(0,\infty)$ Positive

$\left[-\dfrac{5}{4},0\right]$

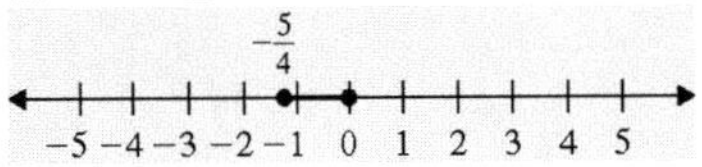

21) $m^2 - 64 < 0$

$(m+8)(m-8) < 0$

$(m+8)(m-8) = 0$

$m+8=0$ or $m+8=0$

$m=-8 \qquad m=8$

Interval A: $(-\infty,-8)$ Positive

Interval B: $(-8,8)$ Negative

Interval C: $(8,\infty)$ Positive

$(-8,8)$

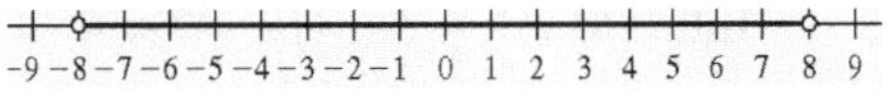

23) $121 - h^2 \le 0$

$(11+h)(11-h) \le 0$

$(11+h)(11-h) = 0$

$11+h=0$ or $11-h=0$

$h=-11 \qquad h=11$

Interval A: $(-\infty,-11)$ Negative

Interval B: $(-11,11)$ Positive

Interval C: $(11,\infty)$ Negative

$(-\infty,-11]\cup[11,\infty)$

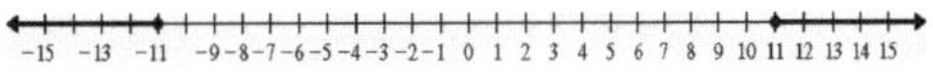

25) $144 \ge 9s^2$

$16 \ge s^2$

$16 = s^2$

$\pm 4 = s$

Interval A: $(-\infty,-4)$ Negative

Interval B: $(-4,4)$ Positive

Interval C: $(4,\infty)$ Negative

$[-4,4]$

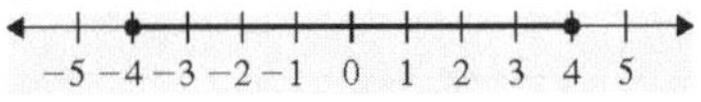

27) $(-\infty,\infty)$ 29) $(-\infty,\infty)$

31) $\varnothing$ 31) $\varnothing$

35) $(r+2)(r-5)(r-1) \le 0$

$(r+2)(r-5)(r-1) = 0$

$r+2=0$ or $r-5=0$ or $r-1=0$

$r=-2 \qquad r=5 \qquad r=1$

Interval A: $(-\infty,-2)$ Negative

Interval B: $(-2, 1)$ Positive

Interval C: $(1, 5)$ Negative

Interval D: $(5,\infty)$ Positive

$(-\infty,-2]\cup[1, 5]$

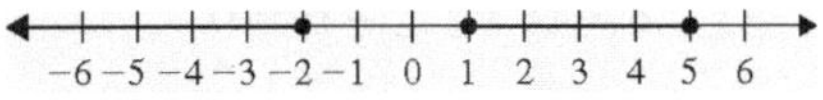

37) $(j-7)(j-5)(j+9) \ge 0$

$(j-7)(j-5)(j+9) = 0$

$j-7=0$ or $j-5=0$ or $j+9=0$

$j=7 \qquad j=5 \qquad j=-9$

Interval A: $(-\infty,-9)$ Negative

Interval B: $(-9,5)$ Positive

Interval C: $(5,7)$ Negative

Interval D: $(7,\infty)$ Positive

$[-9,5]\cup[7,\infty)$

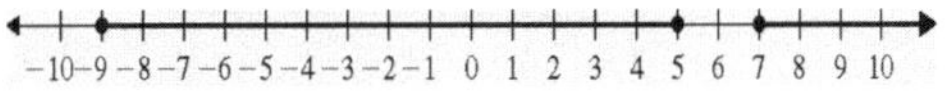

39) $(6c+1)(c+7)(4c-3) < 0$

$(6c+1)(c+7)(4c-3) = 0$

$6c+1=0$ or $c+7=0$ or $4c-3=0$

$6c=-1 \qquad c=-7 \qquad 4c=3$

$c=-\frac{1}{6} \qquad c=\frac{3}{4}$

Interval A: $(-\infty,-7)$ Negative

Interval B: $\left(-7,-\frac{1}{6}\right)$ Positive

Interval C: $\left(-\frac{1}{6},\frac{3}{4}\right)$ Negative

Interval D: $\left(\frac{3}{4},\infty\right)$ Positive

$(-\infty,-7)\cup\left(-\frac{1}{6},\frac{3}{4}\right)$

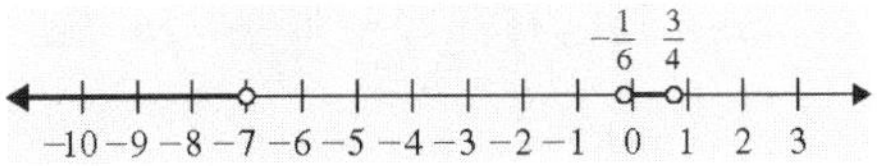

41) $\dfrac{7}{p+6}>0$

Set the numerator and denominator equal to zero and solve for p.

$7\neq 0 \quad p+6=0$

$p=-6$

Interval A: $(-\infty,-6)$ Negative

Interval B: $(-6,\infty)$ Positive

$(-6,\infty)$

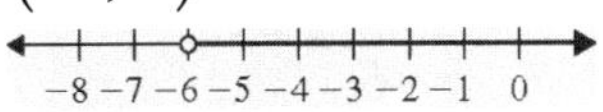

43) $\dfrac{5}{z+3}\leq 0$

Set the numerator and denominator equal to zero and solve for z.

$5\neq 0 \quad z+3=0$

$z=-3$

Interval A: $(-\infty,-3)$ Negative

Interval B: $(-3,\infty)$ Positive

$v\neq -3$ because it makes the denominator equal to zero.

$(-\infty,-3)$

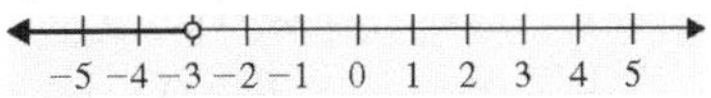

45) $\dfrac{x-4}{x-3}>0$

Set the numerator and denominator equal to zero and solve for x.

$x-4=0 \qquad x-3=0$

$x=4 \qquad x=3$

Interval A: $(-\infty,3)$ Positive

Interval B: $(3,\,4)$ Negative

Interval C: $(4,\infty)$ Positive

$(-\infty,\,3)\cup(4,\infty)$

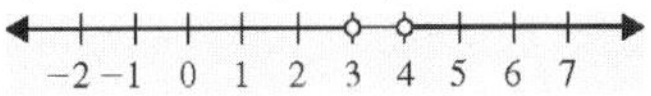

47) $\dfrac{h-9}{3h+1}\leq 0$

Set the numerator and denominator equal to zero and solve for h.

$h-9=0 \qquad 3h+1=0$

$h=9 \qquad 3h=-1$

$h=-\frac{1}{3}$

Interval A: $\left(-\infty,-\frac{1}{3}\right)$ Positive

Interval B: $\left(-\frac{1}{3},\,9\right)$ Negative

Interval C: $(9,\infty)$ Positive

$h\neq -\frac{1}{3}$ because it makes the denominator equal to zero.

$\left(-\frac{1}{3},\,9\right]$

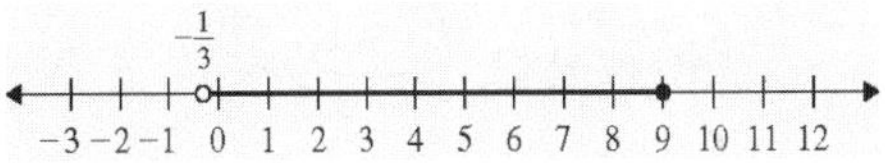

49) $\dfrac{k}{k+3}\leq 0$

Set the numerator and denominator equal to zero and solve for k.

$k=0 \quad k+3=0$

$k=-3$

Interval A: $(-\infty,-3)$ Positive

Interval B: $(-3,\,0)$ Negative

Interval C: $(0,\infty)$ Positive

$k\neq -3$ because it makes the denominator equal to zero.

$(-3,\,0]$

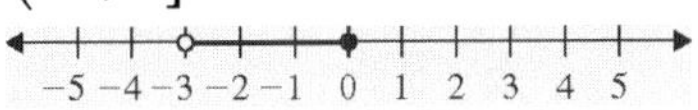

51)
$$\frac{7}{t+6}<3$$
$$\frac{7}{t+6}-3<0$$
$$\frac{7}{t+6}-\frac{3(t+6)}{t+6}<0$$
$$\frac{7-3(t+6)}{t+6}<0$$
$$\frac{7-3t-18}{t+6}<0$$
$$\frac{-3t-11}{t+6}<0$$

Set the numerator and denominator equal to zero and solve for t.

$-3t-11=0 \qquad t+6=0$

$-3t=11 \qquad t=-6$

$t=-\frac{11}{3}$

Interval A: $(-\infty,-6)$ Negative

Interval B: $\left(-6,-\frac{11}{3}\right)$ Positive

Interval C: $\left(-\frac{11}{3},\infty\right)$ Negative

$(-\infty,-6)\cup\left(-\frac{11}{3},\infty\right)$

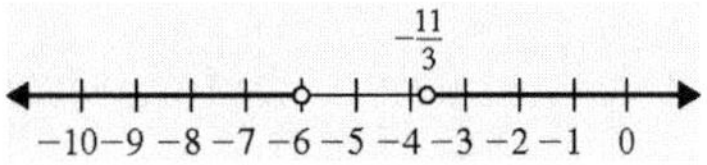

53)
$$\frac{3}{a+7}\geq 1$$
$$\frac{3}{a+7}-1\geq 0$$
$$\frac{3}{a+7}-\frac{(a+7)}{a+7}\geq 0$$
$$\frac{3-(a+7)}{a+7}\geq 0$$
$$\frac{-a-4}{a+7}\geq 0$$

Set the numerator and denominator equal to zero and solve for a.

$-a-4=0 \qquad a+7=0$

$-a=4 \qquad a=-7$

$a=-4$

Interval A: $(-\infty,-7)$ Negative

Interval B: $(-7,-4)$ Positive

Interval C: $(-4,\infty)$ Negative

$a\neq -7$ because it makes the denominator equal to zero.

$(-7,-4]$

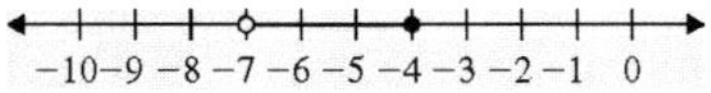

55)
$$\frac{2y}{y-6}\leq -3$$
$$\frac{2y}{y-6}+3\leq 0$$
$$\frac{2y}{y-6}+\frac{3(y-6)}{y-6}\leq 0$$
$$\frac{2y+3(y-6)}{y-6}\leq 0$$
$$\frac{2y+3y-18}{y-6}\leq 0$$
$$\frac{5y-18}{y-6}\leq 0$$

Set the numerator and denominator equal to zero and solve for y.

$5y-18=0 \qquad y-6=0$

$5y=18 \qquad y=6$

$y=\frac{18}{5}$

Interval A: $\left(-\infty,\frac{18}{5}\right)$ Positive

Interval B: $\left(\frac{18}{5},6\right)$ Negative

Interval C: $(6,\infty)$ Positive

$y \neq 6$ because it makes the denominator equal to zero.

$\left[\frac{18}{5}, 6\right)$

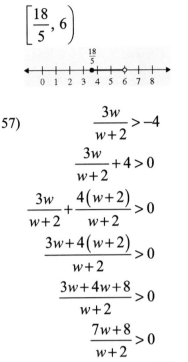

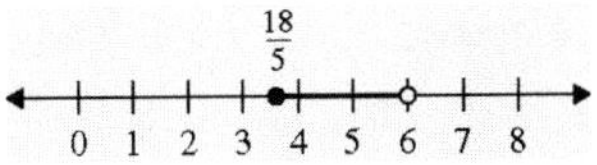

57)
$$\frac{3w}{w+2} > -4$$
$$\frac{3w}{w+2} + 4 > 0$$
$$\frac{3w}{w+2} + \frac{4(w+2)}{w+2} > 0$$
$$\frac{3w+4(w+2)}{w+2} > 0$$
$$\frac{3w+4w+8}{w+2} > 0$$
$$\frac{7w+8}{w+2} > 0$$

Set the numerator and denominator equal to zero and solve for w.

$7w+8=0 \qquad w+2=0$

$7w=-8 \qquad w=-2$

$w=-\frac{8}{7}$

Interval A: $(-\infty,-2)$ Positive

Interval B: $\left(-2,-\frac{8}{7}\right)$ Negative

Interval C: $\left(-\frac{8}{7},\infty\right)$ Positive

$(-\infty,-2)\cup\left(-\frac{8}{7},\infty\right)$

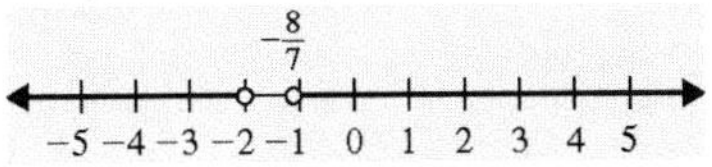

59) $\frac{(6d+1)^2}{d-2} \leq 0$

Numerator will always be positive. Set the denominator equal to zero and solve for d.

$d-2=0$

$d=2$

Interval A: $(-\infty,2)$ Negative

Interval B: $(2,\infty)$ Positive

$d \neq 2$ because it makes the denominator equal to zero.

$(-\infty,2)$

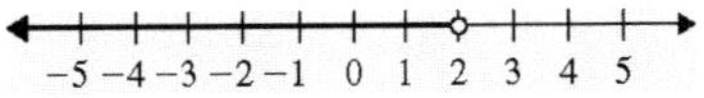

61) $\frac{(4t-3)^2}{t-5} > 0$

Numerator will always be positive. Set the denominator equal to zero and solve for t.

$t-5=0$

$t=5$

Interval A: $(-\infty,5)$ Negative

Interval B: $(5,\infty)$ Positive

$(5,\infty)$

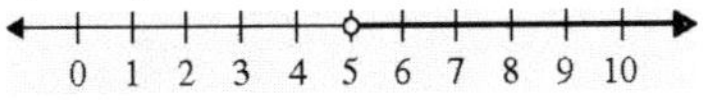

63) $\frac{n+6}{n^2+4} < 0$

Denominator will always be positive. Set the numerator equal to zero and solve for n.

$n+6=0$

$n=-6$

Interval A: $(-\infty,-6)$ Negative

Interval B: $(-6,\infty)$ Positive

$(-\infty,-6)$

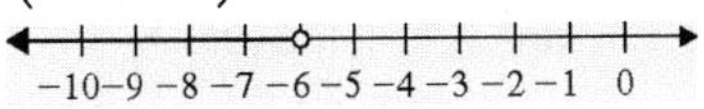

65) $\dfrac{m+1}{m^2+3} \geq 0$

Denominator will always be positive. Set the numerator equal to zero and solve for *m*.

$m+1=0$

$m=-1$

Interval A: $(-\infty,-1)$ Negative

Interval B: $(-1,\infty)$ Positive

$[-1,\infty)$

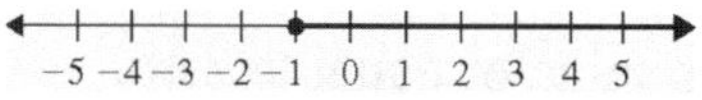

67) $\dfrac{s^2+2}{s-4} \leq 0$

Numerator will always be positive. Set the denominator equal to zero and solve for *s*.

$s-4=0$

$s=4$

Interval A: $(-\infty,4)$ Negative

Interval B: $(4,\infty)$ Positive

$s \neq 4$ because it makes the denominator equal to zero.

$(-\infty,4)$

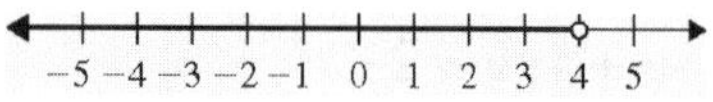

Chapter 14 Review

1) Center: $(-3,5)$; $r=\sqrt{36}=6$

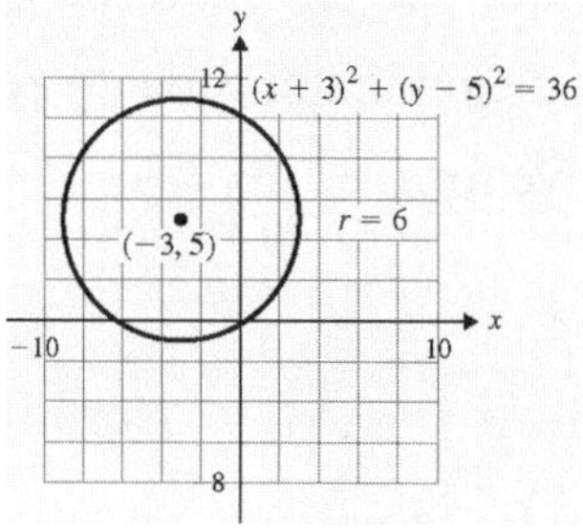

3) $x^2+y^2-10x-4y+13=0$

$(x^2-10x)+(y^2-4y)=-13$

$(x^2-10x+25)+(y^2-4y+4)=-13+25+4$

$(x-5)^2+(y-2)^2=16$

Center: $(5,2)$; $r=\sqrt{16}=4$

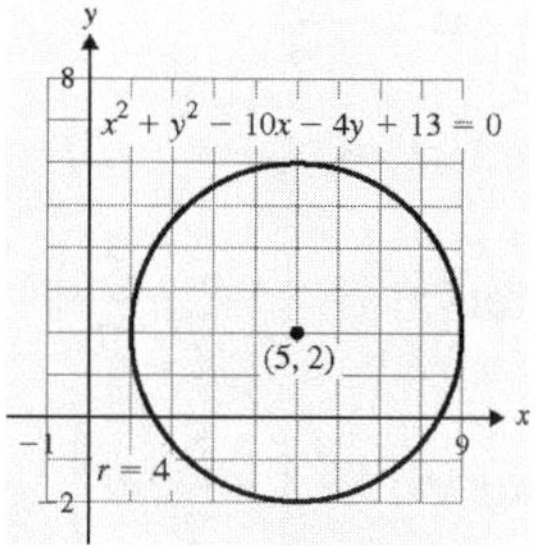

5) $(x-3)^2+(y-0)^2=4^2$

$(x-3)^2+y^2=16$

7) The equation of an ellipse contains the sum of two squares, but the equation of hyperbola contains the difference two squares.

9) Center: $(0,0)$

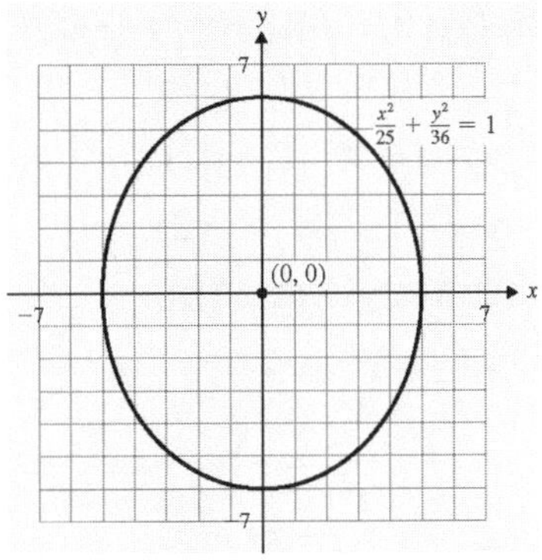

11) Center: $(4,2)$

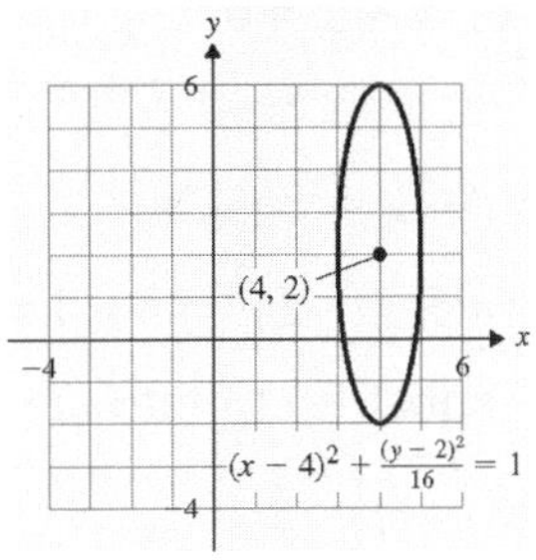

13) Center: $(0,0)$

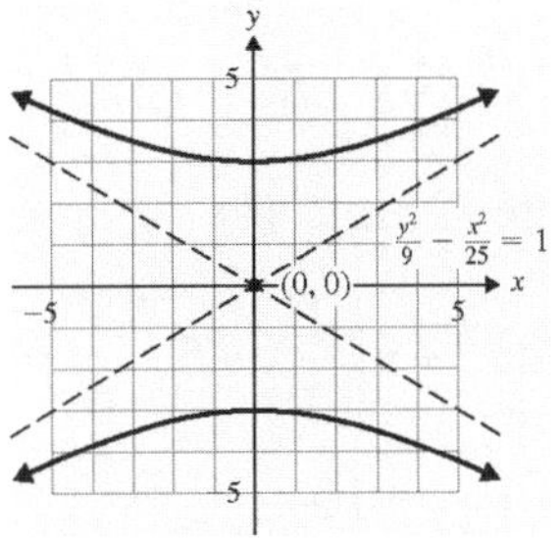

15) Center: $(-1,-2)$

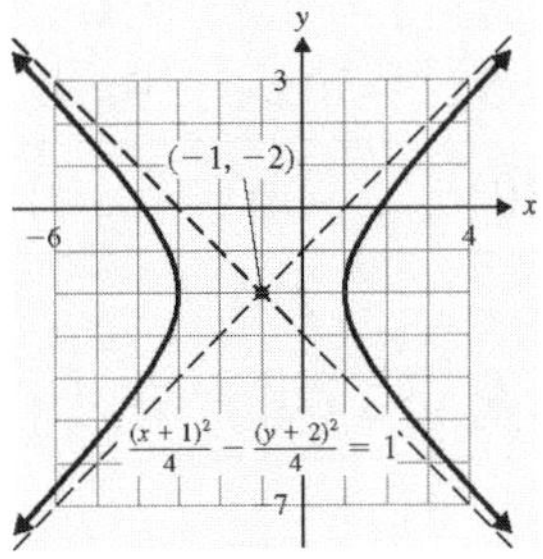

17) $x^2+9y^2=9$

$$\frac{x^2}{9}+\frac{9y^2}{9}=\frac{9}{9}$$

$$\frac{x^2}{9}+y^2=1$$

ellipse

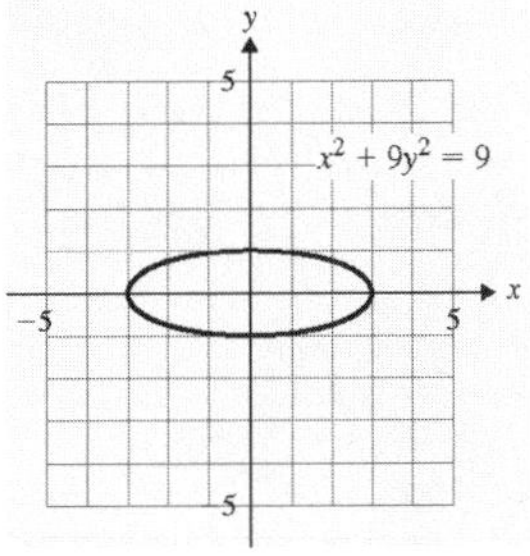

19) parabola

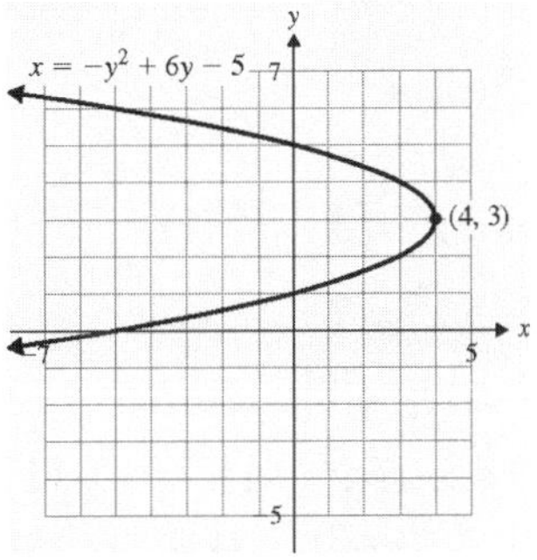

21) hyperbola

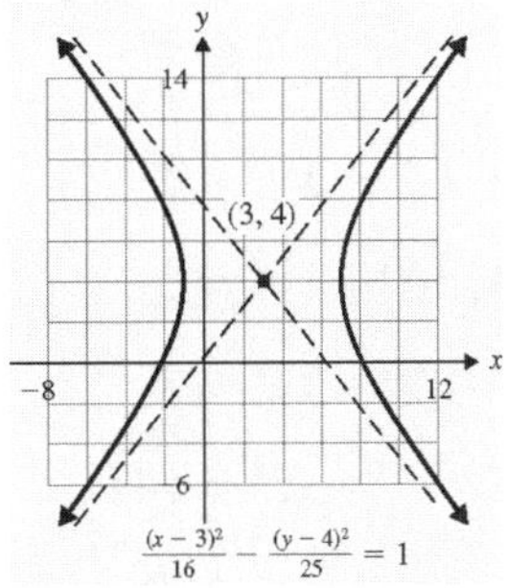

23)

$$x^2+y^2-2x+2y-2=0$$

$$\left(x^2-2x\right)+\left(y^2+2y\right)=2$$

$$\left(x^2-2x+1\right)+\left(y^2+2y+1\right)=2+1+1$$

$$(x-1)^2+(y+1)^2=4$$

circle

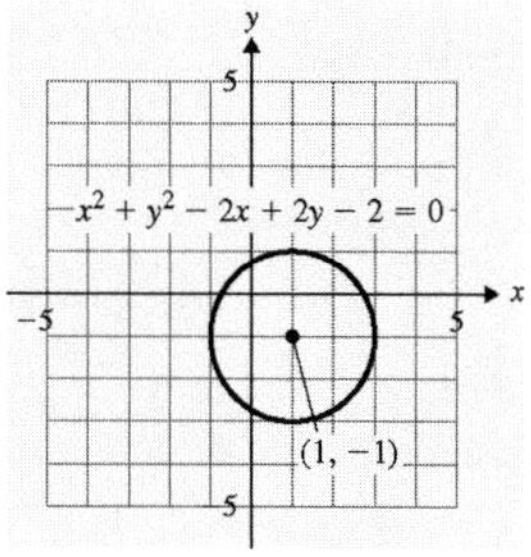

25) parabola

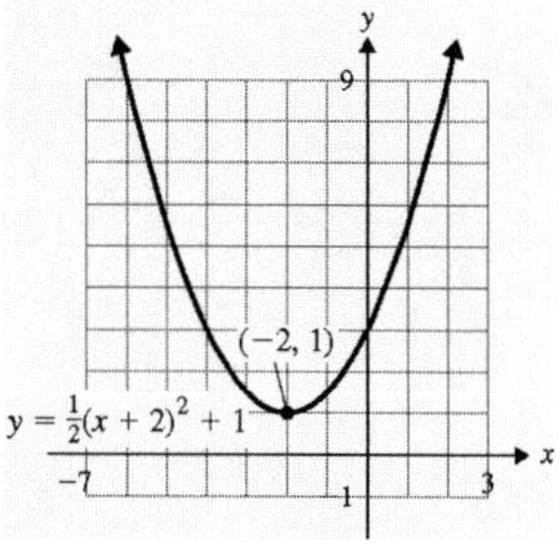

27) $0, 1, 2, 3, \text{or } 4$

29) $-4x^2 + 3y^2 = 3 \quad (1)$

$7x^2 - 5y^2 = 7 \quad (2)$

Solve using elimination.

$$\begin{array}{lr} -20x^2 + 15y^2 = 15 & 5\cdot(1) \\ +\quad 21x^2 - 15y^2 = 21 & 3\cdot(2) \\ \hline x^2 = 36 & \\ x = \pm 6 & \end{array}$$

$$x = 6: -4(6)^2 + 3y^2 = 3$$
$$-4(36) + 3y^2 = 3$$
$$-144 + 3y^2 = 3$$
$$3y^2 = 147$$
$$y^2 = 49$$
$$y = \pm 7$$

$$x = -6: -4(-6)^2 + 3y^2 = 3$$
$$-4(36) + 3y^2 = 3$$
$$-144 + 3y^2 = 3$$
$$3y^2 = 147$$
$$y^2 = 49$$
$$y = \pm 7$$

Verify by substituing into (2).

$$\{(6,7), (6,-7), (-6,7), (-6,-7)\}$$

31) $y = 3 - x^2 \quad (1)$

$x - y = -1 \quad (2)$

Substitute (1) into (2).

$$x - (3 - x^2) = -1$$
$$x^2 + x - 3 = -1$$
$$x^2 + x - 2 = 0$$
$$(x-1)(x+2) = 0$$
$$x - 1 = 0 \text{ or } x + 2 = 0$$
$$x = 1 \qquad x = -2$$
$$x = 1: y = 3 - 1^2 = 3 - 1 = 2$$
$$x = -2: y = 3 - (-2)^2 = 3 - 4 = -1$$

Verify by substituing into (2).

$$\{(1,2), (-2,-1)\}$$

33) $4x^2 + 9y^2 = 36 \quad (1)$

$y = \frac{1}{3}x - 5 \quad (2)$

Substitute (2) into (1).

$$4x^2 + 9\left(\frac{1}{3}x - 5\right)^2 = 36$$
$$4x^2 + 9\left(\frac{1}{9}x^2 - \frac{10}{3}x + 25\right) = 36$$
$$4x^2 + x^2 - 30x + 225 = 36$$
$$5x^2 - 30x + 189 = 0$$
$$x = \frac{-(-30) \pm \sqrt{(-30)^2 - 4(4)(225)}}{2(4)}$$
$$= \frac{30 \pm \sqrt{900 - 3600}}{8}$$
$$= \frac{30 \pm \sqrt{-2700}}{8}$$

does not give real number solutions.

$\varnothing$

35) $x = \text{one number}$

$y = \text{other number}$

$xy = 36$

$x + y = 13$

Substitute $y = 13 - x$ into (1).

$x(13 - x) = 36$

$13x - x^2 = 36$

$0 = x^2 - 13x + 36$

$0 = (x - 9)(x - 4)$

$x - 9 = 0$ or $x - 4 = 0$

$x = 9 \qquad x = 4$

$x = 9: y = 13 - 9 = 4$

$x = 4: y = 13 - 4 = 9$

The numbers are 9 and 4.

37) $a^2 + 2a - 3 < 0$

$(a+3)(a-1) < 0$

$(a+3)(a-1) = 0$

$a + 3 = 0$ or $a - 1 = 0$

$a = -3 \qquad a = 1$

Interval A: $(-\infty, -3)$ Positive

Interval B: $(-3, 1)$ Negative

Interval C: $(1, \infty)$ Positive

$(-3, 1)$

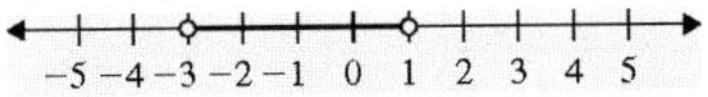

39) $6h^2 + 7h > 0$

$h(6h + 7) > 0$

$h(6h + 7) = 0$

$6h + 7 = 0$ or $h = 0$

$6h = -7$

$h = -\frac{7}{6}$

Interval A: $\left(-\infty, -\frac{7}{6}\right)$ Positive

Interval B: $\left(-\frac{7}{6}, 0\right)$ Negative

Interval C: $(0, \infty)$ Positive

$\left(-\infty, -\frac{7}{6}\right) \cup (0, \infty)$

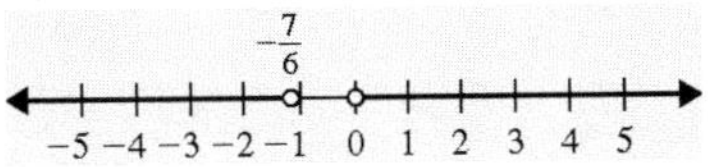

41) $36 - r^2 > 0$

$(6 - r)(6 + r) > 0$

$(6 - r)(6 + r) = 0$

$6 - r = 0$ or $6 + r = 0$

$-r = -6 \qquad r = -6$

$r = 6$

Interval A: $(-\infty, -6)$ Positive

Interval B: $(-6, 6)$ Negative

Interval C: $(6, \infty)$ Positive

$(-6, 6)$

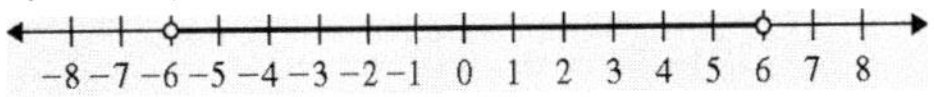

43) $(-\infty, \infty)$

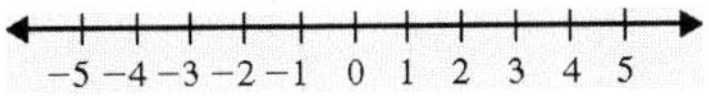

45) $\frac{t+7}{2t-3} > 0$

Set the numerator and denominator equal to zero and solve for t.

$t + 7 = 0 \qquad 2t - 3 = 0$

$t = -7 \qquad 2t = 3$

$t = \frac{3}{2}$

Interval A: $(-\infty,-7)$ Positive

Interval B: $\left(-7,\frac{3}{2}\right)$ Negative

Interval C: $\left(\frac{3}{2},\infty\right)$ Positive

$(-\infty,-7)\cup\left(\frac{3}{2},\infty\right)$

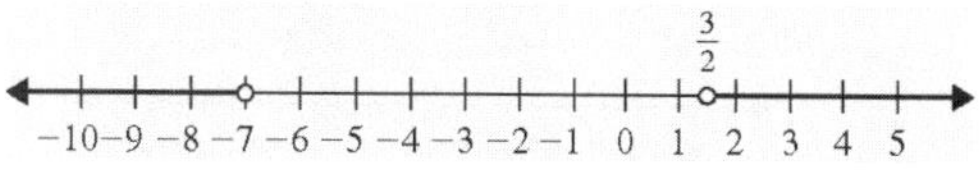

47) $\frac{4w+3}{5w-6}\le 0$

Set the numerator and denominator equal to zero and solve for *w*.

$4w+3=0 \qquad 5w-6=0$

$4w=-3 \qquad 5w=6$

$w=-\frac{3}{4} \qquad w=\frac{6}{5}$

Interval A: $\left(-\infty,-\frac{3}{4}\right)$ Positive

Interval B: $\left(-\frac{3}{4},\frac{6}{5}\right)$ Negative

Interval C: $\left(\frac{6}{5},\infty\right)$ Positive

$w\neq\frac{6}{5}$ because it makes the denominator equal to zero.

$\left[-\frac{3}{4},\frac{6}{5}\right)$

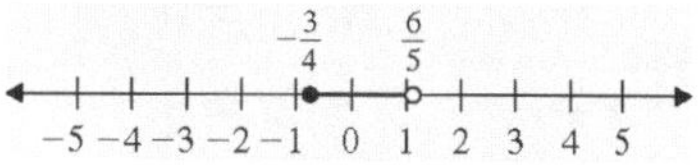

49)

$$\frac{1}{n-4}>-3$$

$$\frac{1}{n-4}+3>0$$

$$\frac{1}{n-4}+\frac{3(n-4)}{n-4}>0$$

$$\frac{1+3(n-4)}{n-4}>0$$

$$\frac{1+3n-12}{n-4}>0$$

$$\frac{3n-11}{n-4}>0$$

Set the numerator and denominator equal to zero and solve for *n*.

$3n-11=0 \qquad n-4=0$

$3n=11 \qquad n=4$

$n=\frac{11}{3}$

Interval A: $\left(-\infty,\frac{11}{3}\right)$ Positive

Interval B: $\left(\frac{11}{3},4\right)$ Negative

Interval C: $(4,\infty)$ Positive

$\left(-\infty,\frac{11}{3}\right)\cup(4,\infty)$

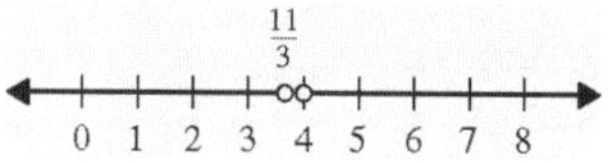

51) $\frac{r^2+4}{r-7}\ge 0$

Numerator will always be positive. Set the denominator equal to zero and solve for *r*.

$r-7=0$

$r=7$

Interval A: $(-\infty,7)$ Negative

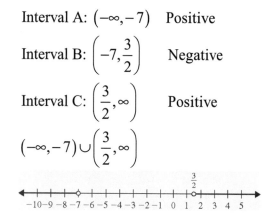

Interval B: $(7,\infty)$ Positive

$r \neq 7$ because it makes the denominator equal to zero.

$(7, \infty)$

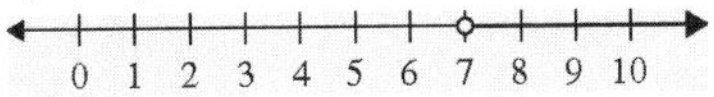

Chapter 14 Test

1) ellipse

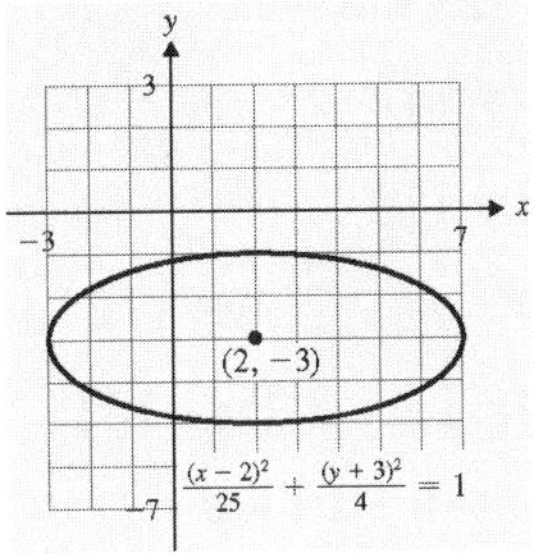

3) $y^2 - 4x^2 = 16$

$$\frac{y^2}{16} - \frac{4x^2}{16} = \frac{16}{16}$$

$$\frac{y^2}{16} - \frac{x^2}{4} = 1$$

hyperbola

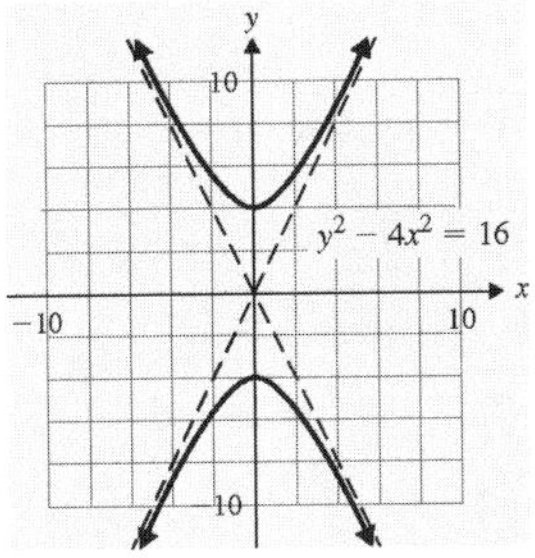

5)
$$x^2 + y^2 + 2x - 6y - 6 = 0$$
$$(x^2 + 2x) + (y^2 - 6y) = 6$$
$$(x^2 + 2x + 1) + (y^2 - 6y + 9) = 6 + 1 + 9$$
$$(x+1)^2 + (y-3)^2 = 16$$

Center: $(-1, 3)$; $r = \sqrt{16} = 4$

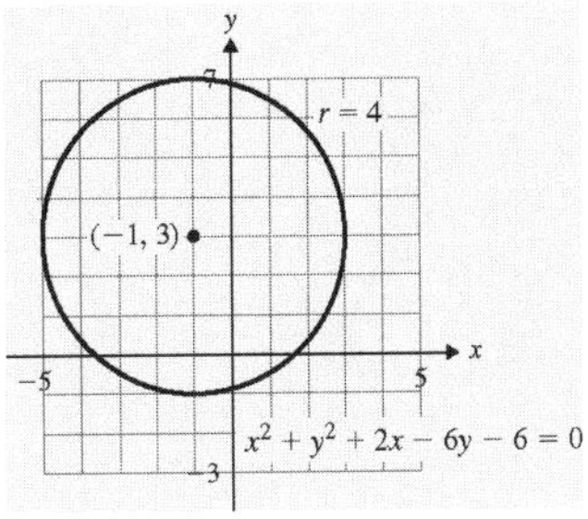

7) a)

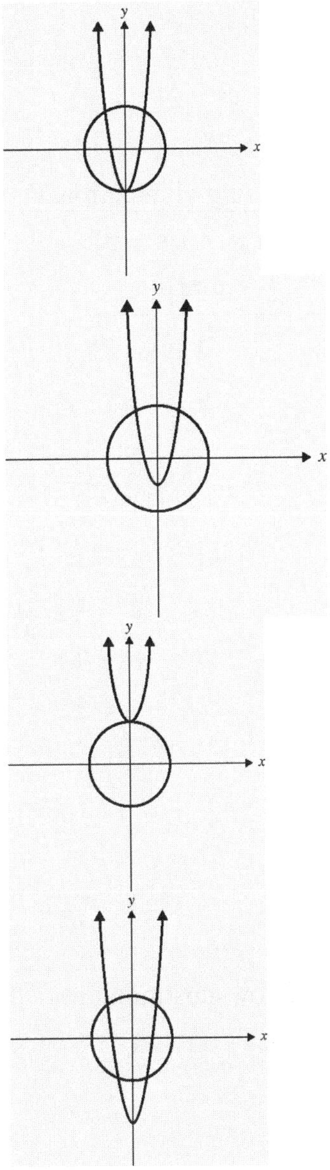

b)

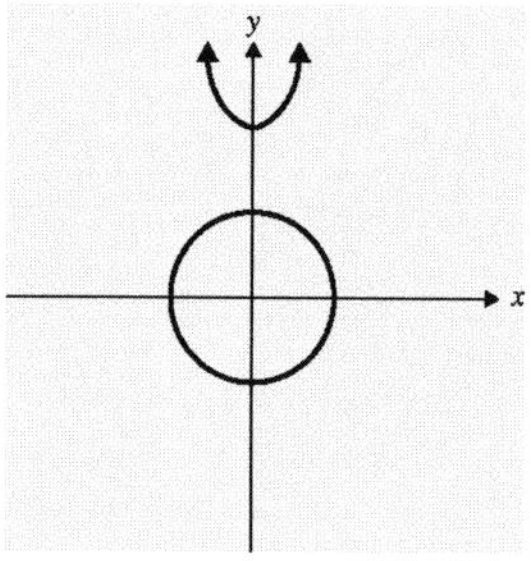

c) 0, 1, 2, 3, or 4

9) $2x^2+3y^2=21$ (1)

$-x^2+12y^2=3$ (2)

Solve using elimination.

$$\begin{array}{rl} 2x^2+3y^2=21 & (1) \\ + \quad -2x^2+24y^2=6 & 2\cdot(2) \\ \hline 27y^2=27 & \\ y^2=1 & \\ y=\pm 1 & \end{array}$$

$y=1: -x^2+12(1)^2=3$

$-x^2+12=3$

$-x^2=-9$

$x^2=9$

$x=\pm 3$

$y=-1: -x^2+12(-1)^2=3$

$-x^2+12=3$

$-x^2=-9$

$x^2=9$

$x=\pm 3$

Verify by substituing into (1).

$\{(3,1),(3,-1),(-3,1),(-3,-1)\}$

11) l = length of frame

w = width of frame

$2l+2w=44$ (1)

$lw=112$ (2)

Solve (1) for l.

$2l=44-2w$

$l=22-w$ (3)

Substitute (3) into (2).

$(22-w)w=112$

$22w-w^2=112$

$0=w^2-22w+112$

$0=(w-8)(w-14)$

$w-8=0$ or $w-14=0$

$w=8$ $\quad$ $w=14$

$w=8: \ l=22-8=14$

$w=14: l=22-14=8$

The dimensions of the window are 8 in×14 in.

13) $2w^2+11w<-12$

$2w^2+11w+12<0$

$(2w+3)(w+4)<0$

$(2w+3)(w+4)=0$

$2w+3=0$ or $w+4=0$

$2w=-3$ $\quad$ $w=-4$

$w=-\dfrac{3}{2}$

Interval A: $(-\infty,-4)$ Positive

Interval B: $\left(-4,-\dfrac{3}{2}\right)$ Negative

Interval C: $\left(-\dfrac{3}{2},\infty\right)$ Positive

$\left(-4,-\dfrac{3}{2}\right)$

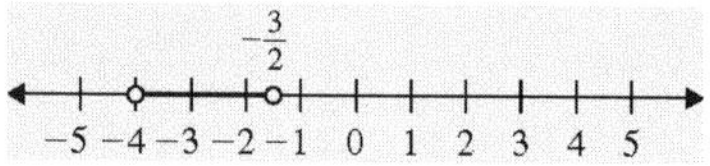

15) $\dfrac{m-5}{m+3} \geq 0$

Set the numerator and denominator equal to zero and solve for m.

$m-5=0 \qquad m+3=0$

$m=5 \qquad m=-3$

Interval A: $(-\infty,-3)$ Positive

Interval B: $(-3, 5)$ Negative

Interval C: $(5,\infty)$ Positive

$m \neq -3$ because it makes the denominator equal to zero.

$(-\infty,-3)\cup[5,\infty)$

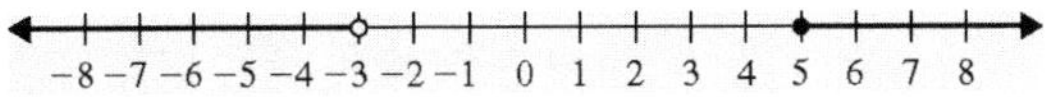

Cumulative Review: Chapters 1-14

1) $\dfrac{1}{6}-\dfrac{11}{12}=\dfrac{2}{12}-\dfrac{11}{12}=-\dfrac{9}{12}=-\dfrac{3}{4}$

3) $A=\dfrac{1}{2}(6)(5)=3(5)=15 \text{ cm}^2$

$P=7+6+5.5=18.5 \text{ cm}$

5) $(-1)^5=-1$

7)
$$\left(\frac{2a^8b}{a^2b^{-4}}\right)^{-3}=\left(\frac{a^2b^{-4}}{2a^8b}\right)^3$$
$$=\left(\frac{1}{2}a^{2-8}b^{-4-1}\right)^3$$
$$=\left(\frac{1}{2}a^{-6}b^{-5}\right)^3$$
$$=\frac{1}{8}a^{-18}b^{-15}=\frac{1}{8a^{18}b^{15}}$$

9)
$$an+z=c$$
$$an=c-z$$
$$n=\frac{c-z}{a}$$

11)
$$n+(n+2)+(n+4)=2(n+4)+13$$
$$3n+6=2n+8+13$$
$$3n+6=2n+21$$
$$n=15$$
$$n=15; n+2=17; n+4=19$$

15, 17, 19

13) 0

15) $m=\dfrac{1-7}{4-(-4)}=\dfrac{-6}{4+4}=\dfrac{-6}{8}=-\dfrac{3}{4}$
$$(y-1)=-\frac{3}{4}(x-4)$$
$$y-1=-\frac{3}{4}x+3$$
$$y=-\frac{3}{4}x+4$$

17) $x=$ milliliters of 8% solution

$20-x=$ milliliters of 16% solution
$$0.08x+0.16(20-x)=0.14(20)$$
$$100(0.08x+0.16(20-x))=100(0.14(20))$$
$$8x+16(20-x)=14(20)$$
$$8x+320-16x=280$$
$$-8x=-40$$
$$x=5$$
$$20-x=15$$

8% solution: 5 ml; 16% solution: 15 ml

19)
$$(4w-3)(2w^2+9w-5)$$
$$=8w^3+36w^2-20w-6w^2-27w+15$$
$$=8w^3+30w^2-47w+15$$

21) $6c^2-14c+8=2(3c^2-7c+4)$
$=2(3c-4)(c-1)$

23) $(x+1)(x+2)=2(x+7)+5x$
$x^2+3x+2=2x+14+5x$
$x^2+3x+2=7x+14$
$x^2-4x-12=0$
$(x-6)(x+2)=0$
$x-6=0$ or $x+2=0$
$x=6 \qquad x=-2 \qquad \{-2,6\}$

25) $\dfrac{\frac{t^2-9}{4}}{\frac{t-3}{24}}=\dfrac{\frac{(t+3)(t-3)}{4}}{\frac{t-3}{24}}$
$=\dfrac{(t+3)(t-3)}{4}\div\dfrac{t-3}{24}$
$=\dfrac{(t+3)\cancel{(t-3)}}{\cancel{4}}\cdot\dfrac{\overset{6}{\cancel{24}}}{\cancel{t-3}}$
$=6(t+3)$

27) $|5r+3|>12$
$5r+3>12$ or $5r+3<-12$
$5r>9 \qquad 5r<-15$
$r>\dfrac{9}{5} \qquad r<-3$
$(-\infty,-3)\cup\left(\dfrac{9}{5},\infty\right)$

29) $\sqrt[3]{48}=\sqrt[3]{8}\cdot\sqrt[3]{6}=2\sqrt[3]{6}$

31) $(16)^{-3/4}=\left(\dfrac{1}{16}\right)^{3/4}=\left(\sqrt[4]{\dfrac{1}{16}}\right)^3$
$=\left(\dfrac{1}{2}\right)^3=\dfrac{1}{8}$

33) $\dfrac{5}{\sqrt{3}+4}=\dfrac{5}{\sqrt{3}+4}\cdot\dfrac{\sqrt{3}-4}{\sqrt{3}-4}$
$=\dfrac{5(\sqrt{3}-4)}{(\sqrt{3})^2+4^2}=\dfrac{5\sqrt{3}-20}{3-16}$
$=\dfrac{5\sqrt{3}-20}{-13}=\dfrac{20-5\sqrt{3}}{13}$

35) $y^2=-7y-3$
$y^2+7y=-3$
$y^2+7y+\dfrac{49}{4}=-3+\dfrac{49}{4}$
$\left(y+\dfrac{7}{2}\right)^2=\dfrac{37}{4}$
$y+\dfrac{7}{2}=\pm\dfrac{\sqrt{37}}{2}$
$y=-\dfrac{7}{2}\pm\dfrac{\sqrt{37}}{2}$
$\left\{-\dfrac{7}{2}-\dfrac{\sqrt{37}}{2},-\dfrac{7}{2}+\dfrac{\sqrt{37}}{2}\right\}$

37)
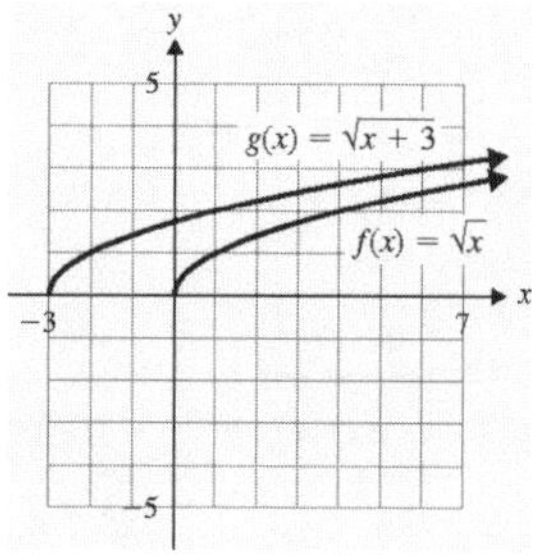

39) a) no b) no

41) $8^{5t}=4^{t-3}$

$(2^3)^{5t}=(2^2)^{t-3}$

$2^{15t}=2^{2t-6}$

$15t=2t-6$

$13t=-6$

$t=-\frac{6}{13}$ $\quad\left\{-\frac{6}{13}\right\}$

43) $\log 100=\log 10^2=2\log 10=2(1)=2$

45) $e^{3k}=8$

$\ln e^{3k}=\ln 8$

$3k\ln e=\ln 8$

$k=\frac{\ln 8}{3}$ $\quad\left\{\frac{\ln 8}{3}\right\};\{0.6931\}$

47)

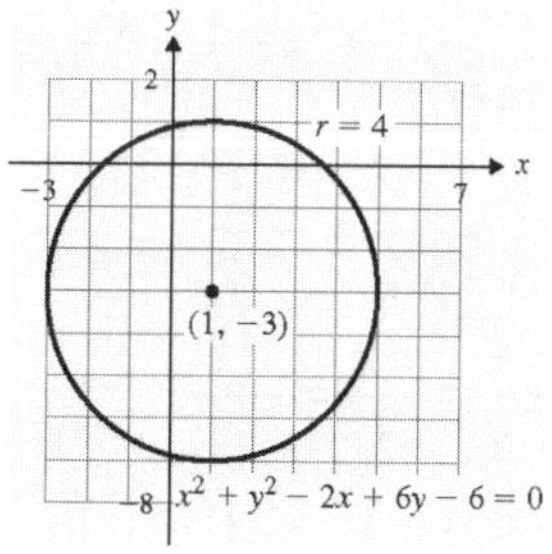

49) $25p^2\le 144$

$25p^2-144\le 0$

$(5p+12)(5p-12)\le 0$

$(5p+12)(5p-12)=0$

$p=\pm\frac{12}{5}$

Interval A: $\left(-\infty,-\frac{12}{5}\right)$ Positive

Interval B: $\left(-\frac{12}{5},\frac{12}{5}\right)$ Negative

Interval C: $\left(\frac{12}{5},\infty\right)$ Positive

$\left[-\frac{12}{5},\frac{12}{5}\right]$

Section 15.1: Exercises

1) $a_1 = 1+2 = 3$
$a_2 = 2+2 = 4$
$a_3 = 3+2 = 5$
$a_4 = 4+2 = 6$
$a_5 = 5+2 = 7$

3) $a_1 = 3(1)-4 = 3-4 = -1$
$a_2 = 3(2)-4 = 6-4 = 2$
$a_3 = 3(3)-4 = 9-4 = 5$
$a_4 = 3(4)-4 = 12-4 = 8$
$a_5 = 3(5)-4 = 15-4 = 11$

5) $a_1 = 2(1)^2 - 1 = 2(1)-1 = 1$
$a_2 = 2(2)^2 - 1 = 2(4)-1 = 8-1 = 7$
$a_3 = 2(3)^2 - 1 = 2(9)-1 = 18-1 = 17$
$a_4 = 2(4)^2 - 1$
$= 2(16)-1 = 32-1 = 31$
$a_5 = 2(5)^2 - 1 = 2(25)-1 = 50-1 = 49$

7) $a_1 = 3^{1-1} = 3^0 = 1$
$a_2 = 3^{2-1} = 3^1 = 3$
$a_3 = 3^{3-1} = 3^2 = 9$
$a_4 = 3^{4-1} = 3^3 = 27$
$a_5 = 3^{5-1} = 3^4 = 81$

9) $a_1 = 5\cdot\left(\frac{1}{2}\right)^1 = 5\cdot\frac{1}{2} = \frac{5}{2}$
$a_2 = 5\cdot\left(\frac{1}{2}\right)^2 = 5\cdot\frac{1}{4} = \frac{5}{4}$
$a_3 = 5\cdot\left(\frac{1}{2}\right)^3 = 5\cdot\frac{1}{8} = \frac{5}{8}$
$a_4 = 5\cdot\left(\frac{1}{2}\right)^4 = 5\cdot\frac{1}{16} = \frac{5}{16}$
$a_5 = 5\cdot\left(\frac{1}{2}\right)^5 = 5\cdot\frac{1}{32} = \frac{5}{32}$

11) $a_1 = (-1)^{1+1}\cdot 7(1) = 1\cdot 7 = 7$
$a_2 = (-1)^{2+1}\cdot 7(2) = -1\cdot 14 = -14$
$a_3 = (-1)^{3+1}\cdot 7(3) = 1\cdot 21 = 21$
$a_4 = (-1)^{4+1}\cdot 7(4) = -1\cdot 28 = -28$
$a_5 = (-1)^{5+1}\cdot 7(5) = 1\cdot 35 = 35$

13) $a_1 = \frac{1-4}{1+3} = \frac{-3}{4} = -\frac{3}{4}$
$a_2 = \frac{2-4}{2+3} = \frac{-2}{5} = -\frac{2}{5}$
$a_3 = \frac{3-4}{3+3} = \frac{-1}{6} = -\frac{1}{6}$
$a_4 = \frac{4-4}{4+3} = \frac{0}{7} = 0$
$a_5 = \frac{5-4}{5+3} = \frac{1}{8}$

15) a) $a_1 = 3(1)+2 = 5$

b) $a_5 = 3(5)+2 = 15+2 = 17$

c) $a_{28} = 3(28)+2 = 84+2 = 86$

17) a) $a_1 = \frac{1-4}{1+6} = \frac{-3}{7} = -\frac{3}{7}$

b) $a_2=\frac{2-4}{2+6}=\frac{-2}{8}=-\frac{1}{4}$

c) $a_{16}=\frac{16-4}{16+6}=\frac{12}{22}=\frac{6}{11}$

19) a) $a_1=10-1^2=10-1=9$

b) $a_6=10-6^2=10-36=-26$

c) $a_{20}=10-20^2=10-400=-390$

21) $a_n=2n$ 23) $a_n=n^2$

25) $a_n=\left(\frac{1}{3}\right)^n$ 27) $a_n=\frac{n}{n+1}$

29) $a_n=(-1)^{n+1}\cdot(5n)$

31) $a_n=(-1)^n\cdot\left(\frac{1}{2}\right)^n$

33) $2592-\frac{1}{3}(2592)=2592-864=1728$

$1728-\frac{1}{3}(1728)=1728-576=1152$

$1152-\frac{1}{3}(1152)=1152-384=768$

$768-\frac{1}{3}(768)=768-256=512$

\$1728, \$1152, \$768, \$512

35) $100+10=110$

$110+10=120$

$120+10=130$

$130+10=140$

$140+10=150$

$150+10=160$ 160 lb

37) A sequence is a list of terms in a certain order, and a series is a sum of the terms of a sequence.

39) $\sum_{i=1}^{6}(2i+1)=(2(1)+1)+(2(2)+1)$

$+(2(3)+1)+(2(4)+1)$

$+(2(5)+1)+(2(6)+1)$

$=3+5+7+9+11+13$

$=48$

41) $\sum_{i=1}^{5}(i-8)=(1-8)+(2-8)+(3-8)$

$+(4-8)+(5-8)$

$=-7-6-5-4-3=-25$

43) $\sum_{i=1}^{4}(4i^2-2i)$

$=\left(4(1)^2-2(1)\right)+\left(4(2)^2-2(2)\right)$

$+\left(4(3)^2-2(3)\right)+\left(4(4)^2-2(4)\right)$

$=(4-2)+(16-4)$

$+(36-6)+(64-8)$

$=2+12+30+56=100$

45) $\sum_{i=1}^{6}\frac{i}{2}=\frac{1}{2}+\frac{2}{2}+\frac{3}{2}+\frac{4}{2}+\frac{5}{2}+\frac{6}{2}=\frac{21}{2}$

47) $\sum_{i=1}^{5}(-1)^{i+1}\cdot(i)=(-1)^{1+1}\cdot1+(-1)^{2+1}\cdot2$

$+(-1)^{3+1}\cdot3+(-1)^{4+1}\cdot4$

$+(-1)^{5+1}\cdot5$

$=1-2+3-4+5=3$

49) $\sum_{i=5}^{9}(i-2)=(5-2)+(6-2)+(7-2)$
$+(8-2)+(9-2)$
$=3+4+5+6+7=25$

51) $\sum_{i=3}^{6}(i^2)=3^2+4^2+5^2+6^2$
$=9+16+25+36=86$

53) $\sum_{i=1}^{5}\frac{1}{i}$ 55) $\sum_{i=1}^{4}(3i)$ 57) $\sum_{i=1}^{6}(i+4)$

59) $\sum_{i=1}^{7}(-1)^i\cdot(i)$ 61) $\sum_{i=1}^{4}(-1)^{i+1}\cdot(3^i)$

63) $\overline{x}=\dfrac{19+24+20+17+23+17}{6}=20$

65) $\overline{x}=\dfrac{8+7+11+9+12}{5}=9.4$

67) $\overline{x}=\dfrac{1431.60+1117.82+985.43}{6}$
$=+\dfrac{1076.22+900.00+813.47}{6}$
$=\$1054.09$

Section 15.2: Exercises

1) It is a list of numbers in a specific order such that the difference between any two successive terms is the same. $1,4,7,10,...$

3) yes, $d=11-3=8$

5) yes, $d=6-10=-4$ 7) no

9) yes, $d=-14-(-17)=3$

11) $a_1=7$
$a_2=7+2=9$
$a_3=9+2=11$
$a_4=11+2=13$
$a_5=13+2=15$
7, 9, 11, 13, 15

13) $a_1=15$
$a_2=15-8=7$
$a_3=7-8=-1$
$a_4=-1-8=-9$
$a_5=-9-8=-17$
15, 7, −1, −9, −17

15) $a_1=-10$
$a_2=-10+3=-7$
$a_3=-7+3=-4$
$a_4=-4+3=-1$
$a_5=-1+3=2$
−10, −7, −4, −1, 2

17) $a_1=6(1)+7=6+7=13$
$a_2=6(2)+7=12+7=19$
$a_3=6(3)+7=18+7=25$
$a_4=6(4)+7=24+7=31$
$a_5=6(5)+7=30+7=37$
13, 19, 25, 31, 37

19) $a_1=5-1=4$
$a_2=5-2=3$
$a_3=5-3=2$
$a_4=5-4=1$
$a_5=5-5=0$
4, 3, 2, 1, 0

21) a) $a_1=4; d=7-4=3$

b) $a_n = 4 + (n-1)3$

$a_n = 4 + 3n - 3$

$a_n = 3n + 1$

c) $a_{24} = 3(24) + 1 = 72 + 1 = 73$

23) a) $a_1 = 4; d = -1 - 4 = -5$

b) $a_n = 4 + (n-1)(-5)$

$a_n = 4 - 5n + 5$

$a_n = -5n + 9$

c) $a_{19} = -5(19) + 9 = -95 + 9 = -86$

25) $a_n = -7 + (n-1)(2)$

$a_n = -7 + 2n - 2$

$a_n = 2n - 9$

$a_{25} = 2(25) - 9 = 50 - 9 = 41$

27) $a_n = 1 + (n-1)\left(\frac{1}{2}\right)$

$a_n = 1 + \frac{1}{2}n - \frac{1}{2}$

$a_n = \frac{1}{2}n + \frac{1}{2}$

$a_{18} = \frac{1}{2}(18) + \frac{1}{2} = \frac{18}{2} + \frac{1}{2} = \frac{19}{2}$

29) $a_n = 0 + (n-1)(-5)$

$a_n = -5n + 5$

$a_{23} = -5(23) + 5 = -115 + 5 = -110$

31) $a_{16} = -5 + (16-1)(4)$

$= -5 + (15)(4) = -5 + 60 = 55$

33) $a_{21} = -7 + (21-1)(-5)$

$= -7 + (20)(-5)$

$= -7 - 100 = -107$

35) $a_3 = a_1 + (3-1)d$

$11 = a_1 + (3-1)d$

$11 = a_1 + 2d \quad (1)$

$a_7 = a_1 + (7-1)d$

$19 = a_1 + (7-1)d$

$19 = a_1 + 6d \quad (2)$

Solve using elimination.

$$\begin{array}{rll} -11 = -a_1 - 2d & & -1\cdot(1) \\ +\quad 19 = a_1 + 6d & & (2) \\ \hline 8 = 4d & & \\ 2 = d & & \end{array}$$

$d = 2: 11 = a_1 + 2(2)$

$11 = a_1 + 4$

$7 = a_1$

$a_n = 7 + (n-1)2$

$= 7 + 2n - 2 = 2n + 5$

$a_{11} = 2(11) + 5 = 22 + 5 = 27$

37) $a_2 = a_1 + (2-1)d$

$7 = a_1 + (2-1)d$

$7 = a_1 + d \quad (1)$

$a_6 = a_1 + (6-1)d$

$-13 = a_1 + (6-1)d$

$-13 = a_1 + 5d \quad (2)$

Solve using elimination.

$$\begin{array}{rll} -7 = -a_1 - d & & -1\cdot(1) \\ +\quad -13 = a_1 + 5d & & (2) \\ \hline -20 = 4d & & \\ -5 = d & & \end{array}$$

$d = -5: \ 7 = a_1 + (-5)$

$12 = a_1$

$$a_n = 12 + (n-1)(-5)$$
$$= 12 - 5n + 5 = -5n + 17$$
$$a_{14} = -5(14) + 17 = -70 + 17 = -53$$

39) $a_4 = a_1 + (4-1)d$

$-5 = a_1 + (4-1)d$

$-5 = a_1 + 3d \qquad (1)$

$a_{11} = a_1 + (11-1)d$

$16 = a_1 + (11-1)d$

$16 = a_1 + 10d \qquad (2)$

Solve using elimination.

$$\begin{array}{rl} 5 = -a_1 - 3d & -1\cdot(1) \\ +\ 16 = a_1 + 10d & (2) \\ \hline 21 = 7d & \\ 3 = d & \end{array}$$

$d = 3:\ -5 = a_1 + 3(3)$

$-5 = a_1 + 9$

$-14 = a_1$

$a_n = -14 + (n-1)(3)$

$= -14 + 3n - 3 = 3n - 17$

$a_{18} = 3(18) - 17 = 54 - 17 = 37$

41) $a_n = 63,\ a_1 = 8,\ d = 13 - 8 = 5$

$63 = 8 + (n-1)(5)$

$63 = 8 + 5n - 5$

$63 = 3 + 5n$

$60 = 5n$

$12 = n$

43) $a_n = -27,\ a_1 = 9,\ d = 7 - 9 = -2$

$-27 = 9 + (n-1)(-2)$

$-27 = 9 - 2n + 2$

$-27 = 11 - 2n$

$-38 = -2n$

$19 = n$

45) S_{15} is the sum of the first 15 terms of the sequence.

47) $S_{10} = \frac{10}{2}(14 + 68) = 5(82) = 410$

49) $S_7 = \frac{7}{2}(3 + (-9)) = \frac{7}{2}(-6) = -21$

51) $S_8 = \frac{8}{2}(-1 + (-29)) = 4(-30) = -120$

53) $S_8 = \frac{8}{2}[2(3) + (8-1)5]$

$= 4(6 + 35) = 4(41) = 164$

55) $S_8 = \frac{8}{2}[2(10) + (8-1)(-6)]$

$= 4(20 - 42) = 4(-22) = -88$

57) $a_1 = -4(1) - 1 = -4 - 1 = -5$

$a_8 = -4(8) - 1 = -32 - 1 = -33$

$S_8 = \frac{8}{2}(-5 + (-33)) = 4(-38) = -152$

59) $a_1 = 3(1) + 4 = 3 + 4 = 7$

$a_8 = 3(8) + 4 = 24 + 4 = 28$

$S_8 = \frac{8}{2}(7 + 28) = 4(35) = 140$

61) a) $\sum_{i=1}^{10}(2i + 7)$

$= (2(1)+7) + (2(2)+7)$

$+ (2(3)+7) + (2(4)+7)$

$+ (2(5)+7) + (2(6)+7)$

$+ (2(7)+7) + (2(8)+7)$

$+ (2(9)+7) + (2(10)+7)$

$= 9 + 11 + 13 + 15 + 17 +$

$19 + 21 + 23 + 25 + 27 = 180$

b) $a_1 = 2(1)+7 = 2+7 = 9$

$a_{10} = 2(10)+7 = 20+7 = 27$

$S_{10} = \frac{10}{2}(9+27) = 5(36) = 180$

c) Answers may vary.

63) $a_1 = 8(1)-5 = 8-5 = 3$

$a_5 = 8(5)-5 = 40-5 = 35$

$S_5 = \frac{5}{2}(3+35) = \frac{5}{2}(38) = 95$

65) $a_1 = -2(1)+7 = -2+7 = 5$

$a_7 = -2(7)+7 = -14+7 = -7$

$S_7 = \frac{7}{2}(5+(-7)) = \frac{7}{2}(-2) = -7$

67) $a_1 = 3(1)-11 = 3-11 = -8$

$a_{18} = 3(18)-11 = 54-11 = 43$

$S_{18} = \frac{18}{2}(-8+43) = 9(35) = 315$

69) $a_1 = 1$

$a_{500} = 500$

$S_{500} = \frac{500}{2}(1+500)$

$= 250(501) = 125,250$

71) $a_1 = \$1500,\ d = \$100$

$a_n = 1500+(n-1)100$

$a_n = 1500+100n-100$

$a_n = 100n+1400$

$a_9 = 100(9)+1400$

$= 900+1400 = \$2300$

73) $a_1 = \$1,\ d = \$1$

$S_{24} = \frac{24}{2}[2(1)+(24-1)(1)]$

$= 12(2+23) = 12(25) = \$300$

75) a) $a_1 = 12,\ d = -1$

$a_n = 12+(n-1)(-1)$

$a_n = 12-n+1$

$a_n = 13-n$

$a_8 = 13-8 = 5$ logs

b) $1 = 13-n$

$n = 12$

The stack has 12 rows of logs.

$S_{12} = \frac{12}{2}(12+1) = 6(13) = 78$ logs

77) $a_1 = 12,\ d = 2$

$a_n = 12+(n-1)(2)$

$a_n = 12+2n-2$

$a_n = 10+2n$

$a_{14} = 10+2(14)$

$= 10+28 = 38$ seats in last row

$S_{14} = \frac{14}{2}(12+38) = 7(50) = 350$ seats

79) $S_n = 860,\ a_1 = 24,\ a_1 = 62$

$860 = \frac{n}{2}(24+62)$

$860 = \frac{n}{2}(86)$

$860 = 43n$

$20 = n$

Section 15.3: Exercises

1) A sequence is arithmetic if the difference between each term is constant, but a sequence is geometric if each term after the first is obtained by multiplying the preceding term by a common ratio.

3) $r=\frac{2}{1}=2$ 5) $r=\frac{3}{9}=\frac{1}{3}$

7) $r=\frac{\frac{1}{2}}{-2}=-\frac{1}{4}$

9) $a_1=2$

$a_2=2(5)=10$

$a_3=10(5)=50$

$a_4=50(5)=250$

$a_5=250(5)=1250$

11) $a_1=\frac{1}{4}$

$a_2=\frac{1}{4}(-2)=-\frac{1}{2}$

$a_3=-\frac{1}{2}(-2)=1$

$a_4=1(-2)=-2$

$a_5=-2(-2)=4$

13) $a_1=72$

$a_2=72\left(\frac{2}{3}\right)=48$

$a_3=48\left(\frac{2}{3}\right)=32$

$a_4=32\left(\frac{2}{3}\right)=\frac{64}{3}$

$a_5=\frac{64}{3}\left(\frac{2}{3}\right)=\frac{128}{9}$

15) $a_n=4(7)^{n-1}$

$a_3=4(7)^{3-1}=4(7)^2=4(49)=196$

17) $a_n=-1(3)^{n-1}$

$a_5=-1(3)^{5-1}=-1(81)=-81$

19) $a_n=2\left(\frac{1}{5}\right)^{n-1}$

$a_4=2\left(\frac{1}{5}\right)^{4-1}=2\left(\frac{1}{5}\right)^3$

$=2\left(\frac{1}{125}\right)=\frac{2}{125}$

21) $a_n=-\frac{1}{2}\left(-\frac{3}{2}\right)^{n-1}$

$a_4=-\frac{1}{2}\left(-\frac{3}{2}\right)^{4-1}=-\frac{1}{2}\left(-\frac{3}{2}\right)^3$

$=-\frac{1}{2}\left(-\frac{27}{8}\right)=\frac{27}{16}$

23) $a_1=5,\ r=\frac{10}{5}=2$

$a_n=5(2)^{n-1}$

25) $a_1 = -3,\ r = \dfrac{-\frac{3}{5}}{-3} = \dfrac{1}{5}$

$a_n = -3\left(\dfrac{1}{5}\right)^{n-1}$

27) $a_1 = 3,\ r = \dfrac{-6}{3} = -2$

$a_n = 3(-2)^{n-1}$

29) $a_1 = \dfrac{1}{3},\ r = \dfrac{\frac{1}{12}}{\frac{1}{3}} = \dfrac{1}{4}$

$a_n = \dfrac{1}{3}\left(\dfrac{1}{4}\right)^{n-1}$

31) $a_1 = 1,\ r = \dfrac{2}{1} = 2$

$a_n = 1(2)^{n-1} = 2^{n-1}$

$a_{12} = 2^{12-1} = 2^{11} = 2048$

33) $a_1 = 27,\ r = \dfrac{-9}{27} = -\dfrac{1}{3}$

$a_n = 27\left(-\dfrac{1}{3}\right)^{n-1}$

$a_8 = 27\left(-\dfrac{1}{3}\right)^{8-1} = 27\left(-\dfrac{1}{3}\right)^{7}$

$= 3^3 \cdot 3^{-7} \cdot (-1)^7 = -3^{-4} = -\dfrac{1}{81}$

35) $a_1 = -\dfrac{1}{64},\ r = \dfrac{-\frac{1}{32}}{-\frac{1}{64}} = 2$

$a_n = -\dfrac{1}{64}(2)^{n-1}$

$a_{12} = -\dfrac{1}{64}(2)^{12-1} = -2^{-6} \cdot 2^{11}$

$= -2^5 = -32$

37) arithmetic

$a_n = 15 + (n-1)(9)$

$a_n = 15 + 9n - 9$

$a_n = 9n + 6$

39) geometric

$a_n = -2(-3)^{n-1}$

41) geometric

$a_n = \dfrac{1}{9}\left(\dfrac{1}{2}\right)^{n-1}$

43) arithmetic

$a_n = -31 + (n-1)(7)$

$a_n = -31 + 7n - 7$

$a_n = 7n - 38$

45) a) $a_1 = 40,000,\ r = 1 - 0.20 = 0.80$

$a_n = 40,000(0.80)^{n-1}$

b) $a_5 = 40,000(0.80)^{5-1}$

$= 40,000(0.80)^4 = \$16,384$

47) a) $a_1 = 500,000,\ r = 1 - 0.10 = 0.90$

$a_n = 500,000(0.90)^{n-1}$

b) $a_3 = 500,000(0.90)^{3-1}$

$= 500,000(0.90)^2 = \$364,500$

49) a) $a_1 = 160,000,\ r = 1 + 0.04 = 1.04$

$a_n = 160,000(1.04)^{n-1}$

b) $a_6 = 160,000(1.04)^{6-1}$

$= 160,000(1.04)^5 \approx \$194,664$

51) $S_6 = \dfrac{9(1-2^6)}{1-2} = \dfrac{9(1-64)}{-1}$

$= -9(-63) = 567$

53) $a_1 = 7,\ r = \dfrac{28}{7} = 4,\ n = 7$

$S_7 = \dfrac{7(1-4^7)}{1-4} = \dfrac{7(1-16,384)}{-3}$

$= \dfrac{7(-16,383)}{-3} = 7(5461)$

$= 38,227$

55) $a_1 = -\dfrac{1}{4},\ r = \dfrac{-\frac{1}{2}}{-\frac{1}{4}} = 2,\ n = 6$

$S_6 = \dfrac{-\frac{1}{4}(1-2^6)}{1-2} = \dfrac{-\frac{1}{4}(1-64)}{-1}$

$= \dfrac{1}{4}(-63) = -\dfrac{63}{4}$

57) $a_1 = 1,\ r = \dfrac{\frac{1}{3}}{1} = \dfrac{1}{3},\ n = 5$

$S_5 = \dfrac{1\left(1-\left(\frac{1}{3}\right)^5\right)}{1-\frac{1}{3}} = \dfrac{\left(1-\frac{1}{243}\right)}{\frac{2}{3}}$

$= \dfrac{3}{2}\left(\dfrac{242}{243}\right) = \dfrac{121}{81}$

59) $a_1 = 18,\ r = 2,\ n = 7$

$S_7 = \dfrac{18(1-2^7)}{1-2} = \dfrac{18(1-128)}{-1}$

$= -18(-127) = 2286$

61) $a_1 = -12,\ r = 3,\ n = 5$

$S_5 = \dfrac{-12(1-3^5)}{1-3} = \dfrac{-12(1-243)}{-2}$

$= 6(-242) = -1452$

63) $a_1 = -\dfrac{3}{2},\ r = -\dfrac{1}{2},\ n = 6$

$S_6 = \dfrac{-\frac{3}{2}\left(1-\left(-\frac{1}{2}\right)^6\right)}{1-\left(-\frac{1}{2}\right)}$

$= \dfrac{-\frac{3}{2}\left(1-\frac{1}{64}\right)}{\frac{3}{2}} = -\left(\dfrac{63}{64}\right) = -\dfrac{63}{64}$

65) $a_1 = 12,\ r = -\frac{2}{3},\ n = 4$

$$S_4 = \frac{12\left(1-\left(-\frac{2}{3}\right)^4\right)}{1-\left(-\frac{2}{3}\right)} = \frac{12\left(1-\frac{16}{81}\right)}{\frac{5}{3}}$$

$$= 12\left(\frac{65}{81}\right)\frac{3}{5} = \frac{52}{9}$$

67) a) $a_1 = 1,\ r = 2$

$$a_n = 1(2)^{n-1}$$

$$a_{10} = 1(2)^{10-1} = 2^9 = 512$$

c) $S_{10} = \frac{1\left(1-2^{10}\right)}{1-2} = \frac{(1-1024)}{-1}$

$$= -(-1023) = 1023 = \$10.23$$

69) $S = \frac{8}{1-\frac{1}{4}} = \frac{8}{\frac{3}{4}} = 8\cdot\frac{4}{3} = \frac{32}{3}$

71) $S = \frac{5}{1-\left(-\frac{4}{5}\right)} = \frac{5}{\frac{9}{5}} = 5\cdot\frac{5}{9} = \frac{25}{9}$

73) $\left|\frac{5}{3}\right| > 1$. The sum does not exist.

75) $a_1 = 8,\ r = \frac{\frac{16}{3}}{8} = \frac{2}{3}$

$$S = \frac{8}{1-\frac{2}{3}} = \frac{8}{\frac{1}{3}} = 8\cdot 3 = 24$$

77) $a_1 = -\frac{15}{2},\ r = \frac{\frac{15}{4}}{-\frac{15}{2}} = -\frac{1}{2}$

$$S = \frac{-\frac{15}{2}}{1-\left(-\frac{1}{2}\right)} = \frac{-\frac{15}{2}}{\frac{3}{2}}$$

$$= -\frac{15}{2}\cdot\frac{2}{3} = -5$$

79) $a_1 = \frac{1}{25},\ r = \frac{\frac{1}{5}}{\frac{1}{25}} = 5$

$|5| > 1$. The sum does not exist.

81) $a_1 = -40,\ r = \frac{-30}{-40} = \frac{3}{4}$

$$S = \frac{-40}{1-\frac{3}{4}} = -\frac{40}{\frac{1}{4}} = -40\cdot 4 = -160$$

83) $a_1 = 3,\ r = 0.75$

$$S = \frac{3}{1-0.75} = \frac{3}{0.25} = 12 \text{ ft}$$

85) a) $a_1 = 27,\ r = \frac{2}{3}$

$$a_n = 27\left(\frac{2}{3}\right)^{n-1}$$

$$a_5 = 27\left(\frac{2}{3}\right)^{6-1} = 27\left(\frac{2}{3}\right)^5$$

$$= 27\cdot\frac{32}{243} = 3\frac{5}{9} \text{ ft}$$

b) $S=\dfrac{27}{1-\dfrac{2}{3}}=\dfrac{27}{\dfrac{1}{3}}=81$

The ball travels this distance twice minus the inital height.

$d=2\cdot 81-27=162-27=135$ ft

Section 15.4: Exercises

1) Answers may vary.

3) $(r+s)^3=r^3+3r^2s+3rs^2+s^3$

5) $(y+z)^5=y^5+5y^4z+10y^3z^2+10y^2z^3+5yz^4+z^5$

7) $(x+5)^4=x^4+4x^3(5)+6x^2(5)^2+4x(5)^3+5^4$
$=x^4+20x^3+150x^2+500x+625$

9) Answers may vary.

11) $2!=2\cdot 1=2$

13) $5!=5\cdot 4\cdot 3\cdot 2\cdot 1=120$

15) $\dbinom{5}{2}=\dfrac{5!}{2!(5-2)!}=\dfrac{5!}{2!3!}=\dfrac{5\cdot 4\cdot \cancel{3!}}{(2\cdot 1)\cancel{3!}}=\dfrac{20}{2}=10$

17) $\dbinom{7}{3}=\dfrac{7!}{3!(7-3)!}=\dfrac{7!}{3!4!}=\dfrac{7\cdot 6\cdot 5\cdot \cancel{4!}}{(3\cdot 2\cdot 1)\cancel{4!}}=\dfrac{210}{6}=35$

19) $\dbinom{10}{4}=\dfrac{10!}{4!(10-4)!}=\dfrac{10!}{4!6!}=\dfrac{10\cdot 9\cdot 8\cdot 7\cdot \cancel{6!}}{(4\cdot 3\cdot 2\cdot 1)\cancel{6!}}=\dfrac{5040}{24}=210$

21) $\dbinom{9}{7}=\dfrac{9!}{7!(9-7)!}=\dfrac{9!}{7!2!}=\dfrac{9\cdot 8\cdot \cancel{7!}}{\cancel{7!}(2\cdot 1)}=\dfrac{72}{2}=36$

23) $\dbinom{4}{4}=1$

25) $\dbinom{6}{1}=\dfrac{6!}{1!(6-1)!}=\dfrac{6!}{1!5!}=\dfrac{6\cdot \cancel{5!}}{1\cdot \cancel{5!}}=6$

27) $\dbinom{5}{0}=1$ 29 10

31) $(f+g)^3$
$=f^3+\dbinom{3}{1}f^{3-1}g+\dbinom{3}{2}f^{3-2}g^2+g^3$
$=f^3+3f^2g+3fg^2+g^3$

33) $(w+2)^4=w^4+\dbinom{4}{1}w^{4-1}(2)+\dbinom{4}{2}w^{4-2}(2)^2+\dbinom{4}{3}w^{4-3}(2)^3+2^4$
$=w^4+4w^3(2)+6w^2(4)+4w(8)+16$
$=w^4+8w^3+24w^2+32w+16$

35) $$(b+3)^5 = b^5 + \binom{5}{1}b^{5-1}(3) + \binom{5}{2}b^{5-2}(3)^2 + \binom{5}{3}b^{5-3}(3)^3 + \binom{5}{4}b^{5-4}(3)^4 + 3^5$$
$$= b^5 + 5b^4(3) + 10b^3(9) + 10b^2(27) + 5b(81) + 243$$
$$= b^5 + 15b^4 + 90b^3 + 270b^2 + 405b + 243$$

37) $$[a+(-3)]^4 = a^4 + \binom{4}{1}a^{4-1}(-3) + \binom{4}{2}a^{4-2}(-3)^2 + \binom{4}{3}a^{4-3}(-3)^3 + (-3)^4$$
$$= a^4 + 4a^3(-3) + 6a^2(9) + 4a(-27) + 81 = a^4 - 12a^3 + 54a^2 - 108a + 81$$

39) $$[u+(-v)]^3 = u^3 + \binom{3}{1}u^{3-1}(-v) + \binom{3}{2}u^{3-2}(-v)^2 + (-v)^3$$
$$= u^3 + 3u^2(-v) + 3uv^2 - v^3 = u^3 - 3u^2v + 3uv^2 - v^3$$

41) $$(3m+2)^4 = (3m)^4 + \binom{4}{1}(3m)^{4-1}(2) + \binom{4}{2}(3m)^{4-2}(2)^2 + \binom{4}{3}(3m)^{4-3}(2)^3 + 2^4$$
$$= 81m^4 + 4(3m)^3(2) + 6(3m)^2(4) + 4(3m)(8) + 16$$
$$= 81m^4 + 4(27m^3)(2) + 6(9m^2)(4) + 4(3m)(8) + 16$$
$$= 81m^4 + 216m^3 + 216m^2 + 96m + 16$$

43) $$[3a+(-2b)]^5$$
$$= (3a)^5 + \binom{5}{1}(3a)^{5-1}(-2b) + \binom{5}{2}(3a)^{5-2}(-2b)^2 + \binom{5}{3}(3a)^{5-3}(-2b)^3$$
$$+ \binom{5}{4}(3a)^{5-4}(-2b)^4 + (-2b)^5$$
$$= 243a^5 + 5(3a)^4(-2b) + 10(3a)^3(4b^2) + 10(3a)^2(-8b^3) + 5(3a)(16b^4) - 32b^5$$
$$= 243a^5 + 5(81a^4)(-2b) + 10(27a^3)(4b^2) + 10(9a^2)(-8b^3) + 5(3a)(16b^4) - 32b^5$$
$$= 243a^5 - 810a^4b + 1080a^3b^2 - 720a^2b^3 + 240ab^4 - 32b^5$$

45) $$(x^2+1)^3 = (x^2)^3 + \binom{3}{1}(x^2)^{3-1}(1) + \binom{3}{2}(x^2)^{3-2}(1)^2 + 1^3$$
$$= x^6 + 3(x^2)^2 + 3(x^2) + 1 = x^6 + 3x^4 + 3x^2 + 1$$

47) $\left[\frac{1}{2}m+(-3n)\right]^4$

$=\left(\frac{1}{2}m\right)^4+\binom{4}{1}\left(\frac{1}{2}m\right)^{4-1}(-3n)+\binom{4}{2}\left(\frac{1}{2}m\right)^{4-2}(-3n)^2+\binom{4}{3}\left(\frac{1}{2}m\right)^{4-3}(-3n)^3+(-3n)^4$

$=\frac{1}{16}m^4+4\left(\frac{1}{2}m\right)^3(-3n)+6\left(\frac{1}{2}m\right)^2\left(9n^2\right)+4\left(\frac{1}{2}m\right)\left(-27n^3\right)+81n^4$

$=\frac{1}{16}m^4+4\left(\frac{1}{8}m^3\right)(-3n)+6\left(\frac{1}{4}m^2\right)\left(9n^2\right)+4\left(\frac{1}{2}m\right)\left(-27n^3\right)+81n^4$

$=\frac{1}{16}m^4-\frac{3}{2}m^3n+\frac{27}{2}m^2n^2-54mn^3+81n^4$

49) $\left(\frac{1}{3}y+2z^2\right)^3=\left(\frac{1}{3}y\right)^3+\binom{3}{1}\left(\frac{1}{3}y\right)^{3-1}\left(2z^2\right)+\binom{3}{2}\left(\frac{1}{3}y\right)^{3-2}\left(2z^2\right)^2+\left(2z^2\right)^3$

$=\frac{1}{27}y^3+3\left(\frac{1}{3}y\right)^2\left(2z^2\right)+3\left(\frac{1}{3}y\right)\left(4z^4\right)+8z^6$

$=\frac{1}{27}y^3+\frac{2}{3}y^2z^2+4yz^4+8z^6$

51) $a=k,\ b=5,\ n=8,\ r=3$

$\frac{8!}{(8-3+1)!(3-1)!}k^{8-3+1}5^{3-1}=\frac{8!}{6!2!}k^6 5^2=28k^6(25)=700k^6$

53) $a=w,\ b=1,\ n=15,\ r=10$

$\frac{15!}{(15-10+1)!(10-1)!}w^{15-10+1}1^{10-1}=\frac{15!}{6!9!}w^6 1^9=5005w^6$

55) $a=q,\ b=-3,\ n=9,\ r=2$

$\frac{9!}{(9-2+1)!(2-1)!}z^{9-2+1}(-3)^{2-1}=\frac{9!}{8!1!}q^8(-3)^1=9q^8(-3)=-27q^8$

57) $a=3x,\ b=-2,\ n=6,\ r=5$

$\frac{6!}{(6-5+1)!(5-1)!}(3x)^{6-5+1}(-2)^{5-1}=\frac{6!}{2!4!}(3x)^2(-2)^4=15\left(9x^2\right)(16)=2160x^2$

59) $a=2y^2,\ b=z,\ n=10,\ r=8$

$\frac{10!}{(10-8+1)!(8-1)!}\left(2y^2\right)^{10-8+1}(z)^{8-1}=\frac{10!}{3!7!}\left(2y^2\right)^3(z)^7=120\left(8y^6\right)\left(z^7\right)=960y^6z^7$

61) $a = c^3,\ b = -3d^2,\ n = 7,\ r = 3$

$$\frac{7!}{(7-3+1)!(3-1)!}\left(c^3\right)^{7-3+1}\left(-3d^2\right)^{3-1} = \frac{7!}{5!2!}\left(c^3\right)^5\left(-3d^2\right)^2 = 21c^{15}\left(9d^4\right) = 189c^{15}d^4$$

63) v^{33} 65) $\binom{n}{n} = \frac{n!}{n!(n-n)!} = \frac{1}{0!} = \frac{1}{1} = 1$

Chapter 15 Review

1) $a_1 = 7(1)+1 = 7+1 = 8$
$a_2 = 7(2)+1 = 14+1 = 15$
$a_3 = 7(3)+1 = 21+1 = 22$
$a_4 = 7(4)+1 = 28+1 = 29$
$a_5 = 7(5)+1 = 35+1 = 36$

3) $a_1 = (-1)^{1+1}\cdot\left(\frac{1}{1^2}\right)$
$= (-1)^2\cdot(1) = 1\cdot 1 = 1$
$a_2 = (-1)^{2+1}\cdot\left(\frac{1}{2^2}\right) = (-1)^3\cdot\left(\frac{1}{4}\right)$
$= -1\cdot\frac{1}{4} = -\frac{1}{4}$
$a_3 = (-1)^{3+1}\cdot\left(\frac{1}{3^2}\right) = (-1)^4\cdot\left(\frac{1}{9}\right)$
$= 1\cdot\frac{1}{9} = \frac{1}{9}$
$a_4 = (-1)^{4+1}\cdot\left(\frac{1}{4^2}\right) = (-1)^5\cdot\left(\frac{1}{16}\right)$
$= -1\cdot\frac{1}{16} = -\frac{1}{16}$
$a_5 = (-1)^{5+1}\cdot\left(\frac{1}{5^2}\right) = (-1)^6\cdot\left(\frac{1}{25}\right)$
$= 1\cdot\frac{1}{25} = \frac{1}{25}$

5) $a_n = 5n$ 7) $a_n = -\frac{n+1}{n}$

9) $8.25 + 0.25 = 8.50$
$8.50 + 0.25 = 8.75$
$8.75 + 0.25 = 9.00$
\$8.25, \$8.50, \$8.75, \$9.00

11) A sequence is a list of terms in a certain order, and a series is a sum of the terms of a sequence.

13) $\sum_{i=1}^{5}\left(2i^2+1\right) = \left(2(1)^2+1\right)+\left(2(2)^2+1\right)$
$+\left(2(3)^2+1\right)+\left(2(4)^2+1\right)$
$+\left(2(5)^2+1\right)$
$= 3+9+19+33+51$
$= 115$

15) $\sum_{i=1}^{4}\frac{13}{i}$

17) $\overline{x} = \frac{18+25+26+20+22}{5} = 22.2$

19) $a_1 = 11$
$a_2 = 11+7 = 18$
$a_3 = 18+7 = 25$
$a_4 = 25+7 = 32$
$a_5 = 32+7 = 39$
11, 18, 25, 32, 39

21) $a_1 = -58$

$a_2 = -58 + 8 = -50$

$a_3 = -50 + 8 = -42$

$a_4 = -42 + 8 = -34$

$a_5 = -34 + 8 = -26$

$-58, -50, -42, -34, -26$

23) a) $a_1 = 6;\ d = 10 - 6 = 4$

b) $a_n = 6 + (n-1)4$

$a_n = 6 + 4n - 4$

$a_n = 4n + 2$

c) $a_{20} = 4(20) + 2 = 80 + 2 = 82$

25) a) $a_1 = -8;\ d = -13 - (-8) = -5$

b) $a_n = -8 + (n-1)(-5)$

$a_n = -8 - 5n + 5$

$a_n = -5n - 3$

c) $a_{20} = -5(20) - 3$

$= -100 - 3 = -103$

27) $a_1 = -15,\ d = -9 - (-15) = 6$

$a_n = -15 + (n-1)(6)$

$a_n = -15 + 6n - 6$

$a_n = 6n - 21$

$a_{15} = 6(15) - 21 = 90 - 21 = 69$

29) $a_n = -4 + (n-1)\left(-\frac{3}{2}\right)$

$a_n = -4 - \frac{3}{2}n + \frac{3}{2}$

$a_n = -\frac{3}{2}n - \frac{5}{2}$

$a_{21} = -\frac{3}{2}(21) - \frac{5}{2} = -\frac{63}{2} - \frac{5}{2} = -34$

31) $a_6 = a_1 + (6-1)d$

$24 = a_1 + (6-1)d$

$24 = a_1 + 5d \qquad (1)$

$a_9 = a_1 + (9-1)d$

$36 = a_1 + (9-1)d$

$36 = a_1 + 8d \qquad (2)$

Solve using elimination.

$-24 = -a_1 - 5d \qquad -1\cdot(1)$

$+\ \ 36 = a_1 + 8d \qquad (2)$

$12 = 3d$

$4 = d$

$d = 4: 24 = a_1 + 5(4)$

$24 = a_1 + 20$

$4 = a_1$

$a_n = 4 + (n-1)4$

$= 4 + 4n - 4 = 4n$

$a_{12} = 4(12) = 48$

33) $a_5 = a_1 + (5-1)d$

$-5 = a_1 + (5-1)d$

$-5 = a_1 + 4d \qquad (1)$

$a_{10} = a_1 + (10-1)d$

$-15 = a_1 + (10-1)d$

$-15 = a_1 + 9d \qquad (2)$

Solve using elimination.

$$\begin{aligned} 5 &= -a_1 - 4d \quad -1\cdot(1) \\ + \; -15 &= a_1 + 9d \quad (2) \\ \hline -10 &= 5d \\ -2 &= d \end{aligned}$$

$d = -2: -5 = a_1 + 4(-2)$

$-5 = a_1 - 8$

$3 = a_1$

$a_n = 3 + (n-1)(-2)$

$= 3 - 2n + 2 = -2n + 5$

$a_{17} = -2(17) + 5 = -34 + 5 = -29$

35) $a_n = 34,\ a_1 = -4,\ d = -2 - (-4) = 2$

$34 = -4 + (n-1)(2)$

$34 = -4 + 2n - 2$

$34 = -6 + 2n$

$40 = 2n$

$20 = n$

37) $a_n = 43,\ a_1 = -8,\ d = -7 - (-8) = 1$

$43 = -8 + (n-1)(1)$

$43 = -8 + n - 1$

$43 = -9 + n$

$52 = n$

39) $S_8 = \frac{8}{2}(5 + (-27)) = 4(-22) = -88$

41) $S_{10} = \frac{10}{2}[2(-6) + (10-1)(7)]$

$= 5(-12 + 63) = 5(51) = 255$

43) $S_{10} = \frac{10}{2}(13 + (-59))$

$= 5(-46) = -230$

45) $a_1 = 2(1) - 5 = 2 - 5 = -3$

$a_{10} = 2(10) - 5 = 20 - 5 = 15$

$S_{10} = \frac{10}{2}(-3 + 15) = 5(12) = 60$

47) $a_1 = -7(1) + 16 = -7 + 16 = 9$

$a_{10} = -7(10) + 16 = -70 + 16 = -54$

$S_{10} = \frac{10}{2}(9 + (-54))$

$= 5(-45) = -225$

49) $a_1 = -11(1) - 2 = -11 - 2 = -13$

$a_4 = -11(4) - 2 = -44 - 2 = -46$

$S_4 = \frac{4}{2}(-13 + (-46))$

$= 2(-59) = -118$

51) $a_1 = 3(1) + 4 = 3 + 4 = 7$

$a_{13} = 3(13) + 4 = 39 + 4 = 43$

$S_{13} = \frac{13}{2}(7 + 43) = \frac{13}{2}(50) = 325$

53) $a_1 = -4(1) + 2 = -4 + 2 = -2$

$a_{11} = -4(11) + 2 = -44 + 2 = -42$

$S_{11} = \frac{11}{2}(-2 + (-42))$

$= \frac{11}{2}(-44) = -242$

55) $a_1 = 15,\ d = 2$

$$a_n = 15 + (n-1)(2)$$
$$a_n = 15 + 2n - 2$$
$$a_n = 13 + 2n$$
$$a_{20} = 13 + 2(20)$$
$$= 13 + 40 = 53 \text{ seats}$$
$$S_{20} = \frac{20}{2}(15 + 53) = 10(68)$$
$$= 680 \text{ seats}$$

57) $a_1 = \$2,\ d = \$2$

$$S_{30} = \frac{30}{2}\left[2(2) + (30-1)(2)\right]$$
$$= 15(4 + 58) = 15(62) = \$930$$

59) $r = \dfrac{20}{4} = 5$

61) $a_1 = 7$

$$a_2 = 7(2) = 14$$
$$a_3 = 14(2) = 28$$
$$a_4 = 56(2) = 56$$
$$a_5 = 24(2) = 112$$

63) $a_1 = 48$

$$a_2 = 48\left(\frac{1}{4}\right) = 12$$
$$a_3 = 12\left(\frac{1}{4}\right) = 3$$
$$a_4 = 3\left(\frac{1}{4}\right) = \frac{3}{4}$$
$$a_5 = \frac{3}{4}\left(\frac{1}{4}\right) = \frac{3}{16}$$

65) $a_n = 3(2)^{n-1}$

$$a_6 = 3(2)^{6-1}$$
$$= 3(2)^5$$
$$= 3(32)$$
$$= 96$$

67) $a_n = 8\left(\dfrac{1}{3}\right)^{n-1}$

$$a_4 = 8\left(\frac{1}{3}\right)^{4-1} = 8\left(\frac{1}{3}\right)^3 = 8\left(\frac{1}{27}\right) = \frac{8}{27}$$

69) $a_1 = 7,\ r = \dfrac{42}{7} = 6$

$$a_n = 7(6)^{n-1}$$

71) $a_1 = -15,\ r = \dfrac{45}{-15} = -3$

$$a_n = (-15)(-3)^{n-1}$$

73) $a_1 = 1,\ r = \dfrac{3}{1} = 3$

$$a_n = (1)(3)^{n-1} = 3^{n-1}$$
$$a_8 = 3^{8-1} = 3^7 = 2187$$

75) $a_1 = 8,\ r = 3,\ n = 5$

$$S_5 = \frac{8(1-3^5)}{1-3} = \frac{8(1-243)}{-2}$$
$$= -4(-242) = 968$$

77) $a_1 = 8,\ r = \dfrac{40}{8} = 5,\ n = 5$

$$S_5 = \frac{8(1-5^5)}{1-5} = \frac{8(1-3125)}{-4}$$
$$= -2(-3124) = 6248$$

79) $a_1 = 8,\ r = \dfrac{4}{8} = \dfrac{1}{2},\ n = 6$

$$S_6 = \frac{8\left(1-\left(\frac{1}{2}\right)^6\right)}{1-\frac{1}{2}} = \frac{8\left(1-\frac{1}{64}\right)}{\frac{1}{2}}$$
$$= 16\left(\frac{63}{64}\right) = \frac{63}{4}$$

81) $a_1 = \frac{7}{2},\ r = \frac{1}{2},\ n = 5$

$$S_5 = \frac{\frac{7}{2}\left(1-\left(\frac{1}{2}\right)^5\right)}{1-\frac{1}{2}} = \frac{\frac{7}{2}\left(1-\frac{1}{32}\right)}{\frac{1}{2}}$$
$$= 7\left(\frac{31}{32}\right) = \frac{217}{32}$$

83) a) $a_1 = 28{,}000,\ r = 1-0.20 = 0.80$

$a_n = 28{,}000(0.80)^{n-1}$

b) $a_3 = 28{,}000(0.80)^{3-1}$

$= 28{,}000(0.80)^2 = \$17{,}920$

85) a) arithmetic

b) $a_1 = 7,\ d = 9-7 = 2$

$a_n = 7+(n-1)2$

$= 7+2n-2 = 2n+5$

c) $S_8 = \frac{8}{2}[2(7)+(8-1)(2)]$

$= 4(14+14) = 4(28) = 112$

87) a) geometric

b) $a_1 = 9,\ r = \frac{\frac{9}{2}}{9} = \frac{1}{2}$

$a_n = 9\left(\frac{1}{2}\right)^{n-1}$

c) $$S_8 = \frac{9\left(1-\left(\frac{1}{2}\right)^8\right)}{1-\frac{1}{2}} = \frac{9\left(1-\frac{1}{256}\right)}{\frac{1}{2}}$$
$$= 18\left(\frac{255}{256}\right) = \frac{2295}{128}$$

89) a) geometric

b) $a_1 = -1,\ r = \frac{-3}{-1} = 3$

$a_n = (-1)(3)^{n-1}$

c) $$S_8 = \frac{-1(1-3^8)}{1-3} = \frac{-(1-6561)}{-2}$$
$$= \frac{-6560}{2} = -3280$$

91) $a_1 = 2000,\ r = 1.5$

$a_n = 2000(1.5)^{n-1}$

$a_5 = 2000(1.5)^{5-1} = 2000(1.5)^4$

$= 10{,}125$ bacteria

93) when $|r| < 1$

95) $S = \frac{-3}{1-\frac{1}{8}} = \frac{-3}{\frac{7}{8}} = -3\cdot\frac{8}{7} = -\frac{24}{7}$

97) $a_1 = -15,\ r = \frac{10}{-15} = -\frac{2}{3}$

$$S = \frac{-15}{1-\left(-\frac{2}{3}\right)} = \frac{-15}{\frac{5}{3}} = -15\cdot\frac{3}{5} = -9$$

99) $a_1 = -4,\ r = \frac{12}{-4} = -3$

$|-3| > 1$. The sum does not exist.

101) $(y+z)^4 = y^4 + 4y^3z + 6y^2z^2 + 4yz^3 + z^4$

103) $6! = 6\cdot5\cdot4\cdot3\cdot2\cdot1 = 720$

105) $\binom{5}{3} = \frac{5!}{3!(5-3)!} = \frac{5!}{3!2!} = \frac{5\cdot4\cdot\cancel{3!}}{\cancel{3!}(2\cdot1)} = \frac{20}{2} = 10$

107) $\binom{9}{1} = \frac{9!}{1!(9-1)!} = \frac{9!}{1!8!} = \frac{9\cdot\cancel{8!}}{1\cdot\cancel{8!}} = \frac{9}{1} = 9$

109) $(m+n)^4 = m^4 + \binom{4}{1}m^{4-1}n + \binom{4}{2}m^{4-2}n^2 + \binom{4}{3}m^{4-3}n^3 + n^4$

$= m^4 + 4m^3n + 6m^2n^2 + 4mn^3 + n^4$

111) $[h+(-9)]^3 = h^3 + \binom{3}{1}h^{3-1}(-9) + \binom{3}{2}h^{3-2}(-9)^2 + (-9)^3$

$= h^3 + 3h^2(-9) + 3h(81) - 729$

$= h^3 - 27h^2 + 243h - 729$

113) $[2p^2 + (-3r)]^5$

$= (2p^2)^5 + \binom{5}{1}(2p^2)^{5-1}(-3r) + \binom{5}{2}(2p^2)^{5-2}(-3r)^2 + \binom{5}{3}(2p^2)^{5-3}(-3r)^3$

$+ \binom{5}{4}(2p^2)^{5-4}(-3r)^4 + (-3r)^5$

$= 32p^{10} + 5(2p^2)^4(-3r) + 10(2p^2)^3(9r^2) + 10(2p^2)^2(-27r^3) + 5(2p^2)(81r^4) - 243r^5$

$= 32p^{10} + 5(16p^8)(-3r) + 10(8p^6)(9r^2) + 10(4p^4)(-27r^3) + 5(2p^2)(81r^4) - 243r^5$

$= 32p^{10} - 240p^8r + 720p^6r^2 - 1080p^4r^3 + 810p^2r^4 - 243r^5$

115) $a = z,\ b = 4,\ n = 8,\ r = 5$

$\frac{8!}{(8-5+1)!(5-1)!}z^{8-5+1}4^{5-1} = \frac{8!}{4!4!}z^4 4^4 = 70z^4(256) = 17{,}920z^4$

117) $a = 2k,\ b = -1,\ n = 13,\ r = 11$

$\frac{13!}{(13-11+1)!(11-1)!}(2k)^{13-11+1}(-1)^{11-1} = \frac{13!}{3!10!}(2k)^3(-1)^{10} = 286(8k^3) = 2288k^3$

Chapter 15 Test

1) $a_1 = 2(1) - 3 = 2 - 3 = -1$
$a_2 = 2(2) - 3 = 4 - 3 = 1$
$a_3 = 2(3) - 3 = 6 - 3 = 3$
$a_4 = 2(4) - 3 = 8 - 3 = 5$
$a_5 = 2(5) - 3 = 10 - 7 = 7$

3) An arithmetic sequence is obtained by adding the common difference to each term to obtain the next term, while a geometric sequence is obtained by muliplying a term by the common ratio to get the next term

5) $d = -11 - (-17) = 6$

7) arithmetic
$a_1 = 5,\ d = -1 - 2 = -3$
$a_n = 5 + (n-1)(-3)$
$= 5 - 3n + 3 = -3n + 8$

9) $\sum_{i=1}^{4}(5i^2 + 6)$
$= (5(1)^2 + 6) + (5(2)^2 + 6)$
$+ (5(3)^2 + 6) + (5(4)^2 + 6)$
$= 11 + 26 + 51 + 86 = 174$

11) $S_{11} = \frac{11}{2}[2(5) + (11-1)(3)]$
$= \frac{11}{2}(10 + 30) = \frac{11}{2}(40) = 220$

13) $S = \dfrac{7}{1 - \frac{3}{10}} = \dfrac{7}{\frac{7}{10}} = 7 \cdot \frac{10}{7} = 10$

15) $a_1 = 400,\ d = 1 + 0.10 = 1.10$
$a_n = 400(1.10)^{n-1}$
$a_4 = 400(1.10)^{4-1}$
$= 400(1.10)^3 \approx 532$

17) $5! = 5 \cdot 4 \cdot 3 \cdot 2 \cdot 1 = 120$

19)
$$(3x+1)^4 = (3x)^4 + \binom{4}{1}(3x)^{4-1}(1) + \binom{4}{2}(3x)^{4-2}(1)^2 + \binom{4}{3}(3x)^{4-3}(1)^3 + (1)^4$$
$$= (3x)^4 + 4(3x)^3 + 6(3x)^2 + 4(3x) + 1$$
$$= 81x^4 + 4(27x^3) + 6(9x^2) + 12x + 1 = 81x^4 + 108x^3 + 54x^2 + 12x + 1$$

Cumulative Review: Chapters 1-15

1) $\frac{5}{8} + \frac{1}{6} + \frac{3}{4} = \frac{15}{24} + \frac{4}{24} + \frac{18}{24} = \frac{37}{24}$

3) a) $(-9k^4)^2 = (-9)^2 k^{4 \cdot 2} = 81k^8$

b) $(-7z^3)(8z^{-9}) = -56z^{3+(-9)}$
$= -56z^{-6} = -\frac{56}{z^6}$

c) $\left(\dfrac{40a^{-7}b^{3}}{8ab^{-2}}\right)^{-3}=\left(\dfrac{8ab^{-2}}{40a^{-7}b^{3}}\right)^{3}$

$$=\left(\frac{1}{5}a^{1-(-7)}b^{-2-3}\right)^{3}$$

$$=\left(\frac{a^{8}}{5b^{5}}\right)^{3}$$

$$=\frac{a^{8\cdot 3}}{5^{3}b^{15}}=\frac{a^{24}}{125b^{15}}$$

5) a) $\dfrac{4}{3}y-7=13$

$$4y-21=39$$
$$4y=60$$
$$y=15 \qquad \{15\}$$

b) $3(n-4)+11=5n-2(3n+8)$

$$3n-12+11=5n-6n-16$$
$$3n-1=-n-16$$
$$4n=-15$$
$$n=-\frac{15}{4}$$

$\left\{-\dfrac{15}{4}\right\}$

7) $x=$ amt invested at 4%

$y=$ amt invested at 6%

$x+y=9000;\ x=9000-y$

$0.04x+0.06y=480$

$0.04(9000-y)+0.06y=480$

$$4(9000-y)+6y=48{,}000$$
$$36{,}000-4y+6y=48{,}000$$
$$2y=12{,}000$$
$$y=6{,}000$$

$x=9000-6000=3000$

\$3000@4% and \$6000@6%

9) $2c+9<3$ or $c-7>-2$

$2c<-6 \qquad c>5$

$c<-3$

$(-\infty,-3)\cup(5,\infty)$

11) $m=\dfrac{2-5}{6-(-3)}=\dfrac{-3}{9}=-\dfrac{1}{3}$

$$y-2=-\frac{1}{3}(x-6)$$
$$y-2=-\frac{1}{3}x+2$$
$$y=-\frac{1}{3}x+4$$

13) $3x+5y=12 \qquad (1)$

$2x-3y=8 \qquad (2)$

Solve using elimination.

$$-6x-10y=-24 \qquad -2\cdot(1)$$
$$+\quad 6x-9y=24 \qquad 3\cdot(2)$$
$$-19y=0$$
$$y=0$$

$y=0:\ 3x+5(0)=12$

$$3x=12$$
$$x=4 \qquad (4,\ 0)$$

15) a) $(9m^{3}-7m^{2}+3m+2)$

$$-(4m^{3}+11m^{2}-7m-1)$$
$$=9m^{3}-7m^{2}+3m+2$$
$$-4m^{3}-11m^{2}+7m+1$$
$$=5m^{3}-18m^{2}+10m+3$$

b) $5(2p+3)+\dfrac{2}{3}(4p-9)$

$$=10p+15+\frac{8}{3}p-6=\frac{38}{3}p+9$$

17) a)

$$\begin{array}{r} 4x^2+5x-9 \\ 3x-2\overline{\smash{)}\,12x^3+7x^2-37x+18} \\ -\ (12x^3-8x^2) \\ \hline 15x^2-37x \\ -(15x^2-10x) \\ \hline -27x+18 \\ -\ (-27x+18) \\ \hline 0 \end{array}$$

b) $$\frac{20a^3b^3-45a^2b+10ab+60}{10ab}$$

$$=\frac{20a^3b^3}{10ab}-\frac{45a^2b}{10ab}+\frac{10ab}{10ab}+\frac{60}{10ab}$$

$$=2a^2b^2-\frac{9a}{2}+1+\frac{6}{ab}$$

19) a) $m^2-15m+54=0$

$(m-6)(m-9)=0$

$m-6=0$ or $m-9=0$

$m=6$ $\quad$ $m=9$ $\quad$ $\{6, 9\}$

b) $x^3=3x^2+28x$

$x^3-3x^2-28x=0$

$x(x^2-3x-28)=0$

$x(x-7)(x+4)=0$

$x-7=0$ or $x+4=0$ or $x=0$

$x=7$ $\quad$ $x=-4$

$\{-4,0,7\}$

21) $$\frac{k}{k^2-11k+18}+\frac{k+2}{2k^2-17k-9}$$

$$=\frac{k}{(k-9)(k-2)}+\frac{k+2}{(2k+1)(k-9)}$$

$$=\frac{k(2k+1)+(k+2)(k-2)}{(2k+1)(k-2)(k-9)}$$

$$=\frac{2k^2+k+k^2-4}{(2k+1)(k-2)(k-9)}$$

$$=\frac{3k^2+k-4}{(2k+1)(k-2)(k-9)}$$

$$=\frac{(3k+4)(k-1)}{(2k+1)(k-2)(k-9)}$$

23) $$\frac{1}{a+1}-\frac{a}{6}=\frac{a-4}{a+1}$$ Multiply by $6(a+1)$

$6-a(a+1)=6(a-4)$

$6-a^2-a=6a-24$

$0=a^2+7a-30$

$0=(a+10)(a-3)$

$a+10=0$ or $a-3=0$

$a=-10$ $\quad$ $a=3$ $\quad$ $\{-10, 3\}$

25) $\left|\frac{1}{4}t-5\right|\le 2$

$-2\le\frac{1}{4}t-5\le 2$

$3\le\frac{1}{4}t\le 7$

$12\le t\le 28$

$[12, 28]$

27) $$\left[\begin{array}{ccc|c} -1 & 3 & 3 & 5 \\ 3 & -2 & -1 & -9 \\ -3 & 1 & 3 & 5 \end{array}\right] \xrightarrow[-3R_1+R_3 \to R_3]{3R_1+R_2 \to R_2} \left[\begin{array}{ccc|c} -1 & 3 & 3 & 5 \\ 0 & 7 & 8 & 6 \\ 0 & -8 & -6 & -10 \end{array}\right] \xrightarrow[-\frac{1}{2}R_3 \to R_3]{-R_1 \to R_1}$$

$$\left[\begin{array}{ccc|c} 1 & -3 & -3 & -5 \\ 0 & 7 & 8 & 6 \\ 0 & 4 & 3 & 5 \end{array}\right] \xrightarrow{-4R_2+7R_3 \to R_3} \left[\begin{array}{ccc|c} 1 & -3 & -3 & -5 \\ 0 & 7 & 8 & 6 \\ 0 & 0 & -11 & 11 \end{array}\right] \xrightarrow[-\frac{1}{11}R_3 \to R_3]{\frac{1}{7}R_3 \to R_3} \left[\begin{array}{ccc|c} 1 & -3 & -3 & -5 \\ 0 & 0 & \frac{8}{7} & \frac{6}{7} \\ 0 & 0 & 1 & -1 \end{array}\right]$$

$$x-3y-3z=-5 \qquad y+\frac{8}{7}z=\frac{6}{7} \qquad x-3y-3z=-5$$
$$y+\frac{8}{7}z=\frac{6}{7} \qquad y+\frac{8}{7}(-1)=\frac{6}{7} \qquad x-3(2)-3(-1)=-5$$
$$z=-1 \qquad y-\frac{8}{7}=\frac{6}{7} \qquad x-6+3=-5$$
$$y=2 \qquad x-3=-5$$
$$x=-2$$
$$(-2,\ 2,-1)$$

29) a) $\sqrt{63}=\sqrt{9}\cdot\sqrt{7}$
$=3\sqrt{7}$

b) $\sqrt[4]{48}=\sqrt[4]{16}\cdot\sqrt[4]{3}$
$=2\sqrt[4]{3}$

c) $\sqrt{20x^2y^9}=2\sqrt{5}\cdot x\cdot y^4\sqrt{y}$
$=2xy^4\sqrt{5y}$

d) $\sqrt[3]{250c^{17}d^{12}}=5\sqrt[3]{2}\cdot c^5\sqrt[3]{c^2}\cdot d^4$
$=5c^5d^4\sqrt[3]{2c^2}$

31) a) $\dfrac{5}{\sqrt{t}}=\dfrac{5}{\sqrt{t}}\cdot\dfrac{\sqrt{t}}{\sqrt{t}}$
$=\dfrac{5\sqrt{t}}{t}$

b) $\dfrac{4}{\sqrt{6}-2}=\dfrac{4}{\sqrt{6}-2}\cdot\dfrac{\sqrt{6}+2}{\sqrt{6}+2}$
$=\dfrac{4\sqrt{6}+8}{6-4}$
$=\dfrac{4\left(\sqrt{6}+2\right)}{2}=2\sqrt{6}+4$

c) $\dfrac{n}{\sqrt[3]{4}}=\dfrac{n}{\sqrt[3]{2^2}}\cdot\dfrac{\sqrt[3]{2}}{\sqrt[3]{2}}=\dfrac{n\sqrt[3]{2}}{\sqrt[3]{2^3}}=\dfrac{n\sqrt[3]{2}}{2}$

33) $\sqrt{-16}=\sqrt{-1}\cdot\sqrt{16}=i\cdot 4=4i$

35) a) $9h^2+2h+1=0$

$$h=\frac{-2\pm\sqrt{2^2-4(9)(1)}}{2(9)}=\frac{-2\pm\sqrt{4-36}}{18}=\frac{-2\pm\sqrt{-32}}{18}=\frac{-2\pm 4i\sqrt{2}}{18}=-\frac{1}{9}\pm\frac{2\sqrt{2}}{9}i$$

$$\left\{-\frac{1}{9}-\frac{2\sqrt{2}}{9}i,\ -\frac{1}{9}+\frac{2\sqrt{2}}{9}i\right\}$$

b) $(w+11)^2+4=0$

$(w+11)^2=-4$

$w+11=\pm 2i$

$w=-11\pm 2i$

$\{-11-2i, -11+2i\}$

c) $(b+4)^2-(b+4)=12$

$(b+4)^2-(b+4)-12=0$

Let $u=b+4$; $u-4=b$

$u^2-u-12=0$

$(u-4)(u+3)=0$

$u-4=0$ or $u+3=0$

$u=4$ $\quad$ $u=-3$

$b=u-4$ $\quad$ $b=u-4$

$=4-4$ $\quad$ $=-3-4$

$=0$ $\quad$ $=-7$

$\{-7,0\}$

d) $k^4+15=8k^2$

$k^4-8k^2+15=0$

$(k^2-5)(k^2-3)=0$

$k^2-5=0$ or $k^2-3=0$

$k^2=5$ $\quad$ $k^2=3$

$k=\pm\sqrt{5}$ $\quad$ $k=\pm\sqrt{3}$

$\{-\sqrt{5},-\sqrt{3},\sqrt{3},\sqrt{5}\}$

37) a) $5x-10=0$

$5x=10$

$x=2$ $\quad$ $(-\infty,2)\cup(2,\infty)$

b) $2x+3\geq 0$

$2x\geq -3$

$x\geq -\frac{3}{2}$ $\quad$ $\left[-\frac{3}{2},\infty\right)$

39) a)

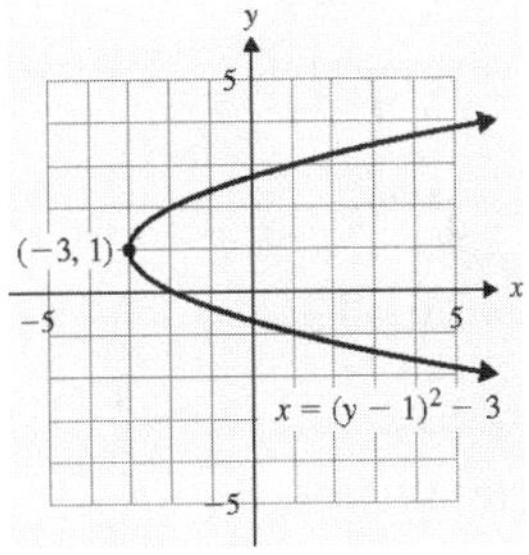

b)

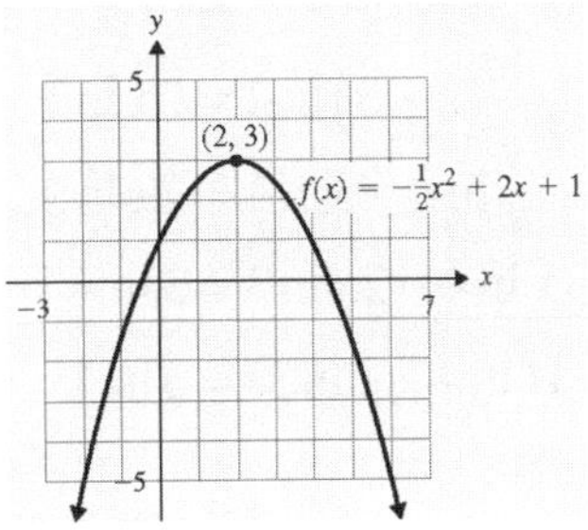

41) a) Let $\log_2 16=x$

$2^x=16$

$2^x=2^4$

$x=4$

b) Let $\log 100=x$

$10^x=100$

$10^x=10^2$

$x=2$

c) $\ln e=1$

d) Let $\log_3\sqrt{3}=x$

$3^x=\sqrt{3}$

$3^x=3^{1/2}$

$x=\frac{1}{2}$

43) a) $5^{2y}=125^{y+4}$

$5^{2y}=\left(5^3\right)^{y+4}$

$5^{2y}=5^{3y+12}$

$2y=3y+12$

$y=-12 \qquad \{-12\}$

b) $4^{x-3}=3^{2x}$

$\ln 4^{x-3}=\ln 3^{2x}$

$(x-3)\ln 4=2x\ln 3$

$x\ln 4-3\ln 4=2x\ln 3$

$x\ln 4-2x\ln 3=3\ln 4$

$x(\ln 4-2\ln 3)=3\ln 4$

$x=\dfrac{3\ln 4}{\ln 4-2\ln 3}$

$\left\{\dfrac{3\ln 4}{\ln 4-2\ln 3}\right\};\{-5.13\}$

c) $e^{-6t}=8$

$\ln e^{-6t}=\ln 8$

$-6t\ln e=\ln 8$

$-6t=\ln 8$

$t=-\dfrac{\ln 8}{6}$

$\left\{-\dfrac{\ln 8}{6}\right\};\{-0.35\}$

45)

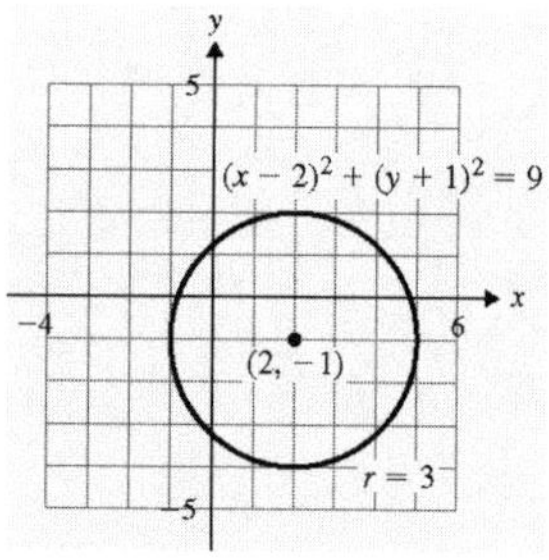

47) a) $w^2+5w>-6$

$w^2+5w+6>0$

$(w+3)(w+2)>0$

$(w+3)(w+2)=0$

$w+3=0$ or $w+2=0$

$w=-3 \qquad w=-2$

Interval A: $(-\infty,-3)$ Positive

Interval B: $(-3,-2)$ Negative

Interval C: $(-2,\infty)$ Positive

$(-\infty,-3)\cup(-2,\infty)$

b) $\dfrac{c+1}{c+7}\le 0$

Set the numerator and denominator equal to zero and solve for *c*.

$c+1=0 \qquad c+7=0$

$c=-1 \qquad c=-7$

Interval A: $(-\infty,-7)$ Positive

Interval B: $(-1,-7)$ Negative

Interval C: $(-7,\infty)$ Positive

$c\ne -7$ because it makes the denominator equal to zero.

$(-7,-1]$

49) a) $S_6=\dfrac{48\left(1-\left(\frac{1}{2}\right)^6\right)}{1-\frac{1}{2}}=\dfrac{48\left(1-\frac{1}{64}\right)}{\frac{1}{2}}$

$=96\left(\dfrac{63}{64}\right)=\dfrac{189}{2}$

b) $S_6=\dfrac{6}{2}(-10+(-25))$

$=3(-35)=-105$

Notes

Notes

Notes

Notes

Notes

Notes